Biology of Conidial Fungi

Volume 1

Biology of Conidial Fungi

Volume 1

Edited by

Garry T. Cole

Department of Botany
University of Texas at Austin
Austin, Texas

Bryce Kendrick

Department of Biology
University of Waterloo
Waterloo, Ontario, Canada

ACADEMIC PRESS   1981
A Subsidiary of Harcourt Brace Jovanovich, Publishers
New York   London   Toronto   Sydney   San Francisco

COPYRIGHT © 1981, BY ACADEMIC PRESS, INC.
ALL RIGHTS RESERVED.
NO PART OF THIS PUBLICATION MAY BE REPRODUCED OR
TRANSMITTED IN ANY FORM OR BY ANY MEANS, ELECTRONIC
OR MECHANICAL, INCLUDING PHOTOCOPY, RECORDING, OR ANY
INFORMATION STORAGE AND RETRIEVAL SYSTEM, WITHOUT
PERMISSION IN WRITING FROM THE PUBLISHER.

ACADEMIC PRESS, INC.
111 Fifth Avenue, New York, New York 10003

United Kingdom Edition published by
ACADEMIC PRESS, INC. (LONDON) LTD.
24/28 Oval Road, London NW1 7DX

Library of Congress Cataloging in Publication Data
Main entry under title:

The Biology of conidial fungi.

   Includes bibliographies and index.
   1. Fungi.   2. Conidia.   I. Cole, Garry T., Date.
II. Kendrick, Bryce.   [DNLM: 1. Fungi. 2. Mycoses.
QW 180 B614]
QK603.B5   589.2'3   80-1679
ISBN 0-12-179501-2 (v. 1)

PRINTED IN THE UNITED STATES OF AMERICA

81 82 83 84   9 8 7 6 5 4 3 2 1

To Heather and Allison

Contents

List of Contributors xi
Foreword xiii
Preface xv
Contents of Volume 2 xvii

I HISTORY

1 The History of Conidial Fungi
Bryce Kendrick
    I. Introduction 3
   II. History 4
        References 16

II SYSTEMATICS

2 The Systematics of Hyphomycetes
Bryce Kendrick
      I. Introduction—The Role of Systematists 21
     II. The Trouble with Anamorphs 22
    III. Classification—The Status Quo 23
   IV. Higher Taxa of Fungi—Heterogeneity 24
    V. Higher Taxa of Fungi—Ephemeral or Protean? 25
   VI. Pseudotaxa 27
  VII. Teleomorph + Anamorph(s) = Holomorph 28
 VIII. Anamorphic and Teleomorphic Holomorphs 30
   IX. Simple Anamorphs—Simplistic Taxa? 31
    X. Guidelines for Species and Genus 32
   XI. The Anamorph-Species and Its Problems 33
  XII. The Anamorph-Genus and Its Problems 35

XIII.	The Intergeneric Hiatus	37
XIV.	The Foundation	38
	References	40

3 Coelomycete Systematics
T. R. Nag Raj

I.	Introduction	43
II.	Techniques	47
III.	Morphological and Ontogenetic Criteria	49
IV.	Some Aspects of Biology of Coelomycetes	75
	Addendum	79
	References	79

4 Systematics of Conidial Yeasts
J. A. von Arx

I.	Introduction and History	85
II.	Generic and Specific Delimitation	86
III.	Classification	88
	References	95

5 Dimorphism
Garry T. Cole and Yoshinori Nozawa

I.	Introduction	97
II.	The Yeast	98
III.	The Hypha	101
IV.	Morphological Changes Associated with Yeast–Hypha Conversion	102
V.	Physical Factors Affecting Yeast–Hypha Conversion	109
VI.	Biochemical Differentiation of Yeast and Hyphal Phases	112
	References	127

6 Pleomorphism
J. W. Carmichael

I.	The Assimilative Stage and Dimorphism	135
II.	The Propagative Stage and Pleomorphism	136
III.	Examples of Pleomorphic Anamorphs	136
IV.	Advantages of Pleomorphism	139
V.	Pleomorphism and Classification	140
VI.	Pleomorphism and Nomenclature	142
	References	142

7 Relations between Conidial Anamorphs and Their Teleomorphs
Emil Müller
I. General Aspects 145
II. Significance of Teleomorph–Anamorph Connections for Systematics 153
III. Conclusions 165
References 165

8 A Survey of the Fungicolous Conidial Fungi
D. L. Hawksworth
I. Introduction 171
II. The Fungicolous Habit...173
III. The Fungicolous Conidial Fungi 180
IV. Discussion..231
References 235

9 Conidial Lichen-Forming Fungi
G. Vobis and D. L. Hawksworth
I. Introduction 245
II. Coelomycetous Anamorphs 246
III. Hyphomycetous Anamorphs 264
IV. Parasymbiotic Conidial Fungi 265
V. Conidial Lichen-Forming Fungi 266
VI. Sterile Lichen-Forming Fungi 267
VII. Discussion..270
References 271

III DISTRIBUTION AND ECOLOGY

10 Ecology of Soil Fungi
Dennis Parkinson
I. Introduction 277
II. Development of Ecological Concepts of Soil Fungi 278
III. Growth and Growth Forms of Soil Fungi..282
IV. Methods for Studying Soil Fungi...287
V. Interaction of Soil Fungi and Soil Microarthropods..............................289
VI. Conclusion 290
References 291

11 Morphology, Distribution, and Ecology of Conidial Fungi in Freshwater Habitats
J. Webster and E. Descals

I.	Introduction	295
II.	Ingoldian Conidial Fungi	296
III.	Aeroaquatic Conidial Fungi	335
	References	348

12 Distribution and Ecology of Conidial Fungi in Marine Habitats
Jan Kohlmeyer

I.	Introduction	357
II.	Morphology and Dispersal of Marine Conidial Fungi	360
III.	Biology of Marine Conidial Fungi	361
IV.	Distribution of Marine Conidial Fungi	368
	References	370

13 The Aerobiology of Conidial Fungi
J. Lacey

I.	Introduction	373
II.	Methods of Study	374
III.	The Aerial Environment	378
IV.	Populations of Airborne Conidia	380
V.	Dispersion of Airborne Conidia	396
VI.	Implications of an Air Spora	406
VII.	Conclusion	409
	References	410

14 Biogeography and Conidial Fungi
D. T. Wicklow

I.	Introduction	417
II.	The Adaptive Value of Conidia	419
III.	Convergent Evolution	426
IV.	Controls of Species Distribution	426
V.	Species Distribution and Environmental Scale	433
VI.	Speciation and Species Diversity	433
VII.	Abundance of Conidial Fungi in Tropical and Temperate Environments	436
VIII.	Conidial Fungi and Ecological Islands	440
	References	442

Subject Index 449
Index to Taxa 457

List of Contributors

Numbers in parentheses indicate the pages on which the authors' contributions begin.

J. W. Carmichael (135), University of Alberta, Mold Herbarium and Culture Collection, Edmonton, Alberta, T6G 2H7, Canada

Garry T. Cole (97), Department of Botany, University of Texas at Austin, Austin, Texas 78712

E. Descals (295), Department of Biological Sciences, Hatherly Laboratories, University of Exeter, Exeter EX4 4PS, England

D. L. Hawksworth* (171, 245), Commonwealth Mycological Institute, Kew, Surrey, England

Bryce Kendrick (3, 21), Department of Biology, University of Waterloo, Waterloo, Ontario N2L 3G1, Canada

Jan Kohlmeyer (357), Institute of Marine Sciences, University of North Carolina at Chapel Hill, Morehead City, North Carolina 28557

J. Lacey (373), Rothamsted Experimental Station, Harpenden, Hertfordshire AL5 2JQ, England

Emil Müller (145), Institut für Spezielle Botanik, Eidg. Technische Hochschule, CH-8092 Zürich, Switzerland

T. R. Nag Raj (43), Department of Biology, University of Waterloo, Waterloo, Ontario N2L 3G1, Canada

Yoshinori Nozawa (97), Department of Biochemistry, Gifu University School of Medicine, Gifu, Japan

Dennis Parkinson (277), Department of Biology, The University of Calgary, Calgary, Alberta T2N 1N4, Canada

G. Vobis (245), Fachbereich Biologie-Botanik der Universität, D-3550 Marburg/Lahn, Germany

*Present address: Commonwealth Agricultural Bureaux, Farnham House, Farnham Royal, Slough SL2 3BN, England.

J. A. von Arx (85), Centraalbureau voor Schimmelcultures, Oosterstraat 1, Baarn, The Netherlands

J. Webster (295), Department of Biological Sciences, Hatherly Laboratories, University of Exeter, Exeter EX4 4PS, England

D. T. Wicklow (417), ARS Culture Collection, Fermentation Laboratory, Northern Regional Research Center, Peoria, Illinois 61604

Foreword

With the publication of these volumes it can be truly said that the conidial fungi—long the stepchildren of mycologists—have finally been legitimized, and the editors are to be congratulated on putting together these authoritative volumes.

No important phase of the biology of these fungi has been neglected and the editors were fortunate to have each topic treated by a recognized authority. Although the discussions vary greatly in length and in detail, the editors have succeeded admirably in intertwining them into a unified whole.

From systematics to biochemistry, from saprobes to human pathogens, from mycotoxins to mycoviruses, every topic that has been explored by mycologists in the last two centuries is included and beautifully summarized. Congratulations to the editors and the authors for a job well done.

Constantine J. Alexopoulos
University of Texas at Austin

Preface

The term Fungi Imperfecti could be broadly defined as including any and all anamorphic (asexual) expressions of true fungi that occur separately in time and/or space from their respective teleomorphs (sexual phases), or are not known to have a teleomorph. This interpretation embraces such phenomena as the sporangial forms of many Zygomycetes, several of the spore forms of the Uredinales, and the Mycelia Sterilia, which do not produce spores. These volumes exclude the Mycelia Sterilia. The conidial fungi, for the purposes of this work, are anamorphic fungi of presumed ascomycetous or basidiomycetous origin (the higher fungi or Dikaryomycota). Sporangial morphs of Zygomycetes, etc., are specifically excluded, though their ecology and biology, but not their phylogeny, closely parallel those of the dikaryomycotan anamorphs. Why, then, are they excluded? The reasons are largely historical. While separate systems of binomials and taxonomic schemes have grown up around dikaryomycotan anamorphs and teleomorphs, respectively, no such dichotomy has occurred in the Zygomycetes. Even if only the sporangial anamorph is known, the basic zygomycotan nature of the organism is assumed by inference, and a teleomorphic, and thus holomorphic, generic name is invariably applied to such anamorphs. Treatments of Zygomycetes have an integrated approach that has been lacking in our dealings with the higher fungi and their various life forms. The history and rationale of this schism between anamorphic and teleomorphic classifications is carefully expounded by Weresub and Pirozynski (1979).* There are other reasons. While several hundred Zygomycetes are known, the Dikaryomycota comprise about 90% of all true fungi, and many thousands of confirmed or presumed dikaryomycotan anamorphs confront mycologists wherever they turn their attention. Such anamorphs affect our lives in many ways. They cause such superficial mycoses as the various kinds of ringworm, and deep-seated mycoses like aspergillosis and histoplasmosis. They cause the majority of important fungal plant

---

*Weresub, L. K., and Pirozynski, K. A. (1979). Pleomorphism of fungi as treated in the history of mycology and nomenclature. *In* The Whole Fungus (B. Kendrick, ed.), Vol. I, pp. 17–26. Nat. Mus., Canada.

diseases, such as southern blight of corn, Panama disease of bananas, and a plethora of wilts, anthracnoses, leaf spots, soft rots, etc. They cause enormous losses of stored processed food (anything from bread to jam), and produce insidious and dangerous toxins such as aflatoxin, vomitoxin, zearalenone, and sporidesmins. They cause deterioration of manufactured natural products including paper, leather, cottons, and even paint.

On the positive side, dikaryomycotan anamorphs produce extremely valuable secondary metabolites, such as antibiotics and organic acids, which we have been able to exploit. They are used in processing gourmet cheeses (camembert, brie, roquefort, gorgonzola, stilton, danish blue); they have potential in biological control of insects, and possibly of nematodes and pathogenic fungi. And perhaps most important of all, they are principally responsible for the conditioning or mineralization of the enormous quantities of plant litter that fall to the ground each year, they comprise the greater part of the microbial biomass in the soil and in many ponds and streams, and they are thus vitally involved in energy flow in many ecosystems. It is probably fair to say that dikaryomycotan anamorphs are the commonest fungal manifestations of all. And yet the only volumes so far that treat them exclusively are concerned with their systematics. We felt it was time that other aspects of their existence were given some concentrated attention. Hence these two volumes. We have tried to bring together in one place detailed considerations of many facets of conidial fungi—some of which have been almost entirely though undeservedly neglected in previous literature. We hope that these volumes will fill some lacunae in our knowledge of anamorphs, and serve as a useful reference to the advanced student who probably encounters many such fungi, but has tended to regard them simply as weeds or contaminants. If we can convey to students our belief that dikaryomycotan anamorphs are among the most successful and versatile fungi of all, and the reasons for that belief, we shall be well satisfied.

Garry T. Cole
Bryce Kendrick

Contents of Volume 2

IV	**CONIDIAL FUNGI AND MAN**
15	**Clinical Aspects of Medically Important Conidial Fungi** *J. W. Rippon*
16	**Mycotoxin Production by Conidial Fungi** *Philip B. Mislivec*
17	**Development of Parasitic Conidial Fungi in Plants** *James R. Aist*
18	**Food Spoilage and Biodeterioration** *John I. Pitt*
19	**Use of Conidial Fungi in Biological Control** *T. E. Freeman*
20	**Predators and Parasites of Microscopic Animals** *G. L. Barron*
21	**Entomogenous Fungi** *Donald W. Roberts and Richard A. Humber*
22	**Food Technology and Industrial Mycology** *William D. Gray*
V	**ULTRASTRUCTURE, DEVELOPMENT, PHYSIOLOGY, AND BIOCHEMISTRY**
23	**Conidiogenesis and Conidiomatal Ontogeny** *Garry T. Cole*
24	**Biochemistry of Microcycle Conidiation** *J. E. Smith, J. G. Anderson, S. G. Deans, and D. R. Berry*

25	**Nuclear Behavior in Conidial Fungi**	
	C. F. Robinow	
26	**Viruses of Conidial Fungi**	
	Paul A. Lemke	
27	**Physiology of Conidial Fungi**	
	Robert Hall	
28	**Cell Wall Chemistry, Ultrastructure, and Metabolism**	
	Jerome M. Aronson	

VI GENETICS

29	**The Genetics of Conidial Fungi**	
	A. C. Hastie	

VII TECHNIQUES FOR INVESTIGATION

30	**Isolation, Cultivation, and Maintenance of Conidial Fungi**	
	S. C. Jong	
31	**Techniques for Examining Developmental and Ultrastructural Aspects of Conidial Fungi**	
	Garry T. Cole	

Subject Index
Index to Taxa

HISTORY

1

The History of Conidial Fungi

Bryce Kendrick

I.	Introduction	3
II.	History	4
	A. Before 1800	5
	B. 1801–1851	6
	C. The Tulasnes, De Bary, Berkeley, and Darwin	8
	D. Saccardo	9
	E. Potebnia, von Höhnel, and Grove	11
	F. Costantin, Vuillemin, and Mason	11
	G. Wakefield and Bisby, Ingold, and Langeron	13
	H. Hughes	13
	I. 1968–1978	14
	J. 1979	15
	References	16

I. INTRODUCTION

I used to think that history was a field which permitted little leeway to its practitioners, since what's done is done and cannot be altered. How naive I was, how unmycologically green! I now know that history is continually being rewritten, that historical personages such as Canada's own Louis Riel can be transmogrified from arch-villain to hero, and that "fact" may come to be regarded as fiction and mere speculation assume the solid flesh of authenticity. Many of these manipulations are politically motivated; some are dictated by changes in the moral tone of society. And although I believe in the perfectibility of neither humans nor mycology, and although the words "enlightenment" and "progress" are subject to many interpretations (remember "newspeak" in George Orwell's

1984), I do believe that mycology has reached a level of sophistication from which we can make some fairly realistic judgments of our forebears.

It would be easy to linger indefinitely among the poppy fields of conidiophores and conidiomata, forgetting (as most specialists occasionally do) that there are other organisms out there and that they too have historical relevance. In order thoroughly to clear the air, and my conscience, I must advise you that there exists an excellent *Introduction to the History of Mycology,* written by the peripatetic scholar Geoffrey Ainsworth and published in 1976 by Cambridge University Press. A perusal of this book should help you place my more specialized and limited account in perspective. If you do not have time to read a whole book, then at least scan the brief survey by the same author that appears in *The Fungi*, Vol. 1, pages 3–20 (1965).

It would also be easy to fall into the trap of believing that the only worthwhile mycological research is that being pursued right now and that the past presents an almost unrelieved panorama of hare-brained hypotheses, off-track observations, erroneous experimental design, and incorrect interpretations.

Of course, mycology has its equivalents of the flat earth and phlogiston hypotheses, and wildly inaccurate phylogenies abound in the literature. Too often, we forget that mycology, as an ongoing cultural artifact, has attained its current level of sophistication largely through the proposal, followed by the testing and rejection of, inadequate hypotheses. Of course, we have an excuse for neglecting the past; so much literature issues from the mouth of the scientific volcano that it is all we can do (we say, casting our eyes piteously to the heavens) to avoid being engulfed.

Yet there is much to be learned from a consideration of the ideas of our forebears, who were just as intelligent as we are, and who, albeit from a narrower data base, were sometimes extremely intuitive. Many scientists labor today to achieve a form of personal immortality; yet if they will not grant this to the pioneers, surely they themselves will ultimately be denied the palm. If some of my readers do not emerge from this chapter with a new respect for some of the historical figures of mycology, I shall not have done my job well.

Those who seek accounts of such historical landmarks as the discovery of penicillin and aflatoxin, and subsequent developments, will find what they seek elsewhere in these volumes. I have contented myself with a consideration of the central, synthesizing discipline of systematics.

II. HISTORY

It is now clear that many of the anamorphs we know today have changed little since the Tertiary period (Pirozynski and Weresub, 1979) and possibly since much earlier times. This means that they considerably antedate human intelligence and must therefore have been noted as molds spoiling fruit and other foods, and sometimes as causing annoying skin diseases such as ringworm and athlete's foot, by the earliest human beings. Their true nature as independent living organisms has been divined and experimentally established beyond doubt only comparatively recently.

Of course, interest in the fungi began with the sporomata of agarics and other macrofungi, and for centuries fungi were simply classified as poisonous or nonpoisonous. Needless to say, this kind of dichotomy did not concern itself with the microfungi (at least, not until very recently, when the word "aflatoxin" entered the language), and the elaboration of a rational classification encompassing all true fungi had to await the elucidation of their microscopic anatomy. The microscope was invented in the middle of the seventeenth century, and at once the world of the microfungi, which includes virtually all conidial fungi, was revealed.

A. Before 1800

In 1665 Hooke published the first illustrations of microfungi, including anamorphic *Mucor*. A few years later Leeuwenhoek provided more information, and in 1680 he observed anamorphic *Saccharomyces cerevisiae* undergoing multiplication by budding. But these were scattered and unsystematic observations.

My average second-year undergraduate student, provided with a concisely written, profusely illustrated textbook and looking at a well-stained slide preparation of a hyphomycete through a twentieth-century microscope, looks me in the eye and says, "Is this anything?" One hardly needs to wonder what he or she would have done in the early 1700s, surrounded by a sea of almost total ignorance and indifference. Yet Micheli, in his *Nova Plantarum Genera* (1729), beautifully illustrated and described the microscopic anamorphs *Aspergillus* and *Botrytis*. I venture to suggest that in his situation most of us would have been nonstarters.

Micheli described no fewer than 900 fungi and must surely be regarded as one of the true founding fathers of mycology. Neither his failure to give magnifications or scales for his illustrations, nor the fact that his classification would not be considered helpful today, can detract from his achievement. The microscopes of the day were plagued with spherical and chromatic aberration, and it is not too surprising that the scientific accuracy of illustrations did not improve much for nearly 200 years. Although the organismic nature of fungi and the role of their spores were not accepted until long after Micheli's death, he was well aware of both concepts—he was simply ahead of his time.

Although Spallanzani (1776) also demonstrated clearly, with anamorphic *Rhizopus,* that it required "seeds" to ensure fungal continuity, it was not until the 1870s that Tyndall and Pasteur finally convinced the world that spontaneous generation (at least in recent geological times), a long-hallowed idea that had originated with Aristotle, was a fallacy. The self-taught Micheli had been vindicated after a lapse of 150 years, Spallanzani after a mere century.

While this controversy still raged, other mycologists were cautiously constructing the framework of fungal taxonomy.

But they were hardly helped by the otherwise great Linnaeus. A taxonomic genius he most certainly was, but his understanding and treatment of the fungi seems to have been almost a calculated slight. He ignored most of them and placed others in the genus *Chaos* under Vermes (worms), having accepted some deplorably bad observations. He disparaged Micheli's magnificent pioneer work and used only two of the many generic names Micheli had proposed. In addition, he arbitrarily changed the application of some well-known Latin names—for example, what we now call *Amanita caesarea,* the Romans are known to have called *Boletus*. Although Linnaeus did not do anything "illegal," I think such name changes represented unfortunate breaks with tradition—hardly an auspicious start for the principle of priority. Despite his blinkered view of the fungi, Linnaeus's system of naming organisms has done much for mycological communication. I cannot improve on the example given by Ainsworth (1965) in which the complex pre-Linnaean appellation *fungus ramosus niger compressus parvus, apicibus albidis* became, after Linnaeus, simply *Xylaria hypoxylon.*

B. 1801–1851

At the beginning of the nineteenth century, recognition of the true nature and significance of the ascus and the basidium lay in the future, and it is hardly surprising that the classifications of the time grouped what we now know to be very disparate elements. Persoon (1801), for example, divided fungi into two main groups on the basis of whether their fructifications were open or closed. Thus Coelomycetes, Ascomycetes, and even Hyphomycetes made uneasy bedfellows. His classification is of interest to us here only because its sixth and last order, Nematothecii (Fungi Byssoidei), was made up of Hyphomycetes (though other Hyphomycetes were disposed in some of his other groups). Although he neither made extensive use of the microscope nor provided good diagnostic illustrations (of the kind later published by Corda), he did something just as important: He laid down dried material of many microfungi, and his herbarium survives to this day at Leiden. Thus we can discover to what actual fungi his names are attached. As a result, many of his generic names are still in use today.

In 1809 Link used the same kind of primary separations as Persoon and thus grouped many Uredinales with sporodochial Hyphomycetes and some acervular Coelomycetes. Von Martius (1817) worked along similar lines but used different group names. What Link had called Epiphytae and Mucedines, Martius called Coniomycetes and Hyphomycetes. The reader will note that the last name obviously caught on, since it has now been in use for 160 years. Note also, however, that Martius placed some fungi we would call Hyphomycetes in his Coniomycetes.

Nees von Esenbeck (1817) followed similar lines in his classification, providing yet another name, Protomyci (long disused) for the Epiphytae of Link and the Coniomycetes of Martius.

The next figure to influence the classification of conidial fungi was Fries. His *Systema Mycologicum* (1821-1832), later chosen as the starting point for nomenclature of conidial fungi, actually did not concern itself with such fungi to nearly the same extent as had Persoon's earlier work. The 1821 starting point means that some of Persoon's names, though perfectly understood through reference to his herbarium (see Hughes, 1958), cannot be legally used. Hughes (1959) advanced a number of persuasive arguments for the adoption of 1801 (Persoon's *Synopsis*) as the starting point for Hyphomycetes, but his proposal was rejected. It is probably time for us to try again.

To return to Fries: His *Systema Mycologicum* outlined a highly structured scheme with four main classes. He adopted Martius's Coniomycetes and Hyphomycetes, adding two new names of his own, Gasteromycetes and Hymenomycetes. Note again that, although these names have stuck and have fairly precise modern meanings, Fries placed both Pyrenomycetes and pycnidial Coelomycetes (the latter being brought into the system for the first time) within Gasteromycetes.

Each of Fries's four classes was subdivided into four orders, and each of these in turn into four tribes. Perhaps the most remarkable thing about this scheme, from our viewpoint, is the amazingly naive theory (based in German romantic philosophy) which underlay it; or at least so Fries would have us believe. I cannot resist a brief digression into this theory. A special reproductive force (A), was supposed to give rise to fungi, after which their development could be influenced by air (B), heat (C), and light (D). The four classes of fungi he ascribed to A acting alone, A + B, A + C, and A + D.

In 1825 he revised his classification considerably, now dividing the class Fungi into two subclasses. The Ascomycetes (chalk another one up for Fries), which included what he called Hymenomycetes and Pyrenomycetes, and the Sporomycetes—Gasteromycetes and Coniomycetes. It was all extremely confusing, and I feel constrained to give an undertaking at this point to content myself henceforth with mentioning only forward steps, ignoring the many lateral and regressive moves that bedevil all fungal classifications before those introduced by De Bary in 1866.

In his *Epicrisis Systematis Mycologia* (1836-1838), Fries redeemed the situation with an improved scheme which recognized six classes, perspicaciously separating Discomycetes from Hymenomycetes, and Pyrenomycetes from Gasteromycetes. Fries is noted for his contribution to our knowledge of the Hymenomycetes rather than the conidial fungi, but the magnitude of his contribution can be seen from the fact that, in *Systema Mycologicum* and its supplement, he described almost 5000 fungi.

About this time some improvements were being made in the design of microscopes, and almost at once the quality of illustrations improved tremendously. Corda's *Icones fungorum hucusque cognitorum*—six volumes published in the years 1837-1854—are still referred to by modern students of Hyphomycetes for their hundreds of idiosyncratic but eminently recognizable drawings. And although he placed a sprinkling of acervular fungi in each of his four orders (1842), he did delineate a new family, Melanconiaceae, within the order Myelomycetes, for some of them, and also put most pycnidial genera in either the Sphaeriacei or Sphaeronemeae of the same order.

Léveillé (1846) used the microscope assiduously and injected a strong element of microanatomy into his scheme of classification. He distributed what we now call Hyphomycetes among three of six major subdivisions, but to his credit he erected the family Sphaeropsidei for some pycnidial fungi.

Bonorden (1851) also removed pycnidial fungi from the Pyrenomycetes, though his segregation was less than perfect, since his Sphaeronemei included *Eurotium* and some other Ascomycetes. The acervular fungi and Hyphomycetes both remained relatively scattered.

C. The Tulasnes, De Bary, Berkeley, and Darwin

And now we come to a dramatic period in the history not only of mycology but of all biology. Two momentous developments occurred within a decade. I will deal with the lesser of the two first. By the middle of the century it had become apparent that some, at least, of the conidial fungi were asexual phases of certain Ascomycetes. This phenomenon was repeatedly demonstrated by the Tulasne brothers in a masterful series of publications, crowned by the three volumes of the *Selecta Fungorum Carpologia* (1861-1865) with their incomparable illustrations. As so often happens in science, the same breakthrough was made at about the same time by De Bary, who in 1854 established that *Aspergillus glaucus* is the anamorph of *Eurotium herbariorum*. Charles Tulasne's magnificent illustrations of teleomorph plus anamorph have never been surpassed, though it must be noted that some of the connections celebrated in his drawings were spurious. This degree of error is hardly surprising when the great Berkeley himself (of whom more later), lacking knowledge of sterile techniques, suggested that yeast cells simply represented an abnormal growth phase of *Penicillium*. Mixed cultures were as common among mycologists in those days as mixed metaphors among politicians today. And even Berkeley's error seems conservative beside the careless rapture of one of his contemporaries, who extravagantly asserted that yeast could give rise not only to *Penicillium* but also to

Mucor, Entomophthora, Isaria, and others. Nor was this the worst example: I shall not even outline Hallier's scandalous excursions into pleomorphism—suffice it to say that they were pure science fiction. Perhaps these misguided claims gave the idea of teleomorph–anamorph connections a bad name. Or perhaps it was simply easier to classify teleomorphs without reference to anamorphs, and vice versa. A gulf opened up between the study of sexual and asexual forms which has persisted until today in the minds of many mycologists (see Weresub and Pirozynski, 1979). Of course, the efforts of the Tulasnes, De Bary and, a generation later, Brefeld, were not in vain. After a century we have become very conscious that the elaborate independent classification systems we have built up for teleomorphs and anamorphs must be integrated and have begun to make strenuous efforts in this direction (see Kendrick, 1979). It is clear that we have far to go before this aim will be achieved, as Müller (this volume, Chapter 7) shows only too well, and we can still marvel at how far knowledge of the fungi had already advanced by 1860 from the ancient folklore peddled by Gerarde (1633) to the effect that poisonous fungi acquired their fateful qualities from the exhalations of snakes (a fantasy that originated with Dioscorides in the first century A.D.).

The second, and much the greater, event of the period was Darwin's ringing proclamation of organic evolution in 1859. His massive documentation of the phenomenon gave it instant authority. No branch of biology was unaffected.

Perhaps Berkeley, who published a classification of fungi in 1860, had not had much time to assimilate Darwin's concepts; nevertheless he was the first to consolidate most acervular fungi in the Melanconiei and to group mononematous, synnematous, and sporodochial forms in the Hyphomycetes.

But a few years later came De Bary again, producing in 1866 a classification that was amazingly close in essentials to that widely accepted until very recently (Table I). I must admit that I find De Bary an almost awe-inspiring figure, with an amazingly sure instinct for unlocking fungal secrets. De Bary's classification embodied, as have all subsequent attempts, important elements of phylogenetic speculation. In his revised second edition (1884) De Bary came even closer to modern ideas, regarding rusts as leading into the climax Basidiomycetes and raising questions about the relationships of the Ustilaginales which are still being debated after almost a century. Note also that he did not include conidial fungi as a main-line group. As I have already suggested, it might have been better to leave things that way and to make strenuous efforts to maintain an integrated approach to teleomorph-anamorph systematics.

D. Saccardo

But it was not to be. Fuckel (1869) divided the fungi into two groups, Fungi Perfecti and Fungi Imperfecti, and Saccardo effectively enshrined the dichotomy

TABLE I

De Bary's 1866 Classification of Fungi

Phycomycetes
  Saprolegnieae
  Peronosporeae
  Mucorini
Hypodermii
  Uredinei
  Ustilaginei
Basidiomycetes
  Tremellini
  Hymenomycetes
  Gasteromycetes
Ascomycetes
  Protomycetes
  Tuberacei
  Onygenei
  Pyrenomycetes
  Discomycetes
Flechten (lichens)
Myxomyceten (slime molds)

in his massive works. Once he had made this decision, it was easy to move further into a convenient marriage of artificiality with convenience. In 1884 he introduced such features as color, shape, and septation of spores to delineate subfamilies and genera. At this point we had the three group names still so widely used: Sphaeropsideae (Sphaeropsidaceae, Sphaeropsidales) for forms with pycnidial conidiomata, Melanconieae (Melanconiaceae, Melanconiales), for those with acervular conidiomata, and Hyphomyceteae (Hyphomycetes) for mononematous, synnematous, and sporodochial forms. His subdivisions of the Hyphomyceteae have been in use until fairly recently. In Volume 4 of his monumental *Sylloge Fungorum* (1886) he divided this group into four families—Mucedineae (mononematous, hyaline, or brightly colored), Dematieae (mononematous, darkly pigmented), Stilbeae (synnematous, light or dark), and Tuberularieae (sporodochia, light or dark). Within the first two families his primary subdivision was based on spore septation (sections Amerosporae, Didymosporae, and so on), and within each of these he recognized subsections Micronemeae and Macronemeae (integrated or morphologically undifferentiated versus discrete or morphologically differentiated conidiogenous structures). In the other two families, the divisions were based first on pigment, then on spore septation or shape.

E. Potebnia, von Höhnel, and Grove

Potebnia (1910) also subdivided Fungi Imperfecti on their respective kinds of fructification, but rather differently. He joined acervular and sporodochial fungi in the Acervulales, placed genera with incomplete or atypical pycnidia in the Pseudopycnidiales, and typical pycnidial forms in the Pycnidiales. Our recent investigations of conidiomatal structure (Kendrick and Nag Raj, 1979) have given me considerable sympathy for Potebnia's ideas; yet they were not widely accepted. Nor were von Höhnel's (1911) rather similar suggestions. It appears that no one could successfully challenge Saccardo at that time. Grove (1935, 1937) juxtaposed the Sphaeropsidales and Melanconiales in the Coelomycetes, a category equivalent in rank to the Hyphomycetes, and thus bisected the conidial anamorphic fungi. I almost wrote "neatly bisected," but of course we have the doubts first raised by Potebnia (1910) and restated by Kendrick and Nag Raj (1979) and by Müller (this volume, Chapter 7). But whatever has gone on in the upper levels of the taxonomic hierarchy, Saccardoan criteria (color, and shape and septation of conidia) remained supreme at the lower levels and were used essentially unmodified in all published compilations until the 1950s.

The mainstream was dominated by characters of mature morphology. But for nearly 100 yr (almost since Saccardo first elaborated his criteria) dissident elements in the mycological community have found this system strangely unsatisfying. It is admittedly an extremely empirical, artificial system. It can often be seen to bring together alien forms and to separate forms which one has a hunch are really very closely related. Genera such as *Arthrobotrys,* for many years arbitrarily defined by the obligate presence of a single septum in the conidium (no more, no less) could be intuited to be related to other similar nematode-trapping forms disposed in other genera simply and solely because their conidium septation was "wrong." Only recently (Schenck *et al.,* 1977) has a more reasonable initiative been taken. A kind of institutionalized rigidity in the application of this and other Saccardoan criteria, actually quite understandable in the absence of other readily accessible characters, has left us with a legacy of irritating paradoxes and inconsistencies scattered throughout the taxonomy of the conidial fungi. Some of these are now being exposed by the effort to connect anamorphs to their teleomorphs, and Müller (this volume, Chapter 7) cites some notable examples.

The question is of course whether we are willing to settle for a convenient, artificial arrangement of pigeonholes or insist on striving for a system, as yet clearly unattained and probably unattainable in the foreseeable future, which more closely represents biological relationships.

F. Costantin, Vuillemin, and Mason

Costantin (1888) was perhaps the first to attempt this. Note how closely his heresy followed on the heels of the proclamation of Saccardoan orthodoxy.

Costantin tried to use the manner in which hyphomycete conidia were attached to their parent hypha as a taxonomic character, but his work did not greatly influence those who followed him.

Vuillemin (1910a,b) recognized that the term "spores" covered structures of very different origins. He proposed that initially two main kinds be distinguished: conidia vera, or spores liberated by a specific mechanism as soon as they have been formed, and thallospores, which remain an integrated part of the hypha that produces them. Vuillemin recognized three kinds of thallospores: arthrospores, arising from fragmentation of hyphae; blastospores, arising as buds in persistent acropetal chains; and chlamydospores, thick-walled resting spores. In 1911 he added a fourth kind of thallospore, aleuriospores, which were intermediate between chlamydospores (no mechanism for release) and conidia vera (specialized immediate release). Unfortunately, as Carmichael has pointed out (in Kendrick, 1971 p. 61), although Vuillemin understood and clearly described the special built-in lateral wall-dissolution mechanism by which aleuriospores are freed, he confused the issue, and many later workers, by calling these spores indehiscent. Vuillemin subdivided his Conidiosporales—Hyphomycetes produced conidia vera—into three groups: the Sporotrichae, forming conidia on undifferentiated hyphae; the Sporophorae, with differentiated conidiophores; and the Phialidae, forming conidia at the tip of flask shaped, septum-delimited structures he called phialides.

I have given some detail on Vuillemin's ideas, because they represented such a break with the tradition of mature morphology (despite the fact that he still used the Saccardoan families) and because they foreshadowed much that has gained currency in the last 25 years.

Vuillemin's papers were published in an obscure journal, but they were also very radical and made little headway at the time. Like Spallanzani and Micheli, he was ahead of his time. His ideas were effectively ignored until Mason (1933), in discussing the then-known anamorphs of the Hypocreales, reworked Vuillemin's definition of the phialide. He extended it to include "the fusiform truncate, fusiform beaked or acuminate terminal portion of a hypha, from the apex of which or from within which thin-walled conidia are abstricted." This definition embraced a number of things we certainly would not call phialides today, but Mason, fortunately, was aware of the heterogeneity of what he had created. What we would call blastic-phialidic conidia, he termed meristem phialospores. What we would call blastic-sympodial conidia, he recognized as terminus phialospores. Although Mason's 1933 paper muddied the waters as far as the term "phialide" was concerned, his recognition of the sympodially proliferating condiogenous cell represented a real advance. Mason also introduced the term "radulaspore" for conidia arising randomly from a hypha on denticles. Then, because he believed that all anamorphs of the Hypocreales ought to have a phialosporic anamorph of some description, he suggested that radulaspores might be homologous with phialospores. We do not accept his excessive expansion of the phialidic concept, which clearly arose from a forlorn hope of arriving at one of those underlying generalizations science is always seeking. But we applaud his observations of modes of conidium ontogeny, which were to bear further fruit.

Later, Mason (1937) shifted his attention to the manner in which spores are dispersed and stressed the difference between the slimy-spored (mainly water- and insect-dispersed) and dry-spored (mainly wind-dispersed) forms. He revised his earlier claim that all hypocrealean anamorphs produced phialospores, now suggesting that pleomorphic forms had one dry-spored and one slimy-spored state and that, if the slimy spores were abstricted from free conidiophores, then these or their apical cells would be phialides.

G. Wakefield and Bisby, Ingold, and Langeron

Although conceding the existence of transitional series (not to mention such dry-slimy twins as *Stachybotrys* and *Memnoniella*), Wakefield and Bisby (1941) in their list of British Hyphomycetes, adopted the two divisions, christening them Xerosporae and Gloiosporae (but still subdividing each on spore morphology, after Saccardo).

I admit that such features as waxy, hydrophobic conidia and conidia embedded in mucilage have considerable importance, but I believe that they represent ecologically imposed differences that have evolved again and again, often between phylogenetically closely related taxa. We also now know that fungal slimes are of various kinds, some merely hydrophilic, some actually surface-active (Bandoni, 1975).

Ingold (1942) discussed a third "biological" spore type, the "aquatic" spore, produced and dispersed underwater. This is an interesting group containing many beautiful examples of parallel or convergent evolution, as shown by Webster and Descals (1979), but it is numerically small (less than 200 species), hence of only minor importance in the general classification.

Langeron (1945), in his often admirable *Précis de mycologie*, tried to apply the terminology of Saccardo, Vuillemin, Mason, and Wakefield and Bisby, but his interpretation of terms such as "phialide" was unacceptably broad, and his work has not been influential.

H. Hughes

Working at the Commonwealth Mycological Institute in the late 1940s, under the tutelage of E. W. Mason, was a young Welsh mycologist, S. J. Hughes. Absorbing Mason's (and Vuillemin's) ideas, Hughes was soon to build on them and develop a stimulating new hypothesis. As one of a series of important papers on hyphomycete taxonomy, Hughes (1951) published a discussion of some phialidic conidial fungi and came up with what is essentially the current developmental definition of the phialide and its unique conidiogenesis. This paper was just a taste of what was to come. In 1953, he produced a remarkable contribution in which he suggested that there are "only a limited number of methods whereby conidia can develop from other cells." He proceeded to describe no fewer than eight such methods and ascribed prime taxonomic importance to them. I refer my readers to his classic paper, but also feel constrained to outline the eight sections he proposed, though using more current terminology to describe them.

Section I: blastic-synchronous (e.g., *Botrytis*) or blastic-acropetal (e.g., *Cladosporium*)
Section II: blastic-sympodial (e.g., *Nodulisporium*)
Section III: blastic-annellidic (e.g., *Scopulariopsis*)
Section IV: blastic-phialidic (e.g., *Penicillium, Phialophora*)
Section V: thallic-meristem (e.g., *Oidium*)
Section VI: blastic-tretic (porospores, e.g., *Alternaria*)
Section VII: thallic-arthric (e.g., *Oidiodendron, Geotrichum*)
Section VIII: basauxic (e.g., *Arthrinium*)

Hughes neither formally named his groups nor attempted an overall classification of conidial fungi using these developmental criteria. Yet he made it clear that he thought these features might well represent more natural bases than those on which the Saccardoan scheme had been erected. He also suggested that, although most of his examples were drawn from the Hyphomycetes, the same developmental criteria could be applied throughout the Coelomycetes as well.

I. 1968–1978

The next 15 yr (1953–1968) seemed to be a hiatus in ontogenetic studies on conidial fungi. However, Hughes's concepts were being internalized by other mycologists and eventually began to receive the attention they deserved. Barron (1968), Ellis (1971, 1976), Kendrick and Carmichael (1973) and, most recently Carmichael, Kendrick, Conners and Sigler (1980) produced successively more comprehensive compilations of Hyphomycetes and based their groupings in large measure on ontogeny.

In two major reviews, Sutton (1971, 1973) set out to apply developmental criteria to Coelomycetes, and in recent years he and Nag Raj have been steadily enriching the literature with painstaking, beautifully illustrated accounts of coelomycete genera (this volume, Chapter 3).

But the hoped-for, inclusive scheme has never materialized. Some of the reasons for this became clear at the first Kananaskis Conference (see Kendrick, 1971). While time-lapse photomicrographic studies had added another mode of conidiogenesis to the list (blastic-retrogressive: Cole and Kendrick, 1968; Kendrick, 1971; e.g., *Basipetospora, Trichothecium*), the conference noted that, the more fungi were subjected to ontogenetic analysis, the more atypical and intermediate kinds of conidiogenesis were found. We have since come to believe that there are fewer basic kinds of conidiogenesis than Hughes supposed and that each is capable of more variation than most of us had suspected.

Perhaps I can exemplify this by looking a little more closely at one very common kind of conidiogenous cell that has already come under scrutiny several times in this account, the phialide. Madelin (1966) considered that at least four different phenomena were lumped in Hughes's 1953 definition of the phialide. Questions regarding the homogeneity of the concept were raised repeatedly at the first Kananaskis Conference (see Kendrick, 1971). In 1972, Morgan-Jones, Nag Raj, and Kendrick reported a further variation on the phialidic theme, which they called percurrently proliferating phialides, in some Coelomycetes. After a consideration of the literature these authors were left wondering whether there was a fundamental difference between the phialide and the annellide, or whether a continuous spectrum of intermediates linked the two. At the second Kananaskis Conference (see Kendrick, 1979) it became clear that mycologists were still of two minds about phialides, some suggesting that all extant phialides (at least in ascomycetous anamorphs) shared common ancestry, and others that the phialide has evolved on at least several occasions. Note that, if the latter position turns out to be correct, then phialidic conidium ontogeny as a prime classificatory character will be revealed as essentially artificial.

J. 1979

We have known for many years that some holomorphs incorporate more than one conidial anamorph. The attempt to apply ontogenetic criteria to such fungi led to the conclusion that sibling anamorphs often display different modes of conidium ontogeny. Now we know that the same conidiogenous cell can function in different modes at different times. And at the second Kananaskis Conference, Madelin (1979) advanced a fascinating and plausible explanation of how one method of conidium ontogeny can be transmuted into another by a combination of temporal and edaphic factors. All of which leads me to an inescapable conclusion: Developmental criteria cannot form the basis for a more natural classification of the conidial fungi. This does not mean that the accumulation of developmental data has been a waste of time. On the contrary, it has increased our understanding of these organisms in many ways. But it does mean that extravagant claims can no longer be made for ontogeny. Development now becomes, in perspective, just one more kind of taxonomic information to be added to those we already possess.

As Müller (this volume, Chapter 7) shows so clearly, both morphologically and developmentally based systems sometimes break down in the face of proven connections between anamorph and teleomorph. For example, the anamorphs of closely related species of *Venturia* exhibit three different kinds of conidium ontogeny. It is hardly likely that these ontogenetic differences in anamorphs could bring about the separation of holomorphs at the generic level or above. [The anamorphs of three species of *Broomella* are disposed in three different anamorph genera in three anamorph families, when they should probably be regarded as congeneric in terms of the holomorphs.]

Fungal systematics is in a state of creative flux. In dealing with conidial fungi we must now consider morphology, development, and connections with teleomorphs. Only by integrating all three kinds of data can be hope to progress toward a more rational system of classification.

ACKNOWLEDGMENT

I am grateful for the award of a Guggenheim Fellowship, during the tenure of which this chapter was written.

REFERENCES

Ainsworth, G. C. (1965). Historical introduction to mycology. *In* "The Fungi" (G. C. Ainsworth, and A. S. Sussman, eds.), Vol. 1, pp. 3–20. Academic Press, New York.

Ainsworth, G. C. (1976). "Introduction to the History of Mycology." Cambridge Univ. Press, London and New York.

Bandoni, R. J. (1975). Surface-active spore slimes. *Can. J. Bot.* **53**, 2543–2546.

Barron, G. L. (1968). "The Genera of Hyphomycetes from Soil." Williams & Wilkins, Baltimore, Maryland.

Bonorden, H. F. (1851). "Handbuch der allgemeinen Mykologie." Stuttgart.

Carmichael, J. W., Kendrick, W. B., Conners, I. L., and Sigler, L. (1980). "Genera of Hyphomycetes." Univ. of Alberta Press, Edmonton.

Cole, G. T., and Kendrick, B. (1968). Conidium ontogeny in Hyphomycetes. The imperfect state of *Monascus ruber* and its meristem arthrospores. *Can. J. Bot.* **46**, 987–992.

Corda, A. K. J. (1837–1854). "Icones fungorum hucusque cognitorum," 6 vols. Prague.

Costantin, J. (1888). "Les Mucédinés simples: Matériaux pour l'histoire des champignons." Klincksieck, Paris.

De Bary, A. (1854). Ueber die Entwicklung und den Zusammenhang von *Aspergillus glaucus* und *Eurotium*. *Bot. Z.* **12**, 425–434, 441–451, 465–471.

De Bary, A. (1866). "Morphologie und Physiologie der Pilze, Flechten und Myxomyceten." Engelmann, Leipzig.

De Bary, A. (1884). "Vergleichende Morphologie und Biologie der Pilze, Mycetozoen und Bacterian." Engelmann, Leipzig.

Ellis, M. B. (1971). "Dematiaceous Hyphomycetes." Commonw. Mycol. Inst., Kew, Surrey, England.

Ellis, M. B. (1976). "More Dematiaceous Hyphomycetes." Commonw. Mycol. Inst., Kew, Surrey, England.

Fries, E. M. (1821–1832). "Systema Mycologicum, Sistens Fungorum Ordines, Genera et Species," 3 vols. Lund and Greifswald.

Fries, E. M. (1836–1838). "Epicrisis Systematis Mycologia." Uppsala and Lund.

Fuckel, L. (1869-1875). "Symbolae Mycologicae," 4 vols. Wiesbaden.
Gerarde, J. (1633). "The Herball or General Historie of Plantes . . . " London.
Grove, W. B. (1935). "British Stem- and Leaf-Fungi (Coelomycetes)," vol. 1. Cambridge Univ. Press, London and New York.
Grove, W. B. (1937). "British Stem- and Leaf-Fungi (Coelomycetes)," Vol. 2. Cambridge Univ. Press, London and New York.
Hughes, S. J. (1951). Studies on micro-fungi XI. Some Hyphomycetes which produce phialides. *Mycol. Pap.* **45**, 1-36.
Hughes, S. J. (1953). Conidiophores, conidia and classification. *Can. J. Bot.* **31**, 577-659.
Hughes, S. J. (1958). Revisiones Hyphomycetum aliquot cum appendice de nominibus rejiciendis. *Can. J. Bot.* **36**, 727-836.
Hughes, S. J. (1959). Starting point of nomenclature of Hyphomycetes. *Taxon* **8**, 96-103.
Ingold, C. T. (1942). Aquatic Hyphomycetes of decaying alder leaves. *Trans. Br. Mycol. Soc.* **25**, 339-417.
Kendrick, W. B., ed. (1971). "Taxonomy of Fungi Imperfecti." Univ. of Toronto Press, Toronto.
Kendrick, W. B., ed. (1979). "The Whole Fungus," vols. 1 and 2, National Museums of Canada, Ottawa.
Kendrick, W. B., and Carmichael, J. W. (1973). The Hyphomycetes. *In* "The Fungi" (G. C. Ainsworth, F. K. Sparrow, and A. S. Sussman, eds.), Vol. 4A, pp. 321-509. Academic Press, New York.
Kendrick, W. B., and Nag Raj, T. R. (1979). Morphological terms in Fungi Imperfecti. *In* "The Whole Fungus" (W. B. Kendrick, ed.), Vol. 1, pp. 43-62. National Museums of Canada, Ottawa.
Langeron, M. (1945). "Précis de mycologie." Masson, Paris.
Léveillé, J. H. (1846). "Considérations mycologiques, suivies d'une nouvelle classification des champignons." Paris.
Link, H. F. (1809). Observationes in ordines plantarum naturales. Dissertatio I. *Mag. Ges. Naturf. Freunde, Berlin* **3**, 3-42.
Madelin, M. F. (1966). The genesis of spores in higher fungi. *In* "The Fungus Spore" (M. F. Madelin, ed.), pp. 15-36. Butterworth, London.
Madelin, M. F. (1979). An appraisal of the taxonomic significance of some different modes of producing blastic conidia. *In* "The Whole Fungus" (W. B. Kendrick, ed.), Vol. 1, pp. 63-79. National Museums of Canada, Ottawa.
Mason, E. W. (1933). Annotated account of fungi received at the Imperial Mycological Institute, List II (Fasc. 2). *Mycol. Pap.* **3**, 1-67.
Mason, E. W. (1937). Annotated account of fungi received at the Imperial Mycological Institute, List II (Fasc. 3, General part). *Mycol. Pap.* **4**, 69-99.
Micheli, P. A. (1729). "Nova Plantarum Genera juxta Tournefortii Methodum Disposita." Florence.
Morgan-Jones, G., Nag Raj, T. R., and Kendrick, B. (1972). Conidium ontogeny in Coelomycetes. IV. Percurrently proliferating phialides. *Can. J. Bot.* **50**, 2009-2014.
Nees von Esenbeck, C. G. (1817). "Das System der Pilze und Schwämme." Würzburg.
Persoon, C. H. (1801). "Synopsis Methodica Fungorum." Göttingen.
Pirozynski, K. A., and Weresub, L. K. (1979). The classification and nomenclature of fossil fungi. *In* "The Whole Fungus" (W. B. Kendrick, ed.), Vol. 2, pp. 653-688. National Museums of Canada, Ottawa.
Potebnia, A. (1910). Beiträge zur Micromyceten Flora Mittel-Russlands. *Ann. Mycol.* **8**, 42-93.
Saccardo, P. A. (1884). "Sylloge Fungorum Omnium Hucusque Cognitorum," Vol. 3. Pavia.
Saccardo, P. A. (1886). "Sylloge Fungorum Omnium Hucusque Cognitorum," Vol. 4. Pavia.
Schenck, S., Kendrick, W. B., and Pramer, D. (1977). A new nematode-trapping Hyphomycete and a re-evaluation of *Dactylaria* and *Arthrobotrys*. *Can. J. Bot.* **55**, 977-985.
Spallanzani, L. (1776). "Opuscoli de Fisca Animale e Vegetabile." Modena.

Sutton, B. C. (1971). Conidium ontogeny in pycnidial and acervular fungi. *In* "Taxonomy of Fungi Imperfecti" (W. B. Kendrick, ed.), pp. 263-278. Univ. of Toronto Press, Toronto.
Sutton, B. C. (1973). Coelomycetes. *In* "The Fungi" (G. C. Ainsworth, F. K. Sparrow, and A. S. Sussman, eds.), Vol. 4A, pp. 513-582. Academic Press, New York.
Tulasne, L. R., and Tulasne, C. (1861-1865). "Selecta Fungorum Carpologia," 3 vols. Paris.
von Höhnel, F. (1911). Zur Systematik der Sphaeropsideen und Melanconieen. *Ann. mycol.* **9**, 258-265.
von Martius, C. F. P. (1817). "Flora Cryptogamica Erlangensis." J. L. Schrag, Nuremburg.
Vuillemin, P. (1910a). Matériaux pour une classification rationelle des Fungi Imperfecti. *C.R. Hebd. Seances Acad. Sci.* **150**, 882-884.
Vuillemin, P. (1910b). Les Conidiosporés. *Bull. Soc. Sci. Nancy* [3] **11**, 129-179.
Vuillemin, P. (1911). Les Aleuriosporés. *Bull. Soc. Sci. Nancy* [3] **12**, 151-175.
Wakefield, E. M., and Bisby, G. R. (1941). List of Hyphomycetes recorded for Britain. *Trans. Br. Mycol. Soc.* **25**, 49-126.
Webster, J., and Descals, E. (1979). The teleomorphs of water-borne Hyphomycetes from fresh water. *In* "The Whole Fungus" (W. B. Kendrick, ed.), Vol. 2, pp. 419-452. National Museums of Canada, Ottawa.
Weresub, L. K., and Pirozynski, K. A. (1979). Pleomorphism of fungi as treated in the history of mycology and nomenclature. *In* "The Whole Fungus" (W.B. Kendrick, ed.), Vol. 1, pp. 17-26. National Museums of Canada, Ottawa.

II
SYSTEMATICS

2

The Systematics of Hyphomycetes*

Bryce Kendrick

I.	Introduction—The Role of Systematists	21
II.	The Trouble with Anamorphs	22
III.	Classification—The Status Quo	23
IV.	Higher Taxa of Fungi—Heterogeneity	24
V.	Higher Taxa of Fungi—Ephemeral or Protean?	25
VI.	Pseudotaxa	27
VII.	Teleomorph + Anamorph(s) = Holomorph	28
VIII.	Anamorphic and Teleomorphic Holomorphs	30
IX.	Simple Anamorphs—Simplistic Taxa?	31
X.	Guidelines for Species and Genus	32
XI.	The Anamorph-Species and Its Problems	33
XII.	The Anamorph-Genus and Its Problems	35
XIII.	The Intergeneric Hiatus	37
XIV.	The Foundation	38
	References	40

I. INTRODUCTION—THE ROLE OF SYSTEMATISTS

Taxonomists inadvertently occupy one of two central positions among biological disciplines, although some of them may not realize it and many of them tacitly reject the responsibility this position involves. Their chief importance to human culture, apart from revealing to their fellow humans the dazzling, almost unbelievable richness and diversity of life on earth, is to collect and collate the information about living organisms acquired by all other disciplines and to construct a system—essentially a four-dimensional web involving space and

*This chapter is dedicated to the memory of Pat Talbot, whose untimely death robbed the mycological fraternity of a distinguished colleague, and me of the chance to benefit from his wisdom and experience during my stay in Australia.

time—which will link all these millions of multifarious organisms, extant and extinct, and show how they are related to one another (not how they relate to one another—that is ecology, the other central discipline).

Although the beginning taxonomist is often initially fascinated by the beauty, complexity, and essential strangeness of the particular group he has chosen to work with, he often eventually awakens to a realization of his deeper role and is sometimes foolhardy enough to embrace it, with all its attendant pitfalls, because it is a challenge worthy of a true Renaissance man. (No sexual bias intended: other possible constructions are too clumsy.)

II. THE TROUBLE WITH ANAMORPHS

But before the great web can be woven and the overall pattern made apparent, our understanding and arrangement of each group must be scientifically and intellectually satisfying. And it is always easy, from one's own vantage point, working with a particular group of organisms and only too aware of the difficulties and deficiencies of its systematics, to assume that the taxonomy of other groups is in much better condition than that of one's own. My later comments on ascomycetous and basidiomycetous teleomorphs will reveal the surprise I felt when I first realized the inadequacy of their systematics. But despite my attempts to compensate for a negative bias in my own thinking, I remain convinced that the systematics of conidial anamorphs is particularly disorganized. This conclusion does not imply that taxonomists of conidial fungi are any less competent, diligent, or inspired than their brethren in other groups: I am sure that this is not the case, since I know from personal acquaintance what penetrating, analytic and original minds the best of them have. No, the truth is that we have been faced with some extremely difficult problems, the most important of which is that we usually work with only part of an organism—only one phase of its life cycle. This problem has recently been extensively and intensively explored in *The Whole Fungus* (Kendrick, 1979). In a nutshell: The anamorphic or asexual phase of a dikaryomycotan fungus and its teleomorphic or sexual phase are capable of metamorphosing, each into the other, but we do not as yet understand the factors which bring about the phase change (Müller, 1979) and so cannot control them. So we know the full dual (or sometimes multiple) nature—the holomorph—of only a small percentage of the Dikaryomycota (Ascomycetes plus Basidiomycetes). And some may well have lost their anamorphic or their teleomorphic expression, although this is still difficult to prove. Because of such factors, I feel that any complacency at the appearance of so many useful books on conidial fungi during the last few years would be ill-founded. All these publications do no more than establish a baseline from which we can measure future progress. The conidial fungi remain an enigmatic—and an artificial—group.

III. CLASSIFICATION—THE STATUS QUO

To paraphrase a passage from *The Peter Principle:* "A classification may be incompetent in itself—that is, unable to do regularly and accurately the work for which it was designed... [and] even when competent in itself, a classification vastly magnifies the results of incompetence in its operators."

One of the most critical phrases in this passage is "the work for which it was designed." A classification may be admittedly artificial, based on only part of what we know about the organisms and designed simply and solely to facilitate identification—application of the "correct" binomial to unknown organisms—or it may be much more subtly constructed, based on everything we know about the organisms, with the aim of being natural and of expressing relationships, hence phylogeny. This prompts me to make two position statements. First, I do not believe that an effective hierarchical classification of conidial fungi now exists at either of these levels of sophistication. Second, since we aim ultimately to attain the higher level, a phylogenetic scheme, I cannot in this chapter discuss conidial anamorphs without some reference to teleomorphs, since only the whole fungus can be properly classified (Kendrick, 1979).

How can I say that no respectable classification exists when I frequently succeed in identifying hyphomycetes myself? The answer is that I do so not by referring to any published hierarchical scheme but rather by searching in my own memory bank (or herbarium specimens or generic monographs or "picture books"). I do not make this claim lightly, since part of the blame for the absence of a proper scheme may be laid at my door: Though in self-defense I must say that I have spent several years working on two comprehensive arrangements of hyphomycete genera (Kendrick and Carmichael, 1973; Carmichael *et al.*, 1980). If these do not represent "good" systems, what excuses or reasons can I offer for their deficiencies? The body of this chapter is devoted to answering this question.

A hierarchical scheme requires the recognition of taxa of higher and higher rank (species, genus, family, order, class, and so on). If the scheme is to withstand scrutiny, so must the taxa at all levels. Unfortunately for any existing schemes, but happily for the future of mycology, the artificiality and the misleading connotations of relatedness that mar many of the widely recognized fungal taxa are now being increasingly perceived and will be discussed below. Suffice it to say at this point that a formal high-ranking anamorph-taxon such as Deuteromycotina is of marginal use, even in a scheme aimed solely at identification, and must be anathema to anyone dedicated to the concept of the whole fungus. Kendrick and Nag Raj (1979) have seriously questioned the procrustean segregation of conidial anamorphs into the currently used anamorph-classes Coelomycetes and Hyphomycetes (and thus incidentally the rationale for dealing with them in separate chapters of this book). And below the class level, one encounters no widely accepted suprageneric categories—the orders and families of Saccardo began to wither away many years ago, and none of the more recently proposed ordinal or familial names (whether morphologically or ontogenetically based) have met with widespread approval.

So we have in effect a large number of anamorph-genera composed of anamorph-species. And even here the taxonomic concepts have been principally morphological, and often applied very broadly or very narrowly, depending on author and group. Compilations such as those by Barron (1968), Ellis (1971), Kendrick and Carmichael (1973), von Arx (1974), Ellis (1976), and Carmichael *et al.* (1980) are very useful and permit identification of an increasing number of anamorphs, but those of us who have made these compilations know only too well how much our confidence in the taxa varies and what an enormous task of generic reassessment and revision remains to be done. And meanwhile, the flood of new anamorph-taxa continues unabated. At their inception, very few of these anamorph-taxa are linked to a known teleomorph, so yet another vital job is usually left to be done later.

Perhaps this would be a good point at which to bring this general survey to a close and to begin analyzing specific areas in detail.

IV. HIGHER TAXA OF FUNGI—HETEROGENEITY

In his key to classes of Basidiomycotina, Ainsworth (1973) separated Hymenomycetes and Gasteromycetes, as has been common practice among mycologists. He noted that the Gasteromycetes are phylogenetically heterogeneous, and we had already been made aware by Heim (1948), Singer and Smith (1960), and Savile (1968) that true phylogenetic relationships in the Holobasidiomycetidae repeatedly cut across the lines separating these classes. Savile compiled no fewer than 14 lineages showing that the family Secotiaceae contains numerous links between Agaricales and gasteromycetous derivatives. Thus both the Secotiaceae and the Gasteromycetes as a whole are revealed as highly polyphyletic. Yet published systems, such as that given in *The Fungi*, Vols. IVA and IVB (Ainsworth *et al.*, 1973a,b), make no serious attempt to replace these categories with more natural ones.

Nobles (1958) showed that many of the families, and even genera, of the higher Basidiomycetes are heterogeneous. Characters of biochemistry and microscopic anatomy effectively dismember older categories based on gross morphology, especially in the polypores. But although many new generic names have been proposed and old ones resurrected as a result of this kind of work, revision of the classification is still incomplete, and its altered nature has not yet been fully appreciated or assimilated by the mycological public.

In his key to classes of Ascomycotina, Ainsworth (1973) recognized the Plectomycetes. Subsequently Malloch (1979) compiled the recent literature on cleistothecial ascomycetes, with the general conclusion that they are also polyphyletic, being derived in the main from four other presumably more natural groups of Ascomycetes—Pleosporales, Diaporthales, Hypocreales, and Pezizales. Note that these ancestors include bitunicate and unitunicate, perithecial, and apothecial forms. They could hardly be more diverse.

Trappe (1979) writes, "Of the 31 genera of hypogeous Ascomycotina accepted, one is assigned to the monotypic order Elaphomycetales ord.nov. and the rest are placed in the Pezizales. . . . I have discarded the artificial, anachronistic order Tuberales. All hypogeous genera that relate morphologically to families of Pezizales are assigned to those families." The Pezizales are commonly placed in a larger group of fungi which usually have apothecial ascomata and are known as Discomycetes. As one becomes familiar with these same Discomycetes, one is made aware that there are two kinds, those with operculate unitunicate asci and those with inoperculate unitunicate asci. Then one discovers the Patellariaceae and the aberrant *Pseudoscypha* Reid & Piroz. (Reid and Pirozynski, 1966) in which an apothecioid ascoma produces bitunicate asci. Thus one learns that the words "Discomycetes" and "apothecium" are morphologically rather than phylogenetically based. Now Trappe has shown that something classified as an "operculate discomycete" may not even produce an ascoma that is recognizably apothecioid, and that its asci may not be overtly operculate. Trappe has bitten the bullet, and we must applaud this. Unfortunately, for those of us who teach mycology, this means that we now have more explaining to do. The major groups of the Ascomycetes can no longer be recognized by simply examining the gross morphology of their ascomata. Nobody promised that a phylogenetically based classification would be simple or easy to cope with.

V. HIGHER TAXA OF FUNGI—EPHEMERAL OR PROTEAN?

Let us now examine the recent history of orders in the bitunicate ascomycetes. Three major efforts to categorize these fungi have appeared in the 1970s. Barr (1972) recognized three orders, Myriangiales, Dothideales, and Pleosporales. Luttrell (1973) accepted five orders—the three given by Barr plus Hysteriales and Hemisphaeriales. Von Arx and Müller (1975) treated all bitunicates in a single order, Dothideales. Since there is no question as to the experience or reputation of any of the workers just mentioned, we must conclude that they have different ideas about what constitutes an order. Von Arx and Müller obviously used the features of the bitunicate ascus itself as diagnostic. Barr used the three common types of centrum development. Luttrell recognized two additional orders. His Hysteriales accommodates forms in which the ascoma is definitely "discomycete-like," and he admits that this order is probably heterogeneous. He suggests, for example, that the Hysteriaceae could be natural members of the Pleosporales. The Dothioraceae, also apothecioid, have affinity with the Dothideales, and Luttrell placed them in this order. The Patellariaceae have no clear affinities, but their apothecioid ascomata allowed them to be placed in the Hysteriales. Luttrell also accepted the order Hemisphaeriales, while recognizing that it too is heterogeneous, since various of its members show internal structure that may eventually lead to their distribution among the other four orders. They are held together for the present by their dimidiate, appressed ascomata, whose evolutionary significance Luttrell felt it was premature to discount.

Von Arx and Müller had earlier (1954) recognized three orders (Pseudosphaeriales, Dothiorales, and Myriangiales) using characters of asci and ascomata, but subsequently found too many intermediates for comfort and ultimately decided that the bitunicates exhibited a virtual continuum of features.

I am sure we have not heard the last of this controversy, and I do not even suggest that such a divergence of opinions is bad for mycology. I simply wish to emphasize that we should not invest the names of some higher taxa with an aura of authority and immutability that they do not deserve. Classes, subclasses, and orders seem to be extraordinarily prone to changes in both name and circumscription.

I can advance two possible explanations for this. The first is that, as our knowledge of such groups grows, we can produce increasingly rational (though altered) taxa—look what electron microscopy and wall chemistry have done for the classification of what we used to call the Phycomycetes. The second is that the highest echelons of fungal taxa (Ascomycotina and Basidiomycotina, for example) are now considered fixed stars, capable only of gaining or losing a little mass (Endomycetes, Ustomycetes), and at the other end of the scale, individual genera and species tend to be revised or monographed by individual specialists only at long intervals and thus often give a possibly misleading appearance of stability. In contrast to the highest and lowest taxa, those at the middle levels of the hierarchy seem to attract attention from many taxonomists. Although it is usually impossible for any one worker to mongraph more than a small proportion of the genera in a given order or class, such workers usually develop a concept of the higher taxa uniquely colored by their own experience, which may change considerably during their working lifetime. So it is hardly surprising that current opinions (which are, after all, what many taxa are based on) are so diverse.

How have we gotten along without families of conidial fungi for so many years? This may well be because the family concept is of questionable usefulness. Perhaps the best way to exemplify this is by quoting from some universally known groups of flowering plants. The family Ranunculaceae is very heterogeneous—anything that will not fit into the Papaveraceae or Magnoliaceae tends to finish up in the Ranunculaceae. And how different *Ranunculus* is from *Aquilegia*. In fact, this example points to a logical corollary regarding the unsatisfactory family: It is usually easy to recognize the genera within it. But consider now the Cruciferae, all members of which share a number of very telling features. This is a good family. But the delimitation of genera within it is fraught with difficulties. It seems, then, that as a general principle, the better defined the family, the more difficult it is to recognize the genera within it (G. Jones, personal communication); ergo, in a large and varied group such as the Hyphomycetes, one may be better off without families.

Perhaps I may be allowed one more comment on high-ranking taxa before returning to my main theme. The higher the rank of a particular taxon, the more indistinct its periphery. I will content myself with a single example. Fungi, a taxon in no danger of disappearing, is variously defined as encompassing or excluding such groups as the Myxomycetes (Mycetozoa?), Labyrinthulomycetes, and Plasmodiophoromycetes. This is one consequence of the high probability that even what we all recognize as true fungi (the hyphal kind) comprise several groups which seem to have evolved separately from protistan ancestors. Other analogous solutions to basic biological problems have evolved many times, so why not the hyphal life style? (After all, we even have a prokaryotic parallel in the Actinomycetes.) Implications of monophyly in particular groups (the avowed goal of most systematists) are therefore limited at several levels of the fungal classification.

VI. PSEUDOTAXA

I have had to begin with all these unsettling comments about the systematics of the Dikaryomycota because the taxonomy of conidial anamorphic fungi is inextricably bound up with that of the sexual alter egos most of them presumably possess. Specialists in anamorphs have usually, with refreshing candor, admitted that their taxa are in fact pseudotaxa. For a long time these were called form-taxa, but this term has been preempted for use with fossils in a sense differing appreciably from that used in discussing dikaryomycotan anamorphs, and we have adopted the term "anamorph-taxa" for our purposes (see Pirozynski and Weresub, 1979). [*Form-taxa* are maintained for fossils of detached organs which lack diagnostic features indicating natural affinity. The application of this term is not governed by Article 59 of the Botanical Code of Nomenclature. *Anamorph-taxa* are maintained for organs of asexual or somatic reproduction in living fungi. The application of the term *is* governed by Article 59 (Weresub and Hennebert, 1979).]

Those who worked with teleomorphs generally did not admit that the taxa in which they arranged their collections were botanical only in name and based on an arbitrary, legalistic decision of a botanical congress. In fact, as I think I have already demonstrated clearly, many such taxa of what used to be called perfect fungi have turned out to be just as artificial as—and possibly more so than—some anamorph-taxa.

In the term "Fungi Imperfecti" we have escaped some of the misconceptions engendered by such misleading terms of convenience as "Discomycetes," "Pyrenomycetes," "Plectomycetes," and "Gasteromycetes." We have always been fully aware that the Fungi Imperfecti are a miscellaneous group, though even here mycological textbooks have erred by relating them more or less exclusively to ascomycetous teleomorphs (we now recognize numerous basidiomycetous anamorphs; see Kendrick and Watling, 1979). We also fully realize that any and all subdivisions that have been proposed within the Fungi Imperfecti have few—or no—phylogenetic pretensions.

As I pointed out in Chapter 1, the term "Coelomycetes" was coined by Grove (1919) for anamorphic fungi that produced conidia inside initially covered conidiomata. Perhaps because this concept provided a good foil, an "either-or" alternative to the Hyphomycetes with their conidiophores exposed *ab initio,* it has been widely adopted. But we have gradually come to see that the apparent homogeneity of the Coelomycetes is an illusion. We now consider that acervular and pycnidial fungi are not closely related and that pycnidial conidiomata can develop in several fundamentally different ways. Acervular conidiomata grade imperceptibly into sporodochial conidiomata, and thus a link is established between the coelomycetous anamorph-order Melanconiales and the hyphomycetous anamorph-order Moniliales. The walls of pycnidial conidiomata may develop in such a way as to leave an internal cavity, or they can develop as a solid structure within which a secondary lysigenous or schizogenous cavity arises. The conidia may escape through a preformed ostiole or through a secondarily developed lysigenous or schizogenous opening. Although we do not know the phylogenetic significance of such differences, it seems likely that the anamorph-order Sphaeropsidales is internally heterogeneous.

VII. TELEOMORPH + ANAMORPH(S) = HOLOMORPH

I recently edited a publication entitled *The Whole Fungus* (1979) which was based largely on the deliberations of the second Kananaskis Conference held in 1977. The theme of the conference and the book is the recognition and realization of a unified conceptual approach to all phenotypic manifestations of each fungal genome. Of course this was far too ambitious—anamorph–teleomorph connections are known or suspected for only about 10% of higher fungal species. And there remains the strong possibility—in my mind virtually a certainty—that many anamorphs will ultimately be shown to be independent entities that have lost the need and the ability to produce the teleomorph they once had (this idea is further discussed in Section VIII). Just what proportion of conidial fungi will turn out to be anamorphic holomorphs is anyone's guess. And how many are teleomorphic holomorphs is equally obscure. But the number of connections will increase dramatically, and the classification will become increasingly based on data derived from both sexual and asexual phases.

But where do we stand now? Our attempt in *The Whole Fungus* to discern patterns of relationship using both morphs had mixed results. In some areas the process has so far generated more heat than light. And even in a group such as the Nectriaceae, whose anamorphs are comparatively well documented, the picture is far from clear. Booth (1978) has pointed out that *Nectria coccinea* has red, smooth-walled perithecial ascomata without a stroma and has a *Cylindrocarpon* anamorph. *Nectria ralfsii* has yellow, rough-walled perithecia on a well-developed stroma and has a dark-spored *Myrothecium* anamorph. These differences at first sight appear to merit generic segregation, until a wide range of *Nectria* species is examined, when the gaps are bridged and the initially clear-cut hiatus disappears.

Anyone who has read this far must by now be aware that, while the extant classification of the fungi is an inadequate reflection of our knowledge, our knowledge is still inadequate to permit the elaboration of a truly satisfactory classification. We have learned enough about the fungi in recent years to realize that they are sending us back to the drawing board; but we have not yet been granted the insight needed to visualize the grand design.

If the linking of anamorphs with teleomorphs does not give us the breakthrough we are looking for, what does it tell us? I think it tells us that, when species of the same genus evolve into different ecological niches (as they must if they are to be separate species), then evolutionary pressures may well mold anamorph or teleomorph (or both, as in the *Nectria* species mentioned above) into new and different shapes. And since anamorph and teleomorph often develop at different times, and serve different purposes, the fact that each may change independently of the other is not difficult to understand. Nevertheless, the evolutionary process springing from this simple principle has led to situations which challenge the most imaginative mycological system builder.

The expectation that there will be one anamorph per teleomorph is soon dispelled when the connections compiled by Kendrick and DiCosmo (1979) are examined. It is equally clear that this concept breaks down when considered from either side. Anamorph-species of the anamorph-genus *Chrysosporium* are, *pro tem*, associated with representatives of no fewer than 17 ascomycetous genera (*Ajellomyces, Anixiopsis, Aphanoascus, Apinisia, Arachniotus, Arthroderma, Ctenomyces, Echinopodospora, Emmonsiella, Gymnoascus, Neoxenophila, Renispora, Rollandina, Thielavia, Xanthothecium, Zendera,* and *Zopfiella*) and 1 basidiomycetous genus (*Phanerochaete*). Species of *Acremonium* are associated with 30 ascomycetous teleomorphs. More of this later. Looking at the situation from the other side, we find equally spectacular examples of such dissonances. Kendrick and DiCosmo (1979) list connections established between species of the ascomycetous (teleomorphic) genus *Mycosphaerella* and no fewer than 27 anamorph-genera. *Nectria* and *Leptosphaeria* both have connections with 19 anamorph-genera, and *Ceratocystis* and *Hypomyces* with 15 each. This list can easily be extended with less spectacular, but nevertheless disturbing, examples. The figures following the teleomorph names are the numbers of anamorph-genera connected with each: *Physalospora,* 14; *Guignardia,* 14; *Pleospora,* 11; *Cordyceps,* 10; *Cucurbitaria,* 10; *Chaetosphaeria,* 8; *Massaria,* 8; *Botryosphaeria,* 7; *Pezicula,* 7; *Ophiobolus,* 7; *Didymella,* 7; *Godronia,* 7. Of course these "multiple-anamorph" genera do not, like *Chrysosporium,* break the cardinal rule of "one teleomorph per holomorph," but they still present a tangled thicket for mycologists to penetrate.

What do such lists mean? (1) They mean that an element of inaccuracy is involved; mycologists have traditionally been too narrow in their knowledge of the fungal groups. Some teleomorph specialists have had inadequate concepts of the anamorphic genera they encountered, and some anamorph specialists insufficient acquaintance with the teleomorphic genera—or each has obtained bad advice from the other—and so some of the identifications will be found on closer scrutiny to be wrong. (2) They mean that, if the connections are rechecked and found (largely) to be well-founded, then the integrity of the single genus (whether teleomorphic or anamorphic) to which the many are connected must be rigorously examined, since it may be in need of splitting. (3) If the single genus involved on one side of the question is sound, then we have to accept that the phase on the other side of the equation has undergone tremendous radiative evolution.

None of the foregoing should be taken to mean that we can abandon our efforts to link teleomorph and anamorph: Our data base is enriched by every authentic connection made. It simply means that we must not expect too much from this source—it will not solve all our problems.

VIII. ANAMORPHIC AND TELEOMORPHIC HOLOMORPHS

When one first encounters the pseudotaxa of anamorphs in the guise of Fungi Imperfecti with appropriate binomials, one is told that these pseudotaxa are ephemera and that an eventual and inevitable epiphany will reunite them with their respective teleomorphs (long arrogantly called perfect states, and still the source of all holomorph names where a teleomorph is known). But even today, after a century of effort, connections remain in the minority. As I have already indicated, I find it difficult to maintain this doctrine of the perfectibility of fungi. Apart from the entire absence of teleomorph associations for many anamorph-genera, we need only look at other groups of organisms to find examples of how under certain circumstances sexual reproduction can apparently be selected against and sometimes lost. As far as is known, some aphids are obligately parthenogenetic; some mosses have never been seen to fruit; among the flowering plants *Lemna minor,* although capable of flowering, normally reproduces indefinitely in the asexual mode with great success; one of the most common and successful weeds around Adelaide, Australia, is *Oxalis pes-caprae,* which is present in Australia only as a completely sterile pentaploid, reproducing by means of bulbils; in addition, many obligate apomicts in the genera *Rubus, Taraxacum, Alchemilla, Poa, Cassia,* and others, though retaining the external trappings of sex, in fact avoid genetic recombination completely and produce fruit only as a highly evolved dispersal mechanism.

Of course, organisms which dispense with genetic recombination are hostages to fortune. Although they have a genome which is superbly attuned to existing conditions, if these conditions should change dramatically, the species is likely to be wiped out, while other more flexible organisms adapt and survive.

But if we now look back at the conidial fungi we find that they are not actually devoid of all opportunity for genetic recombination. The phenomenon of parasexuality may well provide an adequate amount of recombination, and haploid organisms producing the huge numbers of conidia common among molds must obviously give rise to mutant strains. Anastomosis between hyphae of different strains, and the heterokaryosis arising from such encounters, must help to buffer the conidial fungi against environmental fluctuations. In Chapter 14 Wicklow expands on the reasons for the transcendent success of conidial anamorphs. Most of these could also, I am sure, be advanced as reasons for loss of sexuality. All in all, I see few objections to the idea of anamorphic holomorphs, although I know it will be very difficult to prove their existence unequivocally.

It may be equally challenging to establish the existence of teleomorphic holomorphs, but it seems very probable that they too can be found. For example, we know of no anamorphs in entire families of the Ascomycetes—Brefeldiellaceae, Mesnieraceae, Micropeltidaceae, Microthyriaceae, Piedraceae, Zopfiaceae—and it now seems unlikely that they will be found. Negative proof is usually difficult to establish and, from what Müller (1979) has shown us about the factors inducing phase change, these are as variable and unpredictable as we might expect from organisms with such diverse ecological requirements as the fungi. Nevertheless, I am sure that many presumptive single-phase holomorphs will eventually be confirmed as such.

IX. SIMPLE ANAMORPHS—SIMPLISTIC TAXA?

This section could be subtitled, "They all look the same to me." One of the difficulties facing those who would systematize conidial anamorphs is the relative paucity of morphological features available in a fairly large number of anamorph-genera. *Chrysosporium,* the anamorph-genus mentioned earlier in connection with a plurality of teleomorphic genera, is an example of this. It does not produce a highly characteristic, differentiated, branched, or otherwise complex conidiophore, and its kind of generalized "chlamydosporic" propagules may well have arisen on many occasions—in fact, its seeming violation of the "one teleomorph per holomorph" rule is convincing evidence for this. A second case is presented by *Rhizoctonia* (strictly speaking, this is one of the Mycelia Sterilia and not a conidial anamorph, but we usually have to deal with this fringe group by default). This anamorph-genus often produces poorly defined microsclerotia in the soil. Anamorph-species of this pseudotaxon have been connected in culture to six different teleomorphic genera (*Ascophanus, Ceratobasidium, Sebacina, Thanatephorus, Trichophaea, Tulasnella*—Warcup and Talbot, 1966, 1967). Considering that two of these teleomorphs are Ascomycetes, while the other four are Basidiomycetes, it is clear that the current concept of *Rhizoctonia* is insufficiently precise. But from the stature of the mycologists who established the connections, we can be sure that the matter of separating ascomycetous from basidiomycetous anamorphs, even at the anamorph-generic level, will be far from easy.

My last example of an oversimplified anamorph-generic concept arising directly from simple morphology is *Acremonium.* This genus produces simple, often individual, phialides. Although the phialidic mechanism of conidium ontogeny is a fairly complex and highly evolved one (Kendrick, 1971, 1979; Cole and Samson, 1979), the physical reality of single, tapering phialides is outwardly simple in the extreme and conducive neither to the segregation of closely related taxa nor even to the separation of such monophialidic anamorphs evolved as parts of very different holomorphs. If a single anamorph-genus such as *Acremonium* is polyphyletic, as is apparent from its association with no fewer than 30 ascomycetous telemorphic genera (28 unitunicates, 2 bitunicates—see Kendrick and DiCosmo, 1979), then we must also entertain the possibility that morphologically different phialidic genera such as *Penicillium* and *Chalara* may be only distantly related, despite their relatively similar conidium ontogenies. Each of these two anamorph-genera has, as a matter of fact, relatively homogeneous teleomorphs, but both (and especially *Penicillium*) probably include anamorphic holomorphs among their species.

X. GUIDELINES FOR SPECIES AND GENUS

The literature of hyphomycete taxonomy has grown without reference to any overall principles and in a completely unregulated manner. This is good in that it has given workers of talent and initiative the freedom to explore new ground, but bad in that it has allowed some of the worst taxonomy I have ever seen to enter the literature. I doubt whether a single formal university course has ever been given to train specific individuals to perform this task. Lumpers and splitters, cognoscenti and ignorami—all have added their contributions, ranging from the invaluable to the invalid. I have often thought that all descriptions of new taxa should be submitted to an international screening committee which would be able to veto inadequate, iterative, or incorrect manuscripts. No such body exists in mycology, with the result that each year sees a further accretion of bad taxa or bad descriptions which will eventually have to be sorted out by someone whose time would be better spent doing original work. Thus, although there should be no need to produce guidelines to anamorph-taxa, perhaps the following outline will be helpful to the beginner.

1. Collections sharing great similarity in almost all known morphological and developmental features (often virtual congruence), and usually having similar habitat or host preferences, can be regarded as conspecific: members of the same anamorph-species.

2. Collections which are as similar to each other as those grouped under the first heading, and which share most major features with that group, yet exhibit one or more marked and consistent differences from that group, are regarded as congeneric though not conspecific with it. The two groups constitute distinct anamorph-species belonging to the same anamorph-genus.

3. Collections which, while sharing major features with the first two groups, are significantly more different from both of them than they are from each other can be regarded as belonging to a different anamorph-genus. Thus intergeneric hiatuses (Singer, 1975) are greater than those found between species within a genus.

4. Above the generic level there are no particularly helpful familial or ordinal groupings: All those encountered should be treated with suspicion.

XI. THE ANAMORPH-SPECIES AND ITS PROBLEMS

The anamorph-species should not present too much difficulty, once it has been perceived by the individual. The gaining of this perception is a most valuable step in the training of a taxonomist. An important principle is not to make the organism fit the literature. As a fledgling mycologist I found a helicosporous hyphomycete which I tried to identify in Linder's excellent monograph of the group (1929). The closest match I could come up with did not seem perfect, yet because it was the nearest thing in the literature I accepted it and applied the name to the fungus I had found. A subsequent trip to Kew and consultation with Martin Ellis soon convinced me that the species I had found simply was not in the literature, and I proceeded to describe it as a new species. That hurdle is an important one for anamorph specialists to overcome, because we know that many new taxa remain to be described. But the manner in which the hurdle is overcome is important too, in that the beginner should have the benefit of experienced advice and should seek it out wherever it may be found, lest he or she become too cautious or too confident.

In most situations, members of the same anamorph-species may be regarded as being genuinely genetically related to one another. Note that this is the only point in anamorph taxonomy at which we can now claim such a relationship.

But of course even the anamorph-species, with which we should have the least trouble, can be relied upon to cause problems. The first of these arises when we discover that what appears to be a single anamorph-species is connected to more than one teleomorph. Now while it is all right for one teleomorph to have two or more different anamorphs (see example given earlier), it is axiomatic that any given fungal genome is capable of giving rise to one, and only one, teleomorphic phase. So when we read of dual or multiple teleomorphs, we are justifiably skeptical. Clearly, the anamorph-species can betray us by concealing genetic heterogeneity beneath the mask of morphological congruence. This is usually a result of convergent evolution, though it may also indicate that some inadequate taxonomy has been going on. Which brings me to the second problem, that of standardization of circumscription. Although the species is the foundation of all taxonomy, no one has yet come up with the perfect prescription, let alone persuaded everyone else to adopt such a concept. A species is still what a taxonomist says it is. And although I am not among those who are willing to abandon their taxonomic prerogatives to some numerical or computerized—and hence suspiciously procrustean—scheme, I believe hyphomycete taxonomists should communicate with each other much more than they do and try to establish realistic guidelines. This will call for some sacrifice of individual freedom in the decision-making process, but it would probably lend taxonomy an extra measure of credibility in the eyes of the scientific community at large. Considering our central role, I think we should strive for and earn such acceptance instead of maintaining the aloof, almost Olympian, attitude some of our fraternity espouse. Needless to say, while promulgating the idea of a degree of conceptual standardization, I lack the temerity, as well as the enormous breadth of knowledge, needed to prepare such proposals. Perhaps they could be formulated at a future Kananaskis Conference.

A third problem of the anamorph-species concept is bound up with information retrieval, for example, in genera with very large numbers of species. While in Australia I collected fungi on *Eucalyptus* but usually found myself entirely unable to decide on which of the about 500 species of the host genus the fungus was growing. Another Australian genus, *Acacia,* is equally large and difficult for the nonspecialist. As the task of describing the world's mycoflora moves slowly toward completion, fungal examples of this kind of problem will inevitably multiply.

A fourth area in which our current species concept is apparently inadequate is that of economically important or intensively worked genera, in which significant differences (e.g., in biochemistry, or in host–pathogen relationships) have been documented in otherwise almost identical organisms. In some groups such as the rust fungi (which have of course anamorphic phases) this has engendered a trinomial nomenclature—and even this cannot cope with our detailed knowledge of the fungi causing some cereal rusts. In other groups, such as *Penicillium* and *Aspergillus,* this has led to the recognition of infrageneric but supraspecific series, which could easily be regarded as an equivalent of the "aggregate species" recognized by angiosperm taxonomists. The actual anamorph-species of *Penicillium* and *Aspergillus* may well be essentially subspecific taxa and indicate that the groups concerned are speciating rapidly under strong selection pressures.

Just to show that such problems are not universal in conidial fungi, I will cite the example of the anamorph-genus *Arthrinium,* whose species share some very distinctive features—the basauxic growth pattern of their conidiophores; the presence in many species of very thick, darkly pigmented cross-walls in the conidiophores; and the production of often strangely shaped, dark, amerosporous conidia, which often have a well-marked germ slit. Surely this anamorph-genus will eventually be proven to be a good botanical entity (or at least the anamorphic expression of one).

XII. THE ANAMORPH-GENUS AND ITS PROBLEMS

From 1958 to 1965, in a series of six papers, M. B. Ellis described no fewer than 64 species in the anamorph-genus *Sporidesmium*. The breadth of his interpretation of this genus can be appreciated by browsing through these papers (Ellis, 1958, 1959, 1961, 1963a,b, 1965) and perhaps even from studying Fig. 1. He describes the conidiophores as discrete, mononematous, unbranched, and brown; conidiogenous cells as integrated, terminal, determinate, or percurrent; and conidia as solitary, dry, borne apically, unbranched, and of various shapes (straight, curved, sigmoid, cylindrical, fusiform, obclavate, obpyriform, obturbinate, and sometimes beaked). Their color ranges from subhyaline through straw-colored, to pale or dark brown, olivaceous brown, or reddish brown. Their walls may be smooth or rough. They are transversely septate or pseudoseptate. This is obviously a broad circumscription. In fact, in some senses the limits of this anamorph-genus can be best expressed in terms of the features that currently separate other, similar dematiaceous hyphomycetes from it. *Clasterosporium* and *Ceratophorum* both produce hyphopodia, and in *Annellophora* the conidia

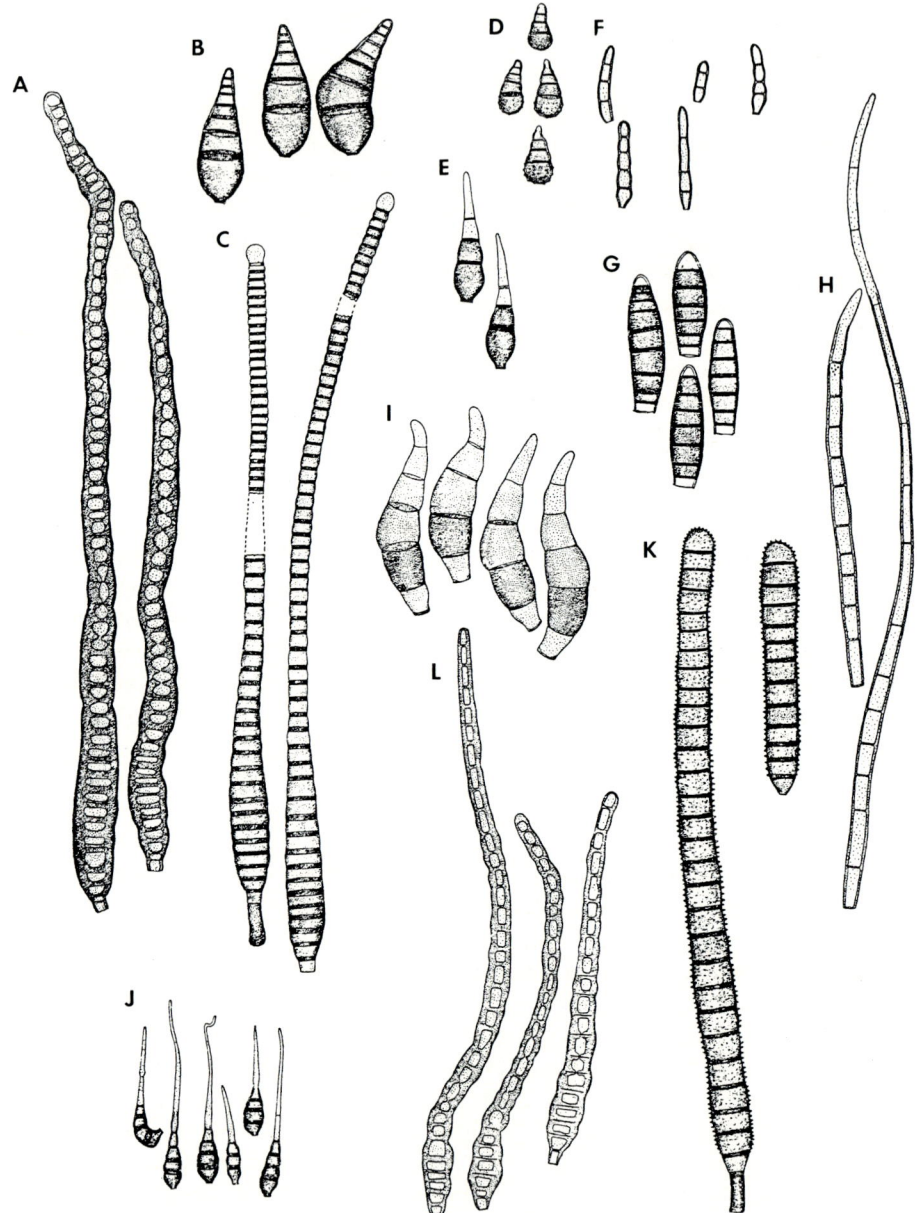

Fig. 1. Conidia of 12 species of *Sporidesmium*, all drawn to the same scale. (A) *Sporidesmium adscendens;* (B) *S. altum;* (C) *S. anglicum;* (D) *S. aturbinatum;* (E) *S. bicolor;* (F) *S. cajani;* (G) *S. cambrense;* (H) *S. eucalypti;* (I) *S. inflatum;* (J) *S. longirostratum;* (K) *S. raphiae;* (L) *S. vagum.* (After M. B. Ellis.)

often proliferate at the apex to form secondary conidiophores and conidia, becoming annellate in the process. *Sporidesmium,* then, lacks hyphopodia, and its conidia do not proliferate. But after these hiatuses have been established, there remains great latitude, which Ellis has clearly made use of. It can be suggested that his generic concept is too broad, that it hinders a ready visualization in the mind's eye of what a *Sporidesmium* looks like. This may be deemed of minor importance, or it may be regarded as a major obstacle to identification of the genus by nonspecialists (think how readily you can visualize some of the common anamorph-genera of the Hyphomycetes—*Aspergillus, Penicillium, Fusarium*—despite the numerous species each contains). This conflict of interests is one of our perennial problems in information storage and retrieval because, when the generic concept is relatively narrow and easily visualized, species differentiation within the genus is often difficult. When the genus is broadly defined, species differentiation within it may be much easier. We should tread carefully between the extremes I have just outlined, and many taxonomists wander far enough in one direction or the other to exacerbate, rather than minimize, problems of identification—the process of information retrieval may reveal the classification as incompetent. This whole business resembles the family–genus conflict mentioned earlier and seems to be a difficulty inherent in hierarchical schemes. The assignment of characters to the various levels of the hierarchy must be done with great insight and with a fine sense of balance and proportion. Often this is only possible through hindsight, after the units of the classification have initially been established in an *ad hoc* manner (as they usually are) by many different workers at different times.

Because much of hyphomycete taxonomy has so far simply been a matter of convenience, we are burdened with many anamorph-genera which are extremely narrowly defined. The generic characters are either arbitrarily chosen (for example, differences in spore septation) or seem inconsequential to the disinterested (unbiased) observer. I will not cite particular examples, but a number of cases may be discerned with ease from a scrutiny of the plates in Carmichael *et al.* (1980), and some have been discussed in detail by Kendrick (1980).

XIII. THE INTERGENERIC HIATUS

Booth (1978) pointed out that the evidence for a particular intergeneric hiatus may be drawn from too few collections. Consider species A and G in Fig. 2. The generic constellation to which these two actually belong will ultimately be found to contain 12 species. The catch is that 10 of these have not been collected or described. So A and G may easily be perceived as having a larger taxonomic distance, or hiatus, between them than is actually the case. This may well lead to their being assigned to different genera. In this and other ways, the literature

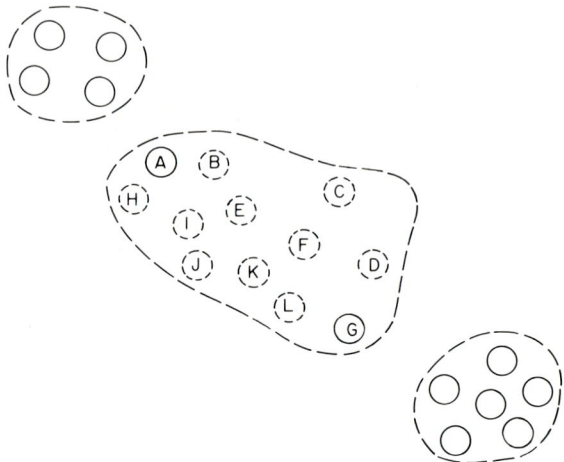

Fig. 2. The intergeneric hiatus.

becomes clogged with genera that will later be recognized as redundant. This process—the description of taxa and their subsequent redundancy—is an inevitable accompaniment of the slow maturation of taxonomy and is not necessarily an indication of any lurking incompetence in those who initially describe these taxa. In some cases, however, the hiatus may be too rigidly and arbitrarily prescribed. Genus X may be defined as having amerosporous conidia, while genus Y, otherwise very similar to X, has didymosporous conidia. Such unthinking reliance on a single generic character is common among the Hyphomycetes and undoubtedly retards taxonomic progress.

XIV. THE FOUNDATION

I have printed a rather bleak picture. You may be left with the impression that the work of more than a century has been singularly unproductive. This would not be a fair assessment. I have focused on problem areas and have not discussed the strong points of hyphomycete taxonomy. Perhaps this is a good point at which to try to redress the balance.

We now have hundreds of extremely well-characterized and illustrated anamorph-genera. We have a large number of reliably established connections between anamorph and teleomorph. And despite the fact that no acceptable system has yet emerged, we have many inklings and indications which will serve as intellectual stimuli to the generation of mycologists now gestating in the universities of the world. Let me exemplify these insights.

Hughes (1953) recognized eight ways in which conidial fungi formed their spores. The proceedings of the first Kananaskis Conference (Kendrick, 1971) reduced the number of basic mechanisms, and we are now aware that, for example, phialidic conidium ontogeny and annellidic conidium ontogeny are more closely related than we believed then. Hughes's initial analytic exercise was parent to all subsequent versions (inevitably it too had antecedents, for which refer to Kendrick, this volume, Chapter 1). Without the initial splitting, the subsequent refinements would never have been realized. I have often stressed the usefulness of an initial splitting process in erecting classifications. This prevents loss of information during the subsequent (and almost inevitable) lumping that ensues when the disguise of superficial differences is penetrated and underlying similarities are perceived. Barron (this series, Vol. II, Chapter 20) suggests that, with one exception, all the adhesive-spored Hyphomycetes which are endoparasites of minute soil animals are probably related to *Verticillium:* this despite the fact that they have been described in *Cephalosporium, Harposporium, Meria, Plesiospora, Cephalosporiopsis,* and *Paecilomyces.* Barron thinks they are probably all derived from verticillate ancestors and are more closely related to each other than to other species of the genera in which they were described.

I had earlier (in Schenck *et al.,* 1977) arrived at similar conclusions about a number of nematode-trapping Hyphomycetes with sympodially proliferating conidiophores, and conidia which were amerosporous, didymosporous, or phragmosporous. Although they were distributed among the anamorph-genera *Arthrobotrys, Dactylaria, Candelabrella,* and *Genicularia,* the last two names may be considered as doubtfully distinct from *Arthrobotrys,* which is much the earliest name, and the nematode-trapping *Dactylaria* species were separated from *Arthrobotrys* only by their phragmosporous conidia. *Arthrobotrys* was traditionally treated as being exclusively didymosporous, but at least two amerosporous species have now been described, and it seems illogical to exclude phragmosporous but otherwise similar *Dactylaria* species, especially when the type species of this genus grows on wood and does not trap nematodes.

For many years, obligately dimorphic anamorph-genera bedeviled our attempts to establish a classification of anamorphs. We recently (Carmichael *et al.,* 1980) took what seems to us the eminently sensible step of applying the appropriate generic name (if one existed) to each of the associated morphs. Thus the dimorphic name *Chalaropsis* is now recognized as a pleomorphic anamorph, with morphs in *Chalara* and in *Humicola.* The pleomorphic anamorph-genus *Diheterospora* is logically dissected into its *Stemphyliopsis* and *Verticillium* morphs.

From ongoing studies of such groups as the nectrioid fungi we are learning more about relationships which override such Saccardoan features as conidiophore aggregation or the lack of it. We have learned, the hard way, that the grouping of anamorphs by morphological similarity alone is often misleading, and we are looking more deeply into this. The anamorph specialist must now learn to spot characters more subtle yet more basic than any used so far. What are these features, and where does the future lie?

I once wrote that I thought the future lay in an understanding of fungal ecology and the impact that ecological pressures had made on fungal phenotypes during the course of evolution. I still believe this will be of enormous help. But I also feel that we need as many other kinds of data as possible to make up for the relative dearth of fossils (or, rather, knowledge about them) and the relative morphological simplicity of most fungi. To follow the lead of botanists, we need to establish our species on a more nearly genetic base. This means, among other things, chromosome counts. I used to think this was a pipedream as far as the conidial fungi were concerned, since their nuclei are small and the chromosomes have been assumed to be extremely contracted or compacted together, and thus very difficult if not impossible to count and characterize. But this kind of information is now within reach, as Robinow shows (this series, Vol. 2, Chapter 25). I am confident that other now apparently inaccessible kinds of data will become available in the years ahead and will ultimately make it possible to produce a truly rational classification of the conidial anamorph, dovetailed into that of the whole fungus.

ACKNOWLEDGMENTS

This chapter was written during my tenure of a Guggenheim Fellowship and while I was distinguished visiting scholar at the Waite Agricultural Research Institute of the University of Adelaide, South Australia. I thank Dr. Glyn Jones of Royal Holloway College, University of London, for sharing many taxonomic insights with me.

REFERENCES

Ainsworth, G. C. (1973). Introduction to keys to higher taxa. *In* "The Fungi" (G. C. Ainsworth, F. K. Sparrow, and A. S. Sussman, eds.), Vol. 4A, pp. 1–7. Academic Press, New York.

Ainsworth, G. C., Sparrow, F. K., and Sussman, A. S., eds. (1973a). "The Fungi" Vol. 4A. Academic Press, New York.

Ainsworth, G. C., Sparrow, F. K., and Sussman, A. S., eds. (1973b). "The Fungi," Vol. 4B. Academic Press, New York.

Barr, M. E. (1972). Preliminary studies on the Dothideales in temperate North America. *Contrib. Univ. Mich. Herb.* **9**, 523–638.

Barron, G. L. (1968). "The Genera of Hyphomycetes from Soil." Williams & Wilkins, Baltimore, Maryland.

Booth, C. (1978). Do you believe in genera? *Trans. Br. Mycol. Soc.* **71**, 1–9.

Carmichael, J. W., Kendrick, W. B., Conners, I. L., and Sigler, L. (1980). "The Genera of Hyphomycetes." Univ. of Alberta Press, Edmonton.

Cole, G. T., and Samson, R. A. (1979). "Patterns of Development in Conidial Fungi." Pitman, London.

Ellis, M. B. (1958). *Clasterosporium* and some allied Dematiaceae-Phragmosporae. I. *Mycol. Pap.* **70,** 16–84.
Ellis, M. B. (1959). *Clasterosporium* and some allied Dermatiaceae-Phragmosporae. II. *Mycol. Pap.* **72,** 73–75.
Ellis, M. B. (1961). Dematiaceous Hyphomycetes. III. *Mycol. Pap.* **82,** 45–46.
Ellis, M. B. (1963a). Dematiaceous Hyphomycetes. IV. *Mycol. Pap.* **87,** 25–31.
Ellis, M. B. (1963b). Dematiaceous Hyphomycetes. V. *Mycol. Pap.* **93,** 25–28.
Ellis, M. B. (1965). Dematiaceous Hyphomycetes. VI. *Mycol. Pap.* **103,** 43–46.
Ellis, M. B. (1971). "Dematiaceous Hyphomycetes." Commonw. Mycol. Inst., Kew, Surrey, England.
Ellis, M. B. (1976). "More Dematiaceous Hyphomycetes." Commonw. Mycol. Inst., Kew, Surrey, England.
Grove, W. B. (1919). Mycological notes. IV. *J. Bot., Br. Foreign* **57,** 206–210.
Heim, R. (1948). Phylogeny and natural classification of macrofungi. *Trans. Br. Mycol. Soc.* **30,** 161–178.
Hughes, S. J. (1953). Conidia, conidiophores and classification. *Can. J. Bot.* **31,** 577–659.
Kendrick, W. B., ed. (1971). "Taxonomy of Fungi Imperfecti." Univ. of Toronto Press, Toronto.
Kendrick, W. B., ed. (1979). "The Whole Fungus," Vols. 1 and 2. National Museums of Canada, Ottawa.
Kendrick, W. B. (1980). The generic concept in Hyphomycetes—a reappraisal. *Mycotaxon.* **11,** 339–364.
Kendrick, W. B., and Carmichael, J. W. (1973). Hyphomycetes. *In* "The Fungi" (G. C. Ainsworth, F. K. Sparrow, and A. S. Sussman, eds.), Vol. 4A, pp. 323–509. Academic Press, New York.
Kendrick, W. B., and DiCosmo, F. (1979). Teleomorph-anamorph connections in Ascomycetes. *In* "The Whole Fungus" (W. B. Kendrick, ed.), Vol. 1, pp. 283–410. National Museums of Canada, Ottawa.
Kendrick, W. B., and Nag Raj, T. R. (1979). Morphological terms in Fungi Imperfecti. *In* "The Whole Fungus" (W. B. Kendrick, ed.), Vol. 1, pp. 43–61. National Museums of Canada, Ottawa.
Kendrick, W. B., and Watling, R. (1979). Mitospores in basidiomycetes. *In* "The Whole Fungus" (W. B. Kendrick, ed.), Vol. 2, pp. 473–545. National Museums of Canada, Ottawa.
Linder, D. H. (1929). A monograph of the helicosporous Fungi Imperfecti. *Ann. Mo. Bot. Gard.* **16,** 227–388.
Luttrell, E. S. (1973). Loculoascomycetes. *In* "The Fungi" (G. C. Ainsworth, F. K. Sparrow, and A. S. Sussman, eds.), Vol. 4A, pp. 135–219. Academic Press, New York.
Malloch, D. (1979). Plectomycetes and their anamorphs. *In* "The Whole Fungus" (W. B. Kendrick, ed.), Vol. 1, pp. 153–165. National Museums of Canada, Ottawa.
Müller, E. (1979). Factors inducing asexual and sexual sporulation in fungi (mainly ascomycetes). *In* "The Whole Fungus" (W. B. Kendrick, ed.), Vol. 1, pp. 265–282. National Museums of Canada, Ottawa.
Nobles, M. K. (1958). Cultural characters as a guide to the taxonomy and phylogeny of the Polyporaceae. *Can. J. Bot.* **36,** 883–926.
Pirozynski, K. A., and Weresub, L. K. (1979). The classification and nomenclature of fossil fungi. *In* "The Whole Fungus" (W. B. Kendrick, ed.), Vol. 2, pp. 653–688. National Museums of Canada, Ottawa.
Reid, J., and Pirozynski, K. A. (1966). A new Loculoascomycete on *Abies balsamea* (L.) Mill. *Can. J. Bot.* **44,** 351–354.
Savile, D. B. O. (1968). Possible interrelationships between fungal groups. *In* "The Fungi" (G. C. Ainsworth and A. S. Sussman, eds.), Vol. 3, pp. 649–675. Academic Press, New York.

Schenck, S., Kendrick, B., and Pramer, D. (1977). A new nematode-trapping hyphomycete and a reevaluation of *Dactylaria* and *Arthrobotrys*. *Can. J. Bot.* **55,** 977–985.

Singer, R. (1975). "The Agaricales in Modern Taxonomy," 3rd ed. Cramer, Weinheim.

Singer, R., and Smith, A. H. (1960). Studies on secotiaceous fungi. IX. The astrogastraceous series. *Mem. Torrey Bot. Club.* **21,** 1–112.

Trappe, J. M. (1979). The orders, families, and genera of hypogeous Ascomycotina (truffles and their relatives). *Mycotaxon* **9,** 297–340.

von Arx, J. A. (1974). "The Genera of Fungi Sporulating in Pure Culture," 2nd ed. Cramer, Vaduz.

von Arx, J. A., and Müller, E. (1954). Die Gattungen der amerosporen Pyrenomyceten. *Beitr. Kryptogamenfl. Schweiz.* **11,** 1–434.

von Arx, J. A., and Müller, E. (1975). A re-evaluation of the bitunicate Ascomycetes with keys to families and genera. *Stud. Mycol.* **9,** 1–159.

Warcup, J. H., and Talbot, P. H. B. (1966). Perfect states of some rhizoctonias. *Trans. Br. Mycol. Soc.* **49,** 427–435.

Warcup, J. H., and Talbot, P. H. B. (1967). Perfect states of rhizoctonias associated with orchids. *New Phytol.* **66,** 631–641.

Weresub, L. K., and Hennebert, G. L. (1979). Anamorph and teleomorph: Terms for organs of reproduction rather than karyological phases. *Mycotaxon* **8,** 181–186.

3

Coelomycete Systematics

T.R. Nag Raj

I.	Introduction	43
II.	Techniques	47
III.	Morphological and Ontogenetic Criteria	49
	A. Conidiomata	49
	B. Conidiophores, Conidiogenous Cells, and Sterile Elements	60
	C. Conidia	63
	D. Conidium Ontogeny	69
	E. Taxonomy of Coelomycetes Growing in Artificial Cultures	71
	F. Discussion	73
IV.	Some Aspects of the Biology of Coelomycetes	75
	Addendum	79
	References	79

I. INTRODUCTION

The terms "anamorph," "teleomorph," and "holomorph" (Hennebert and Weresub, 1977) are already being widely used in mycological literature. The first two terms refer to a particular morph, state, or phase of a holomorph or whole fungus. Teleomorphs represent forms which reproduce with meiospores resulting from plasmogamy and karyogamy of sexually differentiated cells; anamorphs represent forms which propagate by means of mitoconidia resulting from mitosis. In theory, the concept of a holomorph is based on the assumption that it comprises a teleomorph—an ascomycete or a basidiomycete—with or without one or more related anamorphs. In practice, however, one can never be certain if the fungus is in reality a holomorph, particularly when pleomorphic fungi are involved. Such a nagging doubt stems from the fact that in nature teleomorphs and anamorphs are separated spatially or in time; or, more often, an anamorph may not be correlated with a teleomorph, or the latter may no longer exist. (In *Eutypella prunastri,* the causal agent of gummosis or dieback of apricots, the *Cytosporina* anamorph develops first in the infected living branches but is untraceable by the time the teleomorph appears on dead branches, usually 2 or 3 yr after the initial infection.)

In essence, we are left with a motley assemblage of anamorphic fungi on the one hand, and teleomorphic fungi or the so-called perfect fungi in the Ascomycetes and Basidiomycetes on the other hand. The ideas of perfect and imperfect fungi were so fixed and dominant in the minds of mycologists at the time the Code of Botanical Nomenclature was first adopted in fungal systematics that they allowed the teleomorph to outweigh the anamorph for nomenclatural priority. It was expected that the large and heterogeneous assemblage of anamorphic fungi would dwindle as more and more anamorph—teleomorph connections became established and the fungi were disposed in their appropriate slots in a natural scheme of classification reflecting their phylogenies. However, with the passage of time, the anamorphic fungi, rather than disappearing, have increasingly become a group to be reckoned with, new taxa being constantly added to those already known. The provisions of the Code were then modified to allow binomials for the anamorphic fungi as form-genera and form-species (now more correctly called anamorph-genera and anamorph-species) under an artificial subdivision called Deuteromycotina or Fungi Imperfecti, which includes two large subgroups: the Hyphomycetes and Coelomycetes.

The first few coelomycetes—acervular and pycnidial anamorphs—were recognized as early as 1796. In the following 170 years, their number grew to 1115 anamorph-generic names and about 6500 anamorph-species names (Ainsworth, 1967). Considering the time span involved, one might be justified in expecting the systematics of the group to be as advanced as the systematics of the teleomorphic Ascomycetes and Basidiomycetes; but such is not the case. This is mainly because so few mycologists have concentrated on coelomycetous anamorphs. As examples, let us consider *Phyllosticta* Desm. (nom. cons.), type sp.: *Phyllosticta convallariae* Pers., and *Phyllostictina* Syd., type sp.: *Phyllostictina murrayae* Syd. The anamorph-genus *Phyllosticta* was first introduced by Persoon in 1818 and validated by Desmaziéres in 1847. Small, globose pycnidia producing ameroconidia and occurring in more or less characteristic leaf spots were the features stressed in the generic concept by Saccardo (1878, 1884) and accepted by mycologists working with the group until as recently as 5 yr ago. By this time the number of anamorph-species names in the genus exceeded 2000. The anamorph-generic name was conserved at the International Botanical Congress in 1965. The anamorph-genus *Phyllostictina* was proposed in 1916 for *Phyllosticta*-like forms with conidia possessing a mucous sheath and an apical mucoid appendage. By 1973, *Phyllostictina* included 84 anamorph-species. Van der Aa (1973) concluded that the species of *Phyllostictina* belonged to *Phyllos-*

ticta in which he accepted 46 species, mostly as transfers from *Phyllostictina*. Subsequently, Punithalingam (1974) established by type studies that *Phyllostictina murrayae* did indeed belong to *Phyllosticta*. These conclusions affected the work of plant pathologists in two ways: (1) The status of a large number of plant-pathogenic *Phoma*-like forms that were being identified as known species of *Phyllosticta* was in limbo, and (2) the epithets of *Phyllostictina* commonly recognized so far had to be switched to those under *Phyllosticta*, occasionally leading to confusion. There are many more similar examples in the scattered early mycological literature relating to the Coelomycetes.

For lack of an up-to-date and reliable compilation of data comparable to that available for the Hyphomycetes (Kendrick and Carmichael, 1973; Carmichael *et al.*, 1980) a neophyte student or even an applied mycologist of good standing is often frustrated in attempts to establish the identity of a coelomycete collection. This plus the problems until recently inherent in coelomycete systematics account for the relative lack of active interest in coelomycete taxonomy among the mycological fraternity.

The predominant approach to coelomycete systematics has been based on morphological criteria, though pathogenic, cultural, biochemical, and serological characteristics have occasionally been used to differentiate anamorph-species and lower taxa. Following Persoon (1801), several schemes of classification were proposed for the fungi. Many such schemes grouped discordant forms together, acervular and hyphomycetous anamorphs with rusts in one group, and pycnidial anamorphs with ascomycetes in the second. The scheme proposed by Saccardo (1880, 1884), found to be more practicable, was widely accepted and remained static for a considerable length of time. For the anamorphic fungi, Saccardo employed the characteristics of conidium-bearing structures to separate the pycnidial Sphaeropsideae, the acervular Melanconieae, and the mononematous, synnematous, and sporodochial Hyphomyceteae. Familial categories were recognized according to color, texture, dehiscence, and the relationship of the conidium-bearing structures to their substrates: Sphaeropsidaceae, Nectrioidaceae, Leptostromataceae, and Excipulaceae in the Sphaeropsideae, and Melanconiaceae in the Melanconieae. Conidium characteristics such as color, septation, shape, and size aided in recognition of the anamorph-genera, while anamorph-species were recognized largely on the basis of presumed host–parasite relationships. While Saccardo's scheme had its good points, the use of conidium color and septation to separate anamorph-genera had the deleterious effect of separating apparently related anamorphs into different genera (e.g., *Neottiospora* Desm. and *Samukuta* Subram. & Ramakr.) and conversely in aggregating apparently unrelated taxa in a single anamorph-genus [e.g., *Robillarda* (Sacc.) Sacc. and *Pseudorobillarda* Morelet]. Grove (1935, 1937) accepted two categories of anamorphic fungi: the Hyphomycetes, in which the conidia are borne on conidiophores on the exterior of the substrate; and the

Coelomycetes, in which the conidia are borne in a cavity lined by the host tissue of fungal tissue or a combination of both; but his approach to the recognition of genera and species was no different from that of Saccardo.

Disenchanted by the inequities inherent in Saccardo's scheme of classification, many workers embarked on a search for more stable and reliable criteria. Following the leads of Vuillemin (1910a,b, 1911, 1912) and Mason (1933, 1937), Hughes (1953) proposed an experimental classification for the Hyphomycetes based on conidium ontogeny. Extensive research in the years that followed established that Hughes's approach offered better criteria but by itself was not a practicable substitute for that of Saccardo. It became clear that a feasible scheme of classification might be based on an integration of conidium ontogeny and existing morphological data. Such an approach has now made it possible to secure a cleaner delineation of concepts for the Hyphomycetes (Kendrick and Carmichael 1973; Carmichael *et al.*, 1980). In the last 25 yr, a similar approach stressing the correlation of precise information on conidium ontogeny with morphological criteria has been extended to some coelomycete taxa (Boerema and Bollen, 1975; DiCosmo, 1978; Morgan-Jones, 1971a-c, 1973, 1975; Morgan-Jones and Kendrick, 1972; Morgan-Jones *et al.*, 1972a-g; Nag Raj, 1973a,b, 1974, 1975a,b, 1977a-d, 1978a,b, 1979a; Nag Raj and DiCosmo, 1978, 1980; Nag Raj and Kendrick, 1970, 1971, 1972a,b, 1978; Nag Raj and Morgan-Jones, 1973; Nag Raj *et al.*, 1972; Pirozynski and Morgan-Jones, 1968; Pirozynski and Shoemaker, 1971; Sutton, 1961, 1963, 1964a-c, 1967a,b, 1968a, 1969, 1970, 1971a,b, 1973, 1975a,b; Sutton and Chao, 1970; Sutton and DiCosmo, 1977; Sutton and Kobayashi, 1970; Sutton and Pirozynski, 1963, 1965; Sutton and Sandhu, 1969; Sutton and Sellar, 1966).

Despite the seemingly impressive contributions to coelomycete taxonomy in recent years, a considerable amount of work still needs to be completed before a clear and comprehensive overview will be possible. The early achievement of such a goal is hindered by several obstacles, especially since our morphological and ontogenetic approach is tied to the type method. The first of these problems stems from the fact that early mycologists held the view that defining a species on mere morphological grounds was extremely difficult or even impossible and must rest on host differences. This and the tendency on the part of some phytopathologist-mycologists to describe new taxa without reference to the known or validly described species, resulted in unmanageably large genera such as *Phyllosticta, Phoma, Septoria,* and *Hendersonia.* The second problem arose because many early mycologists did not save type specimens. Descriptions, with or without illustrations, were considered the most important. Type specimens of many of the old taxa in European herbaria are inaccessible or untraceable. This in turn leads to the third problem of unknown ontogeny among the type species of some anamorph-genera. An added complication is the fact that species of many of the large anamorph-genera of the Coelomycetes exhibit considerable heterogeneity with regard to conidium ontogeny and morphology (e.g., *Septoria, Ascochyta*). The next problem relates to the fact that in most coelomycetes, particularly the pycnidial and stromatic anamorphs, the conidiogenous cells and the conidia are enclosed in host or fungal tissue or a combination of both, and the ontogenetic details are often obscure. Further, the conidiogenous cells in many such taxa are very small, and techniques for evaluating them are neither well known nor widely used. A problem of interpretation may also arise where small conidiogenous cells bearing catenate conidia with hardly perceptible morphological variations are involved. In such cases it may be difficult to determine if the conidia are being formed from a meristem at the apex of the conidiogenous cell or by fragmentation of the conidiogenous cell. These problems account for the slow pace of critical reevaluation of generic concepts in the Coelomycetes.

II. TECHNIQUES

In recent years, many workers have become fascinated by the modern techniques of scanning electron microscopy (SEM) and transmission electron microscopy (TEM) and have begun to voice the conviction that electron microscopy will provide the ultimate answers to all questions in taxonomy raised by users of conventional light microscope techniques. While it is possible that TEM might facilitate some interpretations in a few problem areas of coelomycete systematics, it is obvious that electron microscope techniques cannot aid in the routine examination and identification of collections. The less expensive, less time-consuming, and simpler techniques of light microscopy still offer an adequate means of studying the group and of making the identifications needed by applied mycologists.

Our optical equipment includes a binocular dissecting microscope, a binocular research microscope fitted with phase optics yielding magnifications of $60\times$ to $2000\times$, a similar research microscope equipped for photomicrography, a camera lucida, and stage and ocular micrometers.

To begin with, one needs to determine the nature of the conidium-bearing structure of the coelomycetous anamorph. It is not rare that overlooking such an essential detail has led some workers to dispose a fungus collection in an alien category. Arrangement of the conidiogenous cells around the cavity, and textural details of the peridium (when present), are some of the data that can be derived from sections of the conidiomata. A freezing microtome is ideal for obtaining reasonably thin sections quickly. [We use a Reichert freezing microtome, the original freezing stage of which has been replaced by an off-the-shelf thermal electric module about $32 \times 32 \times 5$ mm, requiring external water cooling (which is built into the substage) and a dc supply at 3.5 V, 8.5 A (available from Cambion-Cambridge Thermionic Corporation, 445 Concord Avenue, Cambridge, Massachusetts 02138, No. 801–2001–01–00–00; water-cooled unit available as No. 806-1006-01-00-00).] If the collection to be examined is fresh, soft material, presoaking is not essential. On the other hand, if the material is a dried herbarium specimen or consists of lignified material, presoaking in water or water containing a few drops of lactic acid (over a period of 1-48 h depending on the nature of the material involved), or in a 2% aqueous solution of potassium hydroxide for about 15 min, is necessary. The pretreated material is rinsed in water, placed in a drop of water-soluble resin (Cryoform, Damon, IEC Division, 300 Second Avenue, Needham Heights, Massachusetts 02194, No. 3383, obtainable from scientific supply companies) on the freezing stage, oriented suitably, and frozen. Sections cut to a thickness of 10-20 μm are transferred from the cutting edge of the blade to water. Good resin-free sections are transferred with a fine camel's hair brush or a fine pin onto a small drop of water on a microscope slide which is then carefully inverted and carefully lowered onto a clean 18 mm sq coverslip. A drop of 70% aqueous lactic acid is placed at one edge of the coverslip, and the water is drained with a small piece of absorbent paper placed at the other edge. Following a similar procedure, the mountant is completely replaced with concentrated lactic acid. It is imperative that the slide mount be well dehydrated before being sealed; this is achieved by warming the slide gently on a slide warmer, making sure that the mountant does not boil. Excess lactic acid at the edges of the coverslip, if any, should be removed by draining with a strip of absorbent paper or wiping with a moistened facial tissue, taking care not to damage the sections. The slide mount is then sealed with Glyceel (G. D. Gurr, available from Hopkins and Williams, Chadwell Heath, Essex, England), or clear nail lacquer. A second coat of sealant applied after an interval of 24 h will prolong the life of these semipermanent mounts.

Some of the sections left from the preceding step can be used to prepare squash or tease mounts to separate individual or small groups of conidiogenous cells. Quite often, a single conidioma can easily be separated from the covering host tissue or substrate after it has undergone presoaking treatment. Such a conidioma or appropriate sectioned material is placed in a drop of water on a microscope slide, and the tissues of the conidioma are teased apart with the aid of very fine entomological pins. Covering the material on the slide with a coverslip and sliding it sideways while exerting gentle pressure on it may help in making good squash mounts of well-separated conidiogenous cells. In some coelomycetes such as *Chaetospermum,* the gelatinous nature of the conidiomata frustrates efforts to obtain tease mounts. In such cases, soaking the sections of the conidiomata in 1 N hydrochloric acid with a gentle warming will eliminate the mucilaginous matrix; before making lactic mounts, repeated rinsing of such preparations is also necessary to obtain usable preparations. Further steps in preparing lactic mounts of tease or squash mounts are as described for the sections. Conidium mounts are also prepared in a similar manner. The only exception lies in the case of coelomycetous anamorphs where the conidia bear mucoid appendages that become invisible in lactic mounts even under phase optics (e.g., *Coniella*). In such cases, the appendages are best studied and photographed in aqueous mounts containing traces of methylene blue or crystal violet. All the morphological and ontogenetic data used in this chapter and other publications emanating from this laboratory have been derived from the techniques outlined above.

III. MORPHOLOGICAL AND ONTOGENETIC CRITERIA

A. Conidiomata

In coelomycetous anamorphs conidium formation occurs on or in specialized structures that we believe defy simple categorization into either acervuli or pycnidia, because of their morphological diversity. Potebnia (1910) recognized the variability of such structures and placed acervular and sporodochial forms in one category, the pycnidial forms in a second, and the variable types in a third category—pseudopycnidia. In more recent times, the five morphological categories—acervuli, pycnidia, stromata, cupulate, and pycnothyrioid pycnidia—recognized by von Höhnel (1923) have been widely accepted, although some forms could not be satisfactorily placed in any of them. From an analysis of 200 coelomycete genera, Kendrick and Nag Raj (1979) have concluded that the general term "conidiomata" is more appropriate for the conidium-bearing structures of these organisms, with the affinities of the conidiomata being indicated by the terms "acervular," "pycnidial," "pycnothyrial," and "intermediate." Current assessments bear out the conclusion of Potebnia (1910) concerning the variability of the conidiomata in the Coelomycetes. As the data accumulate, it appears that the distinctions between Hyphomycetes and Coelomycetes need to be abandoned because of the apparent continuum which exists among hyphomycetous, acervular, stromatic, and cupulate anamorphs, with the pycnidial anamorphs alone being a relatively distinct group. (This hypothesis seems to derive additional support from the common occurrence of annellidic conidium ontogeny in the hyphomycetous and acervular forms and its rarity among pycnidial forms; see Section III, D).

Acervuli (Plate I, Figs. 1–3) can be subcuticular, intraepidermal, subepidermal, or subperidermal in the host, but lack an enclosing fungal layer. The conidiophores arise from the upper layer of cells of a basal stroma of varying thickness, and the overlying host tissues are eventually ruptured by the developing mass of conidia. Acervuli can be glabrous or setose. The setae are usually many-septate, thick-walled, and pigmented; in some anamorph-genera they are randomly scattered throughout the conidiomata, while in a few they are marginal. In *Plectronidiopsis,* under dry conditions, the densely packed setae are

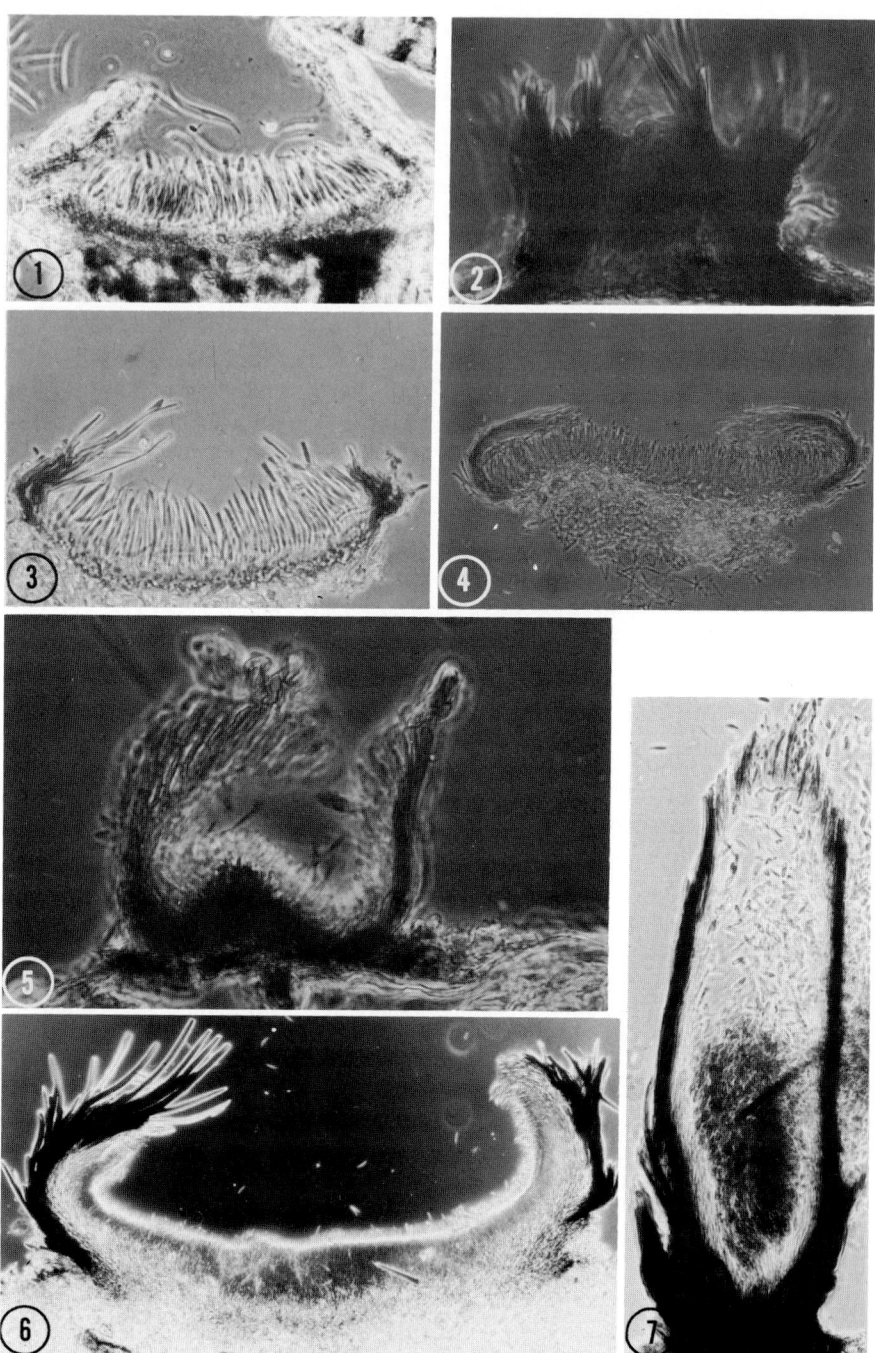

Plate I. Figs. 1–7.

usually incurved over the hymenium, and in this condition the acervuli are easily mistaken for pycnidia. In some fungi, the initially closed pycnidia finally become wide open and are often interpreted as acervuli (Kempton, 1919). At other times cupulate conidiomata have been described as acervuli.

Cupulate conidiomata (Plate I, Figs. 4–7) are discrete, usually superficial, with variable excipular development and adorned with sterile hyphae or setae that are hyaline or pigmented and vairably septate (*Myxormia, Hainesia, Dinemasporium, Polynema, Stauronema, Pseudolachnea*). The variability in conidiomatal morphology—ranging from sporodochial to cupulate and acervular types—exhibited by some anamorphic fungi make it difficult to establish generic limits. A typical example is found in the monographic account of *Myrothecium* (Tulloch, 1972), in which the sporodochia of several species of *Myrothecium*, the synnematous *Saccardoa atra,* and the cupulate forms of *Myxormia atroviride, Myrothecium gramineum,* and *M. leucotricha,* were disposed in a single anamorph-genus. Sporodochia are the typical conidiomata of some hyphomycetes: The conidiophores are closely aggregated and are often formed above the surface of the substrate, characters that differentiate them from acervuli. In cultures, such distinctions are hard to make. It is apparent that acervular, sporodochial, and some cupulate conidiomata have a much closer affinity with each other than with the pycnidial forms.

The conidiomata of Leptostromataceae, Pycnothyriaceae, and Peltasterales are distinct from the pycnidial or stromatic categories. They are generally flat, elongated, hemispherical (*Discosiella, Suttoniella*), or irregular (*Pycnidiopeltis*) and are referred to as pycnothyria (Plate II, Figs. 8–10). They can be subcuticular (*Discosiella*) or superficial (*Poropeltis*). In *Tracyella,* the conidiomata are composed of an upper shield supported by a central column of cells, and conidiogenesis is confined to the resulting concavity. In some others the locule can be simple or divided, and conidiogenesis may occur only at the base (*Acarellina*), only from the lower part of the apical well (*Allothyriella*), or from both (*Schizothyra*). In *Discosiella,* the covering involucre is thick, but there is virtually no underlying stromatic tissue contiguous with it. There is no regular dehiscence to release conidia in *Tracyella*; in others dehiscence may occur by an ostiole (*Discosia, Discosiella*), a slit (*Placella*), irregular tears (*Suttoniella*), or longitudinal fissures (*Leptostroma*).

Simple to complex conidiomata are found in the pycnidial (Plate III, Figs. 11–16; Plate IV, Figs. 17–20) and stromatic (Plate V, Figs. 21–26; Plate VI, Figs. 28–30) categories. Discrete pycnidia are globose to flask-shaped, usually

---

Plate I. Figs. 1–7. Vertical sections through conidiomata, acervular, cupulate, and cornute types. Fig. 1. *Seimatosporium falcatum.* Fig. 2. *Annellolacinia dinemasporiodes.* Fig. 3. *Plectronidiopsis chilensis.* Fig. 4. *Myxormia* sp. Fig. 5. *Satchmopsis brasiliensis.* Fig. 6. *Pseudolachnea bubaki.* Fig. 7. *Hoehneliella perplexa.*

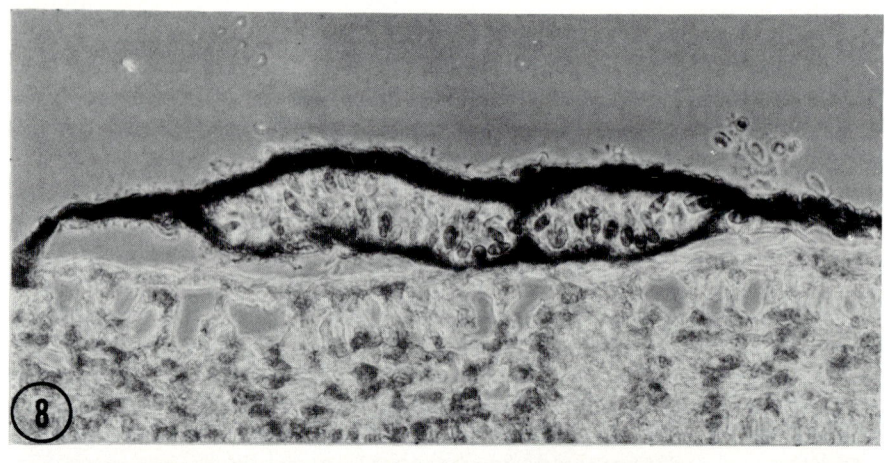

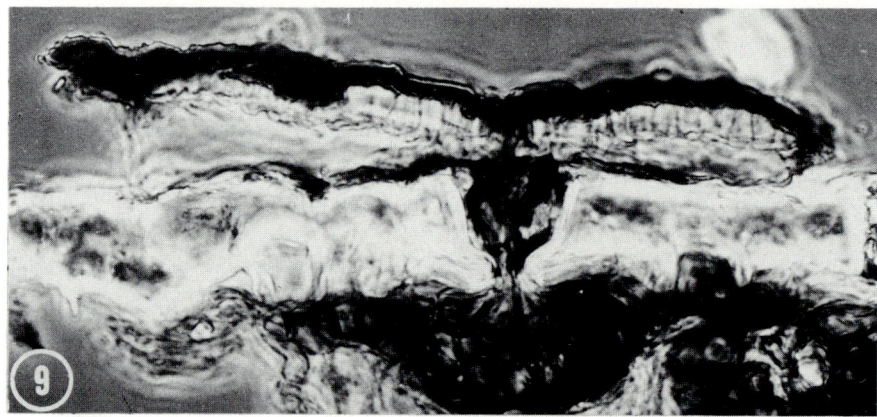

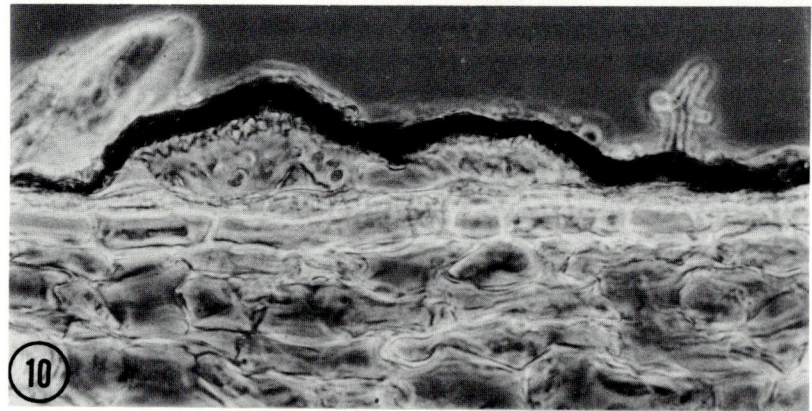

Plate II. Figs. 8–10. Vertical sections through pycnothyrial conidiomata. Fig. 8. *Poropeltis davilliae*. Fig. 9. *Tracyella aristata*. Fig. 10. *Diedickea singularis*.

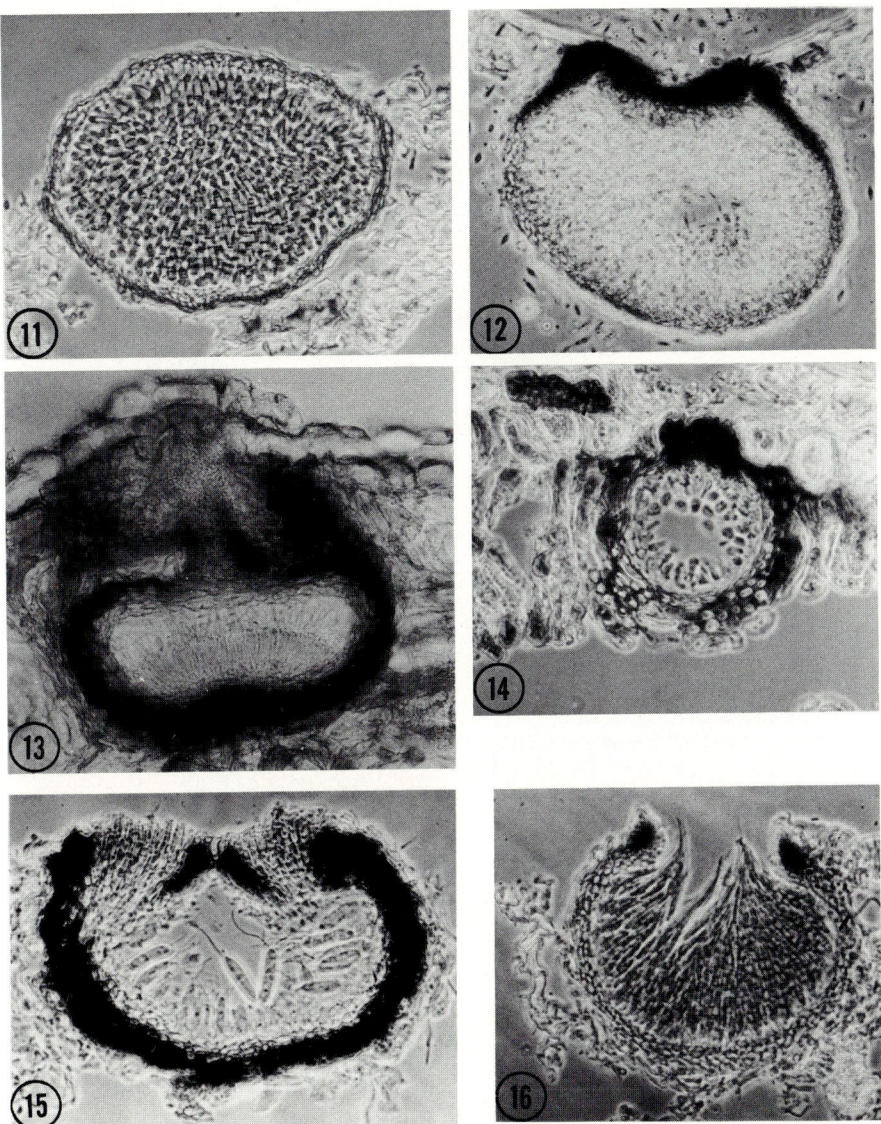

Plate III. Figs. 11–16. Vertical sections through pycnidial conidiomata. Fig. 11. *Doliomyces senegalensis*. Fig. 12. *Neottiospora caricina*. Fig. 13. *Coniella eucalypticola*. Fig. 14. *Phyllosticta* sp. Fig. 15. *Kellermania uniseptata*. Fig. 16. *Scolecosporiella sisyrinchii*.

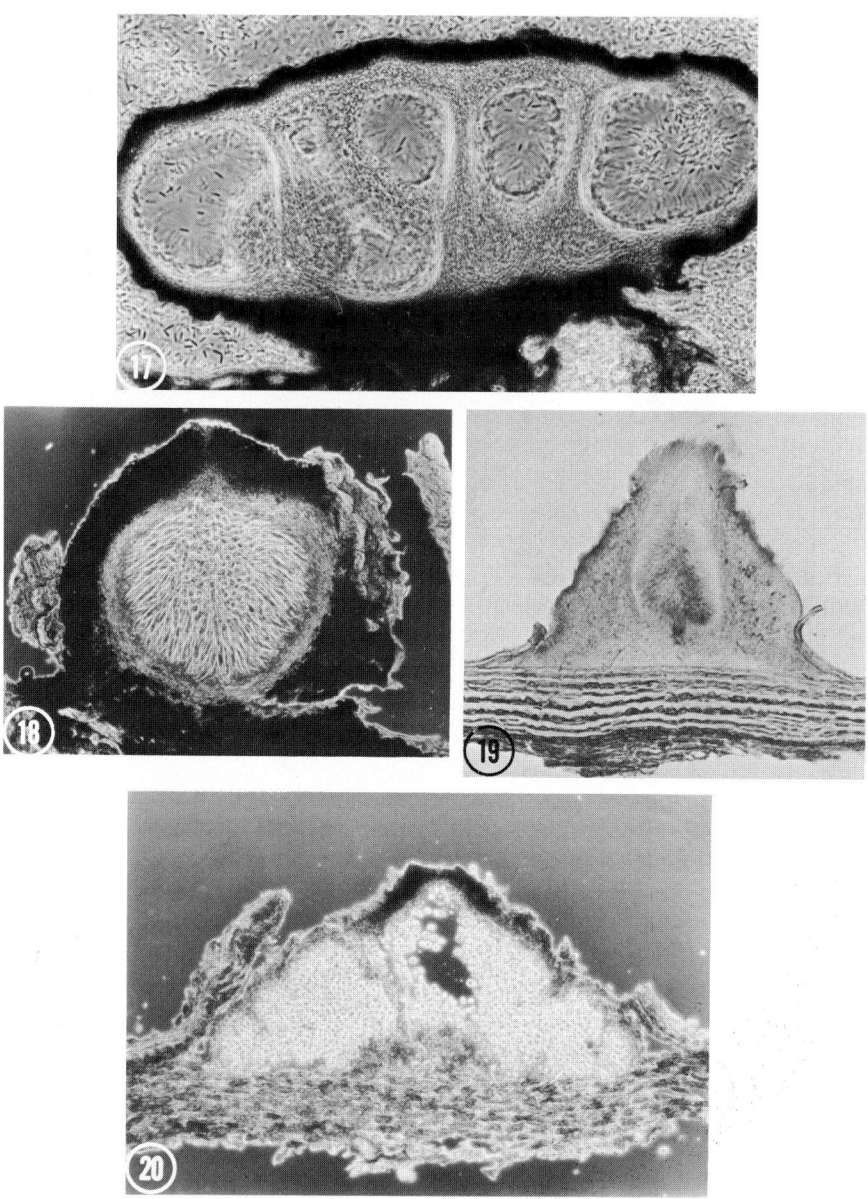

Plate IV. Fig. 17–20. Vertical sections through pycnidial and stromatic conidiomata. Fig. 17. *Ceuthospora* sp. Fig. 18. *Rileya piceae*. Fig. 19. *Dothichiza sorbi*. Fig. 20. *Phacidiopycnis piri*.

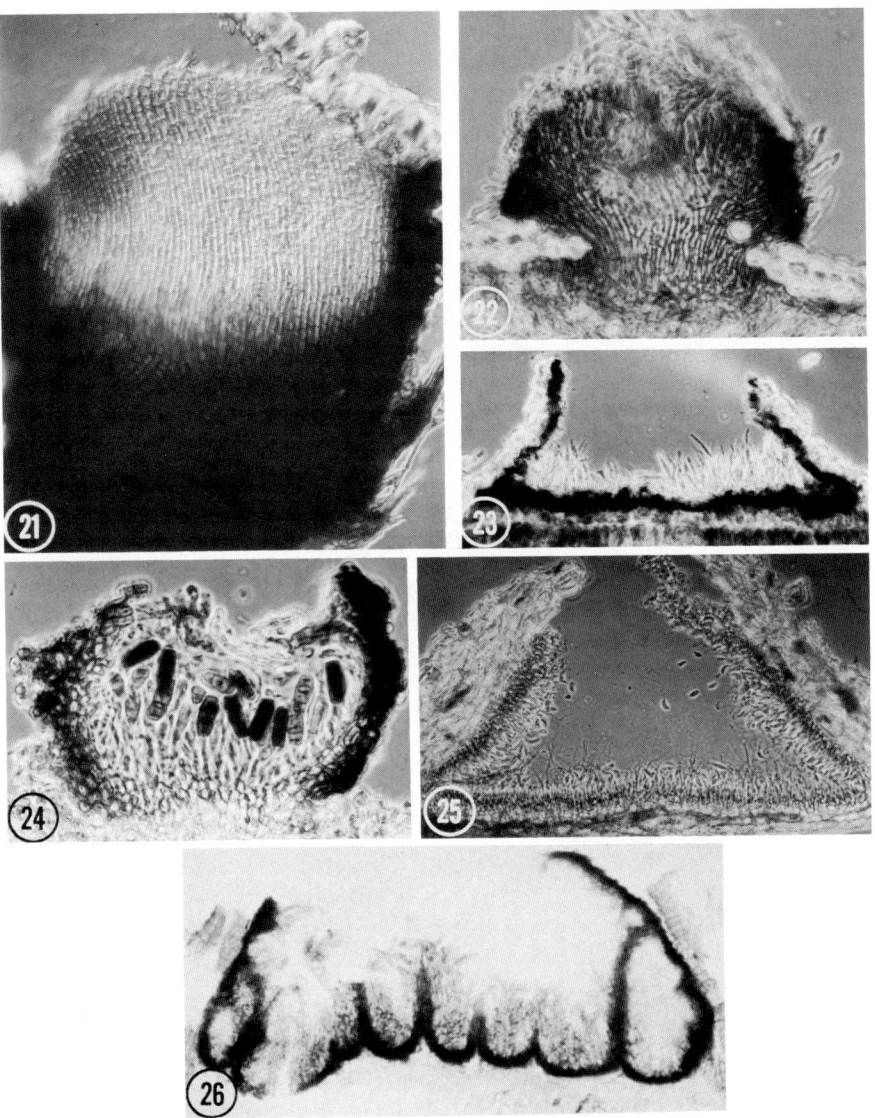

Plate V. Figs. 21–26. Vertical sections through conidiomata of various types. Fig. 21. *Catenophora yuccae*. Fig. 22. *Lecanosticta acicola*. Fig. 23. *Coma circularis*. Fig. 24. *Bleptosporium pleurochaetum*. Fig. 25. *Uniseta flagellifera*. Fig. 26. *Chondroplea populea*.

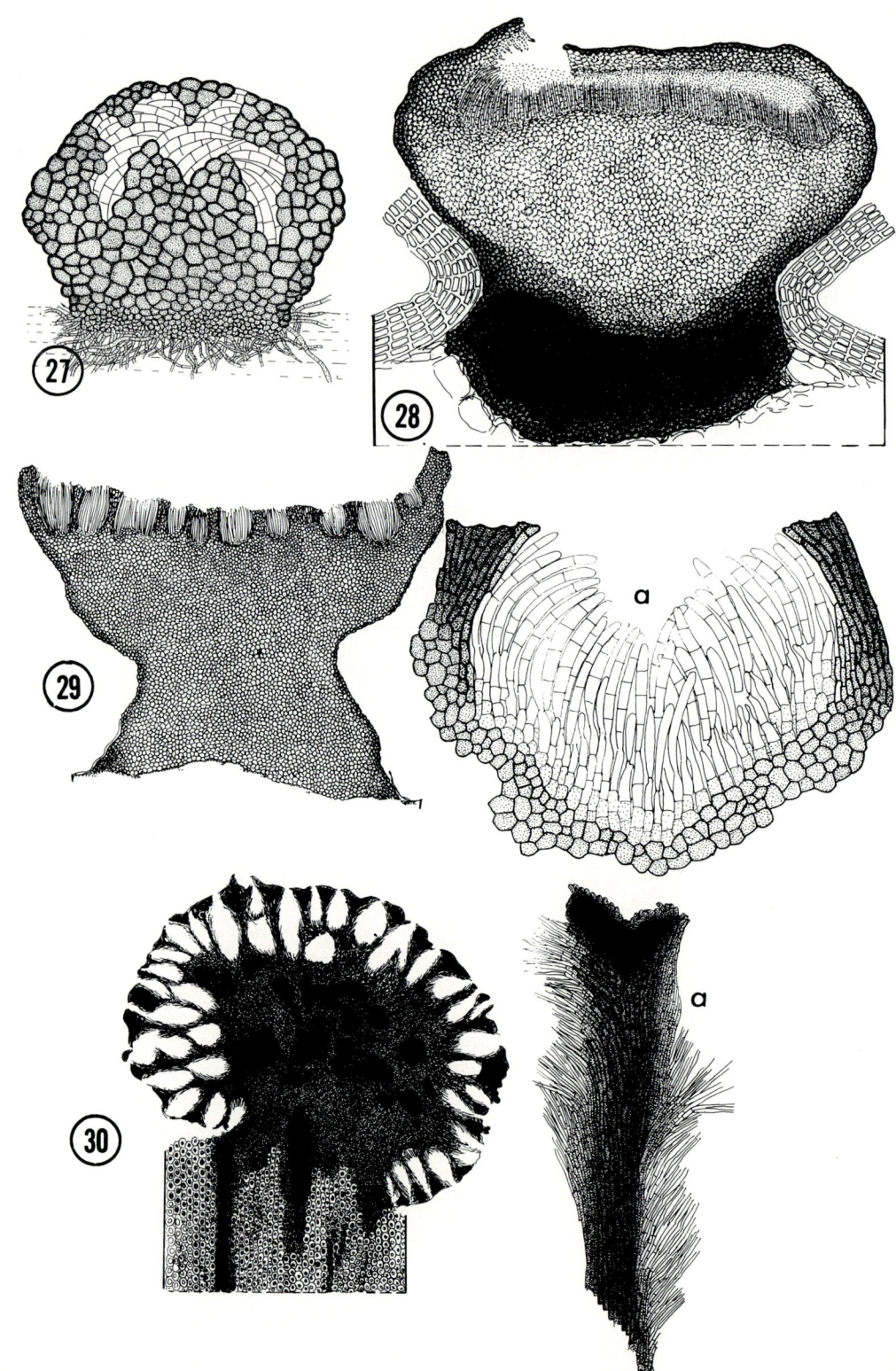

unilocular, glabrous, inostiolate, or with a papillate or rostrate apex and a circular or oval ostiole. In relation to the substrate they may be superficial, immersed (*Phyllosticta*), or erumpent. In some genera the pycnidial conidiomata are adorned with pigmented setae that are septate and simple (*Amerosporium*) or pigmented and stellately branched (*Ceratopycnis*). In *Aristastoma, Chaetoseptoria,* and *Pyrenochaeta* such setae surround the emergent ostiole of the usually immersed conidiomata. In *Chaetomella* the setae occur on either side of the opening, while in *Amerosporium* the setae usually occur in the median part of the conidiomata. In *Hyalopycnis,* very short, nonseptate, hyaline setae surround the ostiole. Linear or botryose aggregations of pycnidia without fungal connective tissues occur in some genera. Stromatic conidiomata are quite distinct from single or aggregated pycnidia and include pulvinate or clavate forms. In the simplest type, the conidiomata generally resemble pycnidia but are variable in shape, normally very thick-walled, and ostiolate or nonostiolate; the cavity may be divided or convoluted (*Mycohypallage, Phaeocytostroma, Phomopsis*). In the anamorph-genera *Ceuthospora, Ciliochora, Placonema,* and *Rileya,* the conidiomata incorporate host cells and fungal elements, while in *Dilophospora* the conidiomata are composed solely of fungus tissue. In *Chondrostroma* and *Haplosporella* individual locules occur at the same level in the conidiomata, and the conidia are released through a common ostiole, while in *Bothrodiscus* (Plate VI, Fig. 29) locules occurring at the same level in the apothecium-like conidioma are separated by connective tissue and each opens by a separate ostiole. In *Dothiorina* (Plate VI, Fig. 30) and *Strasseriopsis,* each of the peripheral locules of the globose conidioma opens by a pore. In *Aschersonia* and *Diplozythia* the pores are very wide. In *Camaropycnis,* an unnamed species of *Ceuthospora* (Plate IV, Fig. 17), and *Fuckelia,* the locules are irregularly divided. In *Pleurophomella* (=*Sirodothis*) several stromatic conidiomata originate from a common base (Sutton, 1973).

So far the wall tissues of the Coelomycetes have usually been categorized simply as pseudoparenchymatous and plectenchymatous. In reality, they are more or less comparable to the tissue types found in the Ascomycetes, and the terminology adopted for describing the apothecial tissues (Korf, 1973) appears to be more appropriate (DiCosmo, 1978; Dyko and Sutton, 1979; Nag Raj, 1977c, 1978a,b). A single tissue type might constitute the wall, as in some pycnidial anamorphs (*Scolecosporiella*), or several tissue types may be involved, as in the case of stromatic conidiomata. In *Scolecosporiella* the wall of textura angularis of pale-brown cells becomes darker near the ostiole. Such dark-celled tissues surround the partly erumpent ostioles in many of the anamorph-genera with

Plate VI. Figs. 27–30. Vertical sections through pycnidial and indeterminate types of conidiomata. Fig. 27. *Geastrumia polystigmatis.* Fig. 28. *Sirodothis populnea.* Fig. 29. *Bothrodiscus berenice.* (a) An individual locule. Fig. 30. *Dothiorina tulasnei.* (a) Tissue detail of the separating wall.

immersed conidiomata (*Kellermania, Neottiospora, Neottiosporina, Parahyalotiopsis, Pseudoneottiospora*). In *Phyllosticta* the wall is several cells thick and is composed of outer textura globulosa of dark-brown, thick-walled cells and inner textura prismatica of hyaline, thin-walled cells. In *Chaetospermum* the wall of the gelatinous pycnidioid conidioma is composed of textura intricata. In *Acarosporium* the outer brittle wall layer is of textura globulosa of dark-brown, thick-walled cells, and the inner light-colored layer is made up of cells immersed in gel. In *Hoehneliella* (Plate VII, Fig. 36), the basal textura globulosa of thick-walled, dark-brown cells gives rise to textura intricata of thinner-walled, brown hyphal elements; the well-developed excipulum is composed of textura intricata in the outer layers, giving rise to setae and inner textura porrecta of thin-walled, subhyaline to hyaline cells. The conidiophores arise from the elements of textura porrecta lining the cavity of the conidioma. Comparable complex wall tissues are encountered in the anamorph-genera *Corniculariella, Gelatinosporium,* and *Ypsilonia* (Plate VII, Fig. 35). In the cupulate, somewhat gelatinous conidiomata of *Pestalotia*, thick, basal and partly lateral tissues of the conidiomata are made up of textura globulosa of dark, thick-walled cells in the outer layers and of less pigmented cells with thinner walls in the inner layers and merge into textura porrecta and textura intricata in the upper levels of the peridium. In *Dinemasporium, Polynema,* and *Pseudolachnea,* the excipular tissues are composed of outer, dark textura porrecta merging into an inner, subhyaline to hyaline textura prismatica that gives rise to the conidiophores or conidiogenous cells. In *Ciliochora* and *Placonema* the stromatic conidiomata possess a well-developed clypeus.

Cavity formation in the pycnidial and stromatic conidiomata is known to result from schizogenous or lysigenous acitivity or a combination of both (Archer, 1926; Baccarini, 1890; Boerema, 1964; de Bary, 1887; Harris, 1935; Punithalingam, 1966). Dodge (1930) ascribed cavity formation in *Chaetomella* to upward and inward growth of the wall of the conidiomata; according to DiCosmo and Cole (1980) such growth is usually initiated at the site of an incipient seta.

In the Coelomycetes, conidia are released from the conidiomata in several ways. In acervuli a definite dehiscence mechanism is absent, the overlying host tissues being ruptured by the developing mass of conidia. In some pycnidial anamorphs the conidiomata are completely enclosed by a peridium. In the case of *Amerosporium* the cavity of the conidioma is filled with masses of conidia embedded in a matrix of slime. Under ideal conditions, moisture accumulates on the conidioma as minute water droplets held in place by the conidiomatal setae. As the conidioma imbibes the moisture, the slimy conidial mass inside also swells

---

Plate VII. Figs. 31–37. Types of textura. Figs. 31 and 32. Basal and upper walls in *Acarosporium sympodialis*. Fig. 33. *Rileya piceae*. Fig. 34. *Chondroplea populea*. Fig. 35. *Ypsilonia* state of *Acanthotheciella barbata*. Fig. 36. *Hoehneliella perplexa*. Fig. 37. *Dothichiza sorbi*.

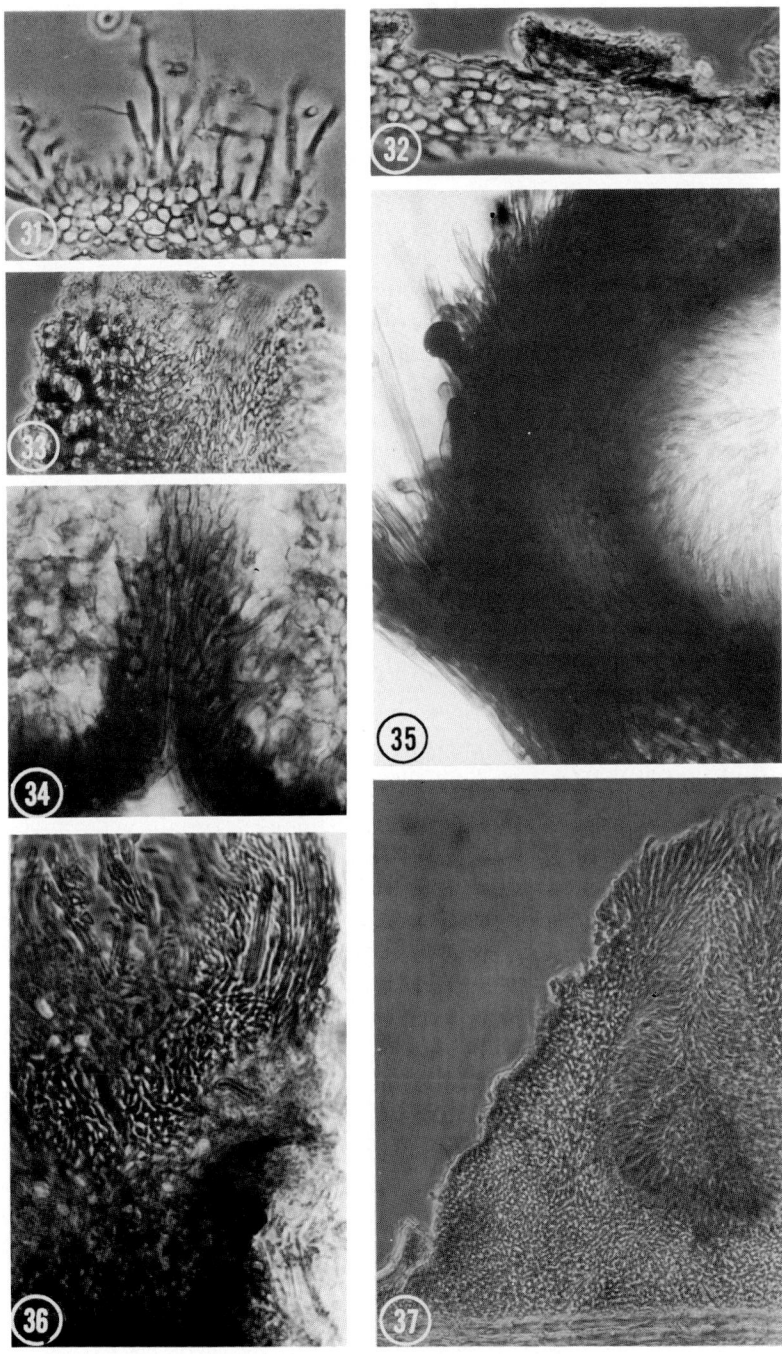

Plate VII. Figs. 31–37.

up, causing an irregular split in the peridium. In *Doliomyces,* the cells of the innermost layer in a small area of the roof of the conidioma are more or less cuboid and embedded in mucus. Such cells aid in the splitting of the peridium after the conidioma has been in contact with moisture. A similar mechanism is involved in *Acarosporium* where mechanical pressure is exerted on the brittle outer wall layer by the swelling mucus surrounding the inner layer of wall cells. In *Neottiospora* the swelling mass of conidia in mucus exerts a pressure on the delicate suture surrounding the operculum. In *Heteropatella,* a similar operculum is released by the swelling of the inner textura porrecta of the peridium. In *Chaetomella* dehiscence occurs along a linear or irregular suture on the upper peridium. In *Petrakomyces,* the swelling mass of conidia in mucus in the cavity and the sterile hyphae lining the roof of the conidioma cause the conidiomata to open along a vertical line of dehiscence in the clypeus. In many anamorph-genera of the Coelomycetes, the most common method of conidium release is through an ostiole which may develop in several ways: by lysis in *Septoria lycopersici* (Archer, 1926) and in *Plenodomus lingam* (Boerema and van Kesteren, 1964); by lysis with mechanical force exerted by the developing mass of conidia in *Septoria* spp. (Punithalingam, 1966); by lysis followed by growth of cells on the inner walls to produce a papillate ostiole in *Ascochyta gossypii* (Chippindale, 1929); and by upward growth of the cells of the upper wall without lysis in *Phoma betae* (Archer, 1926).

B. Conidiophores, Conidiogenous Cells, and Sterile Elements

On account of the emphasis laid on the morphology of conidia and conidomata in Saccardo's scheme of classification for the anamorphic fungi, for many years no serious attempts were made to characterize the conidiophores. Furthermore, in a large number of coelomycetes, the conidiophores are reduced to small conidiogenous cells that are hardly distinguishable from the inner layer of cells of the peridium. As a result, early coelomycete literature lacks accurate information concerning conidiophores and their development.

The different types of conidiophores and conidiogenous cells encountered in coelomycetes are illustrated in Plate VIII, Figs. 38-46 and Plate IX, Figs. 47-56. While in a large number of pycnidial anamorphs the conidiophores line the walls all the way around the cavity, in some anamorph-genera they extend only

---

Plate VIII. Figs. 38-46. Conidium ontogeny. Fig. 38. Thallic mode in *Acarosporium sympodialis.* Fig. 39. Blastic-single mode in *Strasseriopsis tsugae.* Fig. 40. Polyblastic mode in *Libartania laserpiti.* Fig. 41. Blastic-single mode in *Labridella cornu-cervae.* Fig. 42. Blastic-single mode with integrated conidiogenous cells in *Coleophoma taxi.* Fig. 43. Polyblastic mode in *Chaetospermum* sp. Fig. 44. Blastic mode with sympodial proliferations in *Hyalotiella transvalensis.* Fig. 45. Blastic mode with sympodial proliferations in *Heteropatella lacera.* Fig. 46. Blastic mode with limited sympodial proliferations in *Gampsonema exile.*

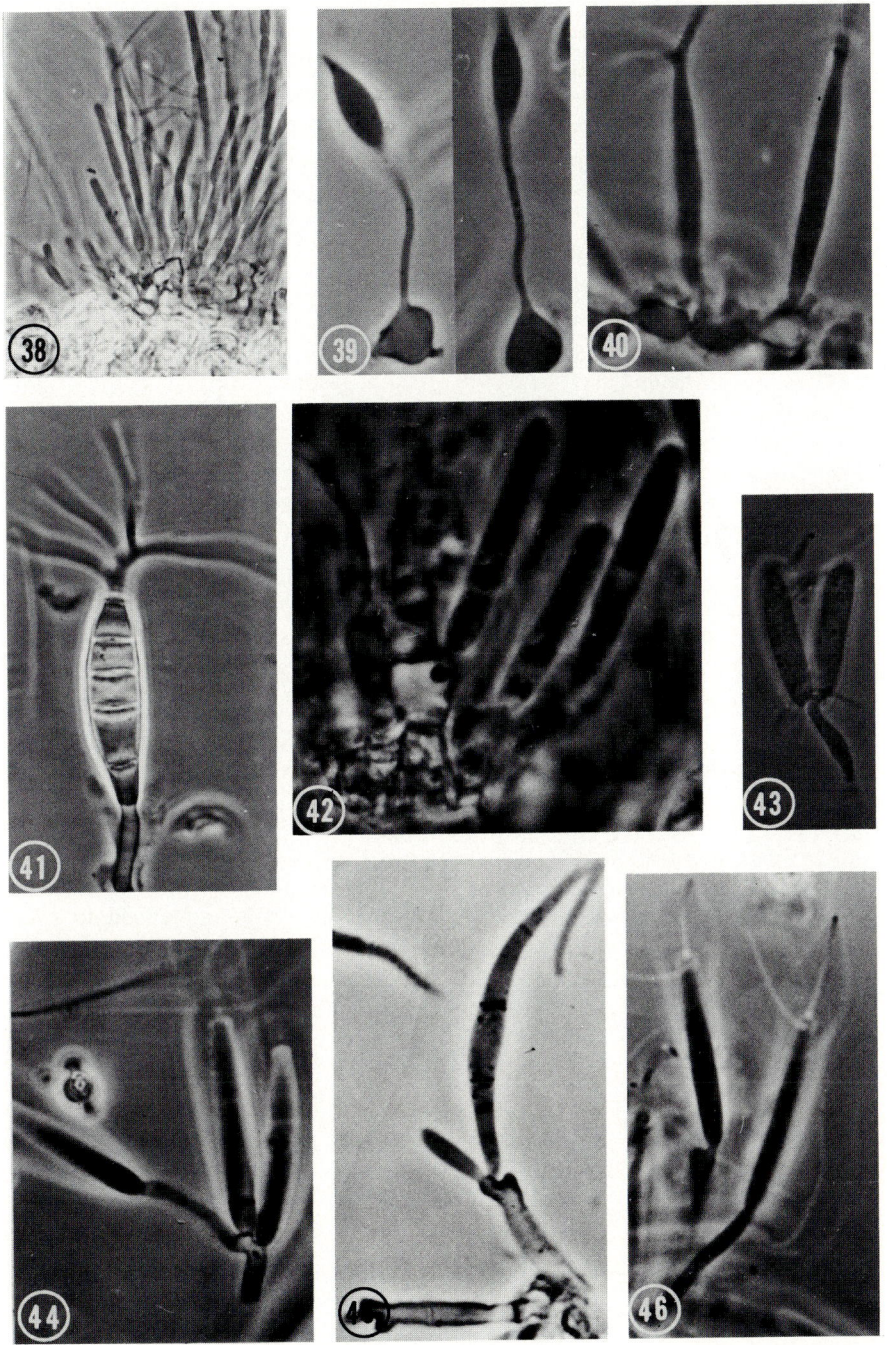

Plate VIII. Figs. 38–46.

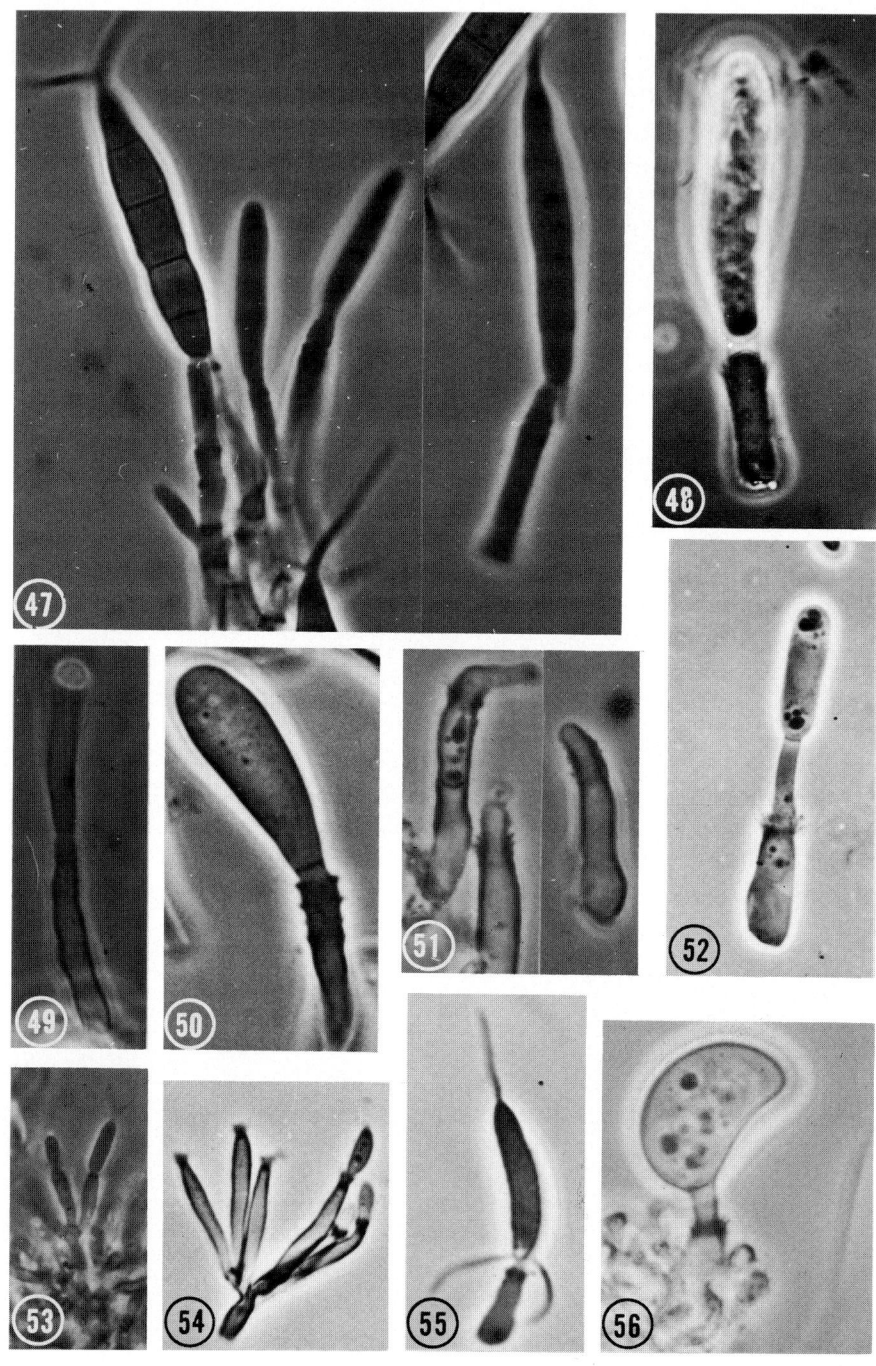

Plate IX. Figs. 47-56.

part way up the sides, and in some (*Acarosporium, Amerosporium, Japonia*) they are restricted to the base. In *Coniella,* the conidiophores are aggregated in a hemispherical, basal cushion typical of the genus (Plate, III, Fig. 13). Conidiophores may often be reduced to conidiogenous cells indistinguishable from the cells of the wall (*Libartania, Sclerophoma*) or may be discrete. In the latter case, they are quite distinct and may be single conidiogenous cells or multicellular structures. Discrete septate conidiophores may be simple or branched with a terminal cluster of conidiogenous cells. Often the conidiophores may be composed of integrated conidiogenous cells, as in Plate X, Figs. 58 and 60 (*Catenophora, Eleutheromyces*). Simple conidiogenous cells, or septate and often branched conidiophores, may be characteristic of individual anamorph-species (*Seimatosporium*), or both types may occur in the same species of an anamorph-genus (*Hyalotiella*). Conidiogenous cells of many coelomycetes are hyaline and smooth-walled; some may be uniformly pigmented; while in a few others they are darker and minutely verruculose near the apical region, often obscuring percurrent proliferations if these are present (*Lecanosticta, Monochaetiella, Phaeopolynema*).

Sterile elements resembling paraphyses often occur interspersed with conidiophores in some coelomycetes (*Massariothea, Mycotribulus, Toxosporiopsis*). Usually they are simple, septate, hyaline, and smooth-walled. In *Mycotribulus* they are irregularly nodulose in the apical part and knobby at the tips. In *Pseudorobillarda* and *Vasudevella,* the sterile elements are much shorter. It is not unusual to find that paraphyses described for some coelomycete taxa are in reality conidiophores. In *Coleophoma,* the terminal cell and occasionally a few of the lateral branches of the conidiophores remain sterile. In *Melanconiopsis* sterile elements line the apical canal of the conidioma. Quite often such sterile, paraphysis-like elements are invested in a mucilaginous matrix. In *Ciliochora,* the floor of the clypeus is lined with septate, pigmented to subhyaline hyphoid elements with their apexes oriented toward the center of the clypeus.

C. Conidia

Conidium morphology had a significant role in determining the generic limits in Saccardo's scheme of classification. The importance attached to the number of

Plate IX. Figs. 47–56. Conidium ontogeny. Fig. 47. Blastic-annellidic mode in *Seimatosporium dilophosporum*. Fig. 48. Blastic-phialidic mode with percurrent proliferations in *Idiocercus pirozynskii*. Fig. 49. Blastic-annellidic mode in *Plectronidium minor*. Fig. 50. Blastic-phialidic mode with percurrent proliferations in *Bleptosporium pleurochaetum*. Fig. 51. Blastic-phialidic mode with percurrent proliferations in *Annellolacinia dinemasporioides*. Fig. 52. Blastic-phialidic mode with percurrent proliferations in *Idiocercus macarangae*. Fig. 53. Blastic-phialidic mode in *Ceuthospora* sp. Fig. 54. Blastic-phialidic mode in *Dinemasporium aberrans*. Fig. 55. Blastic-phialidic mode in *Neobarclaya primaria*. Fig. 56. Blastic-phialidic mode with percurrent proliferations in *Harknessia renispora*.

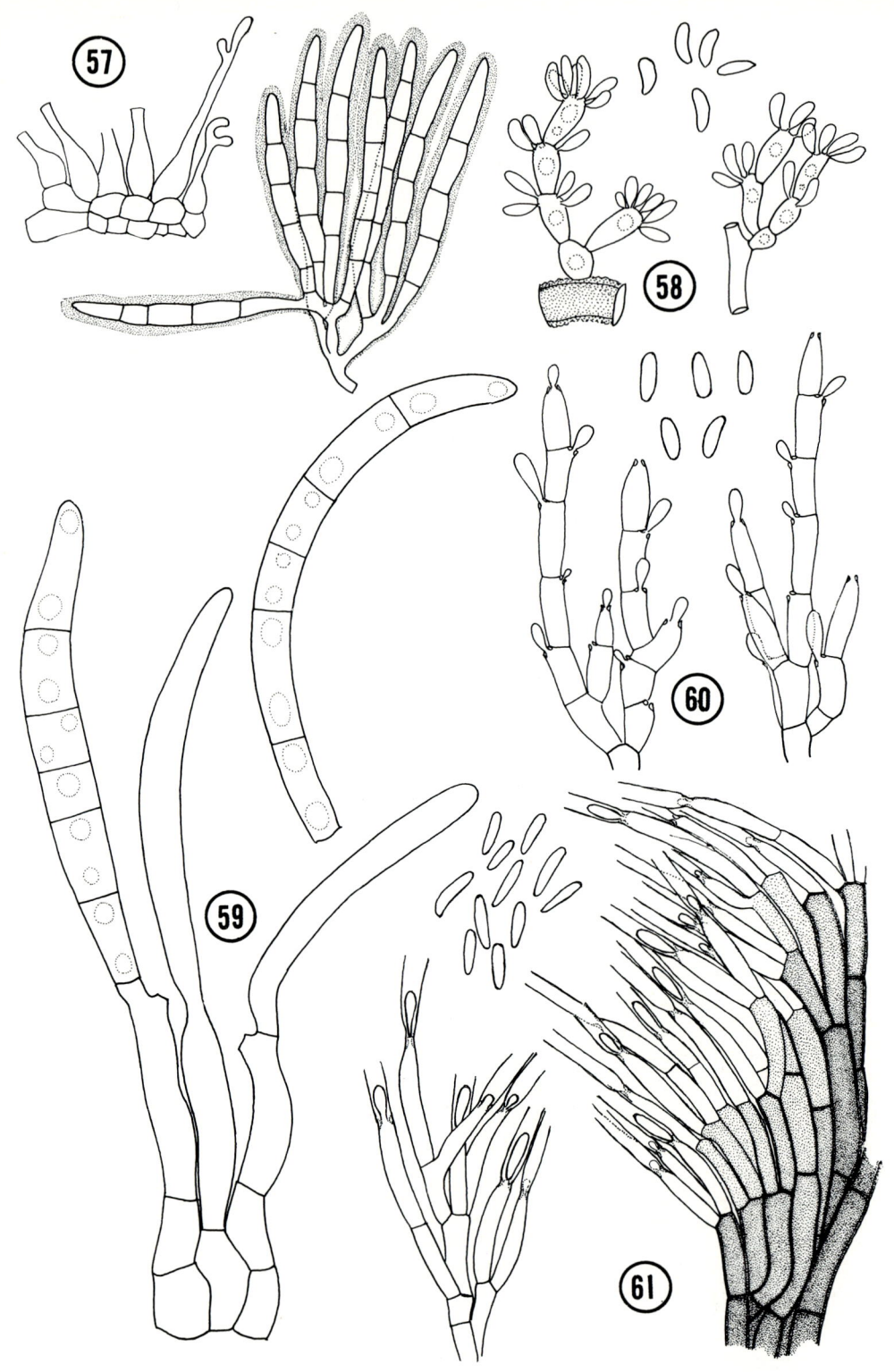

conidial septa and pigmentation was responsible for segregation of not a few coelomycete anamorph-genera now regarded as superfluous. It is now well known that septum formation and development of pigmentation (in taxa with pigmented conidia) can often be delayed until after secession of the conidia (*Botryodiplodia theobromae, Diplodia tumefaciens, Macrophoma pinea*). The current approach is therefore to include anamorph-species possessing varying degrees of septation and pigmentation (but identical in other respects) in a single anamorph-genus (Dorenbosch, 1970; Nag Raj, 1973b; Sutton, 1964b, 1968a; Sutton and Pirozynski, 1965). However, some of Saccardo's terminology for conidium morphology cannot be ignored. The limits of scolecospores, helicospores, and staurospores have recently been reevaluated (Kendrick and Nag Raj, 1979), and these along with the terms "ameroconidia," "didymoconidia," "phragmoconidia," and "dictyoconidia" have been found to facilitate the design and use of keys to the identification of taxa (Kendrick and Carmichael, 1973; Carmichael *et al.*, 1980, Michaelides *et al.*, 1979).

Coelomycete conidia exhibit diversity in form, from the common amerosporous, to the helicosporous in *Helicothyrium riyukyuense,* and even more complex and composite forms in *Geastrumia* (Plate X, Fig. 57), *Tetranacrium, Ypsilonia* (Plate XII, Fig. 77), and others. Some of the conidia have rounded ends, while others have truncate bases which occasionally bear minute marginal frills (*Groveolopsis, Kellermania, Placonema*). Following Luttrell (1963), Sutton (1973) recognized two types of septa in coelomycete conidia: The more common eusepta usually present in hyaline or pigmented conidia and the less common distosepta that occur in the pigmented conidia of *Coryneum, Massariothea, Pestalotia, Prosthemium,* and *Stegonosporium.* Concolorous pigmented conidia are common among the taxa with ameroconidia, the range of pigmentation varying from subhyaline (some species of *Coniella*) to dark brown (*Melanconium*). In phragmoconidia, the end cells may remain hyaline, while the median cells acquire variable degrees of pigmentation (*Pestalotiopsis*). In some taxa, the conidia exhibit a dark band at the septa: *Doliomyces,* species of *Pestalotiopsis, Toxosporiopsis* (Plate XII, Fig. 71) but in *Mycohypallage* and *Poropeltis,* the otherwise pigmented conidia have a hyaline transverse band in the median part (Plate XI, Fig. 63). The pigmented conidia of some species of *Harknessia* have one to several longitudinal light-colored bands. Although reported in *Pilidiella* (Maas *et al.,* 1979), germ slits are not common in coelomycete conidia. In many taxa with hyaline conidia the walls are usually smooth, though some

---

Plate X. Figs. 57–61. Conidium ontogeny. Fig. 57. Blastic-single mode in *Geastrumia polystigmatis*. Fig. 58. polyblastic mode with integrated conidiogenous cells in *Pragmopycnis pithya*. Fig. 59. Blastic mode with limited sympodial proliferations in *Bothrodiscus berenice*. Fig. 60. Blastic-phialidic mode with integrated conidiogenous cells in *Sirodothis populnea*. Fig. 61. Blastic-phialidic mode in *Dothiorina tulasnei*.

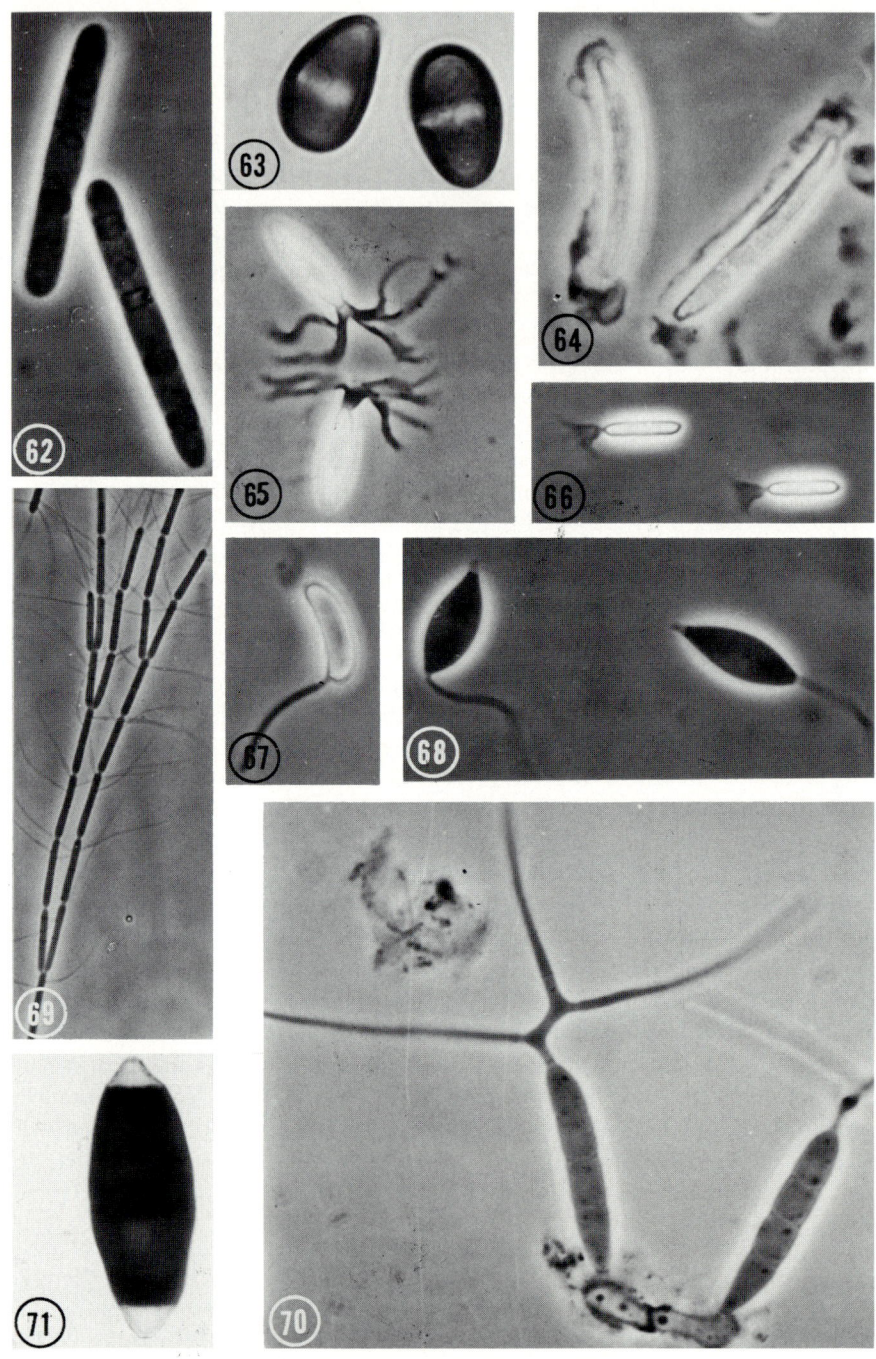

Plate XI. Figs. 62–70 Captions on page 68.

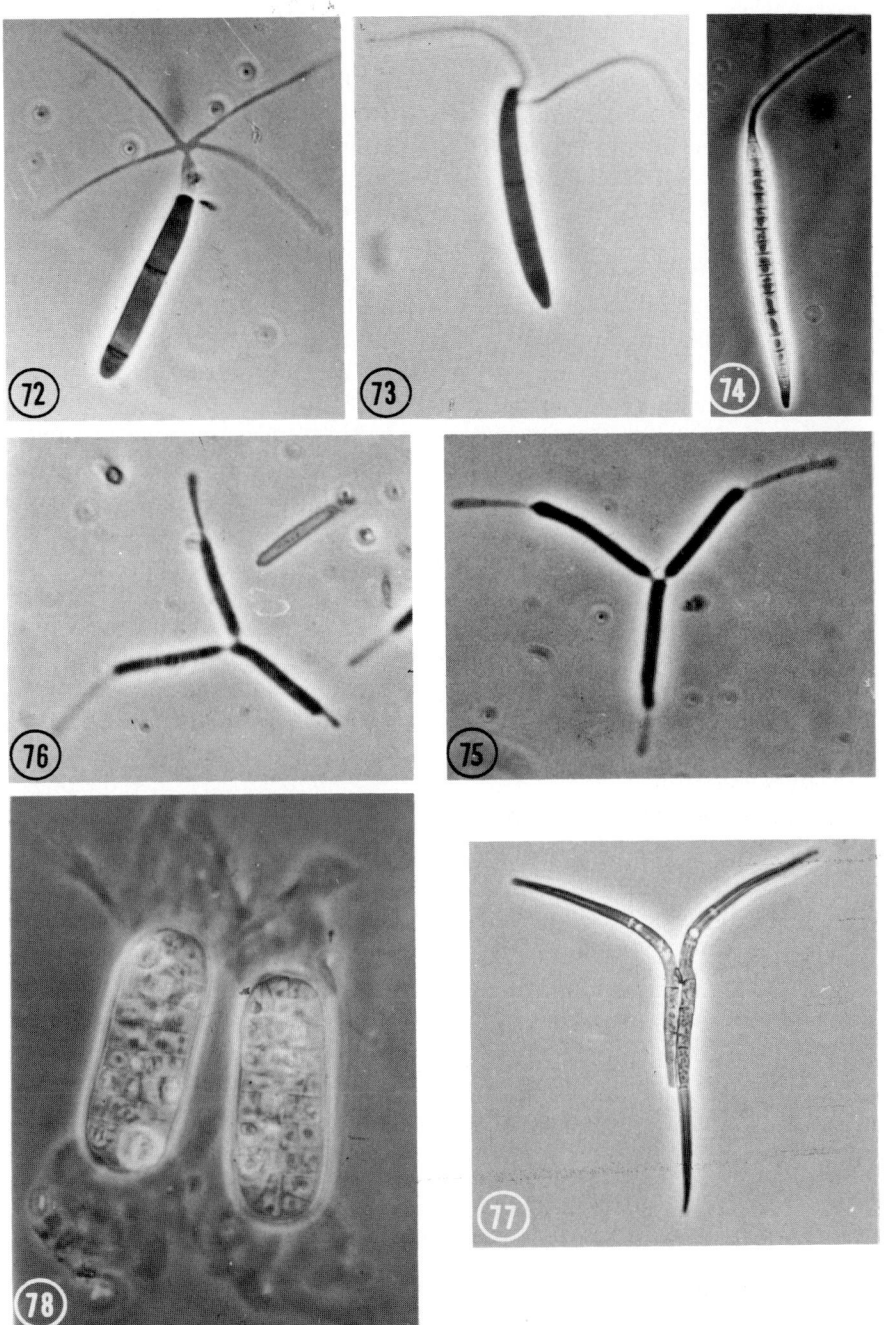

Plate XII. Figs. 72-78 Captions on page 68.

mature conidia in *Dothichiza, Plectronidium,* and *Polynema ornatum* develop pigmentation and ornamentation of the walls.

Other characteristic features of the conidia of many taxa relate to structures that have been described variously as appendages, pedicels, setae, and cilia. These structures are of two basic types: cellular—originating as tubular extensions of the conidium body; and extracellular—arising without protoplasmic continuity with the conidium body. The cellular appendages vary in number from one to several and in their position on the conidia (Plate XI, Figs. 67-69; Plate XII, 70-76): apical (*Obstipipilus*), basal (species of *Harknessia, Strasseria*), apical and basal (*Dinemasporium,* species of *Harknessia*), apical and lateral (*Gampsonema, Pestalozzina*), and apical, lateral, and basal (*Heteroceras, Plectronidiopsis, Polynema, Stauronema*). The apical appendages in some taxa retain their cytoplasmic contents even at maturity (*Colletotrichum caudatum, Diachorella, Monochaetiella*) or are devoid of contents and separated from the conidium body by septa (*Ciliochorella, Japonia, Petrakomyces*). The apical appendages may be branched (*Bartalinia, Chrysalidiopsis, Doliomyces, Mycohypallage, Pestalozziella*), and in some (*Japonia, Libartania themedae*) such branched appendages become septate. In some coelomycetes, development of the basal appendage is followed by formation of the conidium body (*Monodia, Pullospora, Strasseria, Strasseriopsis*). In *Pestalotiopsis* the basal appendage is believed by some workers to arise inside the conidiogenous cell, while in most taxa they are exogenous and excentric (*Placonema*); and in a few they are formed after secession of the conidia (*Heteroceras, Pestalozzina*). The conidia of *Dilophospora* develop their extensively branched apical and basal appendages after seceding from the conidiogenous cells.

Extracellular appendages (Plate XI, Figs. 64-66, and Plage XII, Fig. 78) are mucoid in nature and arise in many ways from a sheath enclosing the developing

---

Plate XI. Figs. 62-70 (*Continued*). Conidia. Fig. 62. Ameroconidia of *Coleophoma taxi.* Fig. 63. Ameroconidia of *Poropeltis davilliae.* Note the median hyaline band. Fig. 64. Ameroconidia with mucoid appendages in *Coniella eucalypticola.* Fig. 65. Ameroconidia with mucoid appendages in *Neottiospora caricina.* Fig. 66. Ameroconidia with mucoid appendages in *Ceuthospora* sp. Fig. 67. Ameroconidium with a mucoid apical appendage and a tubular basal appendage in *Strasseria carpophila.* Fig. 68. Didymoconidia of *Strasseriopsis tsugae* with the basal tubular appendage separated from the conidium body by a septum. Fig. 69. Chains of didymoconidia with tubular appendages in *Acarosporium* sp. Fig. 70. Phragmoconidia with branched, tubular apical appendage in *Libartania themedae.* Note the septa in the basal part of the appendage. Fig. 71. A phragmoconidium of *Toxosporiopsis capitata.* Note the dark bands at the septa.

Plate XII. Figs. 72-78 (*Continued*). Conidia. Fig. 72. Phragmoconidium with branched, tubular apical appendage in *Hyalotiella transvalensis.* Fig. 73. Phragmoconidium with one subapical and one lateral divergent appendages in *Gampsonema exile.* Fig. 74. Phragmo-scolecoconidium of *Rileya piceae.* Figs. 75 and 76. Stauroconidia with tubular appendages (mucoid caps on the appendages are not visible in lactic mounts) in *Furcaspora pini.* Fig. 77. Stauroconidium of *Ypsilonia mirabilis.* Fig. 78. Dictyoconidium with an apical and a basal mucoid appendage in an undescribed coelomycete.

conidium. In *Toxosporiopsis,* the sheath persists as an apical cap on the mature conidia, while in others it splits in different ways, giving rise to apical appendages (*Comatospora, Giulia, Tiarosporella*) or basal appendages (*Neottiospora, Pseudorobillarda*). An apical mucoid appendage and a basal tubular appendage occur on the conidia of *Strasseria,* while tubular appendages bearing mucoid caps occur in *Furcaspora.* The nature of the appendages and their mode of ontogeny, number, and position are fairly consistent for a taxon and aid in taxonomy at the generic or species level.

In coelomycete taxa, as in many other anamorphic fungi, conidium length is more variable than width, and thus the conidium width, and in some instances the conidium length/width ratio, help to differentiate the species in an anamorph-genus. In general, the conidium length ranges from a few micrometers (*Discosiella*) to as much as 110 μm (*Rileya*).

In some coelomycete taxa, small spermatia (often referred to as microconidia) frequently occur intermixed with conidia in the same conidioma or may be formed in separate spermatium-bearing structures resembling and found in association with the conidioma (*Coma, Harknessia, Petrakomyces, Placonema*). The unicellular, invariably hyaline spermatia vary in shape from short cylindrical to ellipsoidal or acerose and generally exhibit the same kind of ontogeny as that of the macroconidial anamorph. From our experience in this laboratory we have begun to accept the occurrence of such spermatia in coelomycete collections as a clue to the presence of the teleomorph of the fungus, which can usually be found after careful scrutiny of the collection.

D. Conidium Ontogeny

Following the basic concepts of conidium ontogeny developed for the Hyphomycetes (Hughes, 1953), an analysis of the published experimental data led to the conclusion that conidia arise in one of two ways: thallic or blastic (Kendrick, 1971). In the thallic mode the conidia arise by the conversion of a preexisting hyphal element, while in the blastic mode the conidia arise *de novo.* Further categories recognized within each of these two major groups were based on the position of the conidiogenous locus and the wall relationships. In the last 25 years, similar concepts have been extended to many coelomycete taxa and are now regarded as of considerable importance in establishing generic concepts. Types of conidium ontogeny in the Coleomycetes are indicated in Scheme 1.

As seen in Scheme 1 basauxic ontogeny has not been reported so far for the Coelomycetes, although members of *Arthrinium,* by virtue of the nature of their conidioma, may be included in this category. Spermatial (*Asteromella*) states of *Mycosphaerella* spp. have been reported to have enterothallic ontogeny but are excluded from this account since spermatial states are not usually regarded as full anamorphs (hence are not usually given binomials). Boerema and van Kesteren

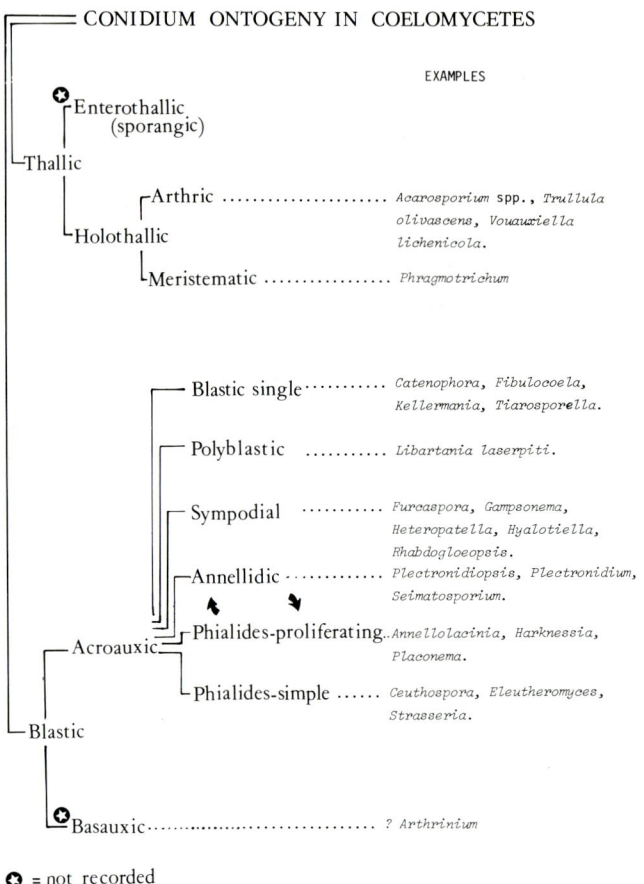

Scheme 1.

(1964) claimed blastic-tretic ontogeny in *Phoma* (unconfirmed), but there is some doubt about the validity of this ontogenetic concept and its taxonomic significance (Brotzman *et al.*, 1975; Carroll, 1972; Carroll and Carroll, 1974).

Despite the simplified scheme of ontogeny outlined in Scheme 1 there are some problem areas of interpretation. In some coelomycetes, it is often difficult to determine whether the chains of conidia result from meristematic-holothallic or retrogressive-blastic ontogeny, since the conidiogenous locus is hard to locate in the latter. A more serious and major point of contention involves the relationship between annellides and phialides. In his reviews, Sutton (1971a, 1973) sought to differentiate annellides with sparse annellations, annellides with flared, cupulate, widely spaced annellations, and retrogressive annellides.

Morgan-Jones *et al.* (1972e) interpreted the last two as proliferating phialides. Such an interpretation helps to reconcile the apparently incongruous report of blastic-single ontogeny in the anamorph and blastic-phialidic ontogeny in the spermatial states of *Harknessia* spp. (Sutton, 1971b). Recent electron microscope studies (Hammill, 1972; Jones, 1977; Reisinger *et al.*, 1977) indicate that the so-called annellides of some coelomycetes are manifestations of blastic-phialidic ontogeny, thus raising serious doubts as to whether real distinctions between phialides and annellides exist. Since the crux of the problem lies in assessing whether or not more than one conidium is formed at each level of proliferation (an assessment that is difficult to accomplish in coelomycetous anamorphs), the difficulty appears insurmountable at present. A partial answer may be found in the years ahead by carrying out extensive TEM studies with a large number of acervular and stromatic coelomycetes, which alone appear by light microscopy to possess annellides comparable to those found in some hyphomycetes.

A critical review of available data points to a preponderance of phialides and proliferating phialides in the Coelomycetes, followed by blastic-single, and blastic-sympodial ontogeny. Blastic-annellidic and polyblastic ontogeny is less common, particularly in pycnidial anamorphs.

Since the Saccardoan scheme of classification for anamorphic fungi was based on the morphology of conidia and conidiomata, it is natural that many of the older and larger genera of the Coelomycetes should turn out to be extremely heterogeneous when the modern criterion of conidium ontogeny is applied to them as a key feature for defining generic concepts. Sutton (1973) has pointed out the extreme degree of heterogeneity in *Marssonina, Septogloeum, Septoria,* and so on, which is also likely to be found in *Ascochyta, Diplodia, Hendersonia, Phoma, Stagonospora,* and others.

E. Taxonomy of Coelomycetes Growing in Artificial Cultures

In the course of their investigations, soil ecologists and plant pathologists have attempted to characterize fungi growing on synthetic media under artificial conditions. They have been concerned not only with establishing the pathogenicity of their isolates but also with cultural behavior and the factors affecting it. Data from such studies have often been used in taxonomy, usually with more disadvantages than advantages in some cases involving coelomycetes.

On the positive side, examples are few and involve such anamorph-genera as *Phoma* and related forms, which have few usable morphological criteria—not enough to separate the large number of species with wide host ranges. Boerema (1969) and Dennis (1946) have shown the need in such cases for a study correlating cultural behavior and pathogenicity with morphological features. Pigment production in cultures has been used to separate many paint-inhabiting species of *Phoma* (Eveleigh, 1961) and varieties of *Phoma exigua* (Boerema and Howeler, 1967). Boerema (1967) used a combination of characters such as pigment production, temperature requirements, and pathogenicity tests on potato to differentiate two varieties of *Phoma exigua*. Species of *Colletotrichum* have been recognized by characters of appressoria and sclerotia in cultures (von Arx, 1957b; Sutton, 1962, 1968b). Similarly, characters of sclerotia have been used in the taxonomy of *Macrophomina* (Ashby, 1927) and *Phoma* (Dorenbosch, 1970).

Examples which cast doubt on the value of morphological features observed in artificial cultures of coelomycetes are numerous. Most coelomycetes exhibit considerable diversity in form and size when grown under artificial conditions. Conidium shape and size can be altered significantly by varying the nature of the medium (Brooks and Searle, 1921; Sutton, 1964a). Similarly, in taxa with appendage-bearing conidia, the number and position of the appendages, number of conidial septa, and even the shape of the conidia, change considerably from the norm when a few drops of lactic acid are added to the culture media. When *Peyronellaea* was proposed as an anamorph-genus distinct from *Phoma,* formation of chains of dictyochlamydospores was considered diagnostic for the genus. Following Chodat (1926) and Lacoste (1955), who showed that chlamydospore production in such fungi was related to the carbon/nitrogen ratio of the medium, and since some of the strains were known to lose the ability to produce chlamydospores, Boerema *et al.* (1965) reduced *Peyronellaea* to synonymy with *Phoma.* In some coelomycetes, the conidiomata that develop in culture differ from those occurring in nature. *Hyalotiella* is a good example. The conidioma of the type species *H. transvalensis,* based on a culture isolated from soil, was described as stromatic. But the conidiomata of another species of *Hyalotiella* observed on leaves of different hosts collected in India and South America are pycnidial (Nag Raj, 1975a, 1979b). Another significant perturbation concerns the affinity between acervuli and sporodochia, which can hardly be distinguished in cultures. A case in point is *Dwayalomella,* conidiomata of which were originally described as acervuli but appear to be sporodochial. In artificial cultures, species of *Discosia* are known to pass through extremes ranging from dematiaceous and tuberculariaceous to acervular. Similar variation is also seen in cultures of species of *Pestalotiopsis*. Recent evidence presented by Madelin (1979) for reciprocal transmutation of conidium ontogeny by relatively simply environmental influences tends to add to the problems of taxonomic decisions based on coelomycete cultures.

I therefore suggest that to propose new anamorph-taxa on the basis of morphological criteria derived from synthetic agar cultures is generally unwise and impedes the progress of coelomycete systematics.

F. Discussion

Mycologists studying the Hyphomycetes now agree that an anatomico-ontogenetic system of classification for this group has yielded somewhat better results than an anatomic system alone. Hyphomycete taxonomy is therefore on a firmer footing than it was in the static post-Saccardoan era. During the last 20 yr a critical reappraisal and redefinition of generic concepts in the Coelomycetes, involving a correlation of morphological data and conidium ontogeny by renewed reference to the type specimens of the innumerable older taxa, has been progressing slowly but steadily. The task confronting the few mycologists who concern themselves with the group is formidable and was aptly compared by Kendrick (1974) to a "generic iceberg." His words, "It is less glamourous, and often more difficult, to reassess old taxa than to describe new ones, but at present the need for such activities should override other considerations," well express the thrust of coelomycete systematics today.

Coelomycete taxonomists have to examine literally hundreds of specimens of old taxa and check whether the names are acceptable or synonyms, *nomina dubia, nomina confusa,* or even illegitimate. This, and the shortage of labor, puts the time of the relatively few workers at a premium. An enormous amount of productive time will be lost if they have to re-examine and redescribe taxa erected in recent years, a loss that could be avoided if students of these fungi would follow what I should like to propose as a procedural quinquelogue (with apologies to L. K. Weresub for irreverently modifying her nomenclatural decalogue for botanists and duodecalogue for mycologists) in the interests of progress toward the goal of an overall developmental-anatomic system of classification (not for the Coelomycetes alone, but the entire anamorphic spectrum).

On pain of its status being left in limbo:

1. **Thou shalt not publish** a new coelomycetous anamorph-taxon without studying vertical sections of the conidioma to determine (a) if it is acervular, pycnidial, or of some other type, (b) the arrangement of conidiogeneous cells or conidiophores in the cavity, and (c) the textura of the periderm, if any. The following problems are typical: (i) In *Amerodiscosiella renispora* Farr nom. dub. f. Nag Raj, 1975b, the nature of the conidioma is unknown, since sections of conidioma are not available for study; conidium ontogeny is unknown; holotype specimen in BPI sterile; type slides in BPI are inadequate to determine nature of conidioma and conidium ontogeny, hence the affinity of the taxon to related anamorph-taxa; isotype specimens in Pnom Penh, Cambodia, inaccessible and apparently lost to posterity. Farr's explanations (*Taxon* **26,** 580–581, 1977) have not alleviated the problem of establishing the status of this taxon. (ii) *Pycnofusarium* and *Thyrsidina* described as coelomycetes but have hyphomycetous conidiomata; see Kendrick and Nag Raj, 1979. (iii) The conidioma of an unde-

scribed *Titaea*-like taxon on bamboo in Brazil appeared to be coelomycetous, but careful study has shown that the hyphal elements and conidiophores aggregate in the host's substomatal cavity, spuriously simulating a pycnidioid conidioma, and there is no peridium or stroma.

2. **Thou shalt not publish** a new coelomycetous anamorph-taxon without ascertaining the type of conidium ontogeny.

3. **Thou shalt not publish** a new coelomycetous anamorph-taxon without appropriate illustrations. "Illustrations serve as a more universal medium of communication than a description or Latin diagnosis.... Photomicrographs lack depth of focus but they cannot be criticized as inaccurate or misleading, as line drawings can be.... A judicious combination of line drawings and photomicrographs not only justifies an author's reasons for describing a new taxon, but also helps the readers to recognize the fungus easily...." These opinions were expressed by members of a symposium entitled, "The Whole Fungus" at Kananaskis, Alberta, Canada, in 1977.

4. **Thou shalt not publish** a new coelomycetous anamorph-taxon without determining if a mycoparasitic relationship is involved. Two undescribed hyphomycetes are known to invade conidiomata of a *Phyllosticta* sp. and the ascomata of the related *Guignardia* sp. In this condition, one can easily interpret the discordant elements of the two fungi as belonging to a new coelomycete taxon if critical observations are not made. *Sporonema nigropunctata* Hino & Katumoto is another example of such association.

5. **Thou shalt not publish** new coelomycetous anamorph-taxa for spermatial states of fungi. (These are not true anamorphs; no purpose is served in applying binomials to such forms, since there are very few morphological criteria available for differentiating them at the species level.)

In view of the utter disregard for the Code with which new taxa are being described even in modern times, and in view of the inaccuracies in some of the features reported, I am tempted to extend the above into a nonalogue, if only to stress the need for good sense.

On pain of being discredited:

6. **Thou shalt not publish** a new taxon, particularly a coelomycetous anamorph-taxon, on the basis of observations on a fungus growing under artificial conditions.

7. **Thou shalt not extrapolate beyond the observed facts** in publishing accounts of a taxon. The features of a fungus should be depicted true to the form; readers will decide whether the interpretations presented are acceptable.

8. **Thou shalt not publish** a new taxon—coelomycetous anamorph or other—without ensuring that the type collections are ample, that the type collections are accessible to other workers through one or more international herbaria (but not

tucked away in a remote corner of the world); see comments on *Amerodiscosiella renispora* on page 73.

9. **Thou shalt not publish** a new taxon without indicating the nomenclatural type. (*Discosiopsis,* Edwards *et al.,* 1974; *Kellermaniopsis,* Edwards *et al.,* 1974, invalidly published.)

IV. SOME ASPECTS OF BIOLOGY OF COELOMYCETES[1]

The phenotypic expression of a fungus, generally stable and distinct in its natural environment, is a reflection of its biology. We know very little of the ecological needs of fungi, but what we do know is that they are denizens of extremely varied ecological niches. Coelomycetes have also adopted diverse habitats as indicated in Table I.

Worldwide in distribution, coelomycetes occur under extremes of climatic conditions ranging from the tropics to arctic regions and even in very arid zones (dead leaves of *Yucca* spp. in the desert regions of the United States are literally covered with coelomycetes). Understandably, tropical regions have a richer and more diverse coelomycete flora than temperate and arctic regions. Reported from living and dead substrates, they are important as saprophytes, recycling carbon and other nutrient elements, and as agents of often serious diseases of agricultural crops and forest trees. In the latter category, some coelomycete pathogens have very wide host ranges, while others are specific to single host species. Some are virulent pathogens; others are wound parasites and weak pathogens attacking only debilitated hosts. Some overwinter in infected plant material, others survive as saprophytes on forest debris (*Diplodia pinea**), and some survive in the soil. In some pathogenic coelomycetes, the conidia act as the primary source of inoculum, while in a few others, as in the *Cytosporina* anamorph of *Eutypella prunastri,* the conidia do not play any role in the infection process.

The success of an anamorphic fungus depends on the viability of its conidia and the efficacy of their dissemination. As in other anamorphic fungi, the conidia of coelomycetes vary in their viability, which is brief in *Marssonina rosae,* or as long as 2 or 3 yr in others; stored pycnidia and conidia of *Coniella diplodiella* have been known to retain their viability and virulence for as long as 16 years*.

Coelomycete conidia are disseminated by wind, water, or arthropods. Under wet conditions, they are usually extruded from the conidiomata as long cirrhi. In *Phaeocytostroma sacchari,* the dried cirrhi break up into small fragments and are carried away by wind. In *Selenophoma donacis,* the conidia in the cirrhi are

---

[1]In This section, some of the statements marked by an asterisk (*) are based on the series: *CMI Descriptions of Pathogenic Fungi and Bacteria,* published by Commonwealth Agricultural Bureaux, England.

TABLE I
Roles of Coelomycetes

I. Plant pathogens
   Agricultural crops
   Ascochyta phaseolorum Sacc.—Wide host range on members of Leguminosae, destructive to pods, stems, and roots of seedlings, serious damage to debilitated hosts
   Diplodia maydis (Berk.) Sacc.—Seedling blight, stalk rot, and white ear rot of maize; up to 18% grain loss in the midwestern United States*
   Macrophomina phaseolina (Tassi) Goid.—Charcoal rot and ashy stem blight; wide host range
   Septoria nodorum (Berk.)*Berk.*—Glume blotch of wheat; up to 50% crop loss reported in Europe
   Horticultural and plantation crops
   Botryodiplodia theobromae Pat.—Plurivorous; facultative wound pathogen
   Colletotrichum falcatum Went—Red rot of sugarcane, widespread in sugarcane-growing areas, often causing considerable losses
   Colletotrichum musae (Berk. & Curt.) von Arx—Anthracnose, finger stalk and main stalk rot, fruit rot of bananas; widespread in banana plantations*
   Coniella diplodiella (Speg.) Petr. & Syd.—White rot of *Vitis vinifera;* 20-80% crop loss*
   Sylvicultural
   Brunchorstia pinea (Karst.) Höhn.—Dieback of pines and spruce, serious nursery losses
   Chondroplea populea (Sacc.) Kleb.—*Dothichiza* dieback and poplar canker
   Diplodia pinea (Desm.)*Kicky.*—Tip and twig blight; bud wilt; seedling collar rot of conifers
   Dothistroma pini Hulbarry—Dothistroma blight and red band of pines; heavy losses reported in New Zealand and Chile
   Phomopsis juniperivora Hahn—Blight and dieback of conifers; serious in the United States*
II. Animal pathogens—*Hendersonula toruloidea* Natrass and *Pyrenochaeta unguis-hominis* Punithalingam & English, though not reported as pathogens, have been isolated in the United Kingdom from infected nails and feet of human patients (Gentle and Evans, *Sabouraudia* **8,** 72-75, 1970; Punithalingam and English, *Trans. Br. Mycol. Soc.* **64,** 539-541, 1975); cases of mycotoxicosis in cattle and sheep and lupinosis in sheep have been attributed to animal feed of maize grains infected by *Diplodia maydis* and lupins infected by *Phomopsis leptostromiformis* (teleomorph: *Diaporthe woodii*), respectively
III. Entomopathogens—Potential sources of biological control of pests
   Aschersonia spp. on scale insects and whiteflies
   Ypsilonia spp. on scale insects
IV. Fungicolous fungi
   Eleutheromyces subulatus (Tode ex Fr.) Fuckel, *Eleutheromycella mycophila* Höhn., on agarics and polypores, respectively
   Cornutispora limaciformis Piroz., on ascomycetes
   Sphaerellopsis (=*Darluca*) spp., on rusts
V. In lichen associations (see Vobis and Hawksworth, this volume, Chapter 9)
VI. In marine and aquatic environments
   Dinemasporium marinum Nilsson on submerged wood in seawater
   Sphaceloma cecidii Kohlm. on *Sargassum natans*
   Robillarda rhizophorae Kohlm. on roots on *Rhizophora mangle* in seawater
   Pseudorobillarda phragmitis (Cunnell) Morelet on submerged dead stems of *Phragmites communis*
VII. In sewage and sludge—*Phoma* spp.
VIII. Coprophilous coelomycetes—*Chaetospora quezeli, Dinemasporium fimeti, Monodia elegans,* some species of *Phoma* or related genera, *Pseudoneottiospora coprophila, Pseudoneottiospora cunicularia,* and *Pullospora tetrachaeta.*
IX. Saprophytes—Most coelomycetes inhabiting plant debris and leaf litter (also includes examples in IV, VII, and VIII above).

disseminated by water films. Some coelomycetes are dispersed by local air currents (*Chondroplea populea, Colletotrichum gloeosporioides, Pestalotiopsis guepini**). Dispersal by water in splashing drops (*Colletotrichum gloeosporioides, Diplodia pinea, Septoria passerini**) or in wind-blown water (*Brunchorstia pinea, Dothistroma pini, Phomopsis juniperivora*) is much more common in coelomycetes than wind dispersal. Another common means of dispersal is by arthropods. Conidia of *Diplodia pinea* are known to be carried in the fluid secreted by the pine spittlebug *Aphrophora parallela.** Colletotrichum musae* is disseminated by ants and other insects,* while mites are known to spread *Sphaceloma fawcetti,* the causal agent of sour orange scab. Entomogenous fungi, such as *Aschersonia* spp. and *Ypsilonia* spp., occurring on scale insects, are spread by ants attracted to the insects by their honeydew secretions.

Slimes associated with conidial masses are believed to have a role in conidium dispersal, but the manner in which they function remained obscure until Bandoni (1975) observed that some behaved as surface-active substances. Bandoni observed that conidia of *Monochaetia* sp., *Pestalotiopsis* sp., and *Robillarda* sp. had highly surface-active slimes. The wet conidial mass in *Monochaetia* spread rapidly throughout the water, while such rapid dispersal of the dry masses in the same species occurred only at the surface of the water film, thus indicating a variability depending on the state of hydration of the slime mass. According to Bandoni (1975), many of the species found to have surface-active slimes are litter-inhabiting forms in which dispersal to new substrates by water is of considerable importance. Ingold (1961) has pointed out that the slimes of different groups of fungi are formed in different ways. In many coelomycetes, such slimes may originate through lysis occurring during cavity formation. In *Coleophoma,* the slime is contributed by gelatinization of some of the sterile cells associated with the conidiophores, while in *Catenophora* spp., the slime originates from the lysis of effete conidiogenous cells in the columnar conidiophores. Unquestionably, slime also plays an indirect role in the dispersal of the conidia by insects. In some instances, the mucoid appendages of conidia may serve to anchor the propagules at their new sites after transportation by water or insects. Mason (1937) suggested that the slimes of fungal spores are of possible taxonomic significance but, according to Bandoni (1975), the mere presence or absence of visible slime is not a reliable indicator of either potential surface activity or dispersal type.

Cupulate forms with setae, such as *Dinemasporium* spp., have conidial masses more or less completely exposed and appear to function as splash-cup dispersal mechanisms. A similar mechanism for dispersal is obvious in *Bothrodiscus.* The conidiomatal setae of many coelomycetes may function as a mechanism for trapping water drops in which the conidia become suspended and may also serve to prevent small arthropods such as mites from feeding on the slimy conidial mass. Bandoni (1975) suggests that structural arrangements may determine the manner in which the encroaching water contacts the slime drops, and that the hydrophobic setae or conidiophores prevent inundation, thus ensuring that the conidia are released onto the water surface. Other biological aspects such as factors affecting sporulation, conidium germination, colonization, nutrition, and so on, have no bearing on the current systematics of the Coelomycetes and are beyond the scope of this chapter.

For quite a long time, mycologists believed that Ascomycetes alone represented the teleomorphs genetically connected to the anamorphic fungi. Early mycological literature is replete with many examples of proven and presumed connections linking Ascomycetes and anamorphic fungi. A recent compilation (Kendrick and DiCosmo, 1979) of such connections shows that ascomycetous teleomorphs correlated with coelomycetes are distributed both in the unitunicate Clavicipitales, Helotiales, Hypocreales, Phacidiales, and Sphaeriales and the bitunicate Dothideales, Hemisphaeriales, Hysteriales, and Pleosporales. No coelomycetes have been correlated with members of Coronophorales, Endomycetales, Erysiphales, Eurotiales, Microascales, Ostropales, Taphrinales, or Tuberales. Recent published data have extended the scope of such connections to include mostly hyphomycetous anamorphs of Basidiomycetes (Kendrick and Watling, 1979). Within the last few years, I have become aware of at least three coelomycetous anamorph-genera that have basidiomycetous affinities. There must be many more such basidiomycetous coelomycetes which will come to light only through careful and critical observations. It is our experience that many of the collections bear both the coelomycetous anamorphs and the teleomorphs, and the two can be correlated (in the absence of cultural evidence) by the kinds of circumstantial evidence listed by Nag Raj (1979c). Often if a teleomorph specialist observes an anamorph in a collection, it is ignored or mentioned so briefly that it is virtually impossible to tell what it is. The converse is also true of some anamorph specialists encountering teleomorphs in their collections. Nag Raj (1979c) stressed the need for fully documenting the features of the morphs present in collections in an effort to advance our knowledge of holomorphs.

Thus, in conclusion, may I add a final commandment, turning the nonalogue into a decalogue:

In the interest of our need for knowledge of the whole fungus:
10. **Thou shalt not fail** to describe and illustrate the morphs found to occur in close association in collections. (Do not simply designate the anamorph as "*Dothichiza*-like." Above all, do not work in isolation from other specialists. If you do not know the anamorph, communicate with an anamorph specialist; conversely, if you do not know the teleomorph, communicate with a teleomorph specialist. Your records might provide a clue to relationships in the future—from recommendations of the Unitunicate Committee at the symposium, "The Whole Fungus," held at Kananaskis, Alberta, Canada, in 1977.)

ADDENDUM

As pointed out earlier, a vast number of coelomycete taxa are yet to be assimilated into a modern anatomico-ontogenetic scheme of classification. In a commentary on the status of generic names proposed for pycnidial and acervular anamorph-taxa, Sutton (1977) accepted 393, rejected 720, and retained a further 223 of these and 149 names of pycnothyrial anamorphs for future evaluation. In view of such inadequacy of data, provision of an overall generic key in this chapter would be impossible. Much of the information now available is presented in the recently published synoptic key to 200 genera of Coelomycetes (Michaelides *et al.*, 1979) based on the first 10 fascicles of *Icones Generum Coelomycetum* (Morgan-Jones, 1974, 1977; Morgan-Jones and Kendrick, 1972; Morgan-Jones *et al.*, 1972b,c,f,g; Nag Raj, 1974, 1977d; Nag Raj and DiCosmo, 1978).

REFERENCES

Ainsworth, G. C. (1967). "Ainsworth and Bisby's Dictionary of Fungi," 5th ed. (reprint). Commonw. Mycol. Inst., Kew, Surrey, England.
Archer, W. A. (1926). Morphological characters of some Sphaeropsidales in culture. *Ann. Mycol.* **24,** 1–84.
Ashby, S. F. (1927). *Macrophomina phaseoli* (Maubl.) comb. nov., the pycnidial stage of *Rhizoctonia bataticola* (Taub.) Butl. *Trans. Br. Mycol. Soc.* **12,** 141–147.
Baccarini, P. (1890). Sullo sviluppo der picnidii. *G. Ital. Bot.* **22,** 150.
Bandoni, R. J. (1975). Surface active spore slimes. *Can. J. Bot.* **53,** 2543–2546.
Boerema, G. H. (1964). *Phoma herbarum* Westend., the type species of the form-genus *Phoma* Sacc. *Persoonia* **3,** 9–16.
Boerema, G. H. (1967). The *Phoma* organisms causing gangrene of potatoes. *Neth. J. Plant Pathol.* **73,** 190–192.
Boerema, G. H. (1969). The use of the term forma specialis for *Phoma*-like fungi. *Trans. Br. Mycol. Soc.* **52,** 509–513.
Boerema, G. H., and Bollen, G. J. (1975). Conidiogenesis and conidial septation as differentiating criteria between *Phoma* and *Ascochyta*. *Persoonia* **8,** 111–144.
Boerema, G. H., and Howeler, L. H. (1967). *Phoma exigua* Desm. and its varieties. *Persoonia* **5,** 15–28.
Boerema, G. H., and van Kesteren, H. A. (1964). The nomenclature of two fungi parasitizing *Brassica*. *Persoonia* **3,** 17–28.
Boerema, G. H., Dorenbosch, M. M. J., and Van Kesteren, H. A. (1965). Remarks on species of *Phoma* referred to *Peyronellaea*. *Persoonia* **4,** 47–68.
Brooks, F. T., and Searle, G. O. (1921). An investigation on some tomato diseases. *Trans. Br. Mycol. Soc.* **7,** 173–197.
Brotzman, H. G., Calvert, O. H., Brown, M. R., and White, J. A. (1975). Holoblastic conidiogenesis in *Helminthosporium maydis*. *Can. J. Bot.* **53,** 813–817.
Carmichael, J. W., Kendrick, B., Conners, I. L., and Sigler, L. (1980). "Genera of Hyphomycetes." Univ. of Alberta Press, Edmonton.
Carroll, F. E. (1972). A fine-structural study of conidium initiation in *Stemphylium botryosum* Wallroth. *J. Cell Sci.* **11,** 33–47.
Carroll, F. E., and Carroll, G. C. (1974). The fine structure of conidium initiation in *Ulocladium atrum*. *Can. J. Bot.* **52,** 443–446.
Chippindale, H. G. (1929). The development in culture of *Ascochyta gossypii* Syd. *Trans. Br. Mycol. Soc.* **14,** 201–215.

Chodat, F. (1926). Recherches expérimentales sur la mutation chez les champignons. *Bull. Soc. Bot. Genève* [2] **18**, 41-144.
de Bary, A. (1887). "Comparative Morphology and Biology of the Fungi, Mycetozoa and Bacteria." Oxford Univ. Press, London and New York.
Dennis, R. W. G. (1946). Notes on some British fungi ascribed to *Phoma* and related genera. *Trans. Br. Mycol. Soc.* **29**, 11-42.
DiCosmo, F. (1978). A revision of *Corniculariella*. *Can. J. Bot.* **56**, 1665-1690.
DiCosmo, F., and Cole, G. T. (1980). Morphogenesis of conidiomata in *Chaetomella acutiseta* (Coelomycetes) *Can. J. Bot.* **58**, 1129-1137.
Dodge, B. O. (1930). Development of the asexual fructifications of *Chaetomella raphigera* and *Pezizella lythri*. *Mycologia* **23**, 446-462.
Dorenbosch, M. M. J. (1970). Key to nine ubiquitous soil-borne *Phoma*-like fungi. *Persoonia* **6**, 1-14.
Dyko, B., and Sutton, B. C. (1979). A revision of *Linodochium, Pseudocenangium, Septopatella* and *Siroscyphella*. *Can. J. Bot.* **57**, 370-385.
Eveleigh, D. E. (1961). *Phoma* spp. associated with painted surfaces. *Trans. Br. Mycol. Soc.* **44**, 573-585.
Grove, W. B. (1935). "British Stem- and Leaf-Fungi (Coelomycetes)," Vol. I. Cambridge Univ. Press, London and New York.
Grove, W. B. (1937). "British Stem- and Leaf-Fungi (Coelomycetes)," Vol. II. Cambridge Univ. Press, London and New York.
Hammill, T. M. (1972). Fine structure of annellophores. V. *Stegonosporium pyriforme*. *Mycologia* **64**, 654-657.
Harris, H. A. (1935). Morphological studies of *Septoria lycopersici*. *Phytopathology* **25**, 790-799.
Hennebert, G. L., and Weresub, L. K. (1977). Terms for states and forms of fungi, their names and types. *Mycotaxon* **6**, 207-211.
Hughes, S. J. (1953). Conidiophores, conidia and classification. *Can. J. Bot.* **31**, 577-659.
Ingold, C. T. (1961). The stalked spore drop. *New Phytol.* **60**, 181-183.
Jones, J. P. (1977). The ultrastructure of conidium ontogeny in *Pestalotiopsis neglecta*. *Can. J. Bot.* **55**, 766-771.
Kempton, F. E. (1919). Origin and development of the pycnidium. *Bot. Gaz. (Chicago)* **68**, 233-261.
Kendrick, W. B., ed. (1971). "Taxonomy of Fungi Imperfecti." Univ. of Toronto Press, Toronto.
Kendrick, W. B. (1974). The generic iceberg. *Taxon* **23**, (5/6); 747-753.
Kendrick, W. B., and Carmichael, J. W. (1973). Hyphomycetes. *In* "The Fungi" (G. C. Ainsworth, F. K. Sparrow, and A. S. Sussman, eds.), Vol. 4A, pp. 323-509. Academic Press, New York.
Kendrick, W. B., and DiCosmo, F. (1979). Teleomorph-anamorph connections in Ascomycetes. *In* "The Whole Fungus" (W. B. Kendrick, ed.), Vol. 1, pp. 283-410. National Museums of Canada, Ottawa.
Kendrick, W. B., and Nag Raj, T. R. (1979). Morphological terms in Fungi Imperfecti. *In* "The Whole Fungus" (W. B. Kendrick, ed.), Vol. 1, pp. 43-61. National Museums of Canada, Ottawa.
Kendrick, W. B., and Walting, W. (1979). Mitospores in Basidiomycetes. *In* "The Whole Fungus" (W. B. Kendrick, ed.), Vol. 2, pp. 473-545. National Museums of Canada, Ottawa.
Korf, R. P. (1973). Discomycetes and Tuberales. *In* "The Fungi" (G. C. Ainsworth, F. K. Sparrow, and A. S. Sussman, eds.), Vol. 4A, pp. 249-319. Academic Press, New York.
La Coste, L. (1955). De la morphologie et de la physiologie de *Peyronellaea stipae* nov. sp. *C.R. Hebd. Seances Acad. Sci.* **241**, 818-820.
Luttrell, E. S. (1963). Taxonomic criteria in *Helminthosporium*. *Mycologia* **55**, 643-674.
Maas, J. L., Pollack, F., and Uecker, F. A. (1979). Morphology and development of *Pilidiella quercicola*. *Mycologia* **71**, 92-102.

3. Coelomycete Systematics

Madelin, M. F. (1979). An appraisal of the taxonomic significance of some different modes of producing blastic conidia. *In* "The Whole Fungus" (W. B. Kendrick, ed.), Vol. 1, pp. 63–80. National Museums of Canada, Ottawa.

Mason, E. W. (1933). Annotated account of fungi received at the Imperial Mycological Institute. List II (fasc. 2). *Mycol. Pap.* **3,** 1–67.

Mason, E. W. (1937). Annotated account of the fungi received at the Imperial Mycological Institute. List II (fasc. 3, general part). *Mycol. Pap.* **4,** 1–99.

Michaelides, J., Hunter, L., Kendrick, B., and Nag Raj, T. R. (1979). "Icones Generum Coelomycetum. Supplement—Synoptic Key to 200 Genera of Coleomycetes," Biol. Ser. University of Waterloo, Waterloo.

Morgan-Jones, G. (1971a). Conidium ontogeny in Coelomycetes. I. Some amerosporous species which possess annellides. *Can. J. Bot.* **49,** 1921–1929.

Morgan-Jones, G. (1971b). Conidium ontogeny in Coelomycetes. II. Some Melanconiales which possess phialides. *Can. J. Bot.* **49,** 1931–1937.

Morgan-Jones, G. (1971c). Conidium ontogeny in Coelomycetes. III. Meristem thalloconidia. *Can. J. Bot.* **49,** 1939–1940.

Morgan-Jones, G. (1973). Genera coelomycetarum. VII. *Cryptocline* Petrak. *Can. J. Bot.* **51,** 309–325.

Morgan-Jones, G. (1974). "Icones Generum Coelomycetum," Fasc. VII, Biol. Ser. University of Waterloo, Waterloo.

Morgan-Jones, G. (1975). Notes on Coelomycetes. I. *Ceratopycnis, Clypeopycnis, Macrodiplodiopsis, Mastigosporella, Paradiscula* and *Septopatella. Mycotaxon* **2,** 167–183.

Morgan-Jones, G. (1977). "Icones Generum Coelomycetum," Fasc. IX, Biol. Ser. University of Waterloo, Waterloo.

Morgan-Jones, G., and Kendrick, B. (1972). "Icones Generum Coelomycetum," Fasc. III, Biol. Ser. University of Waterloo, Waterloo.

Morgan-Jones, G., Nag Raj, T. R., and Kendrick, B. (1972a). Genera coelomycetum. V. *Alpakesa* and *Bartalinia. Can. J. Bot.* **50,** 877–882.

Morgan-Jones, G., Nag Raj, T. R., and Kendrick, B. (1972b). "Icones Generum Coelomycetum," Fasc. I, Biol. Ser. University of Waterloo, Waterloo.

Morgan-Jones, G., Nag Raj, T. R., and Kendrick, B. (1972c). "Icones Generum Coelomycetum," Fasc. II, Biol. Ser. University of Waterloo, Waterloo.

Morgan-Jones, G., Nag Raj, T. R., and Kendrick, B. (1972d). Genera coelomycetum. VI. *Kellermania. Can. J. Bot.* **50,** 1641–1648.

Morgan-Jones, G., Nag Raj, T. R., and Kendrick, B. (1972e). Conidium ontogeny in coelomycetes. IV. Percurrently proliferating phialides. *Can. J. Bot.* **50,** 2009–2014.

Morgan-Jones, G., Nag Raj, T. R., and Kendrick, B. (1972f). "Icones Generum Coelomycetum," Fasc. IV, Biol. Ser. University of Waterloo, Waterloo.

Morgan-Jones, G., Nag Raj, T. R., and Kendrick, B. (1972g). "Icones Generum Coelomycetum," Fasc. V, Biol. Ser. University of Waterloo, Waterloo.

Nag Raj, T. R., (1973a). Genera coelomycetum. IX. *Brycekendrickia* gen. nov. and *Vasudevella. Can. J. Bot.* **51,** 1337–1341.

Nag Raj, T. R., (1973b). Genera coelomycetum. X. *Ellisiella, Samukuta* and *Sakireeta. Can. J. Bot.* **51,** 2463–2472.

Nag Raj, T. R. (1974). "Icones Generum Coelomycetum," Fasc. VI, Biol. Ser. University of Waterloo, Waterloo.

Nag Raj, T. R. (1975a). Genera coelomycetum. XI. *Hyalotia, Hyalotiella* and *Hyalotiopsis. Can. J. Bot.* **53,** 1615–1624.

Nag Raj, T. R. (1975b). Genera coelomycetum. XII. *Tracyella* and *Amerodiscosiella. Can. J. Bot.* **53,** 2435–2442.

Nag Raj, T. R. (1977a). Genera coelomycetum. XIII. *Plectronidium* gen. nov. *Can. J. Bot.* **55**, 625-629.

Nag Raj, T. R. (1977b). Miscellaneous microfungi. II. *Can. J. Bot.* **55**, 757-765.

Nag Raj, T. R. (1977c). *Ypsilonia, Acanthotheciella* and *Kazulia* gen. nov. *Can. J. Bot.* **55**, 1599-1622.

Nag Raj, T. R. (1977d). "Icones Generum Coelomycetum," Fasc. VIII, Biol. Ser. University of Waterloo, Waterloo.

Nag Raj, T. R. (1978a). Genera coelomycetum. XIV. *Allelochaeta, Basilocula, Ceuthosira, Microgloeum, Neobarclaya, Polynema, Pycnidiochaeta,* and *Xenodomus. Can. J. Bot.* **56**, 686-707.

Nag Raj, T. R. (1978b). Genera coelomycetum. XVI. *Fibulocoela* forma-gen. nov., a coelomycete with basidiomycetous affinities. *Can. J. Bot.* **56**, 1485-1491.

Nag Raj, T. R. (1979a). Genera coelomycetum. XVII. New anamorph genera: *Libartania* and *Plectronidiopsis. Can. J. Bot.* **57**, 1389-1397.

Nag Raj, T. R. (1979b). Miscellaneous microfungi. III. *Can. J. Bot.* **57**, 2489-2496.

Nag Raj, T. R. (1979c). Some coelomycetous anamorphs and their teleomorphs. In "The Whole Fungus" (W. B. Kendrick, ed.), Vol. 1, pp. 183-200. National Museums of Canada, Ottawa.

Nag Raj, T. R., and DiCosmo, F. (1978). "Icones Generum Coelomycetum," fasc. X, Biol. Ser. University of Waterloo, Waterloo.

Nag Raj, T. R., and DiCosmo, F. (1980). "Icones Generum Coelomycetum," Fasc. XI, Biol. Ser. University of Waterloo, Waterloo.

Nag Raj, T. R., and Kendrick, W. B. (1970). *Mycotribulus,* a new genus of Sphaeropsidales. *Can. J. Bot.* **48**, 2219-2221.

Nag Raj, T. R., and Kendrick, B. (1971). Genera coelomycetum. I. *Urohendersonia. Can. J. Bot.* **49**, 1853-1862.

Nag Raj, T. R., and Kendrick, B. (1972a). Genera coelomycetum. II. *Doliomyces. Can. J. Bot.* **50**, 45-48.

Nag Raj, T. R., and Kendrick, B. (1972b). Genera coelomycetum. III. *Pestalozziella. Can. J. Bot.* **50**, 607-617.

Nag Raj, T. R., and Kendrick, B. (1978). Genera coelomycetum. XV. *Belaina, Belainopsis* and *Crucellisporium. Can. J. Bot.* **56**, 708-714.

Nag Raj, T. R., and Morgan-Jones, G. (1973). Genera coelomycetum. VIII. *Rhabdogloeopsis* Petrak and *Rhabdogloeum* Sydow. *Can. J. Bot.* **51**, 565-569.

Nag Raj, T. R., Morgan-Jones, G. and Kendrick, B. (1972). Genera coelomycetum. IV. *Pseudorobillarda* gen. nov., a generic segregate of *Robillarda* Sacc. *Can. J. Bot.* **50**, 861-867.

Persoon, D. C. H. (1801). "Synopsis Methodica Fungorum." H. Dietrich, Göttingen.

Pirozynski, K. A., and Morgan-Jones, G. (1968). Notes on microfungi. III. *Trans. Br. Mycol. Soc.* **51**, 185-206.

Pirozynski, K. A., and Shoemaker, R. A. (1971). Some coelomycetes with appendaged conidia. *Can. J. Bot.* **49**, 529-541.

Potebnia, A. (1910). Beiträge zur Micromycetenflora Mittel-Russlands. *Ann. Mycol.* **8**, 42-93.

Punithalingam, E. (1966). Development of the pycnidium in *Septoria. Trans. Br. Mycol. Soc.* **49**, 19-25.

Punithalingam, E. (1974). Studies on Sphaeropsidales in culture. II. *Mycol. Pap.* **136**, 1-63.

Reisinger, O., Morelet, M., and Kiffer, E. (1977). Electron microscopic study of conidium ontogeny in *Coniothyrium cupressacearum* (Coelomycetes). *Persoonia* **9**, 257-264.

Saccardo, P. A. (1878). Fungi veneti novi vel critici. Series 7. *Michelia* **1**, 133-271.

Saccardo, P. A. (1880). Conspectus generum fungorum italiae inferiorum nempe ad Sphaeropsideas, Melanconieas et Hyphomyceteas pertinentium sporologico dispositorum. *Michelia* **2**, 1-38.

Saccardo, P. A. (1884). "Sylloge Fungorum Omnium Hucusque Cognitorum," Vol. 3. Padova.

Sutton, B. C. (1961). Coelomycetes. I. *Mycol. Pap.* **80,** 1-16.
Sutton, B. C. (1962). *Colletotrichum dematium* (Pers. ex Fr.) Grove and *C. trichellum* (Fr. ex Fr.) Duke. *Trans. Br. Mycol. Soc.* **45,** 222-232.
Sutton, B. C. (1963). Coelomycetes. II. *Neobarclaya, Mycohypallage, Bleptosporium* and *Cryptostictis. Mycol. Pap.* **88,** 1-50.
Sutton, B. C. (1964a). *Phoma* and related genera. *Trans. Br. Mycol. Soc.* **47,** 497-509.
Sutton, B. C. (1964b). Coelomycetes. III. *Annellolacinia* gen. nov., *Aristastoma, Phaeocytostroma, Seimatosporium,* etc. *Mycol. Pap.* **97,** 1-42.
Sutton, B. C. (1964c). *Melanconium* Link ex Fr. *Persoonia* **3,** 193-198.
Sutton, B. C. (1967a). Two new genera of Sphaeropsidales and their relationships with *Diachorella, Strasseria* and *Plagiorhabdus. Can. J. Bot.* **45,** 1249-1263.
Sutton, B. C. (1967b). Redescription of *Ajrekarella* Kamat & Kalani. *Mycopathol. Mycol. Appl.* **33,** 76-80.
Sutton, B. C. (1968a). *Kellermania* and its generic segregates. *Can. J. Bot.* **46,** 181-196.
Sutton, B. C. (1968b). The appressoria of *Colletotrichum graminicola* and *C. falcatum. Can. J. Bot.* **46,** 873-876.
Sutton, B. C. (1969). Type studies in *Coniella, Anthasthoopa* and *Cyclodomella. Can. J. Bot.* **47,** 603-608.
Sutton, B. C. (1970). Forest microfungi. III. The heterogeneity of *Pestalotia* de Not. section sexloculatae Klebahn sensu Guba. *Can. J. Bot.* **47,** 2083-2094.
Sutton, B. C. (1971a). Conidium ontogeny in pycnidial and acervular fungi. *In* "Taxonomy of Fungi Imperfecti" (W. B. Kendrick, ed.), pp. 263-278. Univ. of Toronto Press, Toronto.
Sutton, B. C. (1971b). Coelomycetes. IV. The genus *Harknessia* and similar fungi on *Eucalyptus. Mycol. Pap.* **123,** 1-46.
Sutton, B. C. (1973). Coelomycetes. *In* "The Fungi" (G. C. Ainsworth, F. K. Sparrow, and A. S. Sussman, eds.), Vol. 4A, pp. 513-582. Academic Press, New York.
Sutton, B. C. (1975a). *Diploceras,* another synonym of *Seimatosporium. Trans. Br. Mycol. Soc.* **64,** 483-487.
Sutton, B. C. (1975b). Coelomycetes. V. *Coryneum. Mycol. Pap.* **138,** 1-224.
Sutton, B. C. (1977). Coelomycetes. VI. Nomenclature of generic names proposed for coelomycetes. *Mycol. Pap.* **141,** 1-253.
Sutton, B. C., and Chao, R. L. C. (1970). *Leptomelanconium. Trans. Br. Mycol. Soc.* **55,** 37-44.
Sutton, B. C., and DiCosmo, F. (1977). A revision of *Monochaetiella. Can. J. Bot.* **55,** 2535-2543.
Sutton, B. C., and Kobayashi, T. (1970). *Strasseriopsis* gen. nov. based on *Phellostroma tsugae* Kobayashi. *Mycologia* **61,** 1066-1071.
Sutton, B. C., and Pirozynski, K. A. (1963). Notes on British microfungi. I. *Trans. Br. Mycol. Soc.* **46,** 505-522.
Sutton, B. C., and Pirozynski, K. A. (1965). Notes on microfungi. II. *Trans. Br. Mycol. Soc.* **48,** 349-366.
Sutton, B. C., and Sandhu, D. K. (1969). Electron microscopy of conidium development and secession in *Cryptosporiopsis* sp., *Phoma fumosa, Melanconium bicolor* and *M. apiocarpum. Can. J. Bot.* **47,** 745-749.
Sutton, B. C., and Sellar, P. W. (1966). *Toxosporiopsis* n. gen., an unusual member of the Melanconiales. *Can. J. Bot.* **44,** 1505-1513.
Tulloch, M. (1972). The genus *Myrothecium* Tode ex Fr. *Mycol. Pap.* **130,** 1-42.
Van der Aa. H. A. (1973). Studies in *Phyllosticta. I. Stud. Mycol.* **5,** 1-110.
von Arx, J. A. (1957). Die Arten der Gattung *Colletotrichum* Corda. *Phytopathol. Z.* **29,** 413-468.
von Höhnel, F. (1923). System der Fungi Imperfecti Fuckel. *Mykol. Unters.* **1,** (3), 301-369.

Vuillemin, P. (1910a). Matériaux pour une classification rationelle des Fungi Imperfecti. *C.R. Hebd. Seances Acad. Sci.* **150**, 882–884.
Vuillemin, P. (1910b). Les Conidiosporés. *Bull. Soc. Sci. Nancy* [3] **11**, 129–172.
Vuillemin, P. (1911). Les Aleuriosporés. *Bull. Soc. Sci. Nancy* [3] **12**, 151–175.
Vuillemin, P. (1912). "Les Champignons: Essai de classification." Doin, Paris.

4

Systematics of Conidial Yeasts

J.A. von Arx

I.	Introduction and History	85
II.	Generic and Specific Delimitation	86
III.	Classification	88
	A. Candidaceae Windisch	88
	B. Cryptococcaceae Kützing	91
	C. Sporobolomycetaceae Derx	94
	References	95

I. INTRODUCTION AND HISTORY

The taxa of yeasts are few compared with those delimited in the filamentous fungi. About 550 species were accepted in the taxonomic study, *The Yeasts*, edited by Lodder (1970). Nearly 200 species have been described since then, many of which have proved to be unacceptable (von Arx *et al.*, 1977).

The anamorphic yeasts represent the largest group, comprising about 300 species belonging to 22 genera. They were first treated in a comprehensive study by Diddens and Lodder (1942) as anascosporogenous yeasts, as opposed to sporogenous yeasts which form spores in asci and which had been monographed by Stelling-Dekker in 1931. Diddens and Lodder (1942) classified the anamorphic yeasts as a separate subfamily in the Mycotoruloideae and accepted the genera *Candida, Brettanomyces,* and *Trichosporon.*

Lodder and Kreger-van Rij (1952) distinguished three families in the yeasts: the Endomycetaceae for yeasts forming asci, the Sporobolomycetaceae for anamorphic yeasts forming ballistospores on sterigmata, and the Cryptococcaceae for anamorphic yeasts without asci or ballistospores. The genera *Sporobolomyces* and *Bullera* were accepted in the Sporobolomycetaceae, and the genera *Cryptococcus, Torulopsis, Pityrosporum, Brettanomyces, Candida, Kloeckera, Trigonopsis, Trichosporon,* and *Rhodotorula* were treated in the Cryptococcaceae. In Lodder (1970) the genera *Oosporidium, Schizoblastosporion,* and *Sterigmatomyces* were added, while *Phaffia* and *Selenozyma* have been described more recently.

Delimitation of the yeasts from the filamentous fungi, as proposed in the above-mentioned monographs, was unsatisfactory. Von Arx *et al.* (1977) provided a key for all the classic yeast genera and added *Geotrichum, Trichosporonoides,* and *Moniliella.* In Lodder and Kreger-van Rij (1952) and in Lodder (1970) the filamentous genus *Trichosporon* was included, but the similar genus *Geotrichum* was omitted. Both genera usually form arthroconidia. A comparative study (Weijman, 1979) has shown that *Trichosporon sensu* Do Carmo Sousa as treated in Lodder (1970) is heterogeneous; most of the species must be excluded and actually belong to *Geotrichum* and other genera.

II. GENERIC AND SPECIFIC DELIMITATION

The anamorphic yeast genera accepted in Lodder (1970) are distinguished mainly by morphological characters such as the presence or absence of septate hyphae or pseudohyphae, the shape of the yeast cells (round, ellipsoidal, ovate, triangular, seleniform, navicular, or cylindrical), and the kind of conidiogenesis (blastic or arthric). The genera *Cryptococcus* and *Rhodotorula* are distinguished by the fact that the former assimilates inositol while the latter does not. The species of these two genera do not ferment glucose or other sugars, but other genera such as *Candida* and *Torulopsis* include species with and without fermentative abilities.

In recent years it has often been shown that the larger genera of the anamorphic yeasts are particularly heterogeneous. Species classified in *Candida, Torulopsis, Cryptococcus,* and *Trichosporon* proved to be the anamorphs of holomorphic yeasts belonging to either the Saccharomycetaceae (e.g., to *Saccharomyces, Pichia,* or *Kluyveromyces*) or the Basidiomycetes (e.g., *Filobasidium* and *Filobasidiella* of the Filobasidiaceae, or *Rhodosporidium, Aessosporon,* and *Sporidiobolus*).

Such heterogeneous genera are unsatisfactory and need to be revised. Several characteristics are known which easily distinguish the anamorphs of Saccharomycetaceae from those of basidiomycetous yeasts. Relationships among these taxa have been demonstrated by determination of the chemical composition of the cell walls (e.g., Spencer and Gorin, 1969; Weijman, 1979; von Arx and Weijman, 1979), G+C percent of the DNA (e.g., Nakase and Komagata, 1968,

1971; Meyer and Phaff, 1970), coenzyme Q system (Yamada and Kondo, 1973), ultrastructure of the cell wall and bud (Kreger-van Rij and Veenhuis, 1971; van der Walt *et al.*, 1974), ability to use urease, color reactions, formation of pigments, or mode of conidiogenesis. Enteroblastic-basipetal conidia (yeast cells) have been observed only in basidiomycetous yeasts. In the Saccharomycetes the conidia are holoblastic and solitary (multilateral), acropetal, or formed on percurrently elongating cells. The characters which distinguish between anamorphic Saccharomycetes and anamorphic Basidiomycetes are summarized in Table 1.

The species are delimited by mainly physiological characters which generally include the relative ability to ferment glucose, galactose, sucrose, maltose, or lactose and to assimilate about 30 carbon compounds. The keys are usually based on the assimilation of galactose, lactose, or maltose. Morphological characters, such as size and shape of cells, are neglected in the keys provided in Lodder (1970).

It was shown by Scheda and Yarrow (1966, 1968), Neumann (1972), Price *et al.* (1978), and others that the assimilation of carbon compounds could be variable within one species and that it could change in a single strain after several transfers. Therefore, more stable characters had to be found for species delimitation, e.g., the optimum, maximum, and minimum temperatures for growth and sporulation, growth in the presence of certain toxic or other compounds, or growth on high sugar or salt concentrations. Mating and DNA–DNA reassociation experiments resulted in a great reduction in the number of accepted species

TABLE I

Characters of Saccharomycetes and Basidiomycetes

Character	Saccharomycetes	Basidiomycetes
Colonies	Usually unpigmented, not mucoid	Often pigmented and mucoid
Conidiation	Holoblastic, multilateral, acropetal or solitary, or arthric	Enteroblastic-basipetal or holoblastic-sympodial or arthric
Urease activity	Usually negative	Usually positive
Diazonium blue B	No coloration	Reddish
G+C content of DNA	30–52%	48–68%
Fermentation	Present or absent	Absent (weak in some species)
Coenzyme Q group	6, 7, 8, or 9	9 or 10
Cell wall composition	Glucan–mannan, no xylose, no fucose	Chitin–mannan, often also xylose or fucose
Cell wall ultrastructure	A thin outer and a thick, light inner layer	Several layers, mainly around the conidiogenous locus; outer layers dissolving

in genera such as *Saccharomyces, Kluyveromyces, Torulaspora,* and *Pichia.* However, the anamorphic yeasts have not yet been studied in this manner, and the numerous species are still identified by their often unreliable fermentation and assimilation patterns.

III. CLASSIFICATION

Three anamorph-families must be distinguished: the Candidaceae Windisch for the anamorphic genera of the Saccharomycetales, the Cryptococcaceae Kützing and the Sporobolomycetaceae Derx for those of the basidiomycetous yeasts. The Cryptococcaceae comprise the anamorphs of the Filobasidiaceae, a family probably belonging to the Aphyllophorales. The Sporobolomycetaceae comprise the anamorphs of genera such as *Sporidiobolus, Rhodosporidium,* and *Aessosporon* for which no holomorphic family name is available but which may be close to the Ustilaginales (Oberwinkler, 1978).

The three families can be distinguished by the following key:

1. Enteroblastic (phialidic) or sympodial conidiation absent, ballistospores absent ...Candidaceae
1. Enteroblastic or sympodial conidiation present, ballistospores present or absent 2
2. Carotenoid pigments usually absent, occasionally present; conidiation usually sympodial (buds close together) but also enteroblastic; cells roundish, cylindrical, or curved; cell walls contain xylose ....Cryptococcaceae
2. Carotenoid pigments usually present (red yeasts), conidiation enteroblastic (phialidic), cells often pyriform or ovoid; xylose absent ....Sporobolomycetaceae

A. Candidaceae Windisch

The Candidaceae comprise the anamorphs of Saccharomycetaceae, Metschnikowiaceae, Dipodascaceae, Ascoideaceae, and Saccharomycodaceae (von Arx *et al.,* 1977). Nearly all species of yeasts classified in these families include anamorphic states, but in general they have no separate generic and specific names. Many species, however, only form asci under certain conditions, e.g., on special media. Heterothallic species form the teleomorph only in mating experiments, and the haploid states have been described as *Candida, Torulopsis,* and other genera.

The following genera belong to this family:

1. *Candida* Berkhout (*Schimmelgesl. Monilia, Oidium,* etc., Utrecht, p. 72. 1923). Type species: *C. vulgaris* Berkhout = *C. tropicalis* (Cast.) Berkhout (Fig. 1a).

Van Uden and Buckley, in Lodder (1970), restricted the genus to species forming hyphae and/or pseudohyphae and having unpigmented colonies: 81 species were accepted, and approximately 60 further species have been described

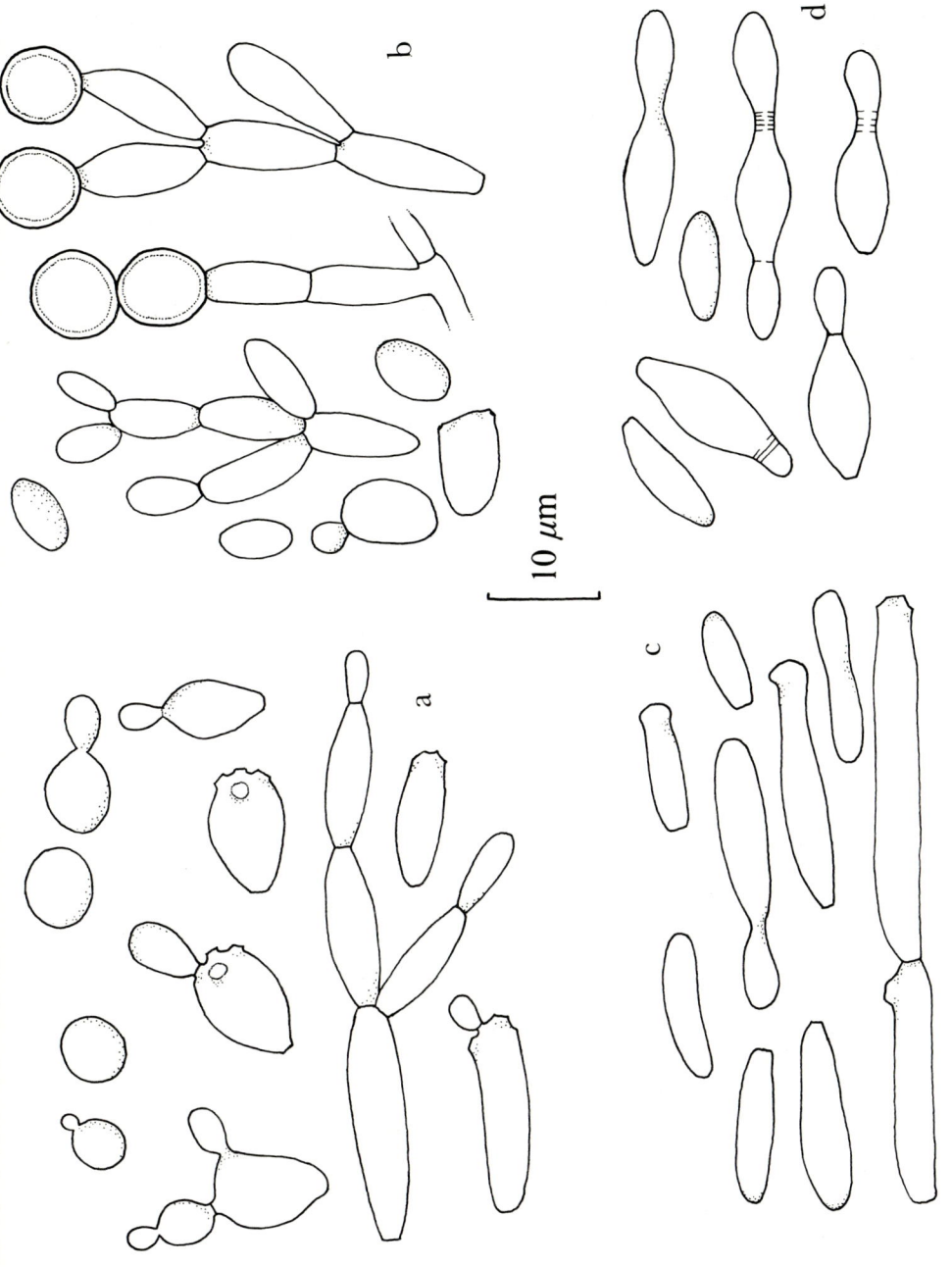

Fig. 1. Conidiation in Candidaceae. (a) *Candida tropicalis*; (b) *Candida albicans*; (c) *Brettanomyces bruxellensis*; (d) *Kloeckera apiculata*.

since. Species without hyphae and pseudohyphae have been classified in *Torulopsis* auct. (non Berlese) which now comprises about 70 species. However, the name *Torulopsis* is illegitimate and must be rejected; most of the species will have to be transferred to *Candida,* and the remainder to *Rhodotorula* and other genera. *Candida* will then comprise species with or without hyphae or pseudohyphae but must be delimited as having holoblastic, multilateral, or acropetal conidiation, and spherical, ovoid, or cylindrical cells (Fig. 1a and b).

A few species are known to form septate hyphae and have often been classified in separate genera. *Candida albicans* (Robin) Berkhout often forms aerial hyphae with terminal or intercalary chlamydospores. It is the type species of the older genus *Syringospora* Quinquand (1868) and is often treated as *Syringospora albicans* (Robin) Dodge. *Candida mesenterica* (Geiger) Diddens and Lodder, the type species of the genus *Pseudomonilia* Geiger (1910), mainly forms septate, aerial hyphae and only a small number of elongate conidia. *Candida chodatii* (Nechitch) Berkhout and *C. ciferrii* Kreger-van Rij develop expanding, septate hyphae, denticulate conidiogenous cells, and small, often pyriform conidia. The former is the haploid anamorph of *Hyphopichia burtonii* (Boidin *et al.*) v. Arx and van der Walt, and the latter that of *Stephanoascus ciferrii* M. T. Smith *et al.*, two heterothallic yeasts of the Ascoideaceae (von Arx and van der Walt, 1976; Smith *et al.*, 1976). A mycelial species often isolated from oil products is *Candida lipolytica* (Harrison) Diddens & Lodder whose teleomorph, *Saccharomycopsis lipolytica* (Wickerham *et al.*) Yarrow, was also obtained in mating experiments.

Many other *Candida* species without septate hyphae represent the anamorphs of Saccharomycetaceae and Metschnikowiaceae. The following connections concern rather common species:

Anamorph	Teleomorph
Candida guilliermondii (Cast.) Langeron & Guerra	*Pichia guilliermondii* Wickerham
C. krusei (Cast.) Berkhout	*Issatchenkia orientalis* Kudrjawzew
C. macedoniensis (Cast. & Chalmers) Berkhout	*Kluyveromyces marxianus* (Hansen) van der Walt
C. mogii Vidal-Leiria	*Debaryozyma castellii* (Capriotti) van der Walt & Johannsen
C. pseudotropicalis (Cast.) Basgal	*Kluyveromyces fragilis* (Jörgensen) van der Walt
C. pulcherrima (Lindner) Windisch	*Metschnikowia pulcherrima* Pitt & Miller
C. sake (Saito & Ota) van Uden & Buckley	*Pichia chambardii* (Ramirez & Boidin) Phaff
C. utilis (Henneberg) Lodder & Kreger-van Rij	*Hansenula jadinii* (Sartory *et al.*) Wickerham
C. valida (Leberle) van Uden & Buckley	*Pichia membranaefaciens* Hansen
Torulopsis candida (Saito) Lodder	*Debaryozyma hansenii* (Zopf) van der Walt
T. domercqii van der Walt & Kerken	*Wickerhamiella domercqii* van der Walt
T. holmii (Jörgensen) Lodder	*Saccharomyces exiguus* Reess & Hansen
T. molischiana (Zikes) Lodder	*Hansenula capsulata* Wickerham

2. *Brettanomyces* Kufferath & van Laer (*Bull. Soc. Chim. Belg.* **30**, 270, 1921). Type species: *B. bruxellensis* Kufferath & van Laer (Fig. 1c).

Van der Walt, in Lodder (1970), included seven species, and two more have been described subsequently. In two species, van der Walt (1964) observed a teleomorph with hat-shaped ascospores, which he classified in a new genus, *Dekkera*. *Brettanomyces* species are rather similar to *Candida* in morphology but produce large amounts of acetic acid on glucose-containing media, so that cultures are short-lived. Subcultures must be made at frequent intervals on media containing calcium carbonate.

3. *Kloeckera* Janke (*Zentralbl. Bakteriol., Parasitenkde, Infektionskr. Hyg., Abt. 2* **59**, 310, 1923). Type species: *K. apiculata* (Reess) Janke (Fig. 1d).

This genus is characterized by fusiform, elongate, ellipsoidal, or limoniform cells which form conidia with rather broad bases at the percurrently proliferating ends (apiculate yeasts). *Kloeckera* includes seven species, all of which are known to have *Hanseniaspora* teleomorphs (Meyer *et al.*, 1978).

Similar anamorphs with apiculate conidiation are known in the genera *Saccharomycodes* Hansen and *Nadsonia* Sydow. Both can be easily distinguished from *Kloeckera* by their much larger cells.

4. *Selenozyma* Yarrow (*Stud. Mycol.* **14**, 29, 1977). Type species: *S. intestinalis* (Krassilnikov) Yarrow.

The genus is characterized by lunate, falcate, or hemispherical cells. The teleomorph of *S. intestinalis* has recently been described as *Metschnikowia lunata* Golubev (1977). A second species has been described as *S. peltata* (Yarrow) Yarrow.

5. *Geotrichum* Link ex Pers. (*Mycol. Eur.* **1**, 26, 1822). Type species: *G. candidum* Link ex Pers.

The genus is usually treated in the Hyphomycetes but should be classified in the yeasts. It is close to the anamorphs of *Saccharomycopsis* and other Ascoideaceae. *Geotrichum* comprises the anamorphs of *Dipodascus* Lagerh. sensu von Arx (1977) and is characterized by arthroconidia which usually develop from aerial hyphae. Von Arx *et al.* (1977) transferred three species of the basidiomycetous genus *Trichosporon* to *Geotrichum*. The genus now comprises eight species, one of which, *G. fermentans* (Diddens & Lodder) v. Arx, also forms blastoconidia with a broad base and has restricted colonies.

The septa of some species of *Geotrichum*, *Dipodascus*, and *Saccharomycopsis* have been examined, and all showed the presence of micropores (plasmodesmata) (Cole, 1975; Kreger-van Rij and Veenhuis, 1973).

B. Cryptococcaceae Kützing

The Cryptococcaceae comprise the anamorphs of Filobasidiaceae, which form basidium-like structures on erect hyphae with clamp connections. The teleomorph is known in only five heterothallic species; the haploid, anamorphic states usually form only yeast cells and no hyphae and are classified in *Cryptococcus*. Septate hyphae are present in the closely related genera *Apiotrichum*, *Trichosporon*, and *Moniliella*.

The following genera belong to this family:

1. *Cryptococcus* Kützing emend. Vuill. (*Rev. Gen. Sci. Pure Appl.* **12,** 741, 1901). Type species: *C. neoformans* (Sanfelice) Vuill. (Fig. 2a).

Phaff and Fell, in Lodder (1970), accepted 17 species, and more have been added since. Typical *Cryptococcus* species have spherical, ovoid, or ellipsoidal cells which propagate by sympodial or basipetal budding (Fig. 2a). The colonies are usually mucoid and grayish, brownish, or reddish.

Filobasidiella neoformans Kwon-Chung, the teleomorph of *Cryptococcus neoformans*, was discovered by Kwon-Chung (1975) in mating experiments. The anamorph of *Filobasidium capsuligenum* (Fell *et al.*) Rodrigues de Miranda has been classified in *Torulopsis* and *Candida* because of its weak fermentation, but in all other respects it is *Cryptococcus*-like. Its conidiation is enteroblastic-basipetal. Some other *Cryptococcus* species are known to have teleomorphs belonging to *Filobasidium* Olive or *Filobasidiella* Kwon-Chung.

2. *Phaffia* Miller *et al.* (*Int. J. Syst. Bacteriol.* **26,** 286, 1976). Type species: *Ph. rhodozyma* Miller *et al.*

This species shows weak fermentation. The colonies are red as a result of carotenoid pigments, and the pyriform or ovoid cells form basipetal conidia. *Phaffia rhodozyma* is close to *Cryptococcus hungaricus* (Zsolt) Phaff & Fell, the type species of the genus *Dioszegia* Zsolt, which also forms carotenoid pigments.

3. *Bullera* Derx (*Ann. Mycol.* **28,** 11, 1930). Type species: *B. alba* Derx (Fig. 2c).

This genus can be distinguished from *Cryptococcus* only by the production of roundish or asymmetric ballistospores on erect sterigmata. Phaff, in Lodder (1970), treated three species, and two have been added more recently, though their position is somewhat doubtful (Stadelmann, 1955).

4. *Sterigmatomyces* Fell (*Antonie van Leeuwenhoek* **32,** 101, 1966). Type species: *S. halophilus* Fell (Fig. 2b).

The genus is close to *Cryptococcus;* the cells are usually roundish, and the conidia are formed singly on narrow, "enteroblastic" stalks (Kreger-van Rij and Veenhuis, 1971). Several species have been described recently (e.g., Rodrigues de Miranda, 1975).

5. *Trichosporon* Behrend (*Biol. Klin. Wochenschr.* **27,** 464, 1890). Type species: *T. beigelii* (Küchenm. & Rabenh.) Vuill.

In the yeast literature the type species is usually treated under the more recent name *Trichosporon cutaneum* (Beurm. *et al.*) Ota. The genus is characterized by expanding, mucoid colonies composed of hyphae which have septa with dolipores and which easily disintegrate into arthroconidia. Only a few species can be accepted in the genus, but *T. cutaneum sensu* Do Carmo-Sousa in Lodder (1970)

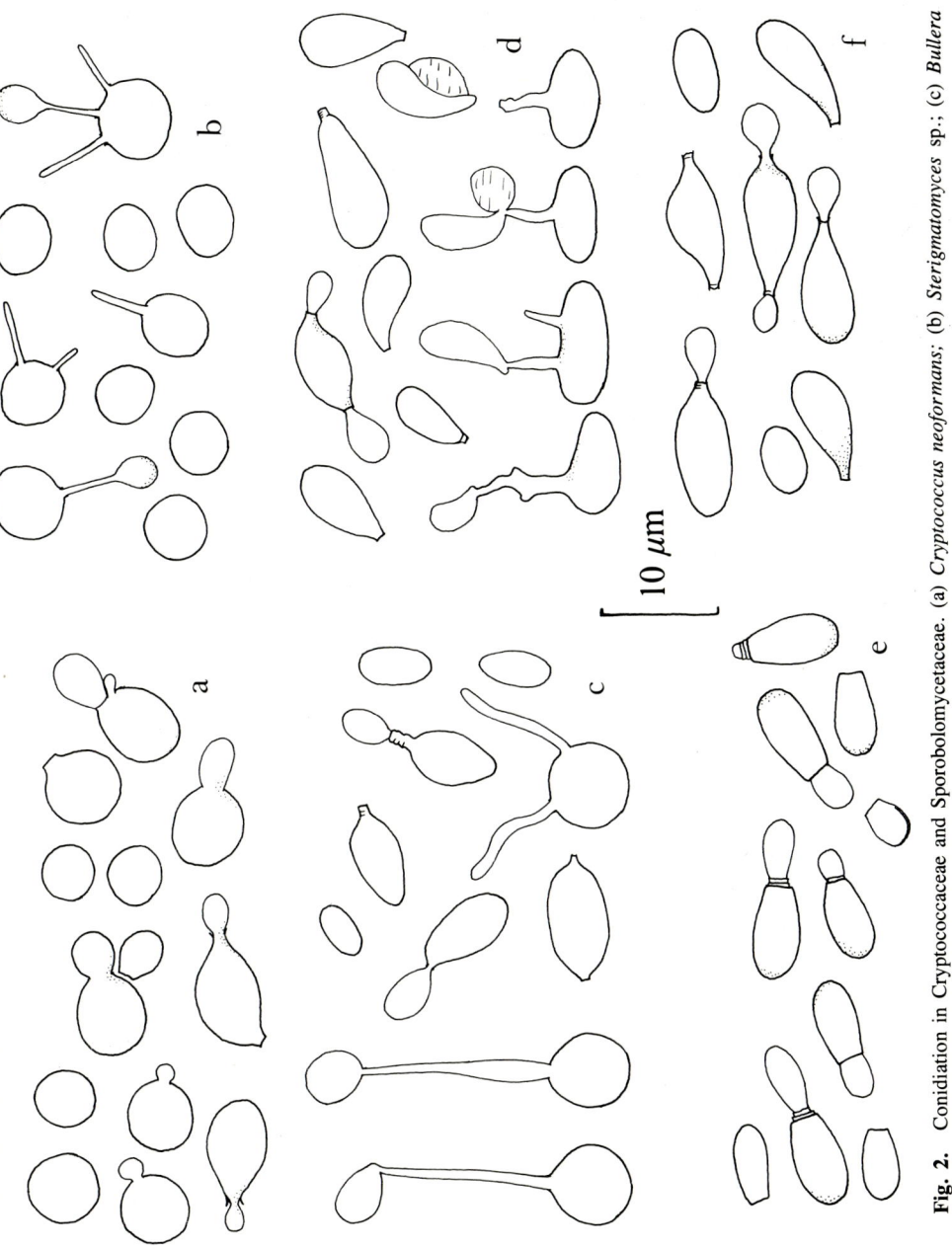

Fig. 2. Conidiation in Cryptococcaceae and Sporobolomycetaceae. (a) *Cryptococcus neoformans*; (b) *Sterigmatomyces* sp.; (c) *Bullera alba*; (d) *Sporobolomyces roseus*; (e) *Pityrosporum ovale*; (f) *Rhodotorula rubra*.

comprises several taxa which can be distinguished by the daily growth rate of the colonies and by the presence or absence of blastoconidia and chlamydospore-like endoconidia (see Weijman, 1979).

6. *Apiotrichum* Stautz (*Phytopathol. Z.* **3**, 163, 1931). Type species: *A. porosum* Stautz.

The genus has been reintroduced by von Arx and Weijman (1979). It is characterized by the formation of septate hyphae and elongate, often curved conidia which are formed sympodially or basipetally. A common species is *A. humicola* (Daszewska) v. Arx & Weijman, hitherto known as *Candida humicola* (Daszewska) Diddens & Lodder.

7. *Sarcinosporon* King & Jong (*Mycotaxon* **3**, 92, 1975). Type species: *S. inkin* (Oho) King & Jong.

This genus was introduced for *Trichosporon inkin* (Oho) Do Carmo-Sousa & Van Uden and was characterized by the formation of sarcina-like clusters of endoconidia in swollen cells. The genus is very close to *Trichosporon*.

8. *Pityrosporum* Sabouraud (*Mal. Cuir Chevel.* **2**, 1904). Type species: *P. ovale* (Bizz.) Cast. & Chalmers (Fig. 2e).

The type species of this genus is often identified as *Malassezia furfur* (Robin) Baillon, the cause of pityriasis versicolor in humans and unknown *in vitro*. *Pityrosporum ovale* and two other species have been isolated only from the skin of humans and animals, and *P. ovale* must be cultivated on media containing oil (olive oil or 1.5% sodium taurocholate). The cells are pyriform, and the buds are formed basipetally from broad bases at the attenuated ends.

Schizoblastosporon starkeyi-henricii Ciferri forms similar blastoconidia, but the cells also form a median septum and subsequently undergo fission.

C. Sporobolomycetaceae Derx

This family comprises the anamorphs of genera such as *Sporidiobolus* Nyland, *Rhodosporidium* Banno, and *Aessosporon* van der Walt, characterized by the formation of chlamydospore-like teliospores which germinate and develop promycelium-producing blastoconidia. These genera have often been classified in or near the Ustilaginales (smut fungi).

The yeasts classified in the Sporobolomycetaceae usually form ovate or pyriform cells with basipetal conidiation at the attenuated ends. Carotenoid pigments are synthesized, and the colonies may become orange, salmon, or red.

The following genera belong here:

1. *Sporobolomyces* Kluyver & van Niel (*Zentralbl. Bakteriol., Parasitenkde, Infektionskr. Hyg., Abt. 2* **63**, 19, 1924). Type species: *S. roseus* Kluyver & van Niel (Fig. 2d).

Phaff, in Lodder (1970), accepted nine species. The genus is characterized by the formation of sterigmata and ballistospores. One to several upright sterigmata are formed on a yeast cell, and the sterigmata may elongate sympodially while forming several ballistospores. The ballistospores are asymmetric, often slightly curved, fusiform, and attenuated at the base. Septate hyphae without clamp connections are present in some species, e.g., in *S. salmonicolor* (Fischer & Brebek) Kluyver & van Niel.

The genus comprises anamorphs of *Aessosporon* van der Walt and *Sporidiobolus* Nyland.

2. *Rhodotorula* Harrison (*Trans. R. Soc. Can.* **3**, (21), 349, 1927). Type species: *R. glutinis* (Fres.) Harrison (Fig. 2f).

Phaff and Ahearn, in Lodder (1970) accepted nine species, to which four more have since been added. Certain other species belonging here are at present wrongly classified in *Cryptococcus, Candida,* and *Torulopsis*. The cells are usually ovoid or pyriform with basipetal, enteroblastic conidiation at the attenuated ends. Conidiation occasionally occurs at both ends. Ballistospores are not formed. The colonies are often reddish, salmon, or yellow as a result of carotenoid pigments, but some species are nearly unpigmented. Species with nearly unpigmented colonies are *R. fujisanensis* (Soneda) Johnson & Phaff and *R. muscorum* (di Menna) v. Arx & Weijman (= *Candida muscorum* di Menna).

The teleomorphs of this genus are classified in *Rhodosporidium* Banno.

REFERENCES

Cole, G. T. (1975). The thallic mode of conidiogenesis in the Fungi Imperfecti. *Can. J. Bot.* **53**, 2983-3001.

Diddens, H. A., and Lodder, J. (1942). "Die anaskosporogenen Hefen," II. Hälfte. Amsterdam.

Geiger, A. (1910). Beiträge zur Kenntnis der Sprosspilze ohne Sporenbildung. *Zentralbl. Bakteriol., Parasitenkde, Infektionskr. Hyg. Abt. 2* **27**, 97-136.

Golubev, W. I. (1977). *Metschnikowia lunata* sp. nov. *Antonie van Leeuwenhoek* **43**, 317-322.

Kreger-van Rij, N. J. W., and Veenhuis, M. (1971). A comparative study of the cell wall structure of basidiomycetous and related yeasts. *J. Gen. Microbiol.* **68**, 87-95.

Kreger-van Rij, N. J. W., and Veenhuis, M. (1973). Electron microscopy of septa in ascomycetous yeasts. *Antonie van Leeuwenhoek* **39**, 481-490.

Kwon-Chung, K. J. (1975). A new genus, *Filobasidiella,* the perfect state of *Cryptococcus neoformans. Mycologia* **67**, 1197-1200.

Lodder, J., ed. (1970). "The Yeasts: A Taxonomic Study." Amsterdam.

Lodder, J., and Kreger-van Rij, N. J. W. (1952). "The Yeasts." Amsterdam.

Meyer, S. A., and Phaff, H. J. (1970). "Taxonomic Significance of the DNA Base Composition in Yeasts." Spectrum, Atlanta, Georgia.

Meyer, S. A., Smith, M. T., and Simione, F. P. (1978). Systematics of *Hanseniaspora* and *Kloeckera. Antonie van Leeuwenhoek* **44**, 79-96.

Nakase, T., and Komagata, K. (1968). Taxonomic significance of base composition of yeast DNA. *J. Gen. Appl. Microbiol.* **14**, 345-357.

Nakase, T., and Komagata, K. (1971). DNA base composition of some species of yeasts and yeast-like fungi. *J. Gen. Appl. Microbiol.* **17**, 363-369.

Neumann, I. (1972). Biotaxonomische und systematische Untersuchungen an einigen Hefen der Gattung *Saccharomyces*. *Beih. Nova Hedwigia* **40**.
Oberwinkler, F. (1978). Was ist ein Basidiomycet? *Z. Mykol.* **44**, 13-29.
Price, C. W., Fuson, G. B., and Phaff, H. J. (1978). Genome comparison in yeast systematics: Delimitation of species within the genera *Schwanniomyces, Saccharomyces, Debaryomyces* and *Pichia*. *Microbiol. Rev.* **42**, 161-193.
Quinquand, M. (1868). Nouvelles recherches sur le muguet. *Arch. Phys. Norm. Pathol.* **1**, 290-305.
Rodrigues de Miranda, L. (1975). Two new species of the genus *Sterigmatomyces*. *Antonie van Leeuwenhoek* **41**, 193-199.
Scheda, R., and Yarrow, D. (1966). The instability of physiological properties used as criteria in the taxonomy of yeasts. *Arch. Mikrobiol.* **55**, 209-225.
Scheda, R., and Yarrow, D. (1968). Variation in the fermentative pattern of some *Saccharomyces* species. *Arch. Mikrobiol.* **61**, 310-316.
Smith, M. T., van der Walt, J. P., and Johannsen, E. (1976). The genus *Stephanoascus* gen. nov. (Ascoideaceae). *Antonie van Leeuwenhoek* **42**, 119-127.
Spencer, J. F. T., and Gorin, P. A. J. (1969). Systematics of the genus *Candida:* Proton magnetic resonance spectra of the mannans and mannose-containing polysaccharides as an aid in classification. *Antonie van Leeuwenhoek* **35**, 33-44.
Stadelmann, F. (1955). A new species of the genus *Bullera*. *Antonie van Leeuwenhoek* **41**, 575-582.
Stelling-Dekker, N. M. (1931). Die sporogenen Hefen. *Verh. K. Akad. Wet., Afd. Natuurkde, Reeks* 2 **28**, 1.
van der Walt, J. P. (1964). *Dekkera,* a new genus of the Saccharomycetaceae. *Antonie van Leeuwenhoek* **30**, 273-280.
van der Walt, J. P., Johannsen, E., and Liebenberg, N. V. D. W. (1974). Cell wall structure, mitosis and urease activity in *Torulopsis* species with high GC content. *Antonie van Leeuwenhoek* **40**, 417-426.
von Arx, J. A. (1977). Notes on *Dipodascus, Endomyces* and *Geotrichum* with the description of two new species. *Antonie van Leeuwenhoek* **43**, 333-340.
von Arx, J. A., and van der Walt, J. P. (1976). The ascigerous state of *Candida chodatii*. *Antonie van Leeuwenhoek* **42**, 309-314.
von Arx, J. A., and Weijman, A. C. M. (1979). Conidiation and carbohydrate composition in *Candida* and *Torulopsis*. *Antonie van Leeuwenhoek* **45**, 547-555.
von Arx, J. A., Rodrigues de Miranda, L., Smith, M. T., and Yarrow, D. (1977). The genera of yeasts and the yeast-like fungi. *Stud. Mycol.* **14**, 1-42.
Weijman, A. C. M. (1979). Carbohydrate composition and taxonomy of *Geotrichum, Trichosporon* and allied genera. *Antonie van Leeuwenhoek* **45**, 119-127.
Yamada, Y., and Kondo, K. (1973). Taxonomic significance of the coenzyme Q system in yeasts and yeast-like fungi. *Proc. 2nd Int. Spec. Symp. Yeasts, 1972,* pp. 62-69.

5

Dimorphism

Garry T. Cole and Yoshinori Nozawa

I.	Introduction	97
II.	The Yeast	98
III.	The Hypha	101
IV.	Morphological Changes Associated with Yeast–Hypha Conversion	102
V.	Physical Factors Affecting Yeast–Hypha Conversion	109
VI.	Biochemical Differentiation of Yeast and Hyphal Phases	112
	A. Chemical Composition	112
	B. Metabolism	122
	C. Mechanisms of Dimorphism	124
	References	127

I. INTRODUCTION

Dimorphic fungi are of special interest to medical mycologists, since the ability of these microbes to alternate between yeast and mycelial phases is a common feature of fungal pathogens of man and animals. In fact, the term "dimorphism" was originally applied only to "fungi pathogenic to man" which "are found in infected tissues in a unicellular yeast-like form, but when cultivated at room temperature grow out in a mycelial form" (Cochrane, 1958). The yeast, however, is not the only cell produced by these pathogens which invade host tissues. In the case of chromomycotic fungi, which cause localized chronic infections of skin and subcutaneous tissues (Emmons *et al.*, 1977), brown hyphae are found in superficial epithelial crusts. *Candida albicans* (Robin.) Berk., an imperfect yeast and a causative agent of mucocutaneous, gastrointestinal, and systemic candidiasis (Odds, 1979), may produce germ tubes, hyphae, or pseudohyphae within host tissues. For fungal pathogens capable of hematogenous and/or lymphatic spread, however, the yeast is usually the disseminating agent, the principal exception being the endosporulating spherules of *Coccidioides immitis* Rixford & Gilchrist. Yeast–hypha conversion, however, is not restricted to pathogenic fungi and, before proceeding further with a discussion of dimorphism, it is appropriate first to examine the ultrastructural, cytological, and biochemical aspects of yeast and hyphal morphogenesis in general. Hyphal growth and development are discussed elsewhere (see Aronson, Volume 2, Chapter 28, and Cole, Volume 2, Chapter 23) and will be only briefly considered here.

II. THE YEAST

Saccharomyces cerevisiae Hansen has been used as a model for extensive morphological (Byers and Goetsch, 1975), biochemical (Cabib, 1975), and genetic (Hartwell, 1974) investigations of yeast growth and development. Although this ascomycetous fungus does not undergo yeast–hypha conversion, it is phylogenetically related to the dimorphic Fungi Imperfecti considered in this chapter. In addition, the elegant experimental systems designed for morphogenetic studies of *S. cerevisiae* are applicable to dimorphic fungi, and the results obtained are relevant to this discussion.

Three morphogenetic stages of the yeast cell cycle are recognized: bud emergence (Fig. 1A and B), bud growth (Fig. 1C–E), and cell separation (Fig. 1F). The bud initial forms by the "blowing out" of a localized region of the parental yeast cell. The outer wall encompassing this fertile locus becomes attenuated, while concomitantly an adjacent inner wall layer is newly synthesized (Fig. 1B). Membrane-bound vesicles located in the cytoplasm juxtaposed to bud emergence are suggested to function in the transport of wall precursors and enzymes from sites of synthesis, such as rough and smooth endoplasmic reticulum, to the plasma membrane with which they fuse, discharging their contents. In addition, these vesicles contribute new membrane to the expanding plasmalemma (Byers and Goetsch, 1974; Cabib, 1975; Farkaš, 1979). In a review of current interpretations of the mechanisms of biosynthesis of fungal cell walls, Farkaš (1979) has summarized that formation of "skeletal polysaccharides is catalyzed by constitutively formed polysaccharide synthases uniformly distributed in the plasmalemma. The unique property of these enzymes is that they can exist in an active or in a temporarily inactive, zymogen state. Only the portion of the polysaccharide synthases located at the growth zone is active during wall growth." Since the cluster of cytoplasmic vesicles appears at the site of bud emergence prior to any recognizable deformation of the parent cell wall (Moor, 1967), it is possible that these organelles also discharge lytic enzymes into the periplasm, which subsequently soften the adjacent region of the preexisting wall (Cortat *et al.*, 1972; Fig. 1B[1]). This pro-

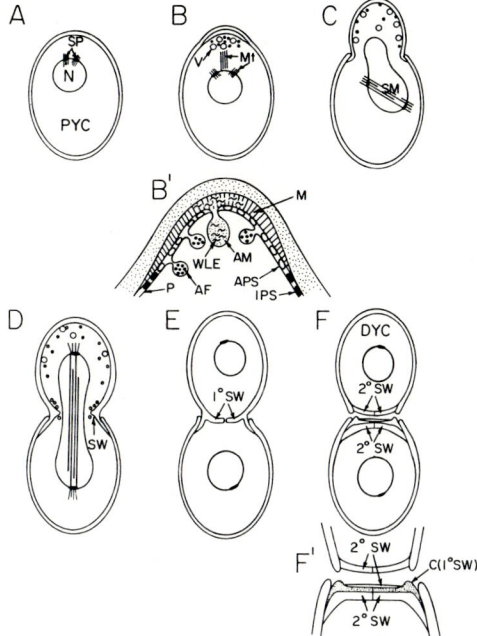

Fig. 1. The yeast cell cycle of *Saccharomyces cerevisiae* demonstrating bud emergence (A, B, and B¹), bud growth (C–E), and cell separation (F and F¹). AF, Activating factors; AM, amorphous material; APS, activated polysaccharide synthase; C, chitin; DYC, daughter yeast cell; IPS, inactivated polysaccharide synthase (zymogen); M, microfilbrils; Mt, microtubules; N, nucleus; P, plasmalemma; PYC, parental yeast cell; SM, spindle microtubules; SP, spindle plaques; SW, septal wall; V, vesicles; WLE, wall lytic enzymes.

cess, combined with the effects of turgor pressure on the parental yeast wall, are suggested to be at least partially responsible for bud emergence (Hartwell, 1974).

It is still unclear what cytological and biochemical events precede bud formation and what factors determine the specific location on the yeast surface where a new bud will emerge (Cabib, 1975). Byers and Goetsch (1974) have suggested that duplication of the spindle plaque or microtubule-organizing center (Pickett-Heaps, 1969) is a necessary precursor of bud initiation (Fig. 1A). In fact, the temporal relationship between events of nuclear division and bud emergence have long been recognized (Guilliermond, 1910; Robinow and Bakerspiegel, 1965; Burnett, 1968). As the spindle plaques continue to move apart, extranuclear microtubules appear, which extend into the bourgeoning initial (Byers and Goetsch, 1975; Fig. 1B). On the basis of examinations of several cell division cycle mutants, Hartwell (1971a, b, 1974) has proposed that spindle plaque duplication, which occurs near the beginning of S phase (DNA synthesis), triggers the events leading to bud emergence. However, initiation of DNA synthesis is not a necessary prerequisite of bud formation (Hartwell, 1971a). Instead, some event which occurs during early S phase may trigger a mechanism which prevents additional budding until completion of cytokinesis; reactivation of events leading to the formation of a new bud occurs only after spindle plaque duplication (Hartwell, 1974).

The nucleus, its envelope remaining intact, first begins to migrate into the bud some time after DNA replication (Hartwell, 1974; Fig. 1C). The two spindle plaques at this stage have separated, but the spindle microtubules are not yet orientated parallel to the long axis of the proliferating cell. Initial migration of the nucleus into the newly formed bud is probably not influenced by the spindle (Byers and Goetsch, 1974; Robinow and Marak, 1966). Later, prior to completion of telophase, the entire spindle abruptly elongates; the plaques are now at opposite poles, and the spindle is parallel to the long axis of the cell (Moens and Rapport, 1971; Robinow and Marak, 1966; Fig. 1D). At this stage, spindle elongation probably influences nuclear migration (Hartwell, 1974).

The parental yeast wall of *S. cerevisiae* is composed of glucans, mannan, protein, and some chitin. The new bud wall consists of newly synthesized glucans which arc added at the tip of the bud distal to the parental cell (Chung *et al.*, 1965; Hartwell, 1974; Tkacz *et al.*, 1971; Tkacz and Lampen, 1972). Wall proteins are also newly synthesized and incorporated into the bud wall. However, the timing and distribution of chitin synthesis is different from that of other wall components. Chitin is found only in the primary septum, which persists after secession as the bud scar on the surface of the parental cell (Cabib and Farkaš, 1971; Cabib, 1975; Cabib *et al.*, 1973). This aspect of wall biosynthesis in *S. cerevisiae* apparently differs markedly in the dimorphic fungi discussed later. The precise stage of the yeast cycle corresponding to the initiation of chitin synthesis is still controversial (Hayashibe and Katohda, 1973; Cabib *et al.*, 1974; Cabib, 1975). The current model suggests that "chitin begins accumulating relatively early in cell division ... in a ring between mother cell and bud, followed by centripetal growth and closure of the primary septum" (Cabib, 1975; Vršanská *et al.*, 1979; Fig. 1D and E). Chitin synthase zymogen, located in the yeast plasmalemma, is converted to active enzyme by an activating factor which has been identified as a protease (Cabib *et al.*, 1973). The activator is presumably synthesized in the endoplasmic reticulum and transported to the plasma membrane in vesicles (Cabib *et al.*, 1973; Duran *et al.*, 1979; Farkaš, 1979). The latter are found frequently at the isthmus between the parental and daughter cells prior to and during septation (Bowers *et al.*, 1974; Byers and Goetsch, 1974; Cortat *et al.*, 1972; Sentandreu and Northcote, 1969; Fig. 1D). An inhibitor of chitin synthase is also present (Cabib, 1975), which "acts as a safety device to inactivate any vacuolar protease (activator) which spills into the cytoplasm preventing random stimulation of the enzyme" (Braun and Calderone, 1979).

Byers and Goetsch (1976) described a "filamentous ring" (microfilaments?) present in the neck of young buds of *S. cerevisiae* which may be associated with formation of the primary septum. The role of microfilaments in septum initiation has been investigated by Girbardt (1979) in an ultrastructural developmental study on septation in the basidiomycete *Trametes versicolor*. He has revealed first the formation of a peripheral belt of 4- to 7-nm microfilaments in a region of the hypha where, a few minutes later, formation of the septal wall begins. It is suggested that associations among the microfilamentous belt, mitochondria, and microvesicles indicate interactions between microfilaments and membranes, which may lead to the trapping of these and other cell organelles. In the budding yeast, invagination of the plasmalemma and septum formation are associated with the synthesis of chitin and development of the annular structure (chitin ring) characteristic of bud growth (Vršanská *et al.*, 1979). The fate of the plasmalemma between the parent and daughter cells during septation (Fig. 1E) is unclear. Although Cabib (1975) has stated that the plasma membranes of the adjacent cells separate, presumably following fusion of the plasmalemma during centripetal growth of the septum, documentation of this event by careful serial sectioning has not been reported (Fig. 1E and F). Upon completion of the primary septum, the secondary septa are laid down using the former as a template (Shannon and Rothman, 1971; Fig. 1F and F^1). The composition of the secondary septal walls is comparable to that of the rest of the yeast cell wall, consisting mainly of glucans and mannan (Bauer *et al.*, 1972; Bush and Horisberger, 1973; Cabib and Bowers, 1971; Cabib 1975). When cell separation finally occurs, a birth scar remains at the base of the daughter cell and a bud scar at the apex of the parent yeast cell (Fig. 1F). The birth scar is apparently composed solely of secondary septal wall material, while the bud scar consists of both primary and secondary septal wall components (Shannon and Rothman, 1971). A raised rim of chitin remains on the surface of the bud scar, which can be demonstrated by fluorescence microscopy using primulin stain (Streiblová and Beran, 1963; Streiblová, 1971). The central region of the bud scar is covered by nonchitinous wall material derived from the secondary septum of the seceded daughter cell (Cabib, 1975; Fig. 1F^1).

III. THE HYPHA

As mentioned earlier, the morphogenesis, ultrastructure, and wall biosynthesis of hyphae are reviewed in Volume 2, Chapters 23 and 28. A comparison of hyphal growth and yeast budding reveals certain basic similarities. Hyphae elongate at their conical tips, demonstrating a polarized type of cell growth comparable to bud emergence. Light microscope examinations of growing hyphae produced by septate fungi show that the hyphal tip contains a cluster of tiny cytoplasmic particles surrounding a dark central body or *Spitzenkörper* (Brunswick, 1924; Girbardt, 1955, 1957; Grove, 1978). Thin sections of hyphal tips disclose that the particles are vesicles (about 0.1 µm in diameter), while the *Spitzenkörper* is in fact a dense cluster of microvesicles (0.05 µm in diameter). Both kinds of vesicles are apparently capable of fusion with the plasmalemma (Grove, 1978; Najim and Turian, 1979). In medial section, the vesicles are shown to reside in a ribosome-free region of the cytoplasm devoid of other organelles (Grove *et al.*, 1970; Grove and Bracker, 1970). The vesicles originate from dictyosomes or their equivalents located at least 1.5–2.0 µm back from the tip. In hyphae of conidial fungi, the dictyosomes are smooth-surfaced cisternae arranged in rings (Grove and Bracker, 1970; Cole and Aldrich, 1971; Cole and Samson, 1979). Evidence for the presence of polysaccharides in the dictyosomes, apical vesicles, and hyphal tip wall has been derived from cytochemical and autoradiographic studies of many filamentous fungi (Grove, 1978). The vesicles at hyphal tips, as in emerging yeast buds, are considered the vehicles which transport enzymes, enzyme activators and inhibitors, wall precursors, and so on, through the cytoplasm from sites of synthesis to the plasmalemma with which they fuse, discharging their contents (Farkaš, 1979). Maintenance of filamentous growth therefore appears to be intimately related to the presence of apical vesicles and a "harmonious balance ... between the processes of wall synthesis and wall lysis" (Bartnicki-Garcia, 1973). It has been shown that a change in direction of hyphal tip growth is preceded by a change in the position of the *Spitzenkörper* (Girbardt, 1957, 1973). The precise location of a new branch of a hypha is also decided by "some event in the cytoplasm (e.g., vesicle accumulation)" (Trinci and Collinge, 1974). The polarity of hyphal growth is dependent on restriction of new wall biosynthesis to the apical dome (Bartnicki-Garcia, 1973). If more than the dome of the elongating hypha is involved in wall intussesception, concomitant with dispersion of the apical cluster of secretory vesicles, the conical tip swells and may differentiate into a terminal, "holoblastic" cell (Cole and Samson, 1979; see Chapter 23). Whether the latter subsequently functions as a conidium or yeast (e.g., in the so-called black yeasts; de Hoog and Hermanides-Nijhof, 1977) is determined by exogenous and endogenous factors which are not well understood. This represents a significant problem for future investigations. The role of the conidium in yeast–hypha conversion has been examined and is discussed below.

IV. MORPHOLOGICAL CHANGES ASSOCIATED WITH YEAST–HYPHA CONVERSION

Candida albicans is a conidial fungus (see Chapter 4, this volume) and has been used as a model for many morphogenetic investigations (Odds, 1979) and will serve here to illustrate certain aspects of yeast–hypha conversion. Yeast cells were used to inoculate glass coverslips coated with cornmeal–Tween 80 agar (Beneke and Rogers, 1970), which were then incubated at 37°C for varying periods and subsequently prepared for scanning electron microscope examination (Cole, 1975). The parent yeast cell in Fig. 2 has produced a daughter cell and a bud after incubation for 2 h. Note the constriction (arrow in Fig. 2) at the base of the latter. Other cultures incubated for the same period demonstrate yeast conversion to the hyphal form (Figs. 3 and 4). Unlike development of the yeast bud, germ tube and hyphal formation does not involve constriction at the junction between the parent yeast cell and the emerging filament. The hyphae later become septate (S in Fig. 4). The coverslip cultures also reveal cells which cannot be classified either as yeast or hyphae; these are elongated ellipsoidal cells or pseudohyphae (PH in Figs. 3 and 4) which arise from parental yeast cells. As in the case of the budding cell in Fig. 2, a constriction is present between the parental cell and the pseudohypha. The ellipsoidal cells do not easily secede and may proliferate apically, producing chains of cells (arrowhead in Fig. 3). In addition, the cylindrical germ tubes may swell apically, giving rise to ellipsoidal cells (arrow in Fig. 3), or the latter may narrow and proliferate to form a hypha (Fig. 5). After incubation for 6 h (Fig. 6), the pseudohyphae undergo further apical proliferation and give rise to new yeast cells at constrictions between adjacent fertile cells. On the other hand, after a 22-h growth on potato dextrose agar at 34°C, septate hyphae form and give rise to yeast cells arranged in chains (Fig. 7). Odds (1979) interpreted the formation of pseudohyphae as a "morphogenetic development intermediate between budding and hyphal growth." The initial development of a pseudohypha from the parental yeast cell is comparable to bud formation. However, the spherical bud soon begins to elongate apically rather than continuing to grow diametrically. Based on earlier discussions of yeast and hyphal morphogenesis, associated cytological and biochemical alterations are suggested by this morphogenetic variation. Perhaps the constriction between the parental cell and both the pseudohypha and the daughter yeast reflects similar processes of wall biosynthesis during early de-

---

Fig. 2. Budding yeast of *Candida albicans* grown for 2 h on cornmeal–Tween 80 agar (CMA-T80) at 37°C. Arrow indicates constriction between parental yeast cell (PYC) and bud. DYC, Daughter yeast cell. ×12,535.

Fig. 3. Yeast of *Candida albicans* which have given rise to both hyphae (H) and pseudohyphae (PH). Arrow indicates germ tube which has converted to a pseudohypha. Arrowhead shows apically proliferating pseudohypha. Two hours on CMA-T80. ×5000.

Fig. 4. Germ tube (GT), hypha (H), and pseudohypha (PH) of *Candida albicans*. Note constriction (arrow) between parental yeast cell and pseudohypha, which is comparable to the constriction at the base of the bud in Fig. 2. S, Septum. Two hours on CMA-T80 at 37°C. ×8750.

Fig. 5. A pseudohypha produced by the parental yeast cell has proliferated to form a hypha. The parental yeast cell has also given rise to a young bud. Two hours on CMA-T80 at 37°C. ×10,240.

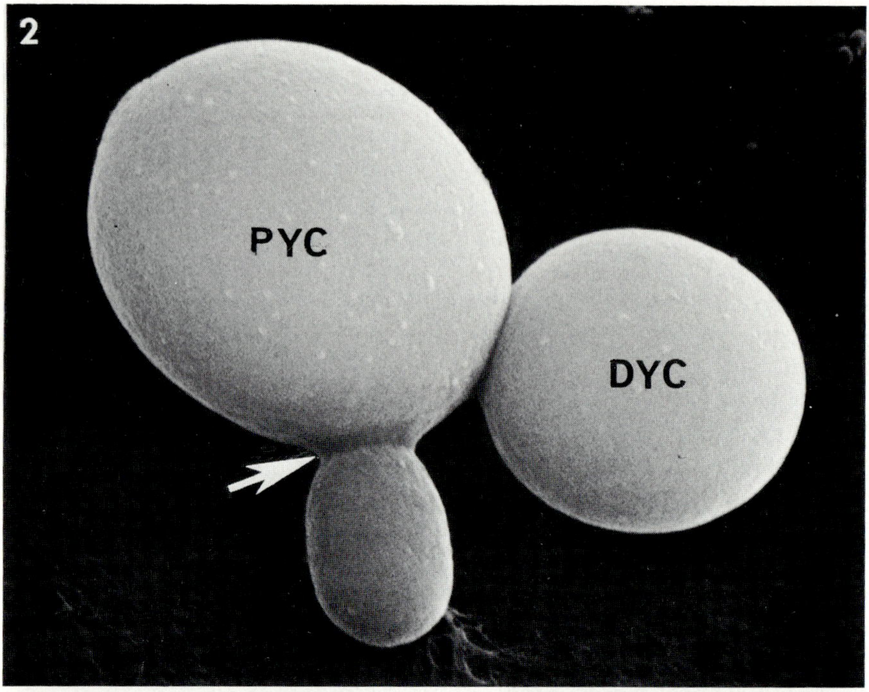

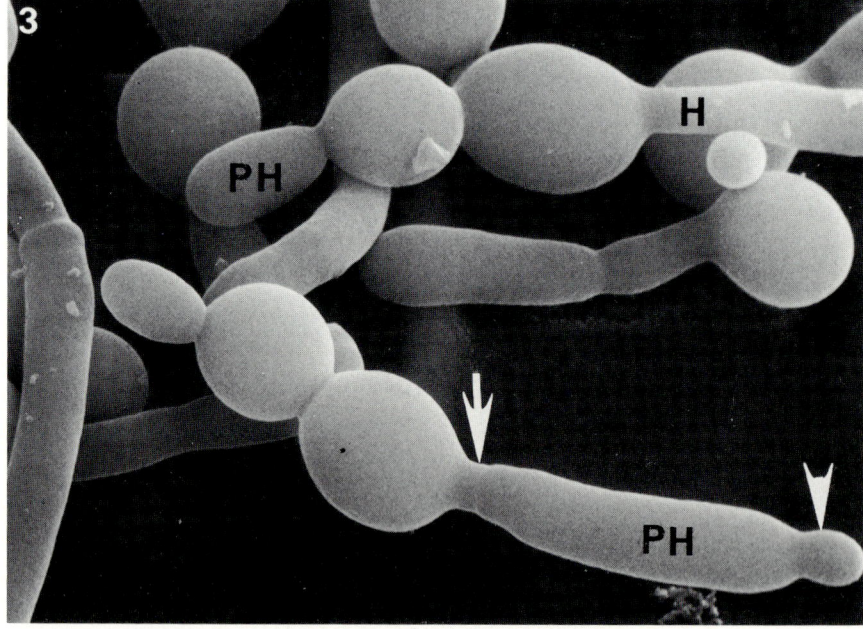

Figs. 2 and 3. Captions on page 103.

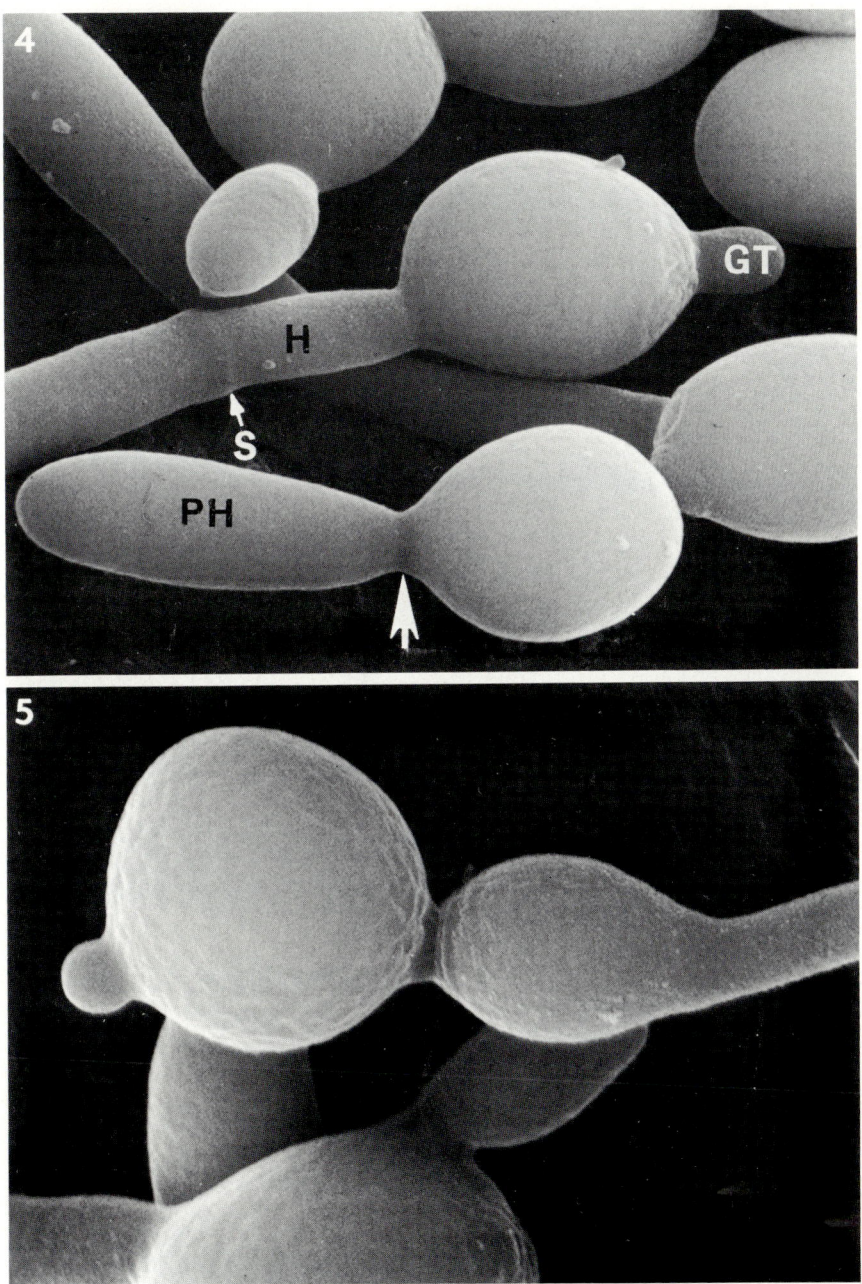

Figs. 4 and 5. Captions on page 103.

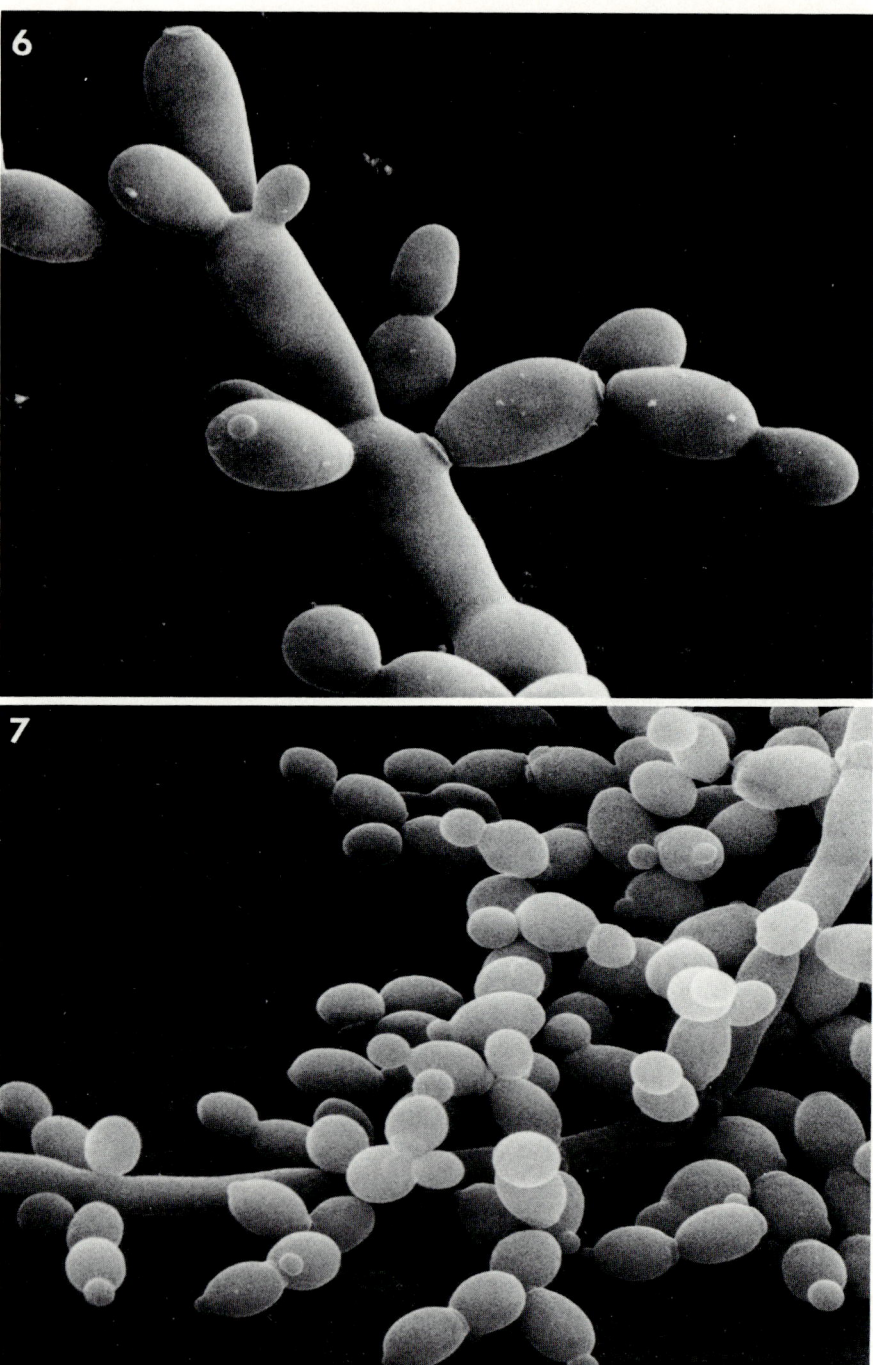

Figs. 6 and 7.

velopment. On the other hand, germ tubes which arise from parental yeast cells and continue to elongate to produce septate hyphae undergo a distinct process of polar growth.

Intermediate states are recognized for most fungi undergoing yeast–hypha conversion. The respiratory pathogen *Cryptococcus neoformans* (San Felice) Vuill. produces primarily a yeast phase in host tissue, but pseudohyphae are occasionally isolated (Stewart and Rogers, 1978; Emmons *et al.*, 1977). The black yeasts, so called because of their dark wall pigmentation (de Hoog and Hermanides-Nijhof, 1977; Ajello, 1978; Cole, 1978), demonstrate several transitional forms. Five different cell types have been characterized in *Exophiala dermatitidis* (Kano) de Hoog [=*Phialophora dermatitidis* (Kano) Emmons, Binford & Utz] in a developmental and ultrastructural study by Oujezdsky *et al.* (1973). In an examination of dimorphism in *Exophiala werneckii* (Horta) v. Arx (=*Cladosporium werneckii* Horta), Hardcastle and Szaniszlo (1974) determined that an increase in temperature from 17° to 30°C resulted in the conversion of yeast to moniliform hyphae in an unenriched medium. Other members of the black yeasts which demonstrate a similar degree of morphogenetic plasticity during yeast–hypha conversion include *E. jeanselmei* (Langer.) McGinnis & Padhye, *E. pedrosoi* (Brumpt) Schol-Schwarz, and *Aureobasidium pullulans* (De Bary) Arn. The morphogenesis of these fungi requires further investigation. Lane and Garrison (1970) suggested that yeast–hypha conversion in *Sporothrix schenckii* Hoktoen & Perkins, *Histoplasma capsulatum* Darling, *Blastomyces dermatitidis* Gilchrist & Stokes, and *Paracoccidioides brasiliensis* (Splendore) Almeida all proceed via a common mechanism. These authors have proposed that each fungus "undergoes phase conversion with the formation of a transitional cell having a mixture of yeast-like and hyphal characteristics." It is apparent from these examinations of yeast–hypha conversion that a developmental continuum exists rather than two exclusive states suggested by the term "dimorphism."

Geotrichum candidum Link and *Coccidioides immitis* are included in most lists of dimorphic fungi (Stewart and Rogers, 1978). However, in both cases budding yeast are not produced, and instead the infectious states are airborne holoarthric and enteroarthric conidia, respectively (Cole and Samson, 1979). Holoarthric conidia of *G. candidum* are formed by septation of vegetative hyphae (Cole and Kendrick, 1969). The conidia produce germ tubes which elongate to form septate hyphae. The latter then undergo fragmentation and are converted back to conidia. In addition to developmental differences between conidiogenesis and yeast production, analyses of the chemical composition of

---

Fig. 6. Pseudohyphae and yeast of *Candida albicans* grown on CMA-T80 for 5 h at 37°C. Note that chains of yeast arise from fertile loci near constrictions between adjacent pseudohyphal cells. ×5060.

Fig. 7. Hyphae and chains of yeast of *Candida albicans* grown on potato dextrose agar at 34°C for 22 h. ×2690.

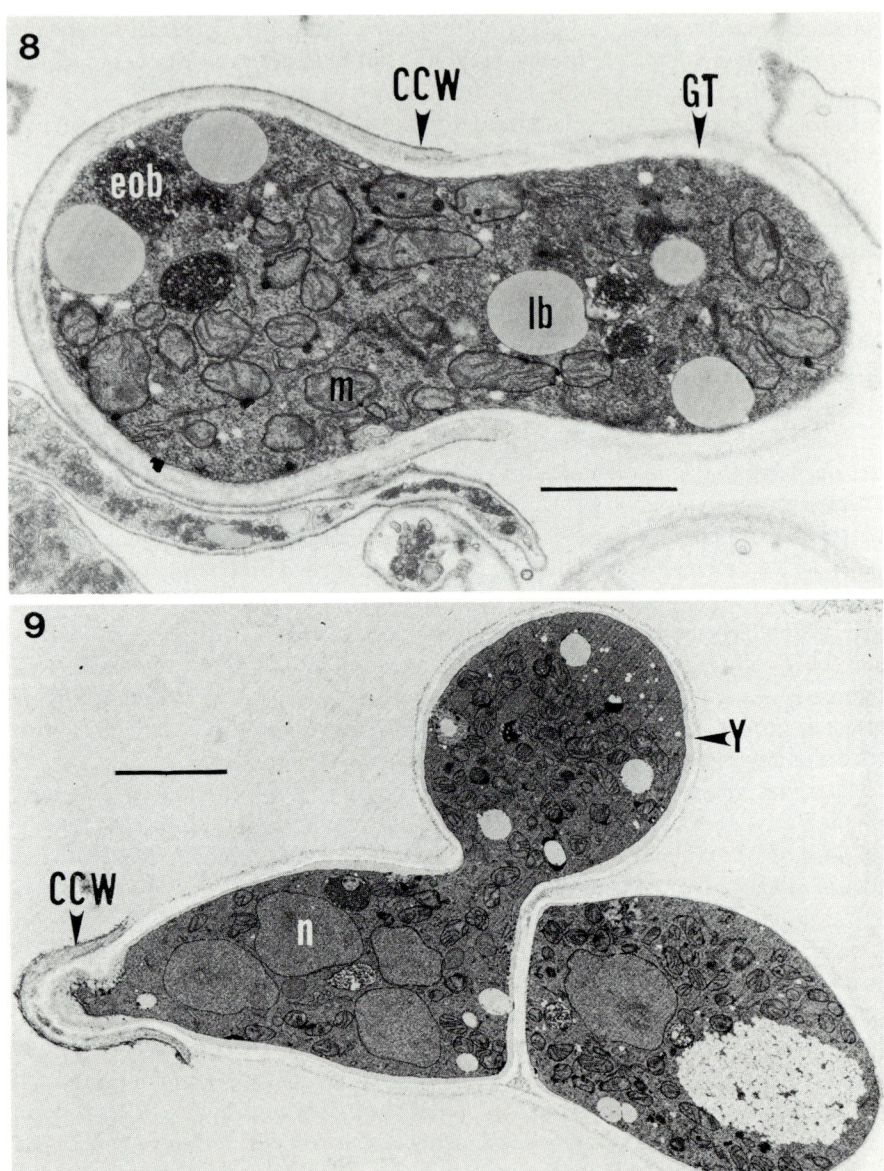

Figs. 8 and 9. *Blastomyces dermatitidis.* Conidia germinate, rupturing their outer wall (CCW) and give rise to germ tubes (GT in Fig. 8). Conidia may subsequently produce yeastlike cells (Y in Fig. 9). eob, Electron opaque body; lb, lipid body; m, mitochondrion; n, nucleus. Bars in Figs. 8 and 9 represent 1 and 2.5 μm, respectively (from Garrison and Boyd, 1978).

isolated conidial and hyphal walls of *G. candidum* reveal comparable amounts of chitin (Ebina *et al.*, 1978; G. T. Cole and Y. Nozawa, unpublished data). In *Coccidiodes immitis*, enteroarthric conidium formation (saprophytic cycle) and spherule endospore development (parasitic cycle) have been examined by Sun and Huppert (1976). The taxonomic position of this fungus remains unresolved, mainly because its teleomorph has not yet been discovered, and development of the spherules resembles both ascal ontogeny (Moore, 1932; Dodge, 1935; Kwon-Chung, 1969) and sporangial formation (Baker *et al.*, 1943; Breslau *et al.*, 1961; Emmons, 1947; O'Hern and Henry, 1956). Neither *Geotrichum candidum* nor *Coccidioides immitis* shows any striking morphogenetic similarity to the dimorphic fungi discussed earlier and should therefore be distinguished from species which undergo yeast–hypha conversion.

When Garrison and Boyd (1978) grew *Blastomyces dermatitidis* in flasks containing Sabouraud's glucose broth agitated continuously at 25°C for 14 days, septate hyphae and conidia were produced. A suspension of both these elements was then inoculated into brain–heart infusion broth and incubated at 37°C. While the hyphae apparently degenerated, the conidia underwent a sequence of morphogenetic changes resulting in the formation of yeastlike cells which were characteristic of this species (Figs. 8 and 9). These authors suggested that the conidium may be more significant than previously suspected in mycelium–yeast conversion in *B. dermatitidis*. Their proposal is supported by the fact that systemic blastomycosis has a respiratory portal of entry, in which case the small airborne condia are probably the primary infective agents (Laskey and Sarosi, 1978). Similar development of yeastlike cells from microconidia of *Histoplasma capsulatum* has been reported by Garrison and Boyd (1978). The same authors (Garrison and Boyd, 1977) had earlier pointed out that "pulmonary histoplasmosis is most likely contracted by the inhalation of . . . airborne" microconidia.

V. PHYSICAL FACTORS AFFECTING YEAST–HYPHA CONVERSION

A summary of the significant environmental conditions affecting yeast–hypha conversion is presented in Fig. 10. Much of the pioneering work on dimorphism has involved *Mucor* Mich. ex Fr. (Bartnicki-Garcia, 1963; Bartnicki-Garcia and Nickerson, 1962a,b,c) and is therefore also considered here. In this zygomycetous fungus, yeast cells are formed under conditions of high carbon dioxide tension and availability of hexose. Mooney and Sypherd (1976), however, reported that the morphology of *M. rouxii* (Cal.) Wehm grown under standardized conditions in yeast extract–peptone–glucose medium was dependent on the nitrogen flow rate and not the glucose concentration. High flow rate of nitrogen maintained the

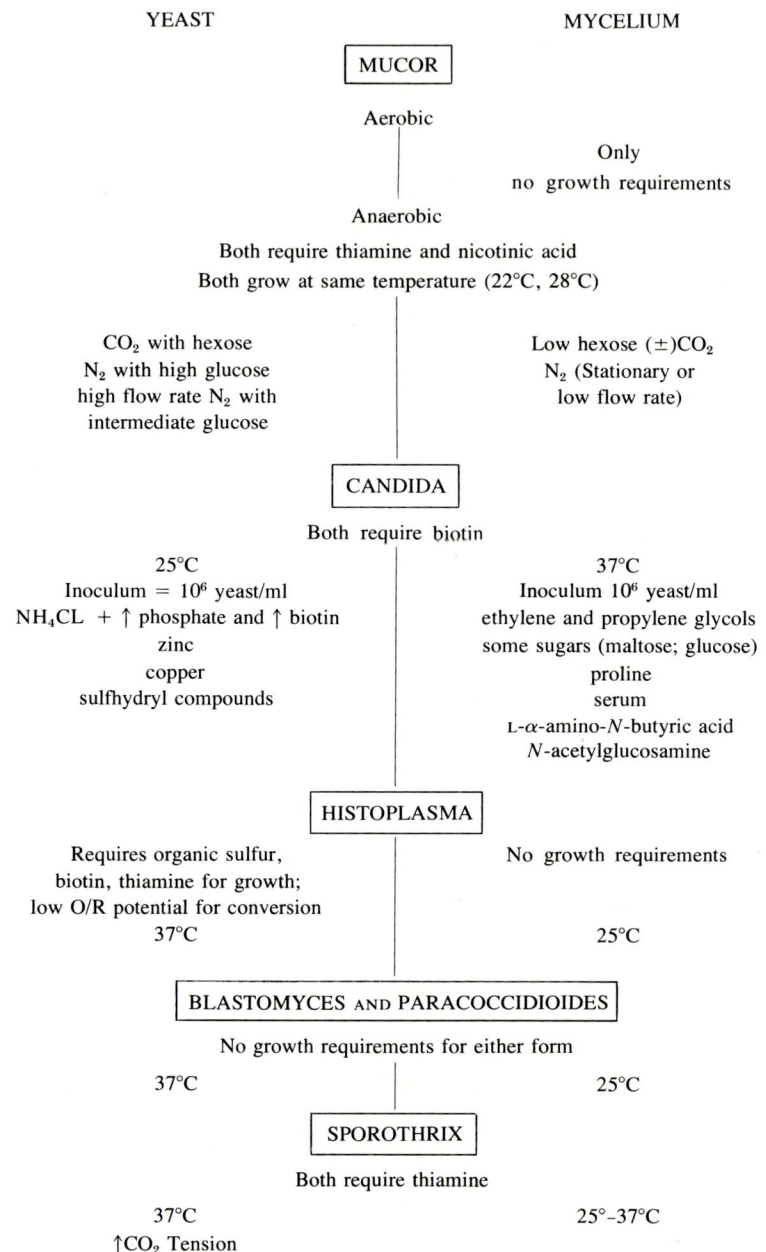

Fig. 10. Summary of environmental factors influencing yeast–hypha conversion in commonly investigated dimorphic fungi. (From J. E. Domer, unpublished.)

yeast phase, while a reduction in the flow rate resulted in conversion to the mycelial phase. These authors suggested that the presence of a volatile compound other than carbon dioxide was pivotal for yeast–hypha conversion in *Mucor*. Larsen and Sypherd (1974) reported that additions of dibutyryl cyclic adenosine monophosphate (dbcAMP) to media which supported growth of the yeast phase of *M. racemosus* Fres. (2% glucose, 100% carbon dioxide) inhibited conversion to hyphae when the culture was exposed to air. Since dbcAMP is the lipophilic derivative of cyclic AMP (cAMP), these authors have stated that addition of this compound to the media influences "the normal morphogenetic patterns of *M. racemosus*, presumably by increasing endogenous levels of cAMP." The lipophilic nature of butyryl derivatives of cAMP facilitates transport of the latter into the cell. Further discussion of the significance of cAMP to dimorphism is presented in Section VI, C.

Many growth conditions have been implicated in the control of dimorphism in *Candida albicans*, and a list of these environmental factors has been compiled by Odds (1979). In contrast to *Mucor*, yeast–hypha conversion in *C. albicans* is less influenced by the carbon dioxide/oxygen ratio and more by nutritional factors. Lee *et al.* (1975) defined a liquid synthetic medium containing six amino acids, biotin, inorganic salts, and glucose, which was successful in supporting the growth of yeast and mycelial phases at 25° and 36°C, respectively. However, we have been unable to maintain our pathogenic strains of *C. albicans* in the yeast phase using this medium, which points out the variation in response of different isolates of this fungus to the same environmental factors. Proline, which was included in the above synthetic medium, has been suggested by several authors to stimulate hyphal formation (Chattaway *et al.*, 1976; Dabrowa *et al.*, 1976; Land *et al.*, 1975a,b). Although cysteine was considered to influence hypha–yeast conversion generally (Nickerson, 1953, 1963; Winsten and Murray, 1956; Taschdjian and Kozinn, 1961), controversy has arisen over its significance in affecting dimorphism in *C. albicans* (Wain *et al.*, 1975) and other fungi. Mitchell and Soll (1979) used pH and temperature shift of the growth medium to examine commitment to pseudomycelium and bud formation. The authors proposed that a germination regulatory substance, capable of stimulating germ tube development, was released from cells upon incubation at 37°C but inactivated at pH values of 4.0 and 9.5 (Hazen and Cutler, 1979). When yeast cells of *C. albicans* in stationary phase at 25°C are used to inoculate fresh medium at 37°C and pH 6.5, the cells synchronously produce pseudomycelia. When the yeast cells are used to inoculate fresh media at either 25°C and pH 6.5 or 37°C and pH 4.5, synchronous formation of new yeast cells results. Mitchell and Soll (1979) determined that, as a result of the pH and temperature shift, the stationary-phase cells became committed either to germ tube or bud formation approximately 1½ h after they were diluted into fresh media. They proposed that "commitment to either growth form during release of cells from stationary phase may involve the

temporal and spatial regulation of septation and the enzyme chitin synthetase.''
The significance of these aspects of dimorphism has been discussed in Section II.

Few growth requirements have been identified for yeast–hypha conversion in *Histoplasma* Darl. and *Sporothrix* Hekt. & Perk. By measuring the oxidation–reduction potential of the liquid medium, Rippon (1968) has determined that the yeast phase of *H. capsulatum* favors a low oxidation–reduction potential. *Blastomyces* Cost. & Roll. and *Paracoccidioides* Almeida demonstrate thermal dimorphism with no specific growth requirements for the yeast or hyphal forms.

Little information is available on environmental factors influencing yeast–hypha conversion in other dimorphic fungi. Barathova *et al.* (1977) examined the effects of numerous carbon and nitrogen sources on dimorphism in *Paecilomyces veridis* Segretain *et al.* (ex Samson). Several species of *Ceratocystis* Ellis & Halst. have been reported to undergo yeast–hypha transformation (Batra and Michie, 1963; Francke-Grosmann, 1967; Mariat, 1975; Tyler and Parker, 1945), and some of the physical factors influencing this morphogenetic process in *C. minor* (Hedgcock) Hunt have been examined (Biel *et al.*, 1977).

VI. BIOCHEMICAL DIFFERENTIATION OF YEAST AND HYPHAL PHASES

The morphological alterations demonstrated during yeast–hypha conversion are associated with marked changes in cell wall composition and metabolism. Some of the significant biochemical differences between yeast and hyphal forms are discussed below.

A. Chemical Composition

1. Whole Cells

a. Nucleic Acids. Nucleic acid metabolism was thought to play a role in yeast–hypha transformation under various cultural conditions. Taylor (1961) measured the DNA and RNA content of yeast (Y) and mycelial (M) cells of *Blastomyces dermatitidis* and found more RNA in the Y phase grown at 37°C than in M-phase cells grown at 25°C after a 6- to 10-day incubation. The RNA content of the M-phase cells was fairly constant during the course of incubation. There was little difference in the DNA content of the two phases, and no significant fluctuation occurred during incubation. Ramirez-Martinez (1970) obtained similar results for *Paracoccidioides brasiliensis*. The Y- to M-form conversion in this species corresponded to an increase in RNA content approximately 48 h after the incubation temperature was reduced from 37° to 23°C (Ramirez-Martinez, 1970). The ratio of RNA to DNA varied from 3.3 (Y form) to 7.2 (M form) during this conversion process. In *Candida albicans,* which demonstrates filamentous growth at temperatures ≥35°C and yeast formation at <35°C (Barlow et al., 1974; Chattaway et al., 1976; Evans et al., 1975; Lee et al., 1975), the RNA and DNA contents were lower in the budding cells (25°C) than in the hyphae (37°C) (Dabrowa et al., 1970). No consistent pattern of fluctuation of nucleic acid content was evident during yeast–hypha conversion in these fungi.

b. Proteins. Variation in total protein content of yeast and mycelial phases has been demonstrated in *B. dermatitidis* (Taylor, 1961). The Y-phase cells contained larger amounts of protein than the M-phase cells. The former revealed a gradual decrease in protein content with age, while that of the latter remained constant during the experiment. In contrast, *C. albicans* shows higher protein content in the filamentous cells than in the yeast (Dabrowa et al., 1970). More RNA and protein in the hyphae of *C. albicans* may reflect increased biosynthesis of these components at the higher growth temperature (37°C). Furthermore, qualitative analysis of water-soluble acidic proteins by disc electrophoresis has revealed differences between budding cells and hyphae of *C. albicans*. Six protein bands were common to both cell types. However, four additional bands, absent from filamentous cells, were found in budding cells. A single protein band observed in filament-producing cells did not occur in yeast (Dabrowa et al., 1970). The significance of these protein differences in yeast–hypha conversion requires further investigation.

2. Cell Walls

Although differences in the chemical composition of yeast and hyphal walls of dimorphic fungi are recognized, difficulties have arisen in obtaining pure cell wall preparations, and detailed knowledge of fungal wall chemistry is still limited. Certain trends are evident, however, from comparisons of data available on isolated walls of *P. brasiliensis* (Carbonell, 1967; Kanetsuna et al., 1969, 1972; Moreno et al., 1969; San-Blas, 1979; San-Blas and San-Blas, 1977; San-Blas et al., 1976), *B. dermatitidis* (Carbonell, 1967; Domer, 1971; Kanetsuna and Carbonell, 1971), *H. capsulatum* (Domer et al., 1967; Domer, 1971; Domer and Hamilton, 1971; San-Blas and Carbonell, 1974), and *Sporothrix schenckii* (Lloyd and Bitoon, 1971; Mendonça et al., 1976; Previato et al., 1979; Travassos et al., 1977). A summary of the whole cell wall composition of these fungi is presented in Table I.

a. Gross Composition. A general similarity is apparent between the corresponding forms of *P. brasiliensis* and *B. dermatitidis*. The predominant constituents of Y-form cell walls of these two fungi are neutral and amino sugars, while neutral sugars and peptides are most abundant in M-form cell walls. No great difference is observed in the lipid content. A similar tendency is found in *H. capsulatum*. *N*-acetylglucosamine, at least some of which is in the polymerized form (i.e., chitin), is a major component of the yeast cell wall. In striking contrast only limited amounts of chitin were detected during yeast

TABLE I

Gross Chemical Composition of Cell Walls of Various Dimorphic Fungi[a]

Component	P. brasiliensis[b]		B. dermatitidis[b]		H. capsulatum[c]		S. schenckii[a]		
	Yeast form (%)	Mycelial form (%)	Yeast form (%)	Mycelial form (%)	Yeast form (%)	Mycelial form (%)	Yeast form (%)	Mycelial form (%)	Conidia (%)
Monosaccharides	37.2	38.0	47.1	43.5	26.5[f]	29.3[e]	61 ± 1.3	44 ± 0.9	47 ± 1.0
N-acetylglucosamine	40.2	11.7	48.0	13.9	33.7[f]	22.2[e]	7.0 ± 0.4[f]	7.0 ± 0.3[f]	8.3 ± 0.6[f]
Peptides	11.2	37.0	7.0	26.8	5.0	10.0	14.4 ± 0.7	21.7 ± 1.0	24.2 ± 0.5
Lipids									
Readily extractable	1.4	2.2	0.9	2.1	0.5	0.6			
Bound	8.6	5.9	4.6	7.2	0.1	1.5	18.0 ± 0.9[g]	26.0 ± 1.3[g]	19.6 ± 1.2[g]

[a] Average values for samples examined.
[b] Data from Kanetsuna et al. (1969).
[c] Data from Domer et al. (1967), Domer (1971), and Kobayashi and Guiliacci, (1967).
[d] Data from Previato et al. (1979).
[e] Average values for whole, unextracted cell walls of chemotype I strains from Domer (1971).
[f] Reported as hexosamine.
[g] Reported as total cell wall lipid.

morphogenesis in *Saccharomyces cerevisiae*. Perhaps the presence of large amounts of chitin in the yeast wall is a significant aspect of the chemical differentiation of the walls of conidial fungi which undergo yeast–hypha conversion. Further discussion of chitin in yeast and hyphal walls is presented below.

The chemical composition of *S. schenckii* is quite different from that of the other dimorphic fungi in Table I. A comparison of the yeast, hyphal, and conidial walls of this fungus demonstrates quantitative and qualitative similarities between mycelial and conidial walls (Previato *et al.*, 1979). This is to be expected considering that conidia arise by a blowing out of the intact fertile hyphal wall (i.e., holoblastic development; Cole, 1976). The yeast walls are distinguished by their higher content of neutral sugars and lower amounts of protein.

b. Carbohydrates. Although there is little difference in the total amount of monosaccharides in yeastlike and mycelial cell walls of *P. brasiliensis* and *B. dermatitidis* (Table I), proportions of specific monosaccharides exhibit marked quantitative and qualitative variations (Kanetsuna and Carbonell, 1971; Kanetsuna *et al.*, 1972), which are shown in Table II. The Y-form cell wall of these two fungi is composed primarily of glucose. The M-form wall, on the other hand, contains glucose, galactose, and mannose in a molar ratio of 1:0.3:0.6 in *P. brasiliensis,* and 1:0.1:0.2 in *B. dermatitidis*. The amounts of α-glucan (alkali-soluble) and β-glucan (alkali-insoluble) in the Y-form walls of each fungus are similar. This, however, is not the case for M-form walls, While no α-glucan is present in *P. brasiliensis,* it is a principal component of *B. dermatitidis*. The higher content of β-glucan in the mycelial walls of both fungi was assumed to result from its increased rate of synthesis at room temperature (Kanetsuna *et al.*, 1972). Changes occurring during temperature-induced yeast–hypha conversion were investigated by examining the incorporation rate of D-[^{14}C]glucose into glucans. As shown in Fig. 11, the temperature shift from 37° to 20°C is followed by a rapid and continuous decrease in α-glucan during a

TABLE II

Carbohydrate Composition of Cell Walls of *Paracoccidioides brasiliensis* and *Blastomyces dermatitidis*[a]

Carbohydrate	*P. brasiliensis*		*B. dermatitidis*	
	Yeast form (%)	Mycelial form (%)	Yeast form (%)	Mycelial form (%)
α-Glucan	38.2	0	32.8	26.3
β-Glucan	6.2	25.1	2.0	17.5
Galactose	Trace	7.5	Trace	4.4
Mannose	Trace	15.0	Trace	8.8

[a] Data from Kanetsuna *et al.* (1972) and Kanetsuna and Carbonell (1971).

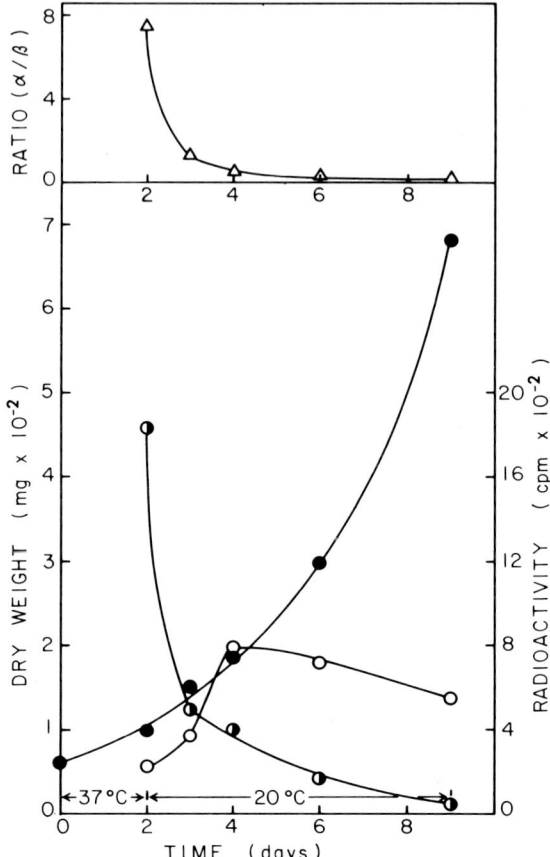

Fig. 11. Change in incorporation rate of [^{14}C]glucose into α- and β-glucans of *Paracoccidioides brasiliensis* resulting from a shift in the temperature of culture from 37° to 20°C. At the time indicated by arrows, [^{14}C]glucose was added to the culture medium and incubated for 24 h. The amount of the fungus grown during 24 h (dry weight in milligrams) was calculated from the growth curve (●, growth). ○, Counts per minute of β-glucan fraction per milligram of fungus grown during 24 h; ◐, counts per minute of α-glucan fraction per milligram of fungus grown during 24 h; △, ratio of radioactivity of α-glucan fraction to that of β-glucan fraction. (After Kanetsuna *et al.*, 1972.)

6-day period. Examination of the change in the α/β ratio over the same period suggests that activities of enzymes involved in β-glucan synthesis are temperature-dependent. Kanetsuna *et al.* (1972) have proposed that β-glucan plays an important role in mycelial wall architecture, while α-glucan is an important structural component of the yeast wall.

Both α-glucan (Y form) and β-glucan (M form) contain 1,3-glycosidic linkages and have been identified as α-1,3- and β-1,3-glucans (Kanetsuna *et al.*, 1972). The β-glucans (mycelial form; Fig. 12) are composed of two chains linked by a 1,6-bond. Each chain consists of approximately 15 glucose units joined by 1,3-bonds, plus two glucose units linked by a 1,6-bond. In addition to glucans, the mycelial wall contains galactose and mannose, probably as a galactomannan complex. The latter was shown to consist of a primary chain of 1,6-linked mannose units possessing galactofuranose residues at the nonreducing

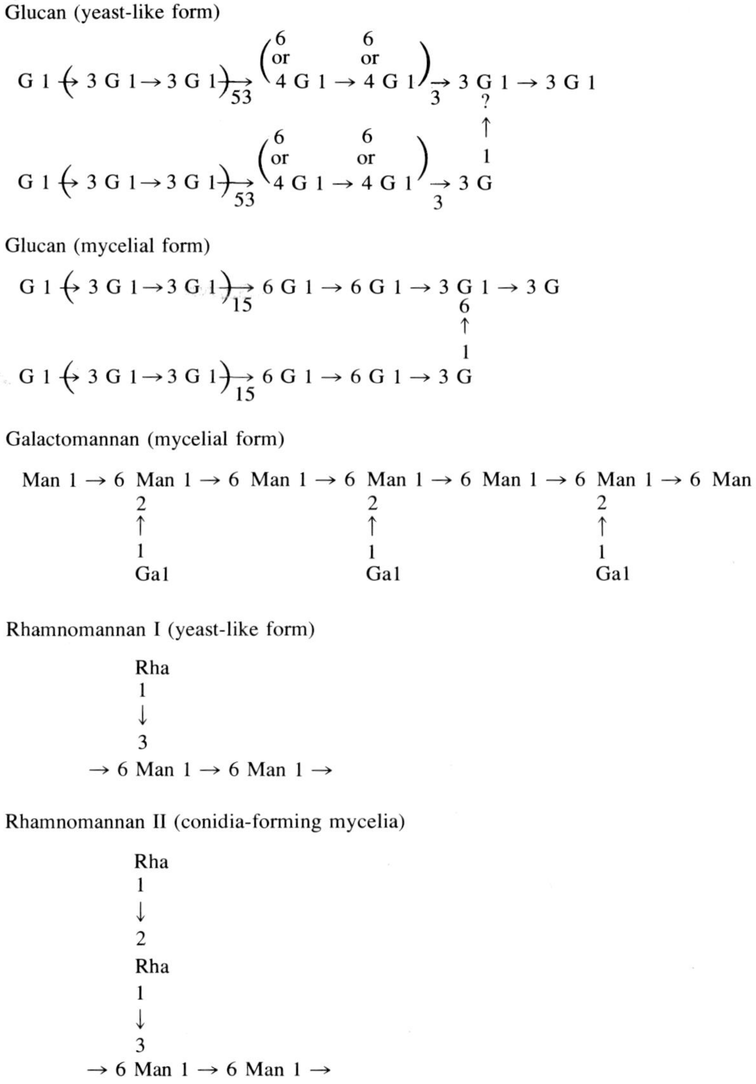

Glucan (yeast-like form)

$$G\,1\,(\to 3\,G\,1\to 3\,G\,1)_{53} \begin{pmatrix} 6 & 6 \\ \text{or} & \text{or} \\ 4\,G\,1 & \to 4\,G\,1 \end{pmatrix}_3 \to 3\,G\,1 \to 3\,G\,1$$

$$G\,1\,(\to 3\,G\,1\to 3\,G\,1)_{53} \begin{pmatrix} 6 & 6 \\ \text{or} & \text{or} \\ 4\,G\,1 & \to 4\,G\,1 \end{pmatrix}_3 \to 3\,G$$

Glucan (mycelial form)

$$G\,1\,(\to 3\,G\,1\to 3\,G\,1)_{15} \to 6\,G\,1 \to 6\,G\,1 \to 3\,G\,1 \to 3\,G$$

$$G\,1\,(\to 3\,G\,1\to 3\,G\,1)_{15} \to 6\,G\,1 \to 6\,G\,1 \to 3\,G$$

Galactomannan (mycelial form)

Man 1 → 6 Man 1 → 6 Man 1 → 6 Man 1 → 6 Man 1 → 6 Man 1 → 6 Man
            2   2   2
            ↑   ↑   ↑
            1   1   1
           Gal   Gal   Gal

Rhamnomannan I (yeast-like form)

Rha
1
↓
3
→ 6 Man 1 → 6 Man 1 →

Rhamnomannan II (conidia-forming mycelia)

Rha
1
↓
2
Rha
1
↓
3
→ 6 Man 1 → 6 Man 1 →

Fig. 12. Proposed structures of glucans and galactomannan of *P. brasiliensis* (Kanetsuna, 1972), and rhamnomannans of *S. schenckii*. (From Mendonca *et al.*, 1976).

terminal positions. Galactomannan was shown to be immunologically active (Malcolm et al., 1979) and to demonstrate cross-reactions with other antigenic polysaccharides extracted from H. capsulatum and B. dermatitidis (Azuma et al., 1974). Such data suggest that these polysaccharides are common serological antigens with the same or similar chemical structures at their antigenic determinants (Azuma et al., 1974). The chemical structures of glucans and galactomannan extracted from the Y- and M-form walls of P. brasiliensis were proposed by Kanetsuna (1972) and are reproduced in Fig. 12. The localization of these polysaccharides, including chitin, is illustrated in Fig. 13. This diagram incorporates data derived from chemical and enzymatic analyses of the walls of P. brasiliensis (F. Kanetsuna, unpublished). The Y-form cell wall is composed principally of alkali-soluble α-glucan in the outer layer, while chitin with small amounts of alkali-insoluble α-glucan and proteins are found in the inner layer. The thinner M-form cell wall does not have an α-glucan layer. Since β-glucan was not solubulized by alkali without chitinase digestion, β-glucan fibers are presumed to be interwoven with chitin and protein. Electron microscopy revealed no distinct wall layers of the developing hypha, although an amorphous galactomannan–protein region on the surface of the young germ tube was found on the basis of enzymatic analyses.

The detection of large amounts of chitin in the yeast cell walls of P. brasiliensis, B. dermatitidis, and H. capsulatum contrasts with the results of Cabib's (1975) study on Saccharomyces cerevisiae in which the small amount of chitin contained in the yeast cell wall was concentrated in the primary septum (Fig. 1F[1]). Several problems for future research can be identified with respect to this marked difference in wall chemistry between nondimorphic (e.g., S. cerevisiae) and dimorphic yeasts. Does the incorporation of chitin occur in a stepwise

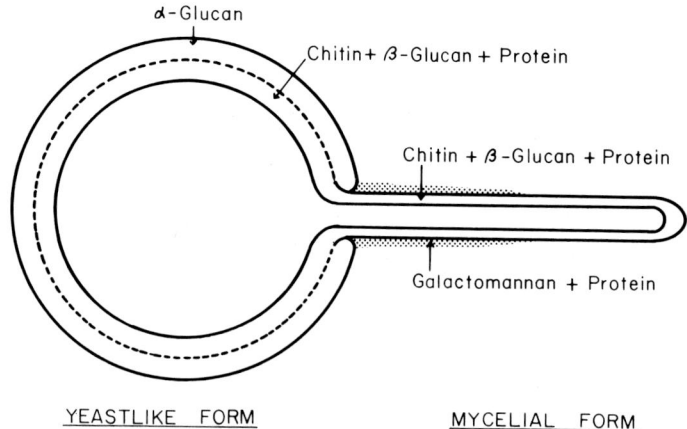

Fig. 13. A proposed model of the cell walls of yeastlike and mycelial forms of P. brasiliensis. (F. Kanetsuna, unpublished.)

fashion in phase with synchronous yeast cell division, as demonstrated for *S. cerevisiae* (Cabib and Farkaš, 1971)? Is chitin preferentially deposited in the region of the primary septum prior to cell separation or is it uniformly (or randomly) distributed throughout the yeast wall? The latter problem could be examined by high-resolution autoradiography of specifically labeled chitin.

Correlation between the formation of specific polysaccharides and morphological differentiation is not clear in *Sporothrix schenckii*. Mendoça *et al.* (1976) observed that, as conidia increased in number, the amount of rhamnomannan released into the medium also increased, suggesting that conidiogenesis was associated with synthesis of this polysaccharide. In general, there is a preferential synthesis of certain polysaccharides by specific cell types (i.e., yeast, conidia, hyphae), resulting in characteristic cell wall composition (Fig. 12). The yeasts produce rhamnomannans with 1,3-α-rhamnopyranosyl side chains, conidia or conidia-forming mycelia form 1,2-α-rhamnopyranosyl-1,3-α-rhamnopyranosyl side chains, and vegetative mycelia synthesize galactomannans and mannans. Little detailed information is available on the polysaccharide structure of cell walls of other dimorphic conidial fungi. Future investigations of total wall carbohydrates and monosaccharides should be directed toward clarification of the role of these components in structural support, morphogenesis and, in the case of pathogenic dimorphic fungi, immunological response of host tissues.

c. Proteins. The amino acid composition of acid hydrolysates of the Y- and M-form cell walls from several dimorphic fungi is shown in Table III. In general, the M-form walls contain larger amounts of peptides than the Y-form (Table I). In *P. brasiliensis* and *B. dermatitidis* the hyphal walls have a lower content of histidine but higher amounts of proline than the yeast cell walls. This same tendency, however, was not observed for *H. capsulatum*. Except in the Y form of *B. dermatitidis* glycine is the most abundant amino acid in the three dimorphic fungi examined, irrespective of morphological type.

There are several lines of evidence supporting the contention that sulfhydryl groups play an important role in conversion from yeast to mycelia (Nickerson and Falcone, 1956; Cortat *et al.*, 1972). As indicated in Table III, the Y-form cell walls of *P. brasiliensis* and *B. dermatitidis* have lower concentrations of sulfhydryl groups than M-form cell walls. The possible roles of disulfide linkages in dimorphism will be discussed later. Although amino acid analysis has been performed for other dimorphic fungi, no clear correlation has been observed between specific amino acid composition and morphological differentiation.

d. Lipids. The content of readily extractable lipids (extracted using chloroform–methanol) and bound lipids (extracted using acidic ethanol) differ in Y and M forms (Table I). The major components of readily extractable lipids in *B. dermatitidis* and *H. capsulatum* (Domer and Hamilton, 1971) were phosphatidylethanolamine (PE), phosphatidylcholine (PC), and phosphatidylserine (PS) representing polar lipids, and triglycerides (TG), diglycerides, sterols,

TABLE III
Amino Acid Composition of Cell Walls of Various Dimorphic Fungi

	P. brasiliensis[a]		B. dermatitidis[a]		H. capsulatum[b]	
Amino acid	Yeast form	Mycelial form	Yeast form	Mycelial form	Yeast form	Mycelial form
Lysine	9.3	8.5	6.4	9.3	0.4	0.6
Histidine	10.3	3.4	5.4	4.9	0.1	0.2
Arginine	2.7	12.7	1.4	6.4	0.3	0.2
Aspartic acid	9.9	34.1	7.6	21.0	0.1	0.7
Threonine	7.9	26.1	2.4	21.4	0.2	1.1
Serine	6.6	23.6	3.2	17.9	0.2	1.0
Glutamic acid	9.9	38.1	5.0	29.1	0.5	1.2
Proline	5.0	42.8	2.7	31.3	Trace	Trace
Glycine	12.8	53.8	4.8	63.0	0.6	1.7
Alanine	7.9	28.8	3.3	16.6	0.3	0.7
Valine	4.4	18.8	1.9	13.0	Trace	0.7
Methionine	1.7	11.3	0.6	1.7	Trace	Trace
Isoleucine	3.9	14.3	1.3	12.5	0.3	0.3
Leucine	4.4	17.1	1.1	14.0	0.4	0.4
Phenylalanine	4.5	7.5	1.2	6.1	0.3	0.1
Half-cystine	0.29	3.40	1.50	2.28		
Total (mg/10 mg[b] or 100 mg[a] cell wall)	10.1	40.8	7.1	30.0	0.5	1.0

[a] Data from Kanetsuna et al. (1969), expressed as micromoles per 100 mg cell wall.
[b] Data from Domer et al. (1967), expressed as micromoles per 10 mg cell wall.

sterol esters, and free fatty acids representing the neutral lipid fractions. Quantitative comparison of the main lipid fraction (PC, PE, TG) revealed little difference between yeast and mycleial cell walls. It is of interest, however, that the major variations found in the fatty acid composition of the two kinds of cell walls primarily involved quantities of oleic ($C_{18:1}$) and linoleic ($C_{18:2}$) acids which accounted for 75–80% of the total fatty acid composition. While oleic acid was predominant in the yeast phase (30–70%), amounts of linoleic acid in the yeast and mycelial walls were comparable (30–40%). The palmitic acid ($C_{16:0}$) level did not show any characteristic trends. The fatty acid pattern of bound lipids was very similar to that of readily extractable lipids.

3. Plasma Membrane

Although evidence indicates that the most distinct differences in chemical composition between the two extreme forms of dimorphic fungi are reflected in the cell wall, the plasma membrane which participates in constructing cell walls can also be expected to reveal significant variations in the two kinds of cells. As pointed out earlier, the plasmalemma is recognized as the location of certain enzyme systems responsible for wall biosynthesis (Farkaš, 1979). However, because of the difficulty in isolating and purifying plasma membranes, only limited information is available at present. The chemical composition of plasma membranes of *C. albicans* has been analyzed in detail and is reviewed in Table IV (Marriot, 1975). The total lipid content was higher in the yeast than in the mycelial membrane, while a significantly larger amount of carbohydrate was found in the M-form membrane. Mycelial plasma membranes contained less sterol but proportionately more triglyceride and free fatty acids than yeast membranes. The most abundant phospholipid in the Y-form membrane was PE (70%), whereas this same phospholipid and PC occurred in equal concentrations in the M form. Ergosterol, which was the major sterol, did not vary significantly in the two types, while the precursors of ergosterol (calciferol and zymosterol) showed a sharp contrast in relative proportions. The fatty acid components were

TABLE IV

Chemical Composition of Plasma Membranes from Yeast and Mycelial Cells of *Candida albicans* [a]

Component	Yeast form (%)	Mycelial form (%)
Gross composition		
Carbohydrate	9.0	25.0
Lipid	43.0	31.0
Protein	52.0	45.0
Carbohydrates		
Hexosamine	3.0	6.5
Mannose	18.0	20.0
Glucose	55.0	63.0
Lipids		
Neutral lipids		
Sterol ester	40	28
Free sterol	19	9
Free fatty acid	17	27
Triglyceride	24	36
Phospholipids		
Phosphatidylethanolamine	70	50
Phosphatidylserine	11	
Phosphatidylcholine	4	50
Sphingolipid	15	
Free sterols		
Calciferol	12.0	Trace
Zymosterol	Trace	16.0
Ergosterol	50.0	42.0

[a] Data from Marriot, 1975.

similar to those of other fungi described in Section II,d. Linoleic acid was more abundant in the Y-form membrane (30%) than in the M-form (18%), and oleic acid was distributed equally between the two forms.

B. Metabolism

Among a variety of environmental factors which affect fungal morphology, the presence of a fermentable hexose is known to be important for maintaining the yeast form of the zygomycetous fungus *Mucor rouxii* (Bartnicki-Garcia, 1963). This suggests that glucose metabolism is closely associated with morphological conversion, either through intermediates or end products. In *M. racemosus* it was revealed that the Y form was consistently correlated with a high flow of glucose carbon through the glycolytic and pentose phosphate pathways, while the mycelial form did not exhibit a consistent pattern in carbon metabolism (Inderlied and Sypherd, 1978).

Comparative studies of enzymatic activities in glycolysis, the pentose phosphate pathway, and the citric acid cycle (Table V) have been carried out for cell-free extracts of the Y and M forms of *P. brasiliensis* (Kanetsuna and Carbonell, 1966). Malic dehydrogenase and glucose-6-phosphate dehydrogenase activities were higher in mycelia than in yeast, but in general enzymatic activities of the M form were lower than those of the Y form. This may be indicative of higher endogenous respiration in the Y form. In *H. capsulatum*, there was no significant difference in enzymatic activities in the two forms with respect to glycolysis and the pentose phosphate pathway, although other enzymes demonstrated higher activities in the yeast than in mycelia (Mahvi, 1965). Land *et al.* (1975b) examined glucose metabolism and respiration in *C. albicans* during hyphal differentiation. Cultures supporting hyphal growth produced more ethanol and less carbon dioxide than yeast cultures, while the former consumed less oxygen than the latter. A shift from aerobic to fermentative metabolism of glucose results in a conversion from the Y to the M form. As yeast cells are transformed to hyphae and fermentative metabolism is initiated, an alteration in electron transport takes place (Nickerson, 1963). It has been suggested that transfer of electrons from flavoprotein is required for maintaining the yeast form, while an increase in the amount of reduced flavoprotein favors hyphal formation. The electron acceptor is not oxygen, since respiration in this case does not involve cytochrome oxidase. Oxidized glutathione and cystine were also excluded as hydrogen acceptors (Romano and Nickerson, 1954), and it was found, instead, that the acceptor was a disulfide linkage associated with a protein bound to the glucomannan complex of the cell wall. In addition, a specific enzyme was demonstrated to catalyze reduction of these disulfide bonds and identified as disulfide reductase. The enzyme was localized exclusively in the mitochondrial fraction (Nickerson and Falcone, 1956). Although disulfide reduc-

TABLE V

Enzymatic Activities[a] of Yeast Form and Mycelial Forms of *Paracoccidiodes brasiliensis*[b]

Enzyme	Yeast form	Mycelial form	Ratio (Y/M)
Phosphorylase	0.43	0.36	1.2
Phosphoglucomutase	+22.30	+13.20	1.7
Hexokinase[c]	+2.10	+1.37	1.5
Phosphoglucose isomerase	+4.98	+4.29	1.2
Phosphofructokinase	+0.30	+0.19	1.6
Aldolase	1.7	1.5	1.1
Glyceraldehyde-3-phosphate dehydrogenase	+44.7	+9.96	4.5
Phosphoglyceromutase	−2.52	−1.42	1.8
Enolase	−6.25	−3.34	1.9
Pyruvate kinase	−4.55	−2.03	2.2
Glucose-6-phosphate dehydrogenase	+2.18	+2.72	0.8
6-Phosphogluconate dehydrogenase	+0.57	+0.46	1.2
Aconitase	+0.87	+0.37	2.4
Isocitric dehydrogenase	+1.09	+0.51	2.1
Succinic dehydrogenase	−0.17	−0.09	1.9
Fumarase	−0.30	−0.18	1.7
Malic dehydrogenase	−10.16	−19.3	0.5

[a] Phosphorylase and aldolase are expressed as micromoles of orthophosphate per minute per 10 mg protein. Other enzymes are expressed as change in absorbance per minute per milligram of protein; (+) indicates increasing and (−) indicates decreasing optical density. (See original article for details of procedure.)
[b] Data from Kanetsuna and Carbonell, 1966.
[c] Based on reaction with glucose.

tase has been suggested to participate in yeast–hypha conversion, Matile *et al.* (1971) has pointed out that this enzyme is not present in all dimorphic fungi.

Chitin has been shown to be a principal component of the cell wall of several dimorphic fungi and probably plays an important role in morphological differentiation (e.g., *H. capsulatum* and *B. dermatitidis*). On the other hand, in *C. albicans* the alkali-insoluble fraction from hyphae was shown to have three times more chitin than the yeast (Chattaway *et al.*, 1968). Chitin synthesis was compared in yeast and hyphae of *C. albicans* by measuring the incorporation rate of N-acetyl[^{3}H]glucosamine into the acid-alkali-insoluble fractions (Braun and Calderone, 1978). The results indicated that incorporation was 10 times greater in hyphal than in yeast forms. In addition, chitin synthase, which was found mostly in the zymogen form, was localized on the cytoplasmic side of the plasma membrane (Braun and Calderone, 1978), as reported for *Saccharomyces cerevisiae* (Duran *et al.*, 1975). Although chitin synthase was activated by preincubation with a protease-containing vacuolar fraction from yeast cells, no stimulation was observed when a vacuolar fraction prepared from hyphae was used with either yeast or hyphal cultures. The reason for this discrepancy has not yet been determined.

C. Mechanisms of Dimorphism

A variety of factors have been shown to induce morphological differentiation in dimorphic fungi. Romano (1966) distinguished three types of dimorphism: temperature-dependent (e.g., *B. dermatitidis, P. brasiliensis*), temperature- and nutrition-dependent (e.g., *H. capsulatum, S. schenckii*), and nutrition-dependent (e.g., *C. albicans*). Many studies have sought to elucidate the biochemical mechanisms by which yeast–hypha conversion occurs, and several hypotheses have been proposed. For example, thermal dimorphism of *B. dermatitidis* was interpreted as an uncoupling of cell division from cell growth (Nickerson and Edwards, 1949). The effect of temperature on oxygen uptake by hyphal and yeast forms was examined, and it was found that the latter consumed several times more oxygen. The respiratory rate of the mycelial cells increased linearly with an increase in growth temperature up to 30°C. However, above 30°C this relationship was no longer observed, suggesting that inactivation of certain enzymes had occurred. This modification of enzyme activity was suggested directly to affect cell division and growth at the onset of yeast morphogenesis. However, such a hypothesis could not be applied to *C. albicans*, since the fungus grows as a yeast form at both 37°C and at room temperature. Many efforts were subsequently directed toward defining specific nutrients which either maintain the yeast form or promote hyphal development. It was found that glucose, a readily utilizable carbon source, was required for maintenance of yeast shape, and that removal of glucose led to the emergence of filamentous growth (Nickerson and Mankowski, 1953). Since hypha formation can be arrested and reversed to yeast development in *C. albicans* by the addition of cysteine, the sulfhydryl groups were suggested to play a central role in cell division and form determination (Nickerson, 1963; Taschdjian and Kozinn, 1961; Winsten and Murray, 1956). Nickerson and Falcone (1956) have presented evidence, from studies on cell division mutants of *Candida albicans*, that protein disulfide reductase is responsible for cleaving disulfide covalent bonds in the cell wall. The activity of this enzyme thereby creates a physically weakened region of the cell wall at the site where budding is initiated. The following scheme was proposed to explain the correlation between morphological conversion and activity of protein disulfide reductase:

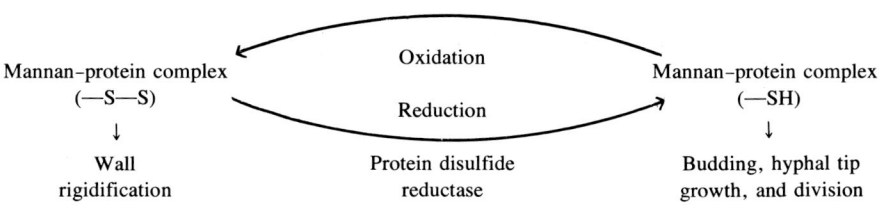

Hartwell (1974) and others have challenged the validity of this concept mainly on the basis of lack of genetic evidence for the interrelationship between synthesis of the enzyme and dimorphism, and the apparent absence of disulfide reductase in many dimorphic fungi. Wain et al. (1975) did not support the view of cysteine-mediated form determination in *C. albicans*. Instead, they suggested that a nonspecific inhibition of hyphal growth occurred, resulting from some toxic action of high concentrations of cysteine or its products of catabolism. These authors also reported that their isolates of *C. albicans* could produce germ tubes in the presence of 10 mM cysteine, although at a slow growth rate. In addition, little or no yeast production occurred under these conditions. On the other hand, Maresca et al. (1978) have reported that organosulfur compounds favor the transition of mycelia to yeast in *H. capsulatum*. Although both forms of this fungus concentrate cystine at comparable rates, only the yeast form was shown to contain NADH-dependent cystine reductase. The authors demonstrated that the enzyme appeared early in the transition from mycelium to yeast and suggested that it may provide reduced sulfhydryl groups which are involved in yeast morphogenesis. A solution to this controversy concerning the significance of cysteine and cystine, protein disulfide reductase, and sulfhydryl groups in yeast–hypha conversion is not yet at hand.

Kanetsuna et al. (1972) presented a hypothesis for thermal dimorphism in *P. brasiliensis* based primarily on differences in α- and β-glucan synthesis and protein disulfide reductase activity in the yeast and mycelial forms. The differentiation of cell wall composition in the Y and M forms was shown in Fig. 12. At 37°C (Y form), α-glucan and chitin are synthesized more actively than β-glucan, while at 20°C (M form) the synthesis of α-glucan decreases. In the M form, a low activity of protein disulfide reductase corresponds with the presence of many disulfide linkages, and interwoven microfibrils of protein, chitin, and β-glucan probably provide rigidity to the hyphal wall. A model was proposed for the β-glucan–chitin complex of hyphal walls of *Schizophyllum commune* Fr. in which amino acids, especially lysine and citrulline, were involved in the linkages between glucan and chitin (Sietsma and Wessels, 1979). Kanetsuna et al. (1972) have suggested that β-glucan exists as ''islets'' in the cell wall and that loss of rigidity around the β-glucans occurs from the combined action of β-glucanase and protein disulfide reductase, resulting in bud formation when the incubation temperature is shifted from 20° to 37°C. Further studies are required to determine how temperature regulates activities of enzymes participating in the synthesis of wall components. It is now well accepted that the physical states of membrane lipids are controlled by membrane fluidity (Kimmelberg, 1977). Therefore one would expect that fluidity-dependent alterations in activities of certain key enzymes associated with wall biosynthesis may have effects on thermal dimorphism.

As pointed out above, cAMP has also been implicated in the control of dimorphism (Larsen and Sypherd, 1974). Endogenous levels of cAMP decrease sharply in yeast cells just prior to the formation of germ tubes. The biosynthesis of cAMP involves a reaction catalyzed by adenylate cyclase, while hydrolysis of cAMP is effected by cAMP phosphodiesterase. Paveto et al. (1975) have shown that the decline in cAMP in yeast corresponds to an increase in phosphodiesterase activity. The requirement of divalent cations for the development of several dimorphic fungi may be associated with modulation of adenylate cyclase and phosphodiesterase activities which result in changes in the intracellular cAMP levels (Bartnicki-Garcia and Nickerson, 1962a; Garrison and Boyd, 1974; Pine and Peacock, 1958; Reiss and Nickerson, 1974; Yamaguchi, 1975; Zorzopulus et al., 1973). Both adenylate cyclase and phosphodiesterase are suggested to be membrane-bound enzymes (Scott et al., 1973; Scott and Solomon, 1975). Stewart and Rogers (1978) have reviewed the possibility that temperature variation during dimorphism may lead to membrane lipid alterations (e.g., linolenic acid) which in turn affect lipid–protein interactions within the biomembrane. The effects of temperature on adenylate cyclase or phosphodiesterase activities would result in changes in the concentration of intracellular cAMP. These authors proposed several roles for cAMP in yeast–hypha conversion, but "the question of how cAMP ultimately generates a morphological effect is left suspended."

The most exploited model of fungal dimorphism is *Mucor,* and current research involving members of this genus has been directed toward unraveling the mechanisms which regulate gene expression during morphogenesis (Sypherd et al., 1979). Developmental mutants have been used to examine the interrelationships between initiation of yeast–hypha conversion (i.e., germ tube formation) and the rapid increase in rate of polypeptide chain elongation. In a conditional morphology mutant of *M. racemosus* which persists as the yeast form when exposed to air and when grown on minimal medium, it has been shown that methionine [precursor of S-adenosylmethionine (SAM)] is required for completing the transition to the hyphal phase (Sypherd et al., 1979). Methionine thereby acts as a "specific stimulus" in this mutant which can be used to examine changes in rates of polypeptide chain elongation and protein synthesis. Such studies may provide information on the mechanisms of transcriptional and translational control of yeast–hypha conversion.

REFERENCES

Ajello, L. (1978). The black yeasts as disease agents: Historical perspective. *Sci. Publ., Pan Am. Health Organ.* **356,** 9-16.
Azuma, I., Kanetsuna, F., Tanaka, Y., Yamamura, Y., and Carbonell, L. M. (1974). Chemical and immunological properties of galactomannans obtained from *H. duboisii, H. capsulatum, P. brasiliensis* and *B. dermatitidis. Mycopathol. Mycol. Appl.* **54,** 111-125.
Baker, E. E., Marak, E. M., and Smith, C. E. (1943). The morphology, taxonomy, and distribution of *Coccidioides immitis* Rixford and Gilchrist 1896. *Farlowia* **1,** 199-244.
Barathova, H., Betina, V., and Ulicky, L. (1977). Regulation of differentiation of the dimorphic fungus *Paecilomyces veridis* by nitrogen sources, antibiotics and metabolic inhibitors. *Folia Microbiol. (Prague)* **22,** 222-231.
Barlow, A. J. E., Aldersley, T. A., and Chattaway, F. W. (1974). Factors present in serum and seminal plasma which promote germ-tube formation and mycelial growth of *Candida albicans*. *J. Gen. Microbiol.* **82,** 261-272.
Bartnicki-Garcia, S. (1963). III. Mold-yeast dimorphism of *Mucor:* Symposium on biochemical bases of morphogenesis in fungi. *Bacteriol. Rev.* **27,** 293-304.
Bartnicki-Garcia, S. (1973). Fundamental aspects of hyphal morphogenesis. *Symp. Soc. Gen. Microbiol.* **23,** 245-267.
Bartnicki-Garcia, S., and Nickerson, W. J. (1962a). Induction of yeast-like development in *Mucor* by carbon dioxide. *J. Bacteriol.* **84,** 829-840.
Bartnicki-Garcia, S., and Nickerson, W. J. (1962b). Nutrition, growth, and morphogenesis of *Mucor rouxii. J. Bacteriol.* **84,** 841-858.
Bartnicki-Garcia, S., and Nickerson, W. J. (1962c). Isolation, composition and structure of cell walls of filamentous and yeast-like forms of *Mucor rouxii. Biochim. Biophys. Acta* **58,** 102-119.
Batra, L. R., and Michie, M. D. (1963). Pleomorphism in some ambrosia and related fungi. *Trans. Kans. Acad. Sci.* **66,** 470-481.
Bauer, H., Horisberger, M., Bush, D. A., and Sigarlakie, E. (1972). Mannan as a major component of the bud scars of *Saccharomyces cerevisiae. Arch. Mikrobiol.* **85,** 202-208.
Beneke, E. S., and Rogers, A. L. (1970). "Medical Mycology Manual." Burgess, Minneapolis, Minnesota.
Biel, A. K., Brand, J. M., Markovetz, A. J., and Bridges, J. R. (1977). Dimorphism in *Ceratocystis minor* var. *barrasii. Mycopathologia* **62,** 179-182.
Bowers, B., Levin, G., and Cabib, E. (1974). Effect of polyoxin D on chitin synthesis and septum formation in *Saccharomyces cerevisiae. J. Bacteriol.* **119,** 564-575.
Braun, P. C., and Calderone, R. A. (1978). Chitin synthesis in *Candida albicans:* Composition of yeast and hyphal forms. *J. Bacteriol.* **135,** 1472-1477.
Braun, P. C., and Calderone, R. A. (1979). Regulation and solubilization of *Candida albicans* chitin synthetase. *J. Bacteriol.* **140,** 666-670.
Breslau, A. M., Hensley, T. J., and Erickson, J. O. (1961). Electron microscopy of cultured spherules of *Coccidioides immitis. J. Biophys. Biochem. Cytol.* **3,** 627-637.
Brunswick, H. (1924). Untersuchungen über Geschlechts und Kernverhaltnisse bei der Hymenomyzetengattung *Coprinus. Bot. Abh.* **5,** (K. Goebel, ed). Gustav Fisher, Jena.
Burnett, J. H. (1968). "Fundamentals of Mycology." Arnold, London.
Bush, D. A., and Horisberger, M. (1973). Mannan of yeast bud scars. *J. Biol. Chem.* **248,** 1318-1320.
Byers, B., and Goetsch, L. (1974). Duplication of spindle plaques and integration of the yeast cell cycle. *Cold Spring Harbor Symp. Quant. Biol.* **38,** 123-131.
Byers, B., and Goetsch, L. (1975). Behavior of spindle plaques in the cell cycle and conjugation of *Saccharomyces cerevisiae. J. Bacteriol.* **124,** 511-523.

Byers, B., and Goetsch, L. (1976). A highly ordered ring of membrane-associated filaments in budding yeast. *J. Cell Biol.* **69,** 717-721.
Cabib, E. (1975). Molecular aspects of yeast morphogenesis. *Annu. Rev. Microbiol.* **29,** 191-214.
Cabib, E., and Bowers, B. (1971). Chitin and yeast budding: Localization of chitin in yeast bud scars. *J. Biol. Chem.* **246,** 152-159.
Cabib, E., and Farkaš, V. (1971). The control of morphogenesis: An enzymatic mechanism for the initiation of septum formation in yeast. *Proc. Natl. Acad. Sci. U.S.A.* **68,** 2052-2056.
Cabib, E., Farkaš, V., Ulane, R. E., and Bowers, B. (1973). Yeast septum formation as a model system for morphogenesis *In* "Yeast, Mould and Plant Protoplasts" (J. R. Villanueva, I. Garcia-Acha, S. Gascón, and F. Uruburu, eds.), pp. 105-116. Academic Press, New York.
Cabib, E., Ulane, R., and Bowers, B. (1974). A molecular model for morphogenesis: The primary septum of yeast. *Curr. Top. Cell. Regul.* **8,** 1-32.
Carbonell, L. M. (1967). Cell wall changes during the budding process of *Paracoccidioides brasiliensis* and *Blastomyces dermatitidis*. *J. Bacteriol.* **94,** 213-223.
Chattaway, F. W., Holmes, M. R., and Barlow, A. J. E. (1968). Cell wall composition of the mycelial and blastospore forms of *Candida albicans*. *J. Gen. Microbiol.* **51,** 367-376.
Chattaway, F. W., O'Reilly, J., Barlow, A. J. E., and Aldersley, T. (1976). Induction of the mycelial form of *Candida albicans* by hydrolysates of peptides from seminal plasma. *J. Gen. Microbiol.* **96,** 317-322.
Chung, K. L., Hawirko, R. Z., and Isaac, P. K. (1965). Cell wall replication in *Saccharomyces cerevisiae*. *Can. J. Microbiol.* **11,** 953-957.
Cochrane, V. W. (1958). "Physiology of Fungi." Wiley, New York.
Cole, G. T. (1975). A preparatory technique for examination of imperfect fungi by scanning electron microscopy. *Cytobios* **12,** 115-121.
Cole, G. T. (1976). Conidiogenesis in pathogenic hyphomycetes. I. *Sporothrix, Exophiala, Geotrichum* and *Microsporum*. *Sabouraudia* **14,** 81-98.
Cole, G. T. (1978). Conidiogenesis in the black yeasts. *Sci. Publ., Pan Am. Health Organ.* **356,** 66-78.
Cole, G. T., and Aldrich, H. C. (1971). Ultrastructure of conidiogenesis in *Scopulariopsis brevicaulis*. *Can. J. Bot.* **49,** 745-755.
Cole, G. T., and Kendrick, W. B. (1969). Conidium ontogeny in Hyphomycetes: The arthrospores of *Oidiodendron* and *Geotrichum* and the endoarthrospores of *Sporendonema*. *Can. J. Bot.* **47,** 1773-1780.
Cole, G. T., and Samson, R. A. (1979). "Patterns of Development in Conidial Fungi." Pitman, London.
Cortat, M., Matile, P., and Wiemken, A. (1972). Isolation of glucanase-containing vesicles from budding yeast. *Arch. Microbiol.* **82,** 189-205.
Dabrowa, N., Howard, D. H., Landau, J. W., and Shechter, Y. (1970). Synthesis of nucleic acids and proteins in the dimorphic forms of *Candida albicans*. *Sabouraudia* **8,** 163-169.
Dabrowa, N., Taxer, S. S. S., and Howard, D. H. (1976). Germination of *Candida albicans* induced by proline. *Infect. Immun.* **13,** 830-835.
de Hoog, G. S., and Hermanides-Nijhof, E. J. (1977). The black yeasts and allied hyphomycetes. C.B.S. *Stud. Mycol.* **15,** 1-222.
Dodge, C. W. (1935). "Medical Mycology." Mosby, St. Louis, Missouri.
Domer, J. E. (1971). Monosaccharide and chitin content of cell walls of *Histoplasma capsulatum* and *Blastomyces dermatitidis*. *J. Bacteriol.* **107,** 870-877.
Domer, J. E., and Hamilton, J. G. (1971). The readily extracted lipids of *Histoplasma capsulatum* and *Blastomyces dermatitidis*. *Biochim. Biophys. Acta* **231,** 465-478.
Domer, J. E., Hamilton, J. G., and Harkin, J. C. (1967). Comparative study of the cell walls of the yeast-like and mycelial phases of *Histoplasma capsulatum*. *J. Bacteriol.* **94,** 466-474.

Duran, A., Bowers, B., and Cabib, E. (1975). Chitin synthetase zymogen is attached to the yeast plasma membrane. *Proc. Natl. Acad. Sci. U.S.A.* **72,** 3952-3955.

Duran, A., Cabib, E., and Bowers, B. (1979). Chitin synthetase distribution on the yeast plasma membrane. *Science* **203,** 363-365.

Ebina, K., Takashita, S., Kamaguchi, A., Yokota, K., and Sakaguchi, O. (1978). Comparison of cell body and cell wall compositions between the yeast-like and mycelial phases of *Geotrichum candidum*. *Jpn. J. Bacteriol.* **33,** 527-538.

Emmons, C. W. (1947). Biology of *Coccidioides*. *In* "Biology of the Pathogenic Fungi" (W. J. Nickerson, ed.), pp. 71-82. Chronica Botanica, Waltham Massachusetts.

Emmons, C. W., Binford, C. H., Utz, J. P., and Kwon-Chung, K. J. (1977). "Medical Mycology." Lea & Febiger, Philadelphia, Pennsylvania.

Evans, E. G. V., Odds, F. C., Richardson, M. D., and Holland, K. T. (1975). Optimum conditions for initiation of filamentation in *Candida albicans*. *Can. J. Microbiol.* **21,** 338-342.

Farkaš, V. (1979). Biosynthesis of cell walls of fungi. *Microbiol. Rev.* **43,** 117-144.

Francke-Grosmann, H. (1967). Ectosymbiosis in wood-inhabiting insects. *In* "Symbiosis" (S. M. Henry, ed.), Vol. 2, pp. 171-180. Academic Press, New York.

Garrison, R. G., and Boyd, K. S. (1974). Ultrastructural studies of induced morphogenesis by *Aspergillus parasiticus*. *Sabouraudia* **12,** 179-187.

Garrison, R. G., and Boyd, K. S. (1977). The fine structure of microconidial germination and vegetative cells in *Histoplasma capsulatum*. *Ann. Microbiol.* **128,** 135-149.

Garrison, R. G., and Boyd, K. S. (1978). Role of the conidium in dimorphism of *Blastomyces dermatitidis*. *Mycopathologia* **64,** 29-33.

Girbardt, M. (1955). Lebendbeobachtungen an *Polystictus versicolor* (L.). *Flora (Jena)* **142,** 540-563.

Girbardt, M. (1957). Der Spitzenkörper von *Polystictus versicolor* (L.). *Planta* **50,** 47-59.

Girbardt, M. (1973). Die Pilzzelle. *In* "Grundlagen der Cytologie" (G. C. Hirsch, H. Ruska, and P. Sitte, eds.), pp. 441-460. Fischer, Jena.

Girbardt, M. (1979). A microfilamentous septal belt (FSB) during induction of cytokinesis in *Trametes versicolor* (L. ex Fr.). *Exp. Mycol.* **3,** 215-228.

Grove, S. N. (1978). The cytology of hyphal tip growth. *In* "The Filamentous Fungi" (J. E. Smith and D. R. Berry, eds.), Vol. 3, pp. 28-50. Wiley, New York.

Grove, S. N., and Bracker, C. E. (1970). Protoplasmic organization of hyphal tips among fungi: Vesicles and Spitzenkörper. *J. Bacteriol.* **104,** 989-1009.

Grove, S. N., Bracker, C. E., and Morré, D. J. (1970). An ultrastructural basis for hyphal tip growth in *Pythium ultimum*. *Am. J. Bot.* **57,** 245-266.

Guilliermond, A. (1910). Remarques sur les récents travaux parus sur la cytologie des levures et quelques nouvelles observations sur ce groupe de champignons. *Zentralbl. Bakteriol., Parasitenkd. Infektionskr. Hyg., Abt. 2* **26,** 577-589.

Hardcastle, R. V., and Szaniszlo, P. J. (1974). Characterization of dimorphism in *Cladosporium werneckii*. *J. Bacteriol.* **119,** 294-302.

Hartwell, L. H. (1971a). Genetic control of the cell division cycle in yeast. II. Genes controlling DNA replication and its initiation. *J. Mol. Biol.* **59,** 183-194.

Hartwell, L. H. (1971b). Genetic control of the cell division cycle in yeast. IV. Genes controlling bud emergence and cytokinesis. *Exp. Cell Res.* **69,** 265-276.

Hartwell, L. H. (1974). *Saccharomyces cerevisiae* cell cycle. *Bacteriol. Rev.* **38,** 164-198.

Hayashibe, M., and Katohda. S. (1973). Initiation of budding and chitin ring. *J. Gen. Appl. Microbiol.* **19,** 23-39.

Hazen, K. C., and Cutler, J. E. (1979). Autoregulation of germ tube formation by *Candida albicans* *Infect. Immun.* **24,** 661-666.

Inderlied, G. B., and Sypherd, P. S. (1978). Glucose metabolism and dimorphism in *Mucor*. *J. Bacteriol.* **133,** 1282-1286.

Kanetsuna, F. (1972). Biochemical characteristics of *Paracocciodioides brasiliensis*. *Sci. Publ., Pan Am. Health Organ.* **254**, 31–37.
Kanetsuna, F., and Carbonell, L. M. (1966). Enzymes in glycolysis and the citric acid cycles in the yeast and mycelial forms of *Paracoccidioides brasiliensis*. *J. Bacteriol.* **92**, 1315–1320.
Kanetsuna, F., and Carbonell, L. M. (1971). Cell wall composition of the yeast-like and mycelial forms of *Blastomyces dermatitidis*. *J. Bacteriol.* **106**, 946–948.
Kanetsuna, F., Carbonell, L. M., Moreno, R. E., and Rodriguez, J. (1969). Cell wall composition of the yeast and mycelial forms of *Paracoccidioides brasiliensis*. *J. Bacteriol.* **97**, 1036–1041.
Kanetsuna, F., Carbonell, L. M., Azuma, I., and Yamamura, Y. (1972). Biochemical studies on the thermal dimorphism of *Paracoccidioides brasiliensis*. *J. Bacteriol.* **110**, 208–218.
Kimmelberg, H. H. K. (1977). The influence of membrane fluidity on the activity of membrane-bound enzymes. *Cell Surf. Rev.* **3**, 205–293.
Kobayashi, G. S., and Guiliacci, P. L. (1967). Cell wall studies of *Histoplasma capsulatum*. *Sabouraudia* **5**, 180–188.
Kwon-Chung, K. J. (1969). *Coccidioides immitis:* Cytological study on the formation of the arthrospores. *Can. J. Bot. Cytol.* **11**, 43–53.
Land, G. A., McDonald, W. C., Stjernholm, R. L., and Friedman, L. (1975a). Factors affecting filamentation in *Candida albicans:* Relationship of the uptake and distribution of proline to morphogenesis. *Infect. Immun.* **11**, 1014–1023.
Land, G. A., McDonald, W. C., Stjernholm, R. L., and Friedman, L. (1975b). Factors affecting filamentation in *Candida albicans:* Changes in respiratory activity of *Candida albicans* during filamentation. *Infect. Immun.* **12**, 119–127.
Lane, J. W., and Garrison, R. G. (1970). Electron microscopy of the yeast to mycelial phase conversion of *Sporotrichum schenckii*. *Can. J. Microbiol.* **16**, 747–749.
Larsen, A. D., and Sypherd, P. S. (1974). Cyclic adenosine 3′,5′-monophosphate and morphogenesis in *Mucor racemosus*. *J. Bacteriol.* **117**, 432–438.
Laskey, W., and Sarosi, G. A. (1978). Endogenous activation in blastomycosis. *Ann. Intern. Med.* **88**, 50–52.
Lee, K. L., Buckley, H. R., and Campbell, C. C. (1975). An amino acid liquid synthetic medium for the development of mycelial and yeast forms of *Candida albicans*. *Sabouraudia* **13**, 148–153.
Lloyd, K. O., and Bitoon, M. A. (1971). Isolation and purification of a peptido-rhamnomannan from the yeast form of *Sporothrix schenckii:* Structural and immunochemical studies. *J. Immunol.* **107**, 663–671.
Mahvi, T. A. (1965). A comparative study of the yeast and mycelial phases of *Histoplasma capsulatum*. I. Pathways of carbohydrate dissimilation. *J. Infect. Dis.* **115**, 226–232.
Malcolm, G. B., Pine, L., Cherniak, R., and Moss, W. (1979). Biochemical and serological characteristics of soluble yeast phase antigens of *Histoplasma capsulatum*. *Mycopathologia* **67**, 3–16.
Maresca, B., Jacobson, E., Medoff, G., and Kobayashi, G. (1978). Cystine reductase in the dimorphic fungus *Histoplasma capsulatum*. *J. Bacteriol.* **135**, 987–992.
Mariat, F. (1975). Observations sur l'écologie de *Sporothrix schenckii* et de *Ceratocystis stenoceras* en Corse et en Alsance, provinces Français indemnes de sporotrichose. *Sabouraudia* **13**, 217–225.
Marriot, M. S. (1975). Isolation and chemical characterization of plasma membranes from the yeast and mycelial forms of *Candida albicans*. *J. Gen. Microbiol.* **86**, 115–132.
Matile, P., Cortat, M., Wiemken, A., and Frey-Wyssling, A. (1971). Isolation of glucanase-containing particles from budding *Saccharomyces cerevisiae*. *Proc. Natl. Acad. Sci. U.S.A.* **68**, 636–640.
Mendonça, L., Gorin, P. A., Lloyd, K. O., and Travassos, L. R. (1976). Polymorphism of *Sporothrix schenckii* surface polysaccharides as a function of morphological differentiation. *Biochemistry* **15**, 2423–2431.

Mitchell, L. H., and Soll, D. R. (1979). Commitment to germ tube or bud formation during release from stationary phase in *Candida albicans*. *Exp. Cell Res.* **120**, 167-179.

Moens, P. B., and Rapport, E. (1971). Spindles, spindle plaques and meiosis in the yeast *Saccharomyces cerevisiae* (Hansen). *J. Cell Biol.* **50**, 344-361.

Mooney, D. T., and Sypherd, P. S. (1976). Volatile factor involved in the dimorphism of *Mucor racemosus*. *J. Bacteriol.* **126**, 1266-1270.

Moor, H. (1967). Endoplasmic reticulum as the initiator of bud formation in yeast. *Arch. Mikrobiol.* **57**, 135-146.

Moore, M. (1932). Coccidioidal granuloma: A classification of the causative agent, *Coccidioides immitis*. *Ann. Missouri Bot. Gdn.* **19**, 397-428.

Moreno, R. E., Kanetsuna, F., and Carbonell, L. M. (1969). Isolation of chitin and glucan from the cell wall of the yeast form of *Paracoccidioides brasiliensis*. *Arch. Biochem. Biophys.* **130**, 212-217.

Najim, L., and Turian, G. (1979). Ultrastructure de l'hyphe végétatif de *Sclerotinia fructigena*. *Can. J. Bot.* **57**, 1299-1313.

Nickerson, W. J. (1953). Reduction of inorganic substances by yeasts. I. Extracellular reduction of sulphite by species of *Candida*. *J. Infect. Dis.* **93**, 43-56.

Nickerson, W. J. (1963). Molecular bases of form in yeasts. *Bacteriol. Rev.* **27**, 305-324.

Nickerson, W. J., and Edwards, G. A. (1949). Studies on the physiological bases of morphogenesis in fungi. I. The respiratory metabolism of dimorphic pathogenic fungi. *J. Gen. Physiol.* **33**, 41-55.

Nickerson, W. J., and Falcone, G. (1956). Identification of protein disulfide reductase as a cellular division enzyme in yeast. *Science* **124**, 722-723.

Nickerson, W. J., and Mankowski, Z. (1953). Role of nutrition in the yeast-shape in *Candida*. *Am. J. Bot.* **40**, 584-592.

Odds, F. C. (1979). "Candida and Candidosis." University Park Press, Baltimore, Maryland.

O'Hern, E. M., and Henry, B. S. (1956). A cytological study of *Coccidioides immitis* by electron microscopy. *J. Bacteriol.* **72**, 632-645.

Oujezdsky, K. B., Grove, S. N., and Szaniszlo, P. J. (1973). Morphological and structural changes during the yeast-to-mold conversion of *Phialophora dermatitidis*. *J. Bacteriol.* **113**, 468-477.

Paveto, C., Epstein, A., and Passeron, A. (1975) Studies on cyclic adenosine 3'-5'-monophosphate levels and adenylate cyclase and phosphodiesterase activities in the dimorphic fungus *Mucor rouxii*. *Arch. Biochem. Biophys.* **169**, 449-457.

Pickett-Heaps, J. D. (1969). The evolution of the mitotic apparatus: An attempt at comparative ultrastructural cytology in dividing plant cells. *Cytobios* **1**, 257-280.

Pine, L., and Peacock, C. L. (1958). Studies on the growth of *Histoplasma capsulatum*. IV. Factors influencing conversion of the mycelial phase to the yeast phase. *J. Bacteriol.* **75**, 167-174.

Previato, J. O., Gorin, P. A. J., and Travassos, L. R. (1979). Cell wall composition in different cell types of the dimorphic species of *Sporothrix schenckii*. *Exp. Mycol.* **3**, 83-91.

Ramirez-Martinez, J. R. (1970). Growth curves and nucleic acid content of mycelial and yeast-like forms of *Paracoccidioides brasiliensis*. *Mycopathol. Mycol.* **41**, 203-210.

Reiss, E., and Nickerson, W. J. (1974). Control of dimorphism in *Phialophora verrucosa*. *Sabouraudia* **12**, 202-213.

Rippon, J. W. (1968). Monitored environmental system to control cell growth, morphology, and metabolic rate in fungi by oxidation-reduction potentials. *Appl. Microbiol.* **16**, 114-121.

Robinow, C. F., and Bakerspiegel, A. (1965). Somatic nuclei and forms of mitosis in fungi. *In* "The Fungi" (G. C. Ainsworth and A. S. Sussman, eds.), Vol. I, pp. 119-142. Academic Press, New York.

Robinow, C. F., and Marak, J. (1966). A fiber apparatus in the nucleus of the yeast cell. *J. Cell Biol.* **29**, 129-151.

Romano, A. H. (1966). Dimorphism. *In* "The Fungi" (G. C. Ainsworth and A. S. Sussman, eds.), Vol. 2, pp. 181-209. Academic Press, New York.

Romano, A. H., and Nickerson, J. W. (1954). Cystine reductase of pea seeds and yeast. *J. Biol. Chem.* **208**, 409-416.

San-Blas, F., San-Blas, G., and Cova, L. J. (1976) A morphological mutant of *Paracoccidioides brasiliensis* strain IVIC Pb9: Isolation and wall characterization. *J. Gen. Microbiol.* **93**, 209-218.

San-Blas, G. (1979). Biosynthesis by subcellular fractions of *Paracoccidioides brasiliensis. Exp. Mycol.* **3**, 249-258.

San-Blas, G., and Carbonell, L. M. (1974). Chemical and ultrastructural studies on the cell walls of the yeastlike and mycelial forms of *Histoplasma farciminosum. J. Bacteriol.* **119**, 602-611.

San-Blas, G., and San-Blas, F. (1977). *Paracoccidioides brasiliensis:* Cell wall structure and virulence. *Mycopathologia* **62**, 77-86.

Scott, W. A., and Solomon, B. (1975). Adenosine 3'-5'-cyclic monophosphate and morphology in *Neurospora crassa:* Drug induced alterations. *J. Bacteriol.* **122**, 454-463.

Scott, W. A., Mishra, N. C., and Tatum, E. L. (1973). Biochemical genetics of morphogenesis in *Neurospora. Brookhaven Symp. Biol.* **25**, 1-18.

Sentandreu, R., and Northcote, D. H. (1969). The formation of buds in yeast. *J. Gen. Microbiol.* **55**, 393-398.

Shannon, J. L., and Rothman, A. H. (1971). Transverse septum formation in budding cells of the yeastlike fungus *Candida albicans. J. Bacteriol.* **106**, 1026-1028.

Sietsma, J. H., and Wessels, J. G. H. (1979). Evidence for covalent linkages between chitin and β-glucan in a fungal wall. *J. Gen. Microbiol.* **114**, 99-108.

Stewart, P. R., and Rogers, P. J. (1978). Fungal dimorphism: A particular expression of cell wall morphogenesis. *In* "The Filamentous Fungi" (J. E. Smith and D. R. Berry, eds.), Vol. 3, pp. 164-196. Wiley, New York.

Streiblová, E. (1971). Cell division in yeasts. *Proc. Symp. Int. Congr. Microbiol., 10th, 1970* pp. 131-140.

Streiblová, E., and Beran, K. (1963). Demonstration of yeast scars by fluorescence microscopy. *Exp. Cell Res.* **30**, 603-605.

Sun, S. H., and Huppert, M. (1976). A cytological study of morphogenesis in *Coccidioides immitis. Sabouraudia* **14**, 185-198.

Sypherd, P. S., Orlowski, M., and Peters, J. (1979). Models of fungal dimorphism: Control of dimorphism in *Mucor racemosus. In* "Microbiology-1979" (D. Schlessinger, ed.), pp. 224-227. Am. Soc. Microbiol., Washington, D.C.

Taschdjian, C. L., and Kozinn, P. J. (1961). Metabolic studies on the tissue phase of *Candida albicans* induced *in vitro. Sabouraudia* **1**, 73-82.

Taylor, J. J. (1961). Nucleic acids and dimorphism in *Blastomyces. Exp. Cell Res.* **24**, 155-158.

Tkacz, J. S., and Lampen, J. O. (1972). Wall replication in *Saccharomyces* species: Use of fluorescein-conjugated concanavalin A to reveal the site of mannan insertion. *J. Gen. Microbiol.* **72**, 243-247.

Tkacz, J. S., Cybulska, E. B., and Lampen, J. O. (1971). Specific staining of wall mannan in yeast cells with fluorescein-conjugated concanavalin A. *J. Bacteriol.* **105**, 1-5.

Travassos, L. R., Sousa, W., Mendonca-Previato, L., and Lloyd, K. O. (1977). Location and biochemical nature of surface components reacting with concanavalin A in different cell types of *Sporothrix schenckii. Exp. Mycol.* **1**, 293-305.

Trinci, A. P. J., and Collinge, A. J. (1974). Occlusion of the septal pore of damaged hyphae of *Neurospora crassa* by hexagonal crystals. *Protoplasma* **80**, 57-67.

Tyler, L. J., and Parker, K. G. (1945). Factors affecting the saprogenic activities of the Dutch elm disease pathogen. *Phytopathology* **35**, 675-687.

Vršanská, M., Krátký, Z., Biely, P., and Machala, S. (1979). Chitin structures of the cell walls of synchronously grown virgin cells of *Saccharomyces cerevisiae*. *Z. Allg. Mikrobiol.* **19,** 357–362.

Wain, W. H., Price, M. F., and Cawson, R. A. (1975). A re-evaluation of the effect of cysteine on *Candida albicans*. *Sabouraudia* **13,** 74–82.

Winsten, S., and Murray, T. J. (1956). Virulence enhancement of a filamentous strain of *Candida albicans* after growth on media containing cysteine. *J. Bacteriol.* **71,** 738.

Yamaguchi, H. (1975). Control of dimorphism in *Candida albicans* by zinc: Effect on cell morphology and composition. *J. Gen. Microbiol.* **86,** 370–372.

Zorzopulus, J., Jabbagy, A. J., and Terenzi, H. F. (1973). Effects of ethylenediaminetetracetate and chloramphenicol on mitochondrial activity and morphogenesis in *Mucor rouxii J. Bacteriol.* **115,** 1198–1204.

6

Pleomorphism

J.W. Carmichael

I.	The Assimilative Stage and Dimorphism	135
II.	The Propagative Stage and Pleomorphism	136
III.	Examples of Pleomorphic Anamorphs	136
	A. Pleomorphism with Conidia of the Same Ontogenetic Type	137
	B. Pleomorphism with Conidia of Different Types	137
	C. Pleomorphism of Conidiomata	139
IV.	Advantages of Pleomorphism	139
V.	Pleomorphism and Classification	140
VI.	Pleomorphism and Nomenclature	142
	References	142

I. THE ASSIMILATIVE STAGE AND DIMORPHISM

One of the most striking features of the higher fungi (Ascomycota and Basidiomycota) is the uniformity of their assimilative stages compared to the diversity of their propagative stages. The assimilative stage consists of either mycelium or yeast cells and, for the most part, one hypha or yeast cell looks pretty much like another. The hyphae of higher fungi are all branched and septate. Some Ascomycota have dark-colored (dematiaceous) hyphal walls. Other Ascomycota and most Basidiomycota have hyaline or brightly colored (moniliaceous) hyphal walls. Under the light microscope, it is difficult to tell whether a moniliaceous mycelium belongs to an ascal or a basidial fungus unless it is a dikaryon with clamp connections. However, the hyphae of a few fungi produce distinctive vegetative organs such as haustoria, hyphopodia, trapping rings, penetrating organs, and rhizomorphs.

Some species can convert from mycelial to yeast form in response to their environment. The yeast form, which is both assimilative and propagative, is better adapted to spread through a liquid medium. The filamentous form is better adapted for the rapid colonization of solid substrates. Fungi which can convert from mycelial to yeast form are often called dimorphic fungi (see Cole and Nozawa, this volume, Chapter 5).

II. THE PROPAGATIVE STAGE AND PLEOMORPHISM

The propagative stages of higher fungi are extremely diverse, both in the kinds of spores and fruiting structures produced by different fungi and in the variety of different spores produced during the life cycle of a single species. We know that many conidial fungi are asexual propagative stages (anamorphs) in the life cycles of higher fungi. For other species, no sexual stage (teleomorph) is known, and we presume that at least some of these species have evolved a completely asexual or parasexual life cycle.

In some species, the propagative stage may be represented (as far as is known) by a single type of reproductive structure and spores. For example, only basidiospores are produced by *Agaricus brunnescens,* only ascospores by *Gelasinospora tetrasperma,* and only phialoconidia by *Aspergillus fumigatus.* Such fungi could be called monomorphic in propagation, but the term is rarely used. Other fungi, which produce more than one type of spore, are called pleomorphic (see Hennebert, 1971). For example, the rusts (Uredinales) may produce two kinds of conidia (aeciospores and urediniospores) and three kinds of spores connected with sexual reproduction (spermatia, teliospores, and basidiospores). The cup fungus *Botryotinia fuckeliana* produces one kind of conidia, plus spermatia and ascospores. *Cephalotrichum stemonitis* produces two or three kinds of conidia, but no teleomorph has yet been found. Pleomorphism consisting of separate anamorphic and teleomorphic propagation is considered by Müller (this volume, Chapter 7). In the remainder of this chapter, we will be concerned primarily with anamorphic pleomorphism—pleomorphism involving two or more kinds of asexual propagules.

III. EXAMPLES OF PLEOMORPHIC ANAMORPHS

In order to provide a framework for listing examples, anamorphic pleomorphism will be divided, somewhat arbitrarily, into three kinds: (1) pleomorphism with conidia of the same ontogenetic type, (2) pleomorphism with conidia of different types, and (3) pleomorphism of conidiomata. To make the text more readable, I have left out author citations, figures, and references for the fungi cited as examples. However, this information is readily available in Kendrick and Carmichael (1973) and Carmichael *et al.* (1980). Diagrams and photographs of conidium ontogeny for many of the fungi cited as examples can be found in Cole and Samson (1979).

A. Pleomorphism with Conidia of the Same Ontogenetic Type

Most strains of the gymnoascaceous fungus *Ajellomyces capsulatus* produce small, smooth aleurioconidia and large, tuberculate aleurioconidia, both of which are roughly globose in shape. Some strains produce only the large, tuberculate conidia. This species can also grow as an intracellular yeast in humans and animals (the *Histoplasma capsulatum* state). Most species of the dermatophytes *Trichophyton* and *Microsporum* produce small, clavate to globose microconidia and large, clavate to fusiform, multiseptate macroconidia, both of which have aleuric dehiscence. In *T. mentagrophytes* and *T. rubrum*, the macroconidia are lacking in many strains. In *T. ajelloi*, the microconidia are often lacking. In the *T. terrestre* group, the large, septate conidia and small, one-celled conidia are accompanied by a complete range of intermediates. In most dermatophytes, the macroconidia and microconidia are quite distinct and no intermediates are seen. In addition to the micro- and macroconidia, the dermatophytes can propagate by arthroconidia formed in keratinized host tissue. In *M. audouinii* and *T. schoenleinii* the ability to produce micro- and macroconidia is limited, and arthroconidia are the most abundant propagules.

Many species of *Fusarium* also have micro- and macroconidia, but here the macroconidia are slimy phialoconidia. The microconidia of some species are also phialoconidia, but in other species they seem to be solitary.

In *Dwayabeeja sundara*, two kinds of septate conidia are produced from specialized *Torula*-like conidiogenous cells: ellipsoidal phragmoconidia and distinctive scolecoconidia that taper smoothly to the distal end.

In *Triadelphia heterospora*, there are large, beaked phragmoconidia borne at the tip of scarcely modified hyphae, and smaller, ellipsoidal phragmoconidia borne singly on somewhat differentiated disjunctor cells. Both kinds of conidia have dark bands around them.

Most conidial fungi produce spores in either powdery or slimy masses. A few phialidic species are pleomorphic in this regard; for example, *Acremonium alternatum*, *Stachybotrys echinata*, and *Gliocladium roseum*. In *Scopulariopsis brevicaulis*, the conidia are usually formed in long, dry chains from the annellate conidiogenous cells, but under moist conditions they sometimes collapse into slimy balls.

B. Pleomorphism with Conidia of Different Types

One of the most common kinds of anamorphic pleomorphism is the production of small, single-celled, phialidic conidia plus larger, thick-walled, solitary conidia which may be septate and pigmented. For example, several *Sepedonium* species have large. globose, tuberculate aleurioconidia plus small, slimy, hyaline phialoconidia borne on *Acremonium-* or *Verticillium*-like conidiophores. *Mycogone* species may have similar tuberculate macroconidia and slimy phialoconidia. *Sepedonium ampullasporum*, however, has unusual, solitary bottle-shaped microconidia borne terminally on penicillate conidiophores. The type species of *Chlamydomyces* and *Harzia* bear their phialidic microconidia on inflated conidiophores of the *Harziella* type.

A few species with *Chalara*-type phialides produce dark, thick-walled macroconidia. In conidial *Ceratocystis paradoxa*, the dark conidia are unicellular and catenate. This combination of macroconidia and *Chalara* conidia was the basis of the anamorph-genus *Thielaviopsis*. Another fungus, later added as *T. basicola*, has *Chalara*-type phialoconidia plus solitary, dark phragmoconidia borne on cymosely branched hyphae. The genus *Chalaropsis* was proposed for a species with *Chalara* phialides and solitary, dark ameroconidia. In *Chalaropsis punctulata*, the dark ameroconidia have a germ slit.

Hyaline phialoconidia of the *Verticillium* or *Paecilomyces* type are associated with large, thick-walled, hyaline dictyoconidia in *Diheterospora*. Similar dictyoconidia or phragmoconidia are associated with phialidic phragmoconidia in some *Cylindrocarpon* species and in the type species of *Stemphyliopsis* Smith.

Small hyaline phialoconidia of the *Acremonium* type are formed in association with larger, aleuric conidia or chlamydospores in species of *Humicola, Botryotrichum, Trichocladium, Desmidiospora*, and other anamorph-genera.

Phialophora- or *Myrioconium*-type conidia are formed in addition to larger, blastic conidia in *Dimorphospora, Botrytis,* and *Monilia*.

Dark dictyoconidia are associated with slimy, hyaline phialoconidia in *Gliocephalotrichum bulbilium, Septosporium bulbotrichum*, and some *Papulaspora* species, although these fungi are otherwise very different from each other. Some species of *Phoma* also produce chains of dark dictyoconidia (*Peyronelia* anamorph) in addition to the ameroconidia formed in pycnidial conidiomata.

Less common are pleomorphic anamorphs where neither the macro- nor microconidia are phialidic. In *Selenosporella*, somewhat slimy ameroconidia are produced from numerous minute denticles on the conidiogenous cell. It is not certain whether these should be considered polyphialides. In some species they are associated with large, dark *Endophragmia* or *Endophragmiella* conidia. Since *Selenosporella* species often parasitize other fungi, caution is necessary in accepting other spore forms as being genetically connected. In *Sympodiophora stereicola*, didymoconidia borne singly on phialide-like branches are accompanied by thick-walled phragmoconidia with aleuric dehiscence.

In *Cephalotrichum stemonitis* the dark synnemata produce ameroconidia from annellidic conidiogenous cells of the *Scopulariopsis* type. In some strains this state may be accompanied by larger, rough, solitary conidia called *Echinobot-*

ryum atrum. In *Wardomyces dimerus,* two-celled *Scopulariopsis*-type conidia are accompanied by *Wardomyces* conidia.

C. Pleomorphism of Conidiomata

Some fungi can produce their conidia either diffusely on the hyphae, or on or in a fruiting body (conidioma). Even though conidium morphology and ontogeny are not altered, the general appearance of the sporulating fungus may be markedly different.

Fungi that are synnematous on their natural substrate often fail to produce synnemata in culture. Examples with different types of conidium ontogeny are species of *Penicillium* (dry, blastic-phialidic conidia), *Coremiella cuboidea* (alternate thallic-arthric conidia), *Scopulariopsis putredinis* (dry, blastic-annellidic conidia), *Beauveria densa* (dry, blastic-sympodial conidia), and the *Graphium* state of *Petriellidium boydii* (slimy, blastic-annellidic conidia).

Fungi that form sporodochial or acervular conidiomata on their natural substrate often fail to sporulate in culture on agar media, but a few produce a diffuse conidial state or produce conidia from scarcely differentiated mats of hyphae. Examples are found in the anamorph-genera *Fusarium, Pestalotia,* and *Epicoccum.* In some *Volutella* species a *Verticillium-* or *Acremonium*-like diffuse conidial state may be present in addition to the sporodochia.

Pycnidial fungi generally will not produce their conidia without a pycnidium, but a few species of *Phoma,* when cultured on agar, will produce masses of slimy pink conidia directly from the assimilative hyphae. The *Aureobasidium* and *Kabatiella* anamorphs of some *Guignardia* and *Dothidea* species are other examples.

IV. ADVANTAGES OF PLEOMORPHISM

In his fascinating book, *The Advance of the Fungi,* E. C. Large (1940) stresses the point that the life cycle of plant-pathogenic fungi often includes a resistant teleomorph that serves as the overwintering stage and a more ephemeral anamorph that produces numerous small, propagative spores in the summer. Pleoanamorphic fungi do not usually show such seasonal succession, but no doubt the kinds of conidia they produce reflect the same compromises relating to the quantity of propagules that can be produced versus their persistence of effectiveness. In the vast majority of cases of pleoanamorphy, one kind of conidium is small and numerous, while the other kind(s) is less numerous and larger, and also often thick-walled, multicellular, and pigmented. In the few exceptions that come to mind (*Triadelphia, Dwayabeeja*), both kinds of conidia are large, dark, and septate.

The conidia of most fungi are either dry and adapted to dispersion by air currents or slimy and adapted to dispersion by rain or insects. Some chlamydospores and aleurioconidia are not readily dispersed in the air, even though they are not obviously slimy. Probably some of these are hitchhikers on insects, mites, and other animals. The so-called amphibious fungi usually have solitary blastic conidia that project in three dimensions by lobes, arms, or twists. There are few examples of pleomorphic aquatics (*Dimorphospora*), but we do not have much knowledge of their life cycles as yet. I mentioned earlier a few examples of fungi with facultatively slimy conidia. However, I can think of even fewer instances where a pleoanamorph produces two kinds of conidia, one in slime and the other adapted for aerial dispersion. The ballistospores and blastospores of *Sporobolomyces* are one exception. The *Scytalidium* state of *Monochaetia* may be another, although it is not certain that the arthroconidia are readily airborne. On the other hand, there are many examples of species that produce slimy spermatia and airborne conidia. These include the spermatial and aecial states of Uredinales, the *Myrioconium* and *Botrytis* or *Monilia* anamorphs of *Botryotinia* and *Monilinia*, the *Harziella* and *Harzia* or *Chlamydomyces* anamorphs of *Melanospora*, and the *Acremonium* and *Humicola* or *Botryotrichum* anamorphs of *Chaetomium*.

The distinction between spermatia and conidia is not always sharp. In some of the above examples spermatia can also germinate to form new colonies, and there are many cases where conidia or conidium-like propagules can effect fertilization. The oidia (arthroconidia) produced by some Hymenomycetes and some heterothallic species of *Chaetomium* and *Myxotrichum* can also serve both functions.

V. PLEOMORPHISM AND CLASSIFICATION

The pleomorphism of higher fungi has long caused problems for taxonomists. These problems are of two basic kinds: first, how to identify an anamorphic state as belonging to a particular teleomorphic species, and second, where to classify the many species that have no known teleomorphic state. These two problems have been "solved" by treating conidial anamorphs as if they belonged to a separate higher taxon of the fungi: the Deuteromycota or Fungi Imperfecti. This solution violates one of the cardinal rules of biological classification: that each taxon can belong to only one taxon of the next higher rank. To avoid this anomaly, the Fungi Imperfecti is called a form-class (or anamorph-class), and the genera based on anamorphs are called form-genera (or anamorph-genera).

Anamorph-genera are convenient pigeonholes for morphologically similar anamorphs. They do not necessarily correspond to genera in the Ascomycota and Basidiomycota. For example, the conidial anamorphs of *Ajellomyces cap-*

sulatus, Renispora flavissima, Corynascus sepedonium, Ctenomyces serratus, and *Arthroderma tuberculatum* are all so similar that they can easily be accommodated in the anamorph-genus *Chrysosporium*. On the other hand, the conidial anamorphs of *Arthroderma curreyi, A. benhamiae,* and *A. tuberculatum* appear very different from each other (although the microconidia of *A. benhamiae* are similar to the only conidia of *A. curreyi*). Similar noncorrespondences are to be found among the anamorphs of *Ceratocystis, Mycosphaerella, Nectria, Ophiostoma,* and a few other genera. It does not appear likely that anamorph-genera can ever be made to correspond exactly to the genera of higher fungi.

Because of the difficulty in identifying anamorphic states with the appropriate teleomorphic species, it has been suggested that all species included in anamorph-genera are anamorph-species. For example, it has been shown that *Cochliobolus lunatus* has a conidial anamorph indistinguishable from the well-known *Curvularia lunata*. There is no proof as yet that all strains of *Curvularia lunata* are anamorphs of *Cochliobolus lunatus*. Therefore, some workers would regard the name *Curvularia lunata* as applying to an anamorph-species that might consist of two or more biological species including the *Curvularia* state of *Cochliobolus lunatus*. Presumably they would also regard any anamorphic holomorphs as anamorph-species, each consisting of an unknown number of biological species. This is one way of solving the species identification problem, but it does not seem appropriate to me to knowingly apply a species name to a mixture of different species without at least adding some qualifier such as "series," "group," or *"sensu lato"* to indicate the uncertainty. I am equally reluctant to accept the proposition that it is a practical impossibility to distinguish one species from another in the absence of a teleomorphic state.

There are two additional problems with anamorph-genera. The first is whether to recognize double anamorph-genera, that is, anamorph-genera based on the co-occurrence of two different anamorphs. A few of these have been proposed (e.g., *Thielaviopsis, Chalaropsis, Diheterospora, Dimorphospora, Triadelphia,* and *Fonsecaea*), but mycologists have generally been reluctant to accept them unless there is no existing anamorph-genus for at least one of the anamorphs. My own opinion is that double anamorph-genera are unnecessary and undesirable. Most of the existing ones can be "lecto-typified" by one of the anamorphs and thus either preserved as single anamorph-genera or reduced to synonyms. The second additional problem is whether to retain anamorph-genera based on bulbils, sclerotia, rhizomorphs, and other nonconidial structures. This practice has a long history and serves the purpose of permitting a species name to be created for a fungus with no known conidia or teleomorph. However, there does not seem to be much point in attaching a binomial to a mycelium with so few distinguishing features that it would be impossible to identify another example of the same species.

VI. PLEOMORPHISM AND NOMENCLATURE

Having a separate cross-classification for the anamorphs of higher fungi means that a single fungus can have one species name in a genus of the Ascomycota or Basidiomycota and one or more additional species names in anamorph-genera of the Fungi Imperfecti. These different names for the same species are not taxonomic synonyms, however, since the name based on the teleomorph may be applied to the whole fungus (holomorph), while the names based on anamorphs apply only to the anamorphic state denoted by the anamorph-generic name. The International Code of Botanical Nomenclature provides that epithets proposed for anamorphs are not transferrable to genera based on teleomorphs, and that anamorphic names do not affect the priority of teleomorphic names (see Hennebert and Weresub, 1977). During the last decade, mycologists have increasingly come to favor the use of cross-reference names to refer to the anamorphs of species with known teleomorphs (see Kendrick and Carmichael, 1973). For example, the anamorphic name *Monilia sitophila* is now usually replaced by the cross-reference name *Monilia* anamorph of *Neurospora sitophila* or *Neurospora sitophila* anam. *Monilia*.

With pleoanamorphic holomorphs, the situation is less settled. The code permits a separate, valid binomial for each anamorph of a species. I recently proposed (Carmichael, 1979) that each recognized species be limited to one valid binomial and that pleoanamorphic holomorphs also be dealt with by the use of cross-reference names. For example, if they are regarded as anamorphs of the same species, then the name *Echinobotryum atrum* would be treated as a synonym of the cross-reference name *Echinobotryum* state of *Cephalotrichum stemonitis*. If (contrary to fact) *E. atrum* had been described before *C. stemonitis*, then *C. stemonitis* would be treated as a synonym of the *Cephalotrichum* state of *E. atrum*. This proposal has met with considerable resistance, both because of its novelty and because cross-reference names are somewhat cumbersome. However, this system does provide a solution for the nomenclatural problems associated with pleoanamorphy. Now that I have used cross-reference names for a while in my own work, I find it difficult to imagine getting along without them.

REFERENCES

Carmichael, J. W. (1979). Cross-reference names for pleomorphic fungi. *In* "The Whole Fungus" (W. B. Kendrick, ed.), pp. 31–41. National Museums of Canada, Ottawa.

Carmichael, J. W., Kendrick, W. B., Conners, I. L., and Sigler, L. (1980). "Genera of Hyphomycetes." University of Alberta Press, Edmonton.

Cole, G. T., and Samson, R. A. (1979). "Patterns of Development in Conidial Fungi." Pitman, London.

Hennebert, G. L. (1971). Pleomorphism in Fungi Imperfecti. *In* "Taxonomy of Fungi Imperfecti" (W. B. Kendrick, ed.), pp. 202-223. Univ. of Toronto Press, Toronto.

Hennebert, G. L., and and Weresub, L. K. (1977). Terms for states and forms of fungi, their names and types. *Mycotaxon* **6,** 207-211.

Kendrick, W. B., and Carmichael, J. W. (1973). Hyphomycetes. *In* "The Fungi" (G. C. Ainsworth, F. K. Sparrow, and A. S. Sussman, eds.), Vol. 4A, pp. 323-509. Academic Press, New York.

Large, E. C. (1940). "The Advance of the Fungi." Jonathan Cape (Dover reprint, 1962, now available).

7

Relations between Conidial Anamorphs and Their Teleomorphs

Emil Müller

I.	General Aspects	145
	A. Introduction	145
	B. Range and Limits	146
	C. Similarities and Dissimilarities	147
	D. Sequence of States	150
II.	Significance of Teleomorph–Anamorph Connections for Systematics	153
	A. Differentiation of Species	153
	B. Differentiation of Genera	160
	C. Differentiation of Higher Taxa	164
III.	Conclusions	165
	References	165

I. GENERAL ASPECTS

A. Introduction

A still unknown number of imperfect fungi (Fungi Imperfecti, Deuteromycetes) do not represent independent organisms but are states of Ascomycetes and Basidiomycetes. Considering the limited information on teleomorph-anamorph connections within these groups, and the enormous morphological variation within and between perfect states (teleomorphs) and imperfect states (anamorphs), we are still forced to maintain two different systems for their classification: one for the perfect states (Ascomycetes and Basidiomycetes), arranged according to the development and the morphology of their sexual fructifications, and a second for the imperfect fungi including the anamorphs of Ascomycetes and Basidiomycetes, based on their asexual fructifications. The nomenclatural difficulties and consequences arising from this situation have been discussed by Hennebert and Weresub (1977).

In most cases the proof that teleomorphs and conidial anamorphs are phases of the same life cycle can only be obtained by cultural work. Since many teleomorphs form only occasionally in pure culture, we usually start our cultural experiments with ascospores. If even in pure culture no anamorphs are formed, environmental conditions should be changed to ensure that the cultural methods being employed do not block the formation of conidia (compare Müller, 1979). Our present consideration therefore must be restricted to connections between teleomorphs and anamorphs proved either by cultural work or—as for obligate parasites—inoculation experiments. Connections not strictly proven may be taken into account if observed regularly. Anamorphs have been positively connected with relatively few of the known species of Ascomycetes and Basidiomycetes. In certain groups, however, enough connections are known for a pattern to be evident, e.g. Erysiphales (Spencer, 1978), Trichocomaceae (Malloch and Cain, 1972), nectrioid fungi (Samuels and Rossman, 1979) within the Ascomycetes, and Uredinales within the Basidiomycetes. For most other groups our information is incomplete. The systematic search for new connections must therefore be considered one of the most important fields of experimental fungal taxonomy. Known connections dramatically increase our basic information about relationships. As members of fungal groups suggested to be related are found to have more features in common, the evidence for their relationship becomes more convincing.

A few attempts were made to link the system for teleomorphs with that for anamorphs (Tubaki, 1958; Müller, 1971) at the ordinal level. These attempts failed, and only a few general rules could be formulated. De Hoog (1977) tried to approach the problem with the aid of cluster analysis. He demonstrated that within a given arrangement of Ascomycetes there was a significant variation of similarities in anamorphs. Connecting the two systems as proposed by Kendrick (editor) and co-workers (1979) is based not on orders but on lower taxa.

B. Range and Limits

Teleomorph–anamorph connections are widespread in fungi, but we have restricted our consideration here to Ascomycetes and Basidiomycetes and their anamorphs. The discussion concerns not only the kinds and the variation of both sexual and asexual forms but also the presence and absence of anamorphs within individual life cycles. We should also remember that ascomycetous anamorphs always occur in the haploid phase, whereas in Basidiomycetes either the haploid or the dikaryotic mycelium may produce conidia; it has even been found that the same fungus can produce conidia in both phases.

In some groups conidium-like spores may function as male gametes (spermatia; either exclusively so or in combination with their function as asexual propagules of Ascomycetes and Basidiomycetes). As our knowledge of the role of such structures is often incomplete, we must include them in our survey.

In the Heterobasidiomycetes (Donk, 1966) the formation of conidia directly from basidiospores is common to all members of the group. Within Ascomycetes this kind of germination behavior is also widespread, but it has never been used for the delineation of larger groups. These "secondary spores" are mostly formed by budding (Heterobasidiomycetes) or on phialides as in *Tympanis* (Ouellette and Pirozynski, 1974), *Nectria* (Booth, 1959; Samuels, 1976), and *Tapesia* (Aebi, 1972). Scattered conidial structures may also occur in the ascomata of *Chaetoscutula juniperi* (Müller, 1958), or even at the tip of paraphyses (phialoconidia in *Cashiella fuscidula,* Müller, 1977). Such anamorphs are usually not named and are therefore not considered within the system of imperfect fungi.

An even closer connection between teleomorph and anamorph in rust fungi has been found by Hiratsuka (1973) for *Gymnosporangium gaeumanni* fa. *albertensis*. Most species of *Gymnosporangium* occurring on Rosaceae in the haploid phase and on Cupressoideae in the dikaryotic phase are opsis-forms (Kern, 1973) which do not form the uredinial anamorph. *Gymnosporangium gaeumanni,* however, is an eu-form. Unicellular urediniospores develop within the same sori as the bicellular teliospores, as in many other rust fungi, but *G. gaeumanni* also forms chimeric bicellular spores with one cell anamorphic (germinating by a germ tube) and the other cell teleomorphic (forming a basidium).

C. Similarities and Dissimilarities

Anamorphs and teleomorphs undergo completely different developments, and there is no reason to expect that these two states should look the same. This is true for the majority of Ascomycetes and Basidiomycetes. However, we know many cases in which at least some characters of the teleomorph and the corresponding anamorph are alike. Similarities may be found in the form of fructification and in the spore morphology. Similarities of ascomata (and basidiomata) to conidiomata may be expected if they begin to develop before formation of the gametangia. If fertilization is suppressed for any reason, conidia may develop in the previously formed fruit bodies. This is true for *Leptosphaeria doliolum* (Lucas and Webster, 1967) whose ascomata and pycnidia (*Phoma* sp., Boerema, 1976) are identical in form, size, ostiolum, and wall texture. The behavior of other *Leptosphaeria* species, e.g., *L. macrospora* and its *Rhabdospora* anamorph (Müller, 1953), is similar to that of *L. doliolum.* Apothecia and apothecioid conidiomata of *Heterosphaeria* species and their *Heteropatella* conidial anamorphs are often macroscopically indistinguishable. We may even observe first the development of conidia and later the replacement of the conidial layer by a hymenium with asci and paraphyses.

Ascosporal and conidial characters may also be strikingly similar. A resemblance among unicellular asco- and basidiospores and unicellular conidia is probably not very significant. Therefore our examples are selected from species having peculiar spore characteris, such as the ascomycetous genera *Discostroma* and *Broomella* (Amphisphaeriaceae), *Melanochaeta* (Sphaeriaceae), *Durandiella* (Leotiaceae), and *Phaeosphaeria* and *Pleospora* (Pleosporaceae). The anamorphs of *Discostroma* (Brockmann, 1976), belonging to the form-genera *Seimatosporium* (conidia with hyaline, filiform appendages) or *Sporocadus* (conidia without appendages) have conidia which may be distinguished from corresponding ascospores only by the thick-walled septa or by the less obvious septal pores (e.g., *D. massarina*, *D. saccardiana*). In *Melanochaeta hemipsila* (Müller et al., 1969) with the anamorph *Sporoschisma saccardoi,* and in *Chaetosphaerella phaeostroma* (Müller and Booth, 1972), ascospores and conidia (*Oedemium*) are transversely septate with brown central cells and paler end cells. Ascospores in both species, however, tend to be more slender or fusoid than the conidia, which are cylindrical or ellipsoid. A strict similarity in form and septation is also found in several species of *Durandiella* (Groves, 1954) whose ascospores and conidia are both long, fusoid or filiform, and transversely pluriseptate.

A most striking likeness occurs in the four known species of *Broomella,* all found on twigs of *Clematis* spp. growing on different continents. As in *Melanochaeta* and *Chaetosphaerella,* both the ascospores and the conidia of the *Pestalotia* anamorph are transversely septate, and the central cells are darker than the end cells. In addition, in both morphs the spores are provided with simple or branched, filiform, colorless appendages (Shoemaker and Müller, 1963). In *Broomella vitalbae* ascospores and conidia are almost identical, differing only in that the conidia often have a shorter, laterally attached, basal appendage. For *Phaeosphaeria herpotricha,* with long, filiform, transversely pluriseptate ascospores, Webster and Hudson (1957) reported a *Rhabdospora*-like anamorph with similarly long, slightly clavate, almost filiform, transversely septate conidia. *Pleospora herbarum,* characterized by brown, dictyosporous ascospores, also produces the muriform, brown, somewhat smaller conidia of the *Stemphylium* anamorph (e.g., Simmons, 1969).

Germ slits or germ pores are not as common in conidia as in ascospores, and we know of no case in which a germ slit is present in both ascospores and conidia. Only the recently described *Porosphaeria sporoschismoides* (Samuels and Müller, 1979a) has ascospores and conidia with germ pores at each end; the conidia are otherwise endophialidic and similar to the ones formed in *Sporoschisma* (examples of similarities are illustrated in Fig. 1).

Germ pores, however, are widespread in rust fungi and often present in the anamorphic urediniospores as well as in the probasidial teliospores, e.g., in

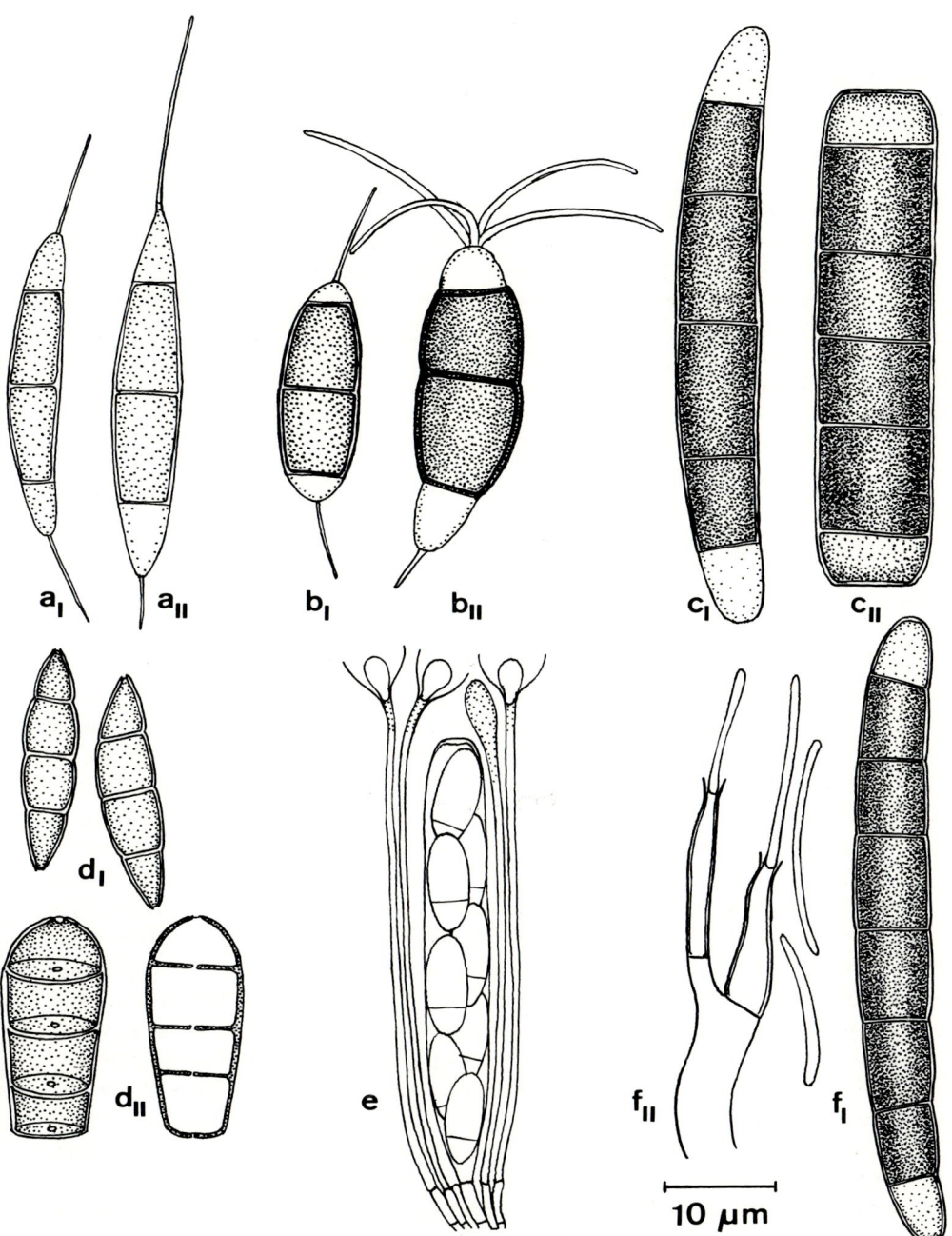

Fig. 1. (a–d) Examples of similar ascospores (a_1, b_1, c_1, and d_1) and conidia (a_{II}, b_{II}, c_{II}, and d_{II}). (a) *Broomella vitalbae*; (b) *Broomella montaniensis*; (c) *Melanochaeta hemipsila*; (d) *Porosphaeria sporoschismoides*; (e) ascus and paraphyses of *Cashiella fuscidula* with paraphyses being phialides; (f) *Melanamphora spinifera* (f_1 ascospore, f_{II} phialidic conidium formation) with ascospores similar to those of *Melanochaeta* but conidia obviously dissimilar.

Puccinia graminis, characterized by urediniospores with three to four germ pores cruciately arranged along the equatorial plane, and by teliospores with one germ pore in each cell.

Often ascospores and conidia do not have the same pigmentation, but a peculiar similarity is characteristic of a number of species belonging to the ascomycetous family Stigmateaceae (=Venturiaceae). These fungi, which have bicellular ascospores, tend to produce an olive-green, olive-gray, or olive-brown pigment in the walls of both ascospores and conidia, the latter belonging to the form-genera *Fusicladium, Karakulinia* (a segregate of *Cladosporium*), *Pollaccia,* and *Spilocaea.*

D. Sequence of States

Anamorphs and teleomorphs which commonly develop simultaneously are comparatively rare. In most cases a definite succession may be observed with the teleomorph following the anamorph(s):

$$\text{Spores} \to \text{Thallus} \to \text{Anamorph(s)} \to \text{Teleomorph}$$

The sequential occurrence of morphs seems to be logical, because anamorphs are simpler and thus develop more rapidly than teleomorphs. Simultaneous development may take place in a number of bitunicate Ascomycetes, pycnidia of the anamorph and ascomata of the teleomorph being morphologically similar. This may be true for a number of *Leptosphaeria* species, e.g., *L. doliolum* with a *Phoma* anamorph (Section I,C) and *L. macrospora* with a *Rhabdospora* anamorph (Table IV), and for *Didymella* species with *Ascochyta* anamorphs. As previously mentioned, the fruit bodies begin their development with a pseudoparenchymatous primordium, and fertilization processes take place only later within these structures. If fertilization is suppressed, the multicellular structures become conidiomata, whereas in *Pyrenophora* species, which only form hyphomycetous *Drechslera* anamorphs, such structures become sclerotia. A succession of anamorph and teleomorph is more typical for both Ascomycetes and Basidiomycetes. Fast-growing saprophytes with immediate sporulation, however, show very little time difference between the maturation of the two states. In many other cases the occurrence of the states is governed by seasonal considerations, most strikingly demonstrated by plant parasites. In temperate zones these parasites often overwinter as slowly developing teleomorphs which mature in early spring so that meiospores will be available at the beginning of the growing season. The developing thallus within or on the infected host then forms the anamorph, its conidia being immediately capable of infecting other host individuals and ensuring spread of the pathogen during the growing season. Some ascomycetous examples are given in Table I.

TABLE I
Seasonal Succession of States of Some Parasitic Ascomycetes

Example	Anamorph				Teleomorph		
	Host	Generic name	Season	Spermatogonia	Ascoma initials	Ascospores	Reference
Venturia inaequalis	*Pirus* or *Sorbus*	*Spilocaea*	Late spring		Autumn	Spring	Aderhold (1897)
Mycosphaerella berberidis	*Berberis*	*Septoria*	Summer	*Asteromella*, autumn	Autumn	Spring	von Arx (1949)
Polystigma rubra	*Prunus*	*Polystigmina*	Summer		Late summer	Autumn, spring	Moreau (1930)
Claviceps purpurea	Gramineae	*Sphacelia*	Summer		Spring	Early summer	Tulasne (1853)
Monilinia fructigena	Rosaceae	*Monilia*	Early summer, autumn	Autumn, spring	Spring	Spring	Aderhold and Ruhland (1905)
Diplocarpon maculatum	Rosaceae	*Entomosporium*	Summer	Autumn	Autumn, winter	Spring	Klebahn (1918)
Rhytisma acerina	*Acer*			*Melasmia*, autumn	Autumn	Spring	Webster (1970)

TABLE II

Seasonal Succession of States in Some Rust Fungi (Uredinales, Basidiomycetes)[a]

Example	Haplophase host	Pycnia	Aecia	Dikaryophase host	Uredinia	Telia	Basidia or basidiospores
Puccinia graminis	*Berberis*	Spring	Spring	Gramineae	Early summer	Late summer	Spring
Uromyces pisi	*Euphorbia*	Spring	Spring	*Pisum*	Early summer	Late summer	Spring
Gymnosporangium sabinae	*Pirus*	Summer	Autumn	*Juniperus*		Early summer	Early summer
Chrysomyxa rhododendri	*Picea*	Early summer	Late summer, autumn	*Rhododendron*	Late spring	Late spring	Late spring

[a] According to Gäumann (1959).

Rust fungi follow a similar pattern with the production of basidiospores in spring, then the anamorphic aecia and uredinia and, toward the end of the growing season, the teleomorphic teliospores, e.g., *Puccinia graminis* on *Berberis* (aecia) and on diverse Gramineae (uredinia and telia). Variations on this theme are not uncommon, e.g., in the genera *Gymnosporangium* and *Chrysomyxa* (Table II).

In the tropics the development of states is often adapted to the succession of dry and wet seasons (Savile, 1976). Under arctic and alpine conditions the whole cycle cannot always be completed during the short summer. Therefore anamorphs and teleomorphs may be found in successive years (Savile, 1972). A succession of morphs which needs more than 1 yr for completion has been observed for *Apiosporina morbosa* occurring in temperate North America on *Prunus* spp. (Luttrell, in Kendrick, 1979). Sometimes teleomorphs are extremely rare, as in some *Cochliobolus* species whose sexual fructifications are known only from laboratory mating tests (Luttrell, 1979).

II. SIGNIFICANCE OF TELEOMORPH–ANAMORPH CONNECTIONS FOR SYSTEMATICS

A. Differentiation of Species

1. Presence or Absence of Anamorphs

Information on the absence of anamorphs may be as helpful in the delimitation of species within fungal groups as information on their presence. In Uredinales, species are commonly characterized not only by their host specialization and host alternation but also by the number and kind of spore forms and the corresponding morphological characteristics. Some examples of closely related rust fungi with differences in the presence of certain fruiting states are given in Table III.

Such differences may also occur in other Basidiomycetes and in Ascomycetes. Unfortunately, for most groups our information is inadequate because many investigators have hesitated to report cases where anamorphs are not formed. A number of species of *Leptosphaeria* (Holm, 1957) were found to produce different kinds of conidial anamorphs: a microconidial state with hyaline, comparatively small amerospores (*Phoma,* Boerema, 1976), and a macroconidial state with cylindrical, fusoid, or almost filiform, hyaline or pigmented, comparatively large phragmospores (belonging to the genera *Stagonospora, Rhabdospora,* and *Camarosporium*). As in the rust fungi, either the microconidial or the macroconidial anamorph, or even both, may be suppressed, as shown in Table IV.

In considering the absence of an anamorph we are confronted with certain problems. Failure to find an anamorph in cultures is ambiguous in that it may

TABLE III

Telemorphs and Anamorphs of Some *Puccinia* Species (Uredinales, Basidiomycetes) on Compositae[a]

Example	Host	Host alternation	Pycnia	Aecidia	Uredinia	Telia	Type of life cycle
Puccinia							
P. aecidii-leucanthemi	*Chrysanthemum or Carex*	+	+	+	+	+	Macrocyclic[b]
P. mulgedii	*Cicerbita*	−	+	+	+	+	Macrocyclic[b]
P. gaeumanni	*Chrysanthemum*	−	−	−	+	+	Brachycyclic[b]
P. tatarica	*Cicerbita*	−	+	+	−	+	Demicyclic[b]
P. leucanthemi	*Chrysanthemum*	−	−	−	−	+	Microcyclic[b]
Uredo							
U. neocomensis	*Chrysanthemum*	−	−	−	+	−	Imperfect rust

[a] According to Gäumann (1959).
[b] Terminology according to Laundon (1967).

TABLE IV
Presence and Absence of Teleomorphs and Anamorphs of Certain *Leptosphaeria* Species (Ascomycetes, Pleosporaceae)

Example	Host	Teleomorph	Anamorph Macroconidial state	Anamorph Microconidial state	Reference
Leptosphaeria					
L. acuta	*Urtica*	+	—	*Phoma*	Müller and Tomasevic (1957)
L. agnita	Compositae	+	—	*Phoma*	Lucas and Webster (1967)
L. doliolum	Plurivorous	+	—	*Phoma*	Lucas and Webster (1967)
L. haematites	*Clematis*	+	—	*Phoma*	Lucas and Webster (1967)
L. maculans	Cruciferae	+	—	*Phoma*	Brefeld (1891), Müller and Tomasevic (1957), Lucas and Webster (1967)
L. millefolium	*Achillaea*	+	*Camarosporium*	*Phoma*	Müller and Tomasevic (1957)
L. ogilviensis	Plurivorous	+	*Camarosporium*	*Phoma*	Müller and Tomasevic (1957)
L. dumentorum	Plurivorous	+	*Stagonospora*	*Phoma*	Webster and Hudson (1957)
L. pratensis	Leguminosae	+	*Stagonospora*	*Phoma*	Lucas and Webster (1967)
L. polygonati	*Polygonatum*	+	*Stagonospora*	—	Müller and Tomasevic (1957)
L. macrospora	Compositae	+	*Rhabdospora*	—	Müller (1953), Lucas and Webster (1967)
L. anemones	*Anemone*	+	*Rhabdospora*	—	Müller (1950)
L. bellynkii	*Polygonatum*	+	—	—	Müller and Tomasevic (1957)
L. fallaciosa	*Satureia*	+	—	—	Müller (1962)
Phoma herbarum	Plurivorous	—	—	*Phoma*	Boerema (1976)

indicate either a genuine lack of conidium production or merely that conidia did not form under the particular conditions imposed on the organisms. But if conidia do not form after repeated attempts under a wide variety of conditions, we can be reasonably sure that the fungus in question has no anamorph.

The significance given to the lack of an anamorph in the life cycle varies from group to group. In rust fungi we tend to consider that morphologically identical teleomorphs differing in the number of connected anamorphs represent independent species. In Ascomycetes similar cases are considered to represent one complex species, e.g., *Mycosphaerella tassiana* which does not produce a *Cladosporium* anamorph under arctic and alpine conditions (von Arx, 1949; Savile, 1972), although one is present in other forms of this heterogeneous species occurring in temperate zones. The same decision may be necessary for cases in which we have morphologically identical conidial fungi, some of which are known to be anamorphs of defined Ascomycetes while others are not. *Ophiostoma stenoceras* and *Sporothrix schenckii* form such a complex (de Hoog, 1974). In this case it has been proved that the long-chain fatty acids of both *Ophiostoma*-connected and independent *Sporothrix schenckii* strains are identical with those of *Ophiostoma stenoceras* (Dart *et al.*, 1976). Similarly, while many strains of *Fusarium solani* cannot form perithecia, either homothallically or heterothallically, others represent the anamorph of *Nectria haematococca* (Samuels, 1976).

A peculiar correlation between compatibility behavior and the formation of anamorphs is found in the ascomycetous genus *Neurospora*. Self-incompatible species (e.g., *N. sitophila, N. crassa*) form an anamorph, whereas self-compatible species do not. A similar correlation is known in *Chaetomium* (Dreyfuss, 1976. *Chaetomium elatum* comprises self-compatible as well as self-incompatible strains (Müller and Sedlar, 1977); the *Acremonium*-like anamorph occurs only in the latter group. The correlation between compatibility behavior and formation of an anamorph, however, cannot be generalized because in many other groups, e.g., in *Nectria* (Booth, 1971; Samuels, 1976), closely related self-compatible and self-incompatible strains do not differ in formation of the anamorph, e.g., *N. haematococca*. An explanation for the correlation observed in *Neurospora* and *Chaetomium* may be the function of the anamorphic spores; in both cases they may be spermatia.

2. Morphology of Teleomorphs and Anamorphs

Morphological characters, which are still the main basis for the differentiation of species, must be considered in both teleomorph and anamorph. In most cases closely related species may be distinguished and identified even if only one state is present.

In some cases, however, either the teleomorphs or the anamorphs of distinct species may be morphologically so similar that differentiation is only possible if

TABLE V

Telemorphs and Anamorphs of the *Guignardia lonicerae* Complex with Morphologically Similar Teleomorphs (Hosts: *Lonicera* species)[a]

	Teleomorph			Anamorph	
Example	Host	Occurrence	Ascospore size (μm)	Name	Conidium size (μm)
G. lonicerae[b]	L. hispidula	Living leaves, North America	12–15 × 6.0–6.5		
G. latemarensis	L. coerulea	Dead leaves, Europe	14–19 × 6.5–8.0	Kabatia lonicerae var. latemarensis	25–43 × 5.5–8.0
G. latemarensis var.[b,c]	L. canadensis	Dead leaves, North America	15–20 × 5–7	Kabatia lonicerae var. americana	14–30 × 5–10
G. mirabilis	L. nigra, L. alpigena	Dead leaves, Europe	12–17 × 6–7.5	Kabatia mirabilis var. mirabilis	21–46 × 6.5–13.0
G. himalayensis[b]	L. qinquelocularis	Dead leaves, Himalayas	14–19 × 6.5–8.0	Kabatia state of G. himalayensis	23–35 × 13–16
G. xylostei	L. xylosteum	Dead leaves, Europe	11–16 × 6.0–7.5	Colletotrichella xylostei	20–30 × 8–12

[a] According to Müller (1953, 1959), Conners (1959), and Reusser (1964).
[b] Connection not proved by cultural work.
[c] Species not named; cf. Reusser (1964).

both states are known. The teleomorphs of some *Guignardia* species (bitunicate Ascomycetes) occurring on *Lonicera* spp. (Caprifoliaceae) are morphologically alike (Müller, 1959; Reusser, 1964), whereas the anamorphs (*Kabatia*), formed within circular necrotic spots on living leaves, may be differentiated easily by the color of the shieldlike conidiomata and by the form, size, and septation of the blastoconidia (Conners, 1959). According to our observations the different species are strictly host-specific (Table V).

Similarly, *Venturia inaequalis* [occurring on some species of *Pirus* and *Sorbus* (Rosaceae)] and *Venturia crataegi* (on *Crataegus*) are very similar in the teleomorph, the ascospores being septate in the wider upper portion. The two species are distinguished by their host range and by the anamorph, which is

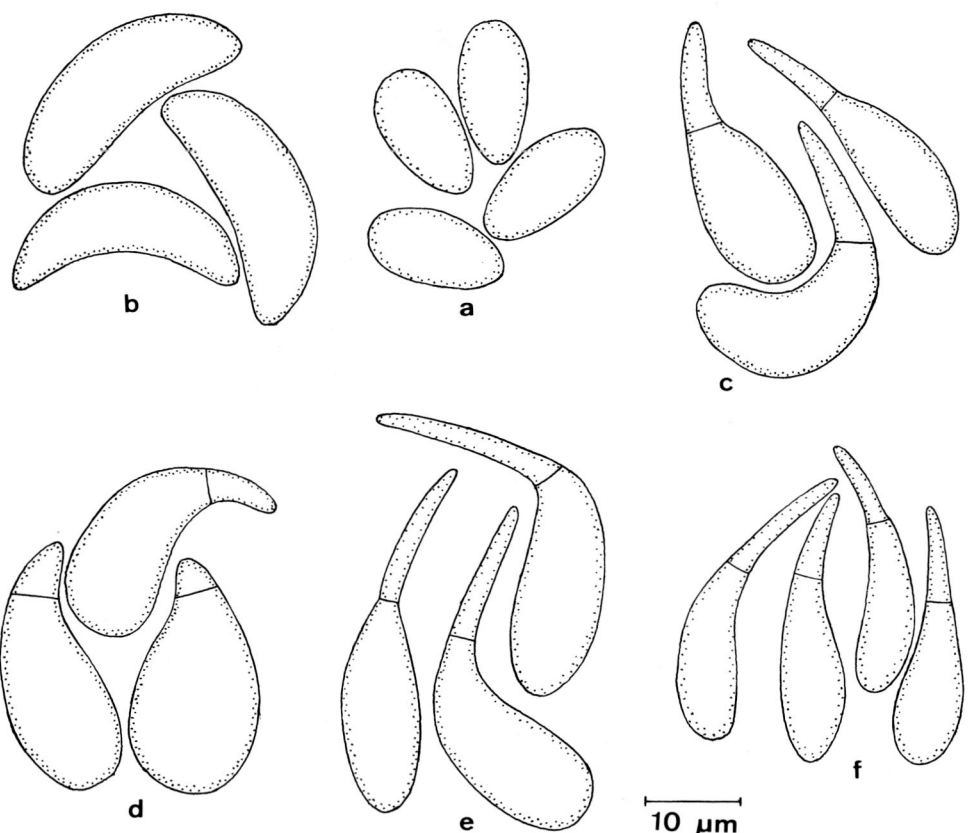

Fig. 2 Differentiation of *Guignardia* spp. on *Lonicera* spp. (Table V). (a) Ascospores of all species. (b–f) Conidia of the different species; (b) *G. xylostei;* (c) *G. latemarensis;* (d) *G. himalayensis;* (e) *G. mirabilis;* (f) *Guignardia* sp. on *Lonicera canadensis*.

TABLE VI

Anamorphs within the Ascomycetous Genus *Tubeufia*

		Anamorph		
			Hyphomycetous	
Example	Pycnidial *Asteromella*	Helicosporous	Dictyosporous	Reference
T. paludosa	−	+	−	Webster (1951), Rossman (1977)
"*T. paludosa*" segregate	−	−	+	Samuels *et al.* (1979)
T. helicoma	−	+	+	Pirozynski (1972)
T. amazonica	+	−	+	Samuels *et al.* (1979)
T. palmarum	−	+	−	Samuels *et al.* (1979)
T. cerea	−	+	−	Booth (1964)

blastic-annellidic in *Venturia inaequalis (Spilocaea pomi)* and blastic-sympodial in *Venturia crataegi (Fusicladium crataegi,* Aderhold, 1897, 1903).

The type species of the pleosporaceous ascomycete genus *Tubeufia, T. paludosa* (Rossman, 1977), occurs in two forms with morphologically indistinguishable teleomorphs but two distinctly different anamorphs (Samuels *et al.*, 1979), one with helicosporous and the other with dictyosporous blastoconidia. According to Table VI both anamorphs are formed in *T. helicoma* (Pirozynski, 1972), whereas the other known species produce only one.

Species of the ascomycetous genus *Dothiora* (Froidevaux, 1972) may be easily distinguished by the arrangement of ascomal loculi in the stromata, the number of ascospores in the asci, and their form, size, and septation. However, the *Dothichiza* anamorphs of all these species are morphologically indistinguishable. Examples of a second anamorph, *Hormonema,* known only in pure culture so far, are likewise so similar that Hermanides-Nijhof (1977) has not been able to differentiate them. Unfortunately, *Dothiora* species may be plurivorous. Therefore *Dothichiza* species occurring independently cannot always be identified.

3. Taxonomic Consequences

The delimitation of a species (holomorph) must be based on all available information. This demand could be interpreted as assigning the same weight to the anamorph as to the teleomorph. Species in which both teleomorph and anamorph are characteristic can therefore be determined on the basis of either morph alone. Within groups of species with similar teleomorphs but diverse anamorphs the differentiation of species has to be based on the anamorph, as in the *Guignardia lonicerae* group, in the *Venturia inaequalis* group, and in the *Tubeufia paludosa* group; this situation is comparable to that in the Mucorales

(Zygomycetes). On the other hand the *Dothichiza* anamorphs of *Dothiora* species are not separable.

Difficulties arise for species with some representatives regularly producing teleomorphs and anamorphs and others that do not form one morph or the other. Since it does not seem practicable to distinguish between these elements, they should be considered representatives of one species (e.g., *Ophiostoma stenoceras* or *Sporothrix schenckii*) unless other characters (e.g., host specificity) indicate real differences, e.g., macrocyclic rust species compared to microcyclic species with similar teleomorphs.

B. Diferentiation of Genera

1. Presence and Absence of Anamorphs

Although it seems extremely difficult to prove with absolute certainty that certain Ascomycetes or Basidiomycetes never produce an anamorph, some genera are known in which no species has ever been reported to produce an anamorph no matter what the cultural conditions. This is true for *Nodulosphaeria, Ophiobolus,* and *Entodesmium* (*sensu* Holm, 1957) belonging to the *Leptosphaeria* complex of the Pleosporaceae.

However, our information based on cultural studies is still inadequate. Within the Pleosporaceae, as delimited by von Arx and Müller (1975) with 77 genera, cultural experiments have been done with at least one species of 46 genera (von Arx *et al.*, 1979), but in 9 genera no anamorphs could be found.

2. Homogeneity and Heterogeneity of Connections

The strict correlation between a certain ascomycetous or basidiomycetous teleomorph and one kind of conidial anamorph represents the ideal situation for connecting these states in a homogeneous holomorph. Within Ascomycetes such cases are known for *Glomerella* (anamorph: *Colletotrichum,* von Arx 1970), *Niesslia (Monocillium,* Gams, 1971), *Pyrenophora (Drechslera,* Luttrell, 1977), *Cochliobolus (Curvularia* including *Bipolaris,* Luttrell, 1977), and *Setosphaeria (Exserohilum,* Luttrell, 1977), and for the Basidiomycetes *Pholiota* (with unnamed, two-celled, *Hansfordia*-like blastic-sympodial conidia, Huebsch, 1978). The connections between *Aspergillus* anamorphs and some ascomycetous genera also represent ideal cases, in that all species of any of these ascomycetous genera include aspergilli from only one group (Raper and Fennell, 1965; Malloch and Cain, 1972, 1973).

Anamorphs disposed in a single anamorph-genus may belong to different genera of Ascomycetes, e.g., *Cytospora* to *Valsa* and *Leucostoma, Dothichiza* and *Hormonema* to *Dothiora* and *Sydowia* (Froidevaux, 1972). This correlation may indicate or confirm a close relationship of the genera involved. For *Dip-*

lodia, however, such a close relationship does not seem likely, because the connected Ascomycetes belong in different families (*Cucurbitaria* and *Otthia* in the Pleosporaceae, *Rhytidhysteron* in the Patellariaceae, and *Botryosphaeria* in the Botryosphaeriaceae, von Arx *et al.,* 1979; Samuels and Müller, 1979c).

In many cases the correlation between anamorph and teleomorph is not so clearly demonstrated. The anamorphs of *Didymella* spp. are mostly found in the genus *Ascochyta,* but sometimes the conidia are not septate and therefore the anamorph is disposed in *Phoma* (Boerema and Dorenbosch, 1973), e.g., *Didymella lycopersici.* The differentiation of genera based merely on conidium septation, although convenient for the determination of such forms, gives too much weight to this character. In such cases we consider the connections overriding and homogeneous. Similarly the different genera mentioned as macroconidial anamorphs of *Leptosphaeria* (Table IV) represent units of several kinds of conidial form and septation but of only one kind of conidial development.

Within the ascomycetous genus *Nectria* (Booth, 1959; Samuels, 1976) the ascomata as well as the conidial states may differ from each other so strongly that heterogeneity seems to be obvious, and the splitting of this genus into smaller units unavoidable. However, Booth (1978) clearly demonstrated why *Nectria* in its present sense should stay. The ascomata of the known species form a continuum in which gaps between distinctly different taxa are bridged by other taxa. Any smaller genus would be separated from others only by a very narrow hiatus. Using both similarities and differences in teleomorphs and anamorphs Booth (1978) outlined a system of developmental pathways that indicated the relationships within the genus.

The majority of Sclerotiniaceae were included in the genus *Sclerotinia sensu lato* with obviously diverse teleomorphs and anamorphs. The splitting of the group into smaller genera, based on differences in stromatic (sclerotial) structures and on the absence or presence and kinds of conidial anamorphs, led to a workable system of closely related genera (Whetzel, 1945; Korf, 1973).

The basidiomycetous genus *Typhula* with two kinds of conidial anamorphs (blastic-sympodial and blastic-annellidic, Berthier, 1976) represents a similar case. Uninuclear, blastic-sympodial conidia have been found in species of the subgenera *Typhula, Gliocoryne, Pistillaria,* and *Microtyphula,* whereas blastic-annellidic conidia are produced by species of the subgenus *Cnazonaria,* by some species not yet allocated to a subgenus, and by all species of the genus *Macrotyphula.* The subgenera are mainly based on sclerotial characters, as in the Sclerotiniaceae, but here no further splitting seems to be desirable.

The separation of the ascomycetous genera *Ophiostoma* and *Ceratocystis sensu stricto* has been debated for a long time (e.g., von Arx, 1974; de Hoog, 1974). The only morphologically expressed difference between the two is in conidiogenesis, with one group of species producing blastic-phialidic conidia (*Ceratocystis*) and the other producing blastic-sympodial conidia. Recent studies on the carbohydrate content of cells (Spencer and Gorin, 1971; Jewell, 1974; Weijman and de Hoog, 1975) indicate that all species with blastic-phialidic conidia lack rhamnose and cellulose, whereas all species with blastic-sympodial conidia have rhamnose, and some also cellulose (in addition to chitin). The two genera are probably unrelated (Samuels and Müller, 1979b). There is very little variation in the outward form of most lignicolous pyrenocarpous Ascomycetes, so the similar morphologies of *Ceratocystis* and *Ophiostoma* are not surprising. Nevertheless it remains quite troublesome that the two genera can only be differentiated on the basis of the anamorph, unless data that call for time-consuming biochemical investigations are available.

Another case of heterogeneity may be found in the *Leptosphaeria* and *Pleospora* groups of the Pleosporaceae. Holm (1957) demonstrated the need to split up *Leptosphaeria sensu lato* on the basis of the teleomorphs. The smaller genera, according to cultural experiments, are in part correlated with certain kinds of pycnidial anamorphs, e.g., *Melanomma* with *Aposphaeria* (Chesters, 1938), *Phaeosphaeria* with *Stagonospora* (e.g., Hedjaroude, 1969), *Paraphaeosphaeria* with *Coniothyrium* (e.g., Webster, 1955), and *Leptosphaeria* with *Phoma, Camarosporium, Rhabdospora,* and *Stagonospora* (Table IV). Dematiacious (*Stemphylium, Alternaria,* and *Dendryphion*) and pycnidial (*Phoma, Stagonospora*) anamorphs are known in *Pleospora*. *Pleospora vagans* with a *Stagonospora* anamorph is—although the ascospores are dictyosporous—closely related to *Phaeosphaeria* (Eriksson, 1967); other species with pycnidial anamorphs have not been examined so far.

3. Taxonomic Consequences

The homogeneity of anamorphs within particular genera of either Ascomycetes or Basidiomycetes supports the conclusion that the species included in these genera are closely related. Diversified anamorphs which exhibit dissimilar features may contain unrelated elements. A further analysis seems to be necessary because, on its own, diversity of anamorphs does not necessitate subdivision. Only additional correlated data, such as distinct differences in the teleomorphs (as in the ascomycetous genus *Melogramma sensu lato,* Laflamme, 1976), or biochemical differences (as in *Ceratocystis;* Section II,B,2), justifies the splitting of such genera. In cases like *Leptosphaeria* (Section I,C) and *Tubeufia* (Section II,A), where some species include two kinds of anamorphs while others have lost one or the other, there is no reason to assume heterogeneity.

On the other hand, known teleomorph–anamorph connections may sometimes call for a more logical arrangement of the anamorphs. The traditional system of Deuteromycetes with Melanconiales, Sphaeropsidales, and Moniliales as the main orders may impose artificial distinctions on otherwise similar forms. According to this system the conidial anamorphs of four species of the genus *Broomella* (Shoemaker and Müller, 1963) belong to four different genera in three

TABLE VII
Conidial Anamorphs of *Broomella* Species (Ascomycetes) on *Clematis* Species[a]

Example, *Broomella*	Host, *Clematis*	Type of conidioma	Type of conidio-genesis	Conidial setae	Anamorph			Proposed arrangement
					Genus	Family		
B. acuta	*C. flammula*	Acervulus	Annellidic	Branched	*Pestalotia*	Melanconiaceae		*Pestalotia*
B. excelsa	*C.* sp. (Pakistan)	Long cylindrical body	Annellidic	Branched	*Ahmadinula*	Pseudographiaceae		*Pestalotia*
B. montaniensis	*C. lugisticifolia*	Acervulus or erect cylindrical body	Annellidic	Branched	*Arthrobotryum*	Stilbaceae		*Pestalotia*
B. vitalbae	*C. vitalba*	Acervulus	Annellidic	Simple	*Pestalotia*	Melanconiaceae		*Pestalotia*

[a] According to Shoemaker and Müller (1963).

different families (Table VII); in fact, they are congeneric. *Heterosphaeria* (inoperculate discomycetes) has anamorphs which belong to the genus *Heteropatella* (Discellaceae) with discoid pycnidia. *Heterosphaeria veratri* (Müller, 1977) has an acervular anamorph which has exactly the same kind of conidiogenesis and shape and septation of conidia.

C. Differentiation of Higher Taxa

1. Differentiation of Families

The principal problems involved in the connection of teleomorphs and anamorphs within families are very similar to those discussed for genera. We must simply be aware of the immature nature of our present family arrangement; statements concerning whole families must be treated with caution. Von Arx *et al.* (1979) compared the families of the Dothideales (von Arx and Müller 1975) in relation to the anamorphs. Several smaller families were found to be quite homogeneous in that their anamorphs possessed obvious common characters. This is true for the Dothideaceae and Botryosphaeriaceae. The Stigmateaceae (=Venturiaceae) are also homogeneous, provided that taxa with pycnidial anamorphs are excluded. In spite of their different kinds of conidiogenesis (sympodial, acropetal, percurrent), the conidia are so similar in size, shape, and pigmentation that we can have no doubt about their relationship. The Xylariaceae also have relatively uniform anamorphs with blastic-sympodial conidia, e.g., *Nodulisporium*. On the other hand, the obviously differing hyphomycetous and pycnidial anamorphs of the Asterinaceae and the Pleosporaceae imply that these are heterogeneous families. Whether or not they will have to be split into smaller units depends on the level of correlation between related groups of anamorphs and connected teleomorphs.

2. Differentiation of Orders

As demonstrated by Tubaki (1958) and Müller (1971) the majority of currently recognized orders of Ascomycetes include taxa with quite diverse anamorphs. Exceptions are the Meliolales, for which no anamorphs are known, and the Erysiphales, with uniform meristem conidial anamorphs. Since information on anamorphs of Basidiomycetes is much more limited than for Ascomycetes, nothing can yet be said about their orders (but see Kendrick and Watling, 1979).

3. Taxonomic Consequences

According to von Arx *et al.* (1979) some families within the Dothideales (=Loculoascomycetes) are characterized by morphologically similar anamorphs, and others by obviously unrelated anamorphs. The former may be considered homogeneous, whereas for the latter the heterogeneous anamorphs may suggest the heterogeneity of the whole family. However, at the family level anamorphs alone are an insufficient basis for decisions concerning the subdivision or amalgamation of currently differentiated families. The same is true at the ordinal level. Since the need for a system of conidial fungi still remains, Luttrell (1979) has proposed—again using the Dothideales as an example—an arrangement in which teleomorph-anamorph connections are considered. This proposal convincingly demonstrates that such a system of Deuteromycetes also includes homogeneous and heterogeneous higher taxa, and that at least with our present information the hoped-for correlation of the two different systems remains a utopian dream.

III. CONCLUSIONS

These reflections on teleomorph-anamorph connections represent an attempt to discover a common base for integrating conidial fungi into the taxonomy of Ascomycetes and Basidiomycetes. The two systems, however, are still too divergent to permit this. On the other hand, anamorphs decisively increase the possibilities for clear delimitation of fungal taxa, for testing the homogeneity of smaller or larger taxa, and for studies on relationships.

The examples discussed demonstrate that the application of anamorphs to the taxonomy of Ascomycetes and Basidiomycetes is not simply helpful—it is now absolutely indispensable. For this reason it is urgent that our investigations of connections be extended and intensified. Only the availability of many more properly worked out connections between teleomorphs and anamorphs can clarify, and ultimately provide solutions to, the many problems which could merely be outlined in the foregoing summary.

ACKNOWLEDGMENT

I wish to thank Dr. Bryce Kendrick and Dr. Gary Samuels for their unselfish help with the English text. To the latter I am grateful for numerous stimulating discussions during his stay at our institute.

REFERENCES

Aderhold, R. (1897). Revision der Species *Venturia chlorospora, inaequalis* und *ditricha autorum*. Hedwigia **36,** 67-83.

Aderhold, R. (1903). Kann das *Fusicladium* von *Crataegus*—und von *Sorbus*—Arten auf den Apfelbaum übergehen? Arb. Biol. Abt. Land- Forstwirtsch.Kaiserl. Gesundheitsamt **4,** 427-442.

Aebi, B. (1972). Untersuchungen über Discomyceten der Gruppe *Tapesia*—*Trichobelonium*. Nova Hedwigia **23,** 49-112.

Berthier, J. (1976). Monographie des *Typhula* Fr., *Pistillaria* Fr. et genres voisins. *Bull. Soc. Linn. Lyon* **45**, Spec. No., 1-213.
Boerema, G. H. (1976). The *Phoma* species studied in culture by Dr. R. W. G. Dennis. *Trans. Br. Mycol. Soc.* **67**, 289-319.
Boerema, G. D., and Dorenbosch, M. M. J. (1973). The *Phoma* and *Ascochyta* species described by Wollenweber and Hochapfel in their study on fruit rotting. *Stud. Mycol.* **3**, 1-50.
Booth, C. (1959). Studies of Pyrenomycetes. IV. *Nectria*. Part. I. *Mycol. Pap.* **73**, 1-115.
Booth, C. (1964). Studies of Pyrenomycetes. VII. *Mycol. Pap.* **94**, 1-16.
Booth, C. (1971). "The Genus *Fusarium*." Commonw. Mycol. Inst. Kew, Surrey, England.
Booth, C. (1978). Presidential address: Do you believe in genera? *Trans. Br. Mycol. Soc.* **71**, 1-9.
Brefeld, O. (1891). "Untersuchungen aus dem Gesammtgebiet der Mykologie," No. X, Ascomyceten II. H. Schöningh Münster i. W. pp. 157-378.
Brockmann, I. (1976). Untersuchungen über die Gattung *Discostroma* Clements (Ascomycetes). *Sydowia* **28**, 275-338.
Chesters, C. G. C. (1938). Studies on British Pyrenomycetes. II. Comparative study of *Melanomma pulvis-pyrius* (Pers.) Fckl., *Melanomma fuscidulum* Sacc. and *Thyridaria rubro-notata* (B. et Br.) Sacc. *Trans. Br. Mycol. Soc.* **22**, 116-150.
Conners, I. L. (1958). Species of *Leptothyrium* and *Kabatia* on *Lonicera*. *Can. J. Bot.* **37**, 419-429.
Dart, R. K., Lee, J. D., and Stretton, R. J. (1976). Classification of *Ceratocystis* and *Sporotrichum* based on their long-chained fatty acids. Trans. Br. Mycol. Soc. **67**, 327-328.
de Hoog, G. S. (1974). The genera *Blastobotrys*, *Sporothrix*, *Calcarisporium* and *Calcarisporiella* gen. nov. *Stud. Mycol.* **7**, 1-84.
de Hoog, G. S. (1977). *Rhinocladiella* and allied genera. *Stud. Mycol.* **15**, 1-140.
Donk, M. A. (1966). Check list of European hymenomycetous Heterobasidiae. *Persoonia* **4**, 145-335.
Dreyfuss, M. (1976). Taxonomische Untersuchungen innerhalb der Gattung *Chaetomium* Kunze. *Sydowia* **28**, 50-132.
Eriksson, O. (1967). On graminicolous pyrenomycetes from Fennoscandia. 1. Dictyosporous species. *Ark. Bot.* **6**, 339-379.
Froidevaux, L. (1972). Contribution à l'étude des Dothioracées (Ascomycètes). *Nova Hedwigia* **23**, 679-734.
Gams, W. (1971). "*Cephalosporium*-artige Schimmelpilze (Hyphomycetes)." Fischer, Stuttgart.
Gäumann, E. (1959). Die Rostpilze Mitteleuropas. *Beitr. Kryptogamen Flora Schweiz* **12**, 1-1407.
Groves, J. W. (1954). The genus *Durandiella*. *Can. J. Bot.* **32**, 116-144.
Hedjaroude, G. A. (1969). Etudes Taxonomiques sur les *Phaeosphaeria* Miyake et leurs formes voisines (Ascomycètes). *Sydowia* **22**, 57-107.
Hennebert, G. L., and Weresub, L. K. (1977). Terms for states and forms of fungi, their names and types. *Mycotaxon* **6**, 207-211.
Hermanides-Nijhof, E. J. (1977). The black yeasts and allied hyphomycetes: *Aureobasidium* and allied genera. *Stud. Mycol.* **15**, 141-177.
Hiratsuka, Y. (1973). Sorus development, spore morphology, and nuclear condition of *Gymnosporangium gaeumanni* ssp. *albertensis*. *Mycologia* **65**, 137-144.
Holm, L. (1957). Etudes taxonomiques sur les Pleosporacées. *Symb. Bot. Ups.* **14**, (3), 1-188.
Huebsch, P. (1978). Nebenfruchtformen bei *Pholiota*-Arten in Reinkultur. *Ceska Mykol.* **32**, 82-86.
Jewell, T. R. (1974). A qualitative study of cellulose distribution in *Ceratocystis* and *Europhium*. *Mycologia* **66**, 139-146.
Kendrick, W. B., ed. (1979). "The Whole Fungus," Vols. 1 and 2. National Museums of Canada, Ottawa.

Kendrick, W. B., and Watling, R. (1979). Mitospores in Basidiomycetes. *In* "The Whole Fungus" (W. B. Kendrick, ed.), pp. 473-545. National Museums of Canada, Ottawa.
Kern, F. D. (1973). "A Revised Taxonomic Account of *Gymnosporangium*." Pennsylvania State Univ. Press, University Park.
Klebahn, H. (1918). "Haupt- und Nebenfruchtformen von Ascomyceten." Leipzig.
Korf, R. P. (1973). Discomycetes and Tuberales. *In* "The Fungi" (G. C. Ainsworth, F. D. Sparrow, and A. S. Sussman, eds.), Vol. 4A, pp. 249-319. Academic Press, New York.
Laflamme, G. (1976). Les genres *Melogramma* Fries et *Melanamphora* gen. nov. (Sphaeriales). *Sydowia* **28**, 237-274.
Laundon, G. F. (1973). Uredinales. *In* "The Fungi" (G. C. Ainsworth, F. K. Sparrow, and A. S. Sussman, eds.), Vol. 4B, pp. 247-279. Academic Press, New York.
Lucas, M. T., and Webster, J. (1967). Conidial states of British species of *Leptosphaeria*. *Trans. Br. Mycol. Soc.* **50**, 85-121.
Luttrell, E. S. (1977). Correlations between conidial and ascigerous state characters in *Pyrenophora, Cochliobolus* and *Setosphaeria*. *Rev. Mycol.* **41**, 271-279.
Luttrell, E. S. (1979). Deuteromycetes and their relationships. *In* "The Whole Fungus" (W. B. Kendrick, ed.), pp. 241-264. National Museums of Canada, Ottawa.
Malloch, D., and Cain, R. F. (1972). The Trichocomataceae: Ascomycetes with *Aspergillus, Paecilomyces* and *Penicillium* imperfect states. *Can. J. Bot.* **50**, 2613-2628.
Malloch, D., and Cain, R. F. (1973). The Trichocomaceae (Ascomycetes). Synonyms in recent publications. *Can. J. Bot.* **51**, 1647-1648.
Moreau, M., and Moreau, F. (1930). Le développement du perithèce chez quelques ascomycètes. *Rev. Gen. Bot.* **42**, 1-34.
Müller, E. (1950). Die schweizerischen Arten der Gattung *Leptosphaeria* und ihrer Verwandten. *Sydowia* **4**, 185-319.
Müller, E. (1953). Kulturversuche mit Ascomyceten. I. *Sydowia* **7**, 325-334.
Müller, E. (1958). Ueber zwei neue Ascomyceten auf *Juniperus*-Arten. Sydowia **12**, 189-196.
Müller, E. (1959). Ueber drei *Guignardia*-Arten und ihre Nebenfruchtformen. *Phytopathol. Z.* **34**, 411-416.
Müller, E. (1962). Kulturversuche mit Ascomyceten. IV. *Sydowia* **16**, 115-120.
Müller, E. (1971). Imperfect-perfect connections in Ascomycetes. *In* "Taxonomy of Fungi Imperfecti" (W. B. Kendrick, ed.), pp. 184-201. Univ. of Toronto Press, Toronto.
Müller, E. (1977). Zur Pilzflora des Aletschwaldreservats (Kt. Wallis, Schweiz). *Beitr. Kryptogamenflora Schweiz* **15**(1), 1-126.
Müller, E. (1979). Factors inducing asexual and sexual sporulation in fungi (mainly ascomycetes). *In* "The Whole Fungus" (W. B. Kendrick, ed.), pp. 265-282. National Museums of Canada, Ottawa.
Müller, E., and Booth, C. (1972). Generic position of *Sphaeria phaeostroma*. *Trans. Br. Mycol. Soc.* **58**, 73-77.
Müller, E., and Sedlar, L. (1977). Compatibilitätsverhältnisse in *Chaetomium*. III. Beziehungen zwischen Selbstcompatibilität und Selbstincompatibilität. *Sydowia* **29**, 352-371.
Müller, E., and Tomasević, M. (1957). Kulturversuche mit einigen Arten der Gattung *Leptosphaeria* Ces. et de Not. *Phytopathol. Z.* **29**, 287-294.
Müller, E., Harr, J., and Sulmont, P. (1969). Deux ascomycètes dont le stade conidien présente des conidies phaeophragmiées endogènes. *Rev. Mycol.* **33**, 369-378.
Ouellette, G. B., and Pirozynski, K. A. (1974). Reassessment of *Tympanis* based on types of ascospore germination within asci. *Can. J. Bot.* **52**, 1889-1911.
Pirozynski, K. A. (1972). Microfungi of Tanzania. I. Miscellaneous fungi on oil palm. *Mycol. Pap.* **129**, 1-139.

Raper, K. B. (1965). "The Genus *Aspergillus*." Williams & Wilkins, Baltimore, Maryland.
Reusser, F. A. (1964). Ueber einige Arten der Gattung *Guignardia* Viala et Ravaz. *Phytopathol. Z.* **51**, 205-240.
Rossman, A. Y. (1977). The genus *Ophionectria* (Euascomycetes, Hypocreales). *Mycologia* **69**, 355-391.
Samuels, G. J. (1976). A revision of the fungi formerly classified as *Nectria* subgenus *Hyphonectria*. *Mem. N. Y. Bot. Gard.* **26**(3), 1-126.
Samuels, G. J., and Müller, E. (1979a). Life-history studies of Brazilian ascomycetes. 1. Two new genera of the Sphaeriaceae having respectively *Sporoschisma*-like and *Codinaea*-like anamorphs. *Sydowia* **31**, 126-136.
Samuels, G. J., and Müller, E. (1979b). Life-history studies in Brazilian ascomycetes. 5. Two new species of *Ophiostoma* and their *Sporothrix* anamorphs. *Sydowia* **31**, 169-179.
Samuels, G. J., and Müller, E. (1979c). Life-history studies in Brazilian ascomycetes. 7. *Rhytidhysteron* and the genus *Eutryblidiella*. *Sydowia* **32**, 277-292.
Samuels, G. J., and Rossman, A. Y. (1979). Conidia and classification of the nectroid fungi. *In* "The Whole Fungus" (W. B. Kendrick, ed.), pp. 167-182. National Museums of Canada, Ottawa.
Samuels, G. J., Rossman, A. Y., and Müller, E. (1979). Life-history studies on Brazilian ascomycetes. 6. Three species of *Tubeufia* with or without dictyosporous pycnidial and helicosporous anamorphs. *Sydowia* **31**, 180-192.
Savile, D. B. O. (1972). Arctic adaptions in plants. *Can. Dep. Agric., Monogr.* **6**, 1-81.
Savile, D. B. O. (1976). Evolution of the rust fungi (Uredinales) as reflected by their ecological problems. *Evol. Biol.* **9**, 137-207.
Shoemaker, R. A., and Müller, E. (1963). Generic correlations and concepts: *Broomella* and *Pestalotia*. *Can. J. Bot.* **41**, 1235-1243.
Simmons, E. G. (1969). Perfect states of *Stemphylium*. *Mycologia* **61**, 1-26.
Spencer, D. M., ed. (1978). "The Powdery Mildews." Academic Press, New York.
Spencer, J. F. T., and Gorin, P. A. J. (1971). Systematics of the genera *Ceratocystis* and *Graphium*. Proton magnetic resonance of the mannose-containing polysaccharides as an aid in classification. *Mycologia* **63**, 387-402.
Tubaki, K. (1958). Studies on the Japanese Hyphomycetes. V. *J. Hattori Bot. Lab.* **20**, 142-244.
Tulasne, L. R. (1853). Mémoire sur l'ergot des Glumacées. *Ann. Sci. Nat., Bot. Biol. Veg.* [3] **20**, 5-56.
von Arx, J. A. (1949). Beiträge zur Kenntnis der Gattung *Mycosphaerella*. *Sydowia* **3**, 28-100.
von Arx, J. A. (1970). A revision of the fungi classified as *Gloeosporium*. *Bibl. Bot.* **24**, 1-203.
von Arx, J. A. (1974). "The Genera of Fungi Sporulating in Pure Culture," 2nd rev. ed. Cramer, Vaduz.
von Arx, J. A., and Müller, E. (1975). A re-evaluation of the bitunicate Ascomycetes, with keys to families and genera. *Stud. Mycol.* **9**, 1-159.
von Arx, J. A., Müller, E., Luttrell, E. S., Pirozynski, K. A., and DiCosmo, F. (1979). Report of the Bitunicate Committee. *In* "The Whole Fungus" (W. B. Kendrick, ed.), pp. 396-410. National Museums of Canada, Ottawa.
Webster, J. (1951). Graminicolous Pyrenomycetes. I. The conidial stage of *Tubeufia helicomyces*. *Trans. Br. Mycol. Soc.* **34**, 304-308.
Webster, J. (1955). Graminicolous Pyrenomycetes. V. Conidial states of *Leptosphaeria michotii, L. microscopica, Pleospora vagans,* and the perfect state of *Dinemasporium graminum*. *Trans. Br. Mycol. Soc.* **38**, 347-365.
Webster, J. (1970). "Introduction to Fungi." Cambridge Univ. Press, London and New York.
Webster, J., and Hudson, H. J. (1957). Graminicolous Pyrenomycetes. IV. Conidia of *Ophiobolus*

herpotrichus, Leptosphaeria luctuosa, L. fuckelii, L. pontiformis and *L. eustomoides*. *Trans. Br. Mycol. Soc.* **40,** 509-522.

Weijman, A. C. M., and de Hoog, G. S. (1975). On the subdivision of the genus *Ceratocystis*. *Antonie van Leeuwenhoek* **41,** 353-360.

Whetzel, H. H. (1945). A synopsis of the genera and species of the Sclerotiniaceae, a family of stromatic inoperculate discomycetes. *Mycologia* **37,** 648-714.

8

A Survey of the Fungicolous Conidial Fungi

D.L. Hawksworth

I.	Introduction	171
II.	The Fungicolous Habit	173
	A. Physical Interactions	173
	B. Chemical Interactions	177
	C. Genetic Interactions	178
	D. Nutritional Interactions	179
III.	The Fungicolous Conidial Fungi	180
	A. Ubiquitous Species	180
	B. Myxomycota	183
	C. Mastigomycotina	186
	D. Zygomycotina	187
	E. Basidiomycotina	188
	F. Ascomycotina	203
	G. Deuteromycotina	224
IV.	Discussion	231
	A. Evolutionary Considerations	232
	B. Practical Considerations	234
	References	235

I. INTRODUCTION

As conidial fungi are able to exploit a vast range of substrates, that some would have developed the ability to utilize other fungi was to be expected. About 1100 species of conidial fungi are only known on other fungi, and approximately 2500 fungi have conidial fungi recorded as growing on them. These figures demonstrate that other fungi constitute a very significant, although relatively little studied, ecological niche for conidial fungi. This chapter aims to provide an overview of the current state of our knowledge of fungicolous conidial fungi both from the biological and systematic standpoints and also to stimulate further interest in this fascinating mode of life.

Fungi occurring on other fungi are commonly termed "mycoparasites" and subdivided into two major biological groups (Barnett, 1963; Barnett and Binder, 1973; Boosalis, 1964; Cooke, 1977): "necrotrophic" (destructive) and "biotrophic" (forming balanced relationships). Cooke (1977) considered the term "mycoparasite" inappropriate, as it could be used for a fungus parasitic on any organism and not only on other fungi, but it is retained here for necrotrophic species. The term "hyperparasite" was employed interchangeably with "mycoparasite" by Boosalis (1964) and has been adopted by several mycologists (e.g., Deighton, 1969; Deighton and Pirozynski, 1972); it was considered unacceptable by Cooke (1977) in strictly implying only a species parasitic on an already parasitic organism. Cooke (1977) speaks of fungi "antagonistic" toward other fungi, but this cannot be commended for general usage as many fungi reported from other fungi may not form actively antagonistic relationships at all. "Mycophilic" (fungus-loving), used for example by Rudakov (1978), may be an acceptable term where the association is obligate, but the precise physiological relationship is unclear.

"Fungicolous" has sometimes been used to refer to fungi on macromycetes (e.g., Tubaki, 1955; Nicot, 1967) but also more widely to embrace a very wide range of fungus–fungus relationships (e.g., Gilman and Tiffany, 1952). "Fungicolous" has been used as a general term by Barnett (1963) and Barnett and Binder (1973) for cases where a definite nutritional relationship has not been demonstrated but is here accepted to include also associations where such relationships have been established. From the following sections, it will be apparent that so little experimental work has been done on most fungicolous fungi that the terms "mycoparasite," "biotrophic," and "necrotrophic" can only exceptionally be applied with confidence. Rudakov (1978) considered that mycophilic fungi could be subdivided into at least six physiological groups, but intermediates undoubtedly exist.

Fungi occurring on lichen-forming (lichenized) fungi are referred to as "lichenicolous." Like "fungicolous," "lichenicolous" does not imply any proven nutritional relationship but is used as an all-embracing term. Some lichenicolous fungi are evidently necrotrophic, because the host is killed, but others are biotrophic, forming very stable associations. Apparently biotrophic lichenicolous fungi have commonly been referred to as "parasymbionts" (Zopf, 1897), but it is conceivable (no experimental information is available) that in some instances they obtain carbohydrates directly from the algal cells present as well as, or instead of, from the lichen-forming fungus (mycobiont); parasymbiotic lichenicolous fungi may consequently also be interpreted as algicolous, lichenized, or constituting a three-membered symbiosis (Hawksworth, 1979; Poelt, 1977). As the concept of lichen itself embraces a wide range of quite different types of fungus–alga relationships (Hawksworth, 1978), it is not surprising that it is difficult to generalize about other fungi entering into these associations. For the purposes of this chapter, lichenicolous fungi are regarded as fungicolous on lichen-forming fungi; elegant physiological and ultrastructural investigations will be needed in order to ascertain the precise nature of such associations.

Fungicolous fungi can be "facultatively" fungicolous (occurring fortuitously or regularly on fungi but also known from other habitats), or "obligately" fungicolous (known only on fungi). Barnett and Binder (1973) considered that "obligate" was undesirable as merely indicating that the fungus was not known in pure culture, but I think this view is unnecessarily pedantic. The terms "facultative" and "obligate" are used here in the sense of the known habitat or host ranges, regardless of whether the fungus is able to grow in culture or not. Some fungi currently appearing to be obligately fungicolous by this definition will certainly prove not to be so.

Pertinent reviews have been prepared by several authors over the last two decades (De Vay, 1956; Barnett, 1963, 1964; Boosalis, 1964; Madelin, 1968; Barnett and Binder, 1973; Hashioka, 1973a; Cooke, 1977), but most concentrate on biological and physiological data in a few selected examples. Surprisingly, no comprehensive survey of fungicolous fungi has been prepared in recent years, although a few regional accounts, host lists, and keys are available for restricted taxonomic areas. (These are cited at appropriate points in Section III.)

II. THE FUNGICOLOUS HABIT

The number of fungus–fungus associations studied in the laboratory is meager compared to the number known in nature. Furthermore, many of the cultural and physiological studies published concern mycoparasitism in the Chytridiales and Mucorales, *not* the conidial fungi. Any account of the features of the fungicolous habit in conidial fungi is consequently inevitably biased by the scarcely representative better studied examples. Consequently, generalizations should only be made from the following synopses with considerable caution.

A. Physical Interactions

1. Appressoria

Appressoria were defined for plant pathogenic fungi by Emmett and Parbery (1975) simply as the "expression of the genotype during the final phase of germination," whether or not they were morphologically differentiated from the vegetative hyphae, as long as they adhered to and penetrated the host. No appressoria, as defined in this way, are known among the conidial fungicolous fungi. Penetrating hyphae do not generally arise directly from a germinating spore and so are not strictly appressoria.

2. Coiling

The coiling of the hyphae of mycoparasitic fungi around the hyphae or spores of their host(s) is widespread and has been repeatedly demonstrated in studies with pure cultures. This is particularly well documented for *Trichoderma* Pers. ex Gray, species of which can coil on the hyphae of a very wide range of other fungi (Weindling, 1932; Komatsu and Hashioka, 1964; Hashioka *et al.*, 1961; Hashioka and Fukita, 1969, Dennis and Webster, 1971c). Hyphal penetration may follow coiling (Aytoun, 1953; Komatsu and Hashioka, 1964). Coiling in *Acremonium* Link ex Fr. sp. on *Pellicularia sasakii* (Shir.) S. Ito has also been studied in detail (Hashioka, 1973a,b).

Rudakov (1978) considered coiling to occur only in the necrotrophic, facultatively necrotrophic, and facultatively biotrophic fungi.

3. Contact Cells

Specialized short globose protuberances which develop from the hyphae of a mycoparasite and clasp the host hyphae have been termed contact cells or, more rarely, buffer (absorptive) cells (Hashioka, 1973a); ultrastructural studies have established that plasmodesmata (Hoch, 1977a, 1978) or pores (Hoch, 1977b) develop in these cells between the host and mycoparasite. These specialized cells are currently known only in six species which have consequently been referred to as contact mycoparasites: *Hansfordia parasitica* (Barnett) de Hoog (*Calcarisporium parasiticum* Barnett) (Barnett and Lilly, 1958), *Gonatobotryum fuscum* (Sacc.) Sacc. (Shigo, 1960), *Gonatobotrys simplex* Corda (Whaley and Barnett, 1963), *Nematogonium ferrugineum* (Pers.) S. Hughes (*N. aurantiacum* Desm., *Gonatorrhodiella highlei* A. L. Sm.) (Gain and Barnett, 1970), *Sphaerulomyces coralloides* Marvanová (Marvanová, 1977), and *Stephanoma phaeospora* Butler & McCain (Rakvidhyasastra and Butler, 1973). *Hansfordia parasitica* is atypical in that an additional buffer cell forms between the contact cell and the unmodified hypha of the mycoparasite. The earlier work on these fungi is reviewed in detail by Barnett and Binder (1973).

4. Haustoria

Differentiated haustoria occur in some mucoraceous fungi parasitic on other Mucorales but are generally considered absent in other mycoparasitic fungi (Barnett, 1963; Barnett and Binder, 1973). The irregularly lobate feeding hyphae developed in the oospores of *Pythium* Pringsh. species by several Hyphomycetes (Drechsler, 1937 on; see Section III,C), however, perhaps most nearly provide an exception to this tenet.

5. Hyphopodia

Specialized lateral cells termed hyphopodia, arising on horizontally spreading hyphae, are well-known in some groups of tropical foliicolous ascomycetes. Their function does not appear to have been investigated in detail, but on vascular plants as penetration through the host cuticle into the epidermal cells rarely (if ever) occurs, they presumably have a primary role as attachment rather than absorptive organs. Hyphopodia are restricted among the fungicolous conidial fungi to certain obligately fungicolous dematiaceous Hyphomycetes on foliicolous ascomycetes and foliicolous lichenized fungi.

6. Intrahyphal Hyphae

In some instances the hyphae of the fungicolous fungus may penetrate and grow within the hyphae of the host (Fig. 1). In some cases the act of penetration

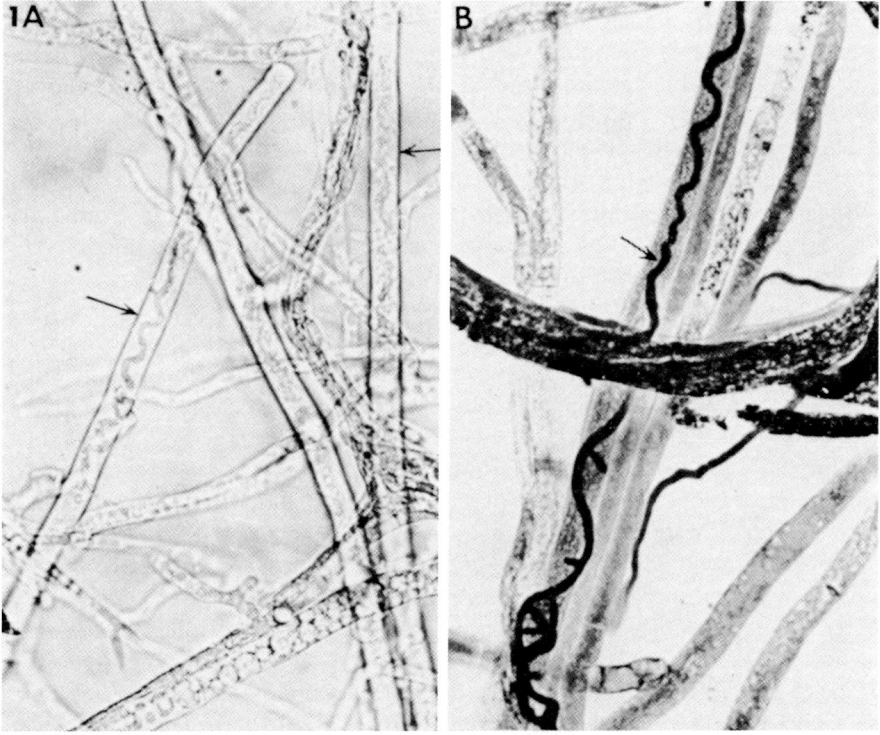

Fig. 1. (A) Hyphae of *Trichoderma viride* growing intracellularly in hyphae of *Pythium ultimum* (arrow). ×600. (B) Hyphae of *T. viride* following hyphae of *P. ultimum* on their surface (arrow). Stained with cotton blue lactic acid. ×600. (From Dennis and Webster, 1971c, reproduced with permission.)

may involve special "penetration pegs" (Boosalis, 1956) which also occur in mucorealean mycoparasites (Manocha and Lee, 1971), while in others, weakened areas such as conidial scars provide points of entry (Barnett and Lilly, 1962), or penetration only occurs after the cell wall has started to decompose either naturally or as a result of enzymatic action or metabolites from the mycoparasite (see Section II,B). Intrahyphal penetration is particularly common in *Aspergillus* Mich. ex Fr., *Penicillium* Link ex Gray, and certain other Hyphomycetes on Mucorales (see Section II,D) but is also known in many other instances: for example, *Trichoderma viride* Pers. ex Gray on *Pythium ultimum* Trow (Dennis and Webster, 1971c), Mucorales (Durrell, 1968), and *Pellicularia sasakii* (Hashioka and Fukita, 1969); *Gliocladium roseum* Bain. on various fungi (Barnett and Lilly, 1962); *Eriomycopsis biseptata* Chavaug. on *Meliola capensis* (Kalchbr. & Cooke) Theiss. (only the conidiogenous cells are sometimes exposed) and *E. flagellata* Hansf. on *M. chlorophorae* Hansf. (Deighton and Pirozynski, 1972); and *Ampelomyces quisqualis* Ces. on *Erysiphe cichoracearum* DC. ex Mérat where the host cells are killed as the invading hypha proceeds (Emmons, 1930). *Aspergillus flavus* Link ex Gray is of interest in that it is able to grow intrahyphally and apparently parasitically on its own hyphae (Boller and Schroeder, 1972). The lichenized basidiomycete *Dictyonema* Ag. ex Kunth. was considered to have intrahyphal hyphae of an unidentified, perhaps parasymbiotic, fungus not causing the host hypha any damage but this proves to be elaborate haustoria (Slocum, 1980).

Rudakov (1978) considered intrahyphal hyphae to occur only in biotrophic, facultatively biotrophic, and facultatively necrotrophic mycophilic fungi.

The penetrating hyphae of some species, for example, *Trichoderma longibrachiatum* Rifai on *Pellicularia sasakii,* become severely restricted (to 0.1-0.2 μm) at the point of penetration and evidently penetrate the *Pellicularia* hypha mechanically (Hashioka, 1973a). In this instance the host forms an internal thickened area of cell wall, an infection papilla, in response to the invading fungus. This is subsequently penetrated by the *Trichoderma* hypha (Hashioka and Fukita, 1969); once inside the host hypha the *Trichoderma* coexists for a while before protoplast breakdown of the *Pellicularia* commences (Hashioka and Fukita, 1969). Similar constrictions in penetrating hyphae sometimes occur in *Verticillium lecanii* (Zimm.) Viégas on *Hemileia vastatrix* Berk. & Br., though lysis of the outer spore wall may be more important in this case (Locci *et al.,* 1971). Some mucoraceous hosts form callosities around invading hyphae (see Section III,D).

The entry of intrahyphal hyphae by enzymatic action is seen, for example, in the action of the *Acremonium* anamorph of *Hypocrea austrograndis* Hashioka & Komatsu against *Cochliobolus miyabeanus* (Ito & Kur.) Drechsler where dissolution of the cell wall occurs and the cytoplasm becomes disorganized in advance of invading hyphae (Hashioka, 1973a).

7. Smothering

The overgrowth of one fungus colony by another is seen in several ubiquitous mycoparasites, for instance, *Gliocladium roseum* and *Trichothecium roseum* Pers. ex Gray, and has been documented both in culture and in nature (Barnett and Lilly, 1962; Shigo, 1958). The ability to grow very rapidly is undoubtedly important in determining the degree of pathogenicity of mycoparasites.

B. Chemical Interactions

Chemical interactions between fungi, particularly in culture, have long been recognized (see Porter and Carter, 1938), and their importance in soil has been discussed by Brian (1960). These can be important in substrate or spatial competition, attraction, penetrating, and also in killing a species to be lived off saprophytically. Barnett (1964) considered that necrotrophic mycoparasites characteristically killed their hosts by enzymes or other toxic compounds of high molecular weight, and Rudakov (1978) found "antibiotics" to be a feature of necrotrophic, facultatively biotrophic, and many semisaprophytic mycophilic fungi. As chemical antagonism is evidently well developed in some of the most destructive mycoparasites, it is pertinent to draw attention to it here, but its significance in most fungicolous fungi is currently unknown.

Trichoderma viride sensu lato filtrates are reported to be particularly rich in antifungal compounds, and mycoparasitic activity has been related to the presence of toxins (Weindling, 1938). However, some of the earlier work requires reexamination because of confusion with other fungi, particularly *Gliocladium virens* J. H. Miller *et al.* (Hashioka and Fukita, 1969). *Trichoderma viride sensu lato* forms both nonvolatile and volatile toxins (Dennis and Webster, 1971a,b) active against a wide range of other fungi (Hashioka *et al.*, 1961; Domsch *et al.*, 1980), and seed treatment with this species or trichodermin can reduce damage by seed-borne pathogens (Upitis, 1956; Fedorinchik, 1961). Rubratoxin B from *Penicillium rubrum* Stoll interferes with cell wall synthesis and leads to hyphal deformation in numerous fungi (Reiss, 1972), and griseofulvin from *P. griseofulvum* Dierckx is extremely toxic to both dermatophytic and plant-pathogenic soil fungi (Domsch and Gams, 1968).

The ability to resist such compounds may also be significant in mycoparasitism. *Gliocladium roseum*, for example, is one of the few fungi not susceptible to products of *Trichothecium roseum* (Barnett and Lilly, 1962), and *Aspergillus ustus* (Bain.) Thom & Church was one of the few 52 tested species resistant to an antibiotic from *Stachybotrys atra* Corda (Butt and Ghaffar, 1972).

Chemical products may also serve as attractants. *Gliocladium roseum* is attracted straight to host hyphae, presumably by products diffusing outward from them. Conversely, both *Gonatobotrys simplex* and *Hansfordia parasitica* stimulate their hosts to branch or grow toward the mycoparasite spores from a distance

Fig. 2. Hyphal interference between *Stilbella erythrocephala* (right-hand colony) and *Pilaira anomala* (Ces.) Schröt. ×2. (Photograph by J Webster.)

of about 50 μm (Barnett and Lilly, 1958; Whaley and Barnett, 1963). Once contact is made, chemical action may also be important in effecting hyphal penetration (see Section II,A,6).

Chemical antagonism at a distance or in close proximity, "hyphal interference" (Ikediugwu and Webster, 1970a; Fig. 2), may be especially important in nature among coprophilous fungi. The action is probably due to changes in membrane permeability and subsequent death of the hyphae. *Stilbella erythrocephala* (Ditm. ex Fr.) Lindau produces a compound that stops sporogenesis in *Pilobolus crystallinus* (Wigg.) Tode, inhibits *Coprinus heptemerus* M. Lange & A. H. Sm. on rabbit dung (Ikediugwu and Webster, 1970a), and is active against *Chaetomium cochliodes* Pall. and *Volutella ciliata* Alb. & Schwein. ex Fr. (Singh and Webster, 1973); *Ascobolus crenulatus* P. Karst. ceased growth in culture when the *Stilbella* was at least 5 mm away (Ikediugwu and Webster, 1970b). Hyphal interference can be interpreted as a highly effective and precise weapon in interspecific competition (Webster, 1970).

There are numerous other reports of chemical interactions between fungi in culture, particularly plant-pathogenic and soil-inhabiting species, which it would be superfluous to enumerate here. Experiments carried out with selected soil fungi are, however, compiled in Domsch *et al.* (1980).

C. Genetic Interactions

Genetic interactions between hosts and pathogenic fungi have been investigated in considerable detail for some economically important crops (Day, 1974).

Types of interactions involve the pathogen developing means of overcoming resistance of the host, hosts evolving methods of resistance to invasion by the pathogen, or even the interdependence of host and formerly harmful organism (biotrophic relationships). While the types of interactions can be seen or assumed to occur in fungicolous conidial fungi, there appears to be no case where interactions at the genetic level have been worked out. Of especial interest in this connection is the apparent ability of some fungicolous fungi to attack or form biotrophic (or parasymbiotic) associations with some species, but not others, of the same fungal genus or family.

D. Nutritional Interactions

Remarkably little experimental work on the nutritional requirements of fungicolous fungi is available. Host extract agar has proved especially suitable to their culture in some cases (Backus and Stowell, 1953; Tubaki, 1955), but most (however, mainly less host-specific species) will grow on usual media and some can germinate in distilled water or water agar (Calpouzos *et al.*, 1957; Emmons, 1930; Shigo, 1960).

Contact mycoparasites (see Section II,A,3) all require the unidentified "mycotrophein" (Barnett and Binder, 1973). Tubaki (1955) found that some agaricicolous species assimilated trehalose and/or mannite as well as or better than glucose. In addition to a requirement for particular metabolites, the ability to utilize a limited number of carbohydrates, or to take them up preferentially, may thus be an important factor. The absence of a particular compound in a fungus might also render it immune to attack by species needing this substance (Barnett, 1964), though its presence might not imply susceptibility if the mycoparasite is unable to make the compound accessible by, for example, changing the permeability of cell membranes to release it.

In the case of balanced biotrophic interactions, mutual nutritional benefit may be gained. This is seen in the classic case of the mutual growth of *Mortierella ramanniana* (Möll.) Linnem. and *Rhodotorula rubra* (Demme) Lodder (Müller, 1941); the *Mucor* obtained thiazole, in which it was deficient, from the yeast, and the yeast obtained the pyrimidine it required from the *Mortierella*, the two forming stable intermixed colonies. In the case of parasymbiotic lichenicolous fungi, certain polyalcohols produced by the phycobiont (Richardson, 1974), perhaps surplus to the mycobiont's needs, are presumably used by the parasymbiont.

As nutritional factors have a role in some fungus–fungus interelationships, the nature of a culture medium can affect the extent to which hosts are susceptible (Boosalis, 1964). Similarly, in nature, whether a facultative fungicolous species assumes this habit might be determined by the availability or otherwise of nutrients where its spores alight.

III. THE FUNGICOLOUS CONIDIAL FUNGI

The following survey of the fungicolous conidial fungi has been arranged according to the major divisions of the fungal kingdom adopted in Ainsworth *et al.* (1973a,b), with a few minor modifications. This proved suitable as (1) the majority of fungicolous fungi have restricted ranges, and (2) those wishing to determine fungicolous fungi will be provided with a synopsis of the known taxa under the group to which the host belongs.

In the course of the preparation of this section it was necessary to compile lists of the described conidial fungicolous fungi from the standard catalogs of fungal taxa (see Hawksworth, 1974, pp. 108-109). These data were supplemented by numerous other publications, those of Hansford (1946), Oudemans (1918-1924), and Viégas (1961) meriting particular mention. This revealed that large numbers of published names have never been critically reassessed, and that they remain in genera that have subsequently been remodeled. A definitive survey of the fungicolous fungi will be impossible until a reappraisal of such names, based on a study of the original material, has been carried out. Generic names which seem particularly inappropriate, and names which are not validly published, are placed in quotation marks here.

The literature survey was supplemented by the host index to the herbarium of the Commonwealth Mycological Institute, Kew (IMI). I was also fortunate in being able to examine both the manuscript of Domsch *et al.* forthcoming compendium of selected soil fungi (1980), and to carry out an on-line computerized search of the Commonwealth Agricultural Bureaux DIALOG data base (1973 on). I have endeavored to survey the extant information as fully as is appropriate here but, because of the particularly scattered literature and the vast amount of data, selection was inevitable.* Selection has been both conscious (from information compiled) and unconscious (from sources not seen). Limitations of space preclude the citation of sources for all occurrences mentioned. Fungi described from completely unidentified hosts are deliberately omitted.

Numbers of potential host genera and species are indicated and, with a few exceptions, are from Ainsworth (1971; Table I).

A. Ubiquitous Species

No fungicolous fungi are recorded from all classes and orders of fungi, but some are known from such disparate hosts that they must be assumed to tend toward this situation. *Gliocladium roseum* is perhaps the most destructive broad-spectrum mycoparasite so far recognized. Barnett and Lilly (1962) demonstrated its activity in culture against *Ceratocystis fimbriata* Ellis and Halst., "*Helminthosporium*" *sativum* Pammel *et al.* [*Cochliobolus sativus* (Ito and

---

*Publications received at CMI after 31 December 1978 are not covered in this chapter.

TABLE I

Numbers of Conidial Fungi Only Known from Particular Fungal Groups[a]

Group	Potential host species[b]	Conidial Fungi			Genera/ hosts	Species/ hosts
		Genera	Species	Genera/ species		
Myxomycota	564	2	11	1:5	1:282	1:51
Mastigomycotina	1,025	0	10			1:103
Zygomycotina	710	0	3			1:236
Basidiomycotina						
Teliomycetes	5,850	7	42	1:6	1:835	1:139
Hymenomycetes	6,300	12	88	1:7	1:525	1:72
Gasteromycetes	700	0	20			1:35
	13,000	19	150	1:8	1:676	1:85
Ascomycotina						
Hemiascomycetes	325	0	6			1:54
Plectomycetes	200	0	0			
Pyrenomycetes	8,000	19	219	1:11	1:421	1:36
Discomycetes	3,000	2	30	1:15	1:1500	1:100
Loculoascomycetes	2,000	12	94	1:8	1:166	1:21
Laboulbeniomycetes	500	0	0			
Lichen-forming	20,000	22	95	1:4	1:909	1:210
	35,000	55	444	1:8	1:636	1:79
Deuteromycotina						
Blastomycetes	200	0	0			
Coelomycetes	7,000	1	45	1:45	1:7000	1:155
Hyphomycetes	7,500	5	104	1:21	1:1500	1:72
Mycelia Sterilia	200	0	7			1:28
	15,000	6	156	1:26	1:2500	1:96

[a] Obligately fungicolous genera and species in more than a single group are omitted here.
[b] Numbers mainly after Ainsworth (1971) with minor modifications.

TABLE II

Examples of Genera of Conidial Fungi Which Include Species Known Only from Particular Fungal Groups[a]

Group	Acremonium	Cladosporium	Endophragmiella	Fusarium	Gliocladium	Mycogone	Phoma	Sympodiophora	Verticillium
Myxomycota	+								+
Mastigomycotina		+							
Zygomycotina									
Basidiomycotina									
Teliomycetes	+								
Hymenomycetes	+	+							+
Gasteromycetes									+
Ascomycotina									
Hemiascomycetes		+							
Plectomycetes									
Pyrenomycetes	+	+	+	+		+	+	+	
Discomycetes		+				+	+		
Loculoascomycetes		+		+			+	+	
Laboulbeniomycetes									
Lichen-forming	+	+	+	+			+		
Deuteromycotina									
Blastomycetes									
Coelomycetes		+	+				+		
Hyphomycetes		+					+		
Mycelia Sterilia							+		
Other substrates	+	+	+	+	+		+		+

[a] See text under group names for examples of the species of the genera cited here.

Kuribayashi) Drechsler ex Dastur *nomen holomorphosis* (nom. hol.)], *Rhinotrichum macrosporum* Farlow, *Thamnidium elegans* Link, and *Trichothecium roseum*; later work has shown it to be active against numerous other fungi (Domsch *et al.*, 1980). *Trichoderma* species have a particularly wide range of hosts and can attack numerous soil fungi, sclerotia, Agaricales, Hyphomycetes, Polyporaceae, Uredinales, and so on (Weindling, 1932; Aytoun, 1953; Boosalis, 1956; Wu, 1977; Domsch *et al.*, 1980). *Stilbella erythrocephala* and *Trichothecium roseum* also affect other fungi by chemical action. The methods of action of these ubiquitous mycoparasites have been reviewed in the preceding section.

Several genera include some species which have adapted to growth on particular fungi or groups of fungi. The best examples of this are seen in *Acremonium*, *Cladosporium* Link ex Fr., *Cylindrocarpon* Wollenw., *Fusarium* Link ex Fr., and *Verticillium* Nees ex Wallr. (see Table II; Section IV,A).

In addition to mycoparasitic fungi, some ubiquitous saprophytes are not uncommonly encountered growing on or intermixed with other fungi, sometimes parasitically, for example, *Alternaria alternata* (Fr.) Keissl., *Acremonium strictum* W. Gams (particularly on Erysiphales and Uredinales), *Aureobasidium pullulans* (de Bary) Arnaud, *Botrytis cinerea* Pers. ex Pers., *Cladosporium herbarum* (Pers. ex Link) Gray (perhaps the most plurivorous fungus known), *Epicoccum purpurascens* Ehrenb. ex Schlecht., and species of *Penicillium* (especially *P. brevicompactum* Dierckx).

B. Myxomycota

No conidial fungicolous fungi are known on the Acrasiomycetes (4 genera, 9 species), Dictyosteliomycetes (4 genera, 19 species), Hydromyxomycetes (4 genera, 13 species), Labyrinthulomycetes (6 genera, 30 species), Plasmodiophoromycetes (9 genera, 35 species), or Protosteliomycetes (10 genera, 18 species). This may be mainly because these slime molds are often mobile and have short-lived, nonpersistent sporangial phases, or are mainly within host tissue; they are consequently scarcely open to attack by true fungi.

The Myxomycetes (70 genera, 400 species) are similarly not colonized in their plasmodial phase, but their exposed sporangial phases do provide a substrate open to colonization. However, rather few obligately myxomyceticolous conidial fungi are known, and these must be able to tolerate the chemical environment of the myxomycete sporangium, which is often very rich in calcium carbonate. Most obligately myxomyceticolous conidial fungi smother the sporangia, giving the whole a white color, but whether most should be viewed as parasites is doubtful, because they colonize an already terminal phase and flourish after a great many spores have already been liberated. A valuable key to the

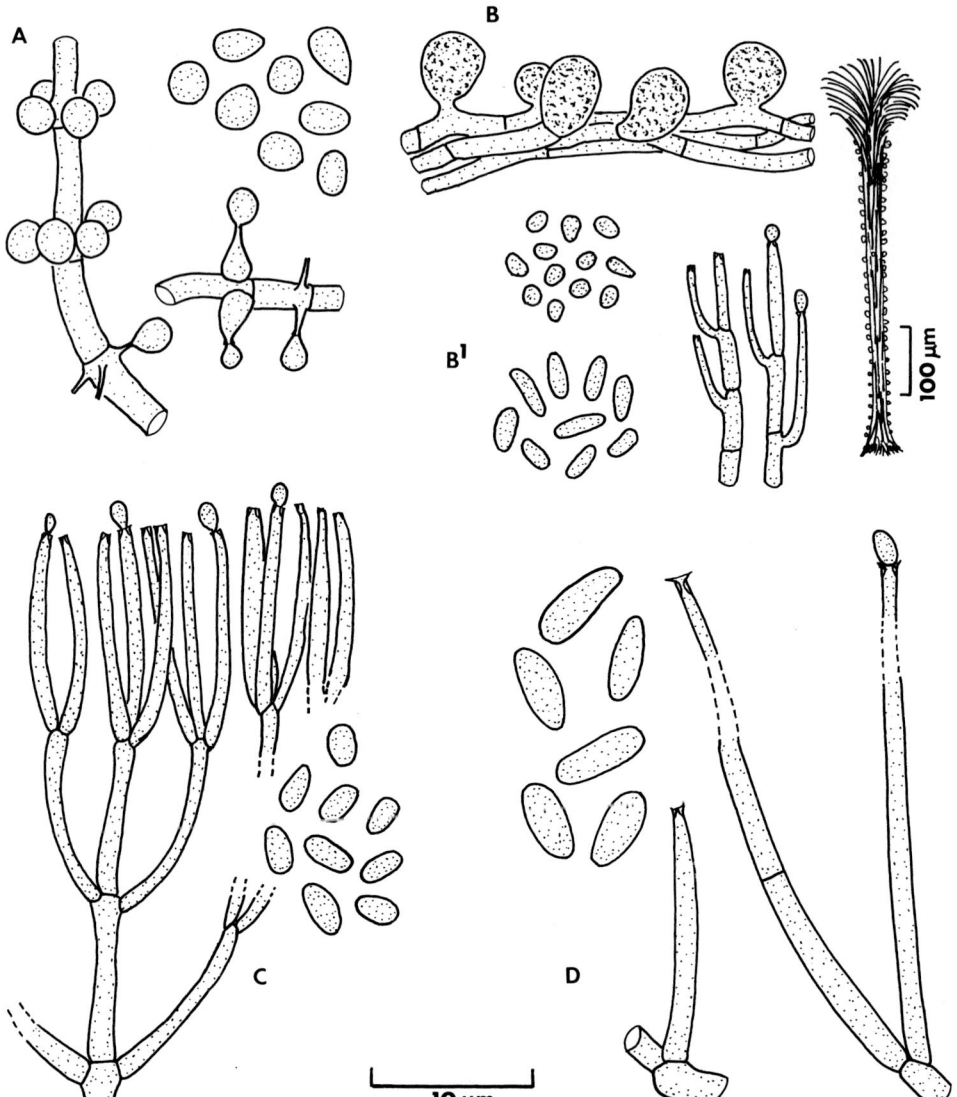

Fig. 3. Conidial fungi occurring on Myxomycetes. (A) *Aphanocladium album*; (B) *Blistum tomentosum*; (B[1]) *B. ovalisporum* conidia; (C) *Gliocladium album*; (D) anamorph of *Nectria candicans*.

myxomyceticolous fungi was prepared by Ing (1974); although this was designed primarily for British students, many of the species are widespread. Ing treated 13 conidial fungi, and it is remarkable that all are phialidic, moniliaceous Hyphomycetes and most are known to be, or are similar to, conidial anamorphs of hypocrealean fungi.

Two exclusively myxomyceticolous genera of conidial fungi are known; the stilboid *Blistum* B. Sutton which comprises *B. ovalisporum* (A. L. Sm.) B. Sutton (Fig. 3B) and the common *B. tomentosum* (Schrader ex Grev.) B. Sutton *Stilbella tomentosa* (Schrader ex Fr.) Bresad., *Byssostilbe stilbigera* (Berk. & Br.) Petch nom. hol.; Fig. 3B, and the recently described monotypic *Leucopenicillifera gracilis* G. Arnold on *Fuligo septica* (L.) Web. with penicillate phialides and chains of narrowly cylindrical conidia (Arnold, 1971).

The majority of myxomyceticolous species, however, have representatives on other substrates and are also catholic with respect to their host range within the Myxomycetes, for example, *Gliocladium album* (Preuss) Petch (Fig. 3C) and *Verticillium rexianum* (Sacc.) Sacc. (*Nectria myxomyceticola* Samuels nom. hol.). The conidial anamorphs of *Nectria candicans* (Plowr.) Samuels (Fig. 3D) and *N. violacea* (Schmidt ex Fr.) Fr. *Acremonium fungicola* Samuels *nomen anamorphosis* [nom. an.], both of which were formerly placed in *Nectriopsis* Maire, have affinities with both *Acremonium* and *Cylindrocarpon* Wollenw. (Gams, 1971; Samuels, 1973). *Aphanocladium album* (Preuss) W. Gams (Fig. 3A) is especially common on Myxomycetes, and some other saprophytic conidial fungi, particularly *Sesquicillium microsporum* (Jaap) Veenbas-Rijks & W. Gams and *Verticillium catenulatum* (Kamyschko ex Barron & Onions) W. Gams, also have a predilection for them.

The only holoblastic conidial fungi described from Myxomycetes appear to have been *Calcarisporium pallidum* Tubaki (on *Stemonitis fusca* Gmelin) which is conspecific with the ubiquitous *Sporothrix schenckii* Hektoen & Perkins (*Ceratocystis stenoceras* (Robak) C. Moreau nom. hol.; de Hoog, 1974), *Helicosporium binale* (Berk. & Curt.) Sacc. (on *Reticularia fuliginosa* Sacc.) which Pirozynski (1966) found to be identical to the wood-inhabiting *Xenosporium berkeleyi* (Curt.) Piroz., and the doubtful *Septocylindrium myxophagum* (Javoron.) Sacc. on *Didymium difforme* (Pers.) Gray. One dematiaceous hyphomycete has been described, *Cladosporium argillaceum* Minoura (on an undetermined myxomycete); its identity requires confirmation, as does the single report of a myxomyceticolous coelomycete, *Colletotrichum palinhae* Gonz. Frag. on *Lamproderma echinulatum* (Berk.) Rostr.

Ubiquitous saprophytic species of several genera (e.g., *Aspergillus* Micheli ex Fr., *Dendryphiella* Bubák & Ranojevic, *Penicillium, Scopulariopsis* Bain., *Sporothrix* Hektoen & Perkins, *Trichoderma* Pers. ex Fr.) are occasionally found on myxomycete sporangia but usually only when these have started to decay.

C. Mastigomycotina

No conidial fungi are known to attack the Chytridiomycetes (93 genera, 460 species) or Hyphochytridiomycetes (6-7 genera, 15 species), perhaps because they are almost all aquatic and (or) immersed within the cells of their host for the nonmotile phases of their life cycles. However, a considerable number of chytrids parasitize other chytrids (e.g., Karling, 1942, 1960; Willoughby, 1956).

The Oomycetes (70 genera, 550 species) includes the primarily aquatic Lagenidiales, Leptomitales, and Saprolegniales, and the mainly terrestrial Peronosporales. [Surprisingly, the order Lagenidiales also includes the lichenicolous fungus *Lagenidium hyphinicola* Moser-Rohrh. which is able to live inside lichen hyphae (Moser-Rohrhofer, 1975).] The three aquatic orders are probably not attacked by conidial fungi, at least in nature. Moreau (1939) reported two from Saprolegniales cultures; "*Macrosporium gemmivorum*" Moreau (on *Achyla conspicua* Coker), perhaps merely a *Stemphylium* Wallr.-like contaminant, and "*Cephalosporium saprolegniae*" Moreau (on *Saprolegnia delica* Coker), probably a member of *Verticillium* sect. *Prostrata* W. Gams (Gams, 1971).

The Peronosporales are generally very delicate fungi parasitizing leaves, roots, or shoots of vascular plants and would perhaps not be expected to be sufficiently robust to support conidial fungi except during their resting stages (oospores and zoosporangia) or on their mycelia. Very few conidial fungi have, however, been able to exploit the Peronosporales. The most fascinating group of these was discovered by seeding growing *Pythium* Pringsh. colonies with leaf mold and other plant debris. This revealed several moniliaceous Hyphomycetes able to invade *Pythium* oospores, assimilating their contents by the formation of lobed intracellular hyphae (Drechsler, 1938, 1943, 1952, 1962, 1963); five species of the mainly nematophagous genus *Dactylella* Grove (*D. anisomeres* Drechs., *D. helminthodes* Drechs., *D. spermatophaga* Drechs., *D. stenocrepis* Drechs., and *D. stenomeres* Drechs.), *Trichothecium arrhenopum* Drechs., *T. polyctonum* Drechs., and the also entomogenous and saprophytic *Trinacrium subtile* Riess. Drechsler (1938) failed to isolate *Dactylella* species from plant tissue affected by *Pythium* and does not appear to have tested these against a wide range of other fungi, only 1-11 *Pythium* species. A further species, *D. tenuis* Drechs., was partially parasitic, not developing on the *Pythium* alone (Drechsler, 1937). A few other conidial taxa have been described from Oomycetes: "*Cicinnobolus*" *heraclei* Dejeva [on *Plasmopara viticola* (Berk. & Curt. ex de Bary) Berl. & de Toni; this genus mainly confined to Erysiphales], *C. bremiphagus* Naoumoff (on *Bremia graminicola* Naoumoff), *Cladosporium colocasiae* Sawada (on *Phytophthora colocasiae* Racib.), "*Macrosporium*" *parasiticum* Thüm. (on *Peronospora schleideni* Unger) and *Trichothecium plasmoparae* Viala [on *Plasmopara nivea* (Unger) Schröt., mycelium.

Other reports of conidial fungi on Oomycetes in nature are: *Cladosporium cladosporioides* (Fres.) de Vries and *C. uredinicola* Speg. (both on *Peronospora* Corda species), *Fusarium semitectum* Berk. & Rav. [on *Sclerospora graminicola* (Sacc.) Schröt., parasitic on the oospores; see Raghavendra Rao and Pavgi, 1977; and *Dactylella spermatophaga, Alternaria alternata, Fusarium oxysporum* Schlecht., *Humicola fuscoatra* Traaen, and *Verticillium chlamydosporium* Godd., all on oospores of *Aphanomyces euteiches* Drechs., *Phytophthora megasperma* var. *sojae* Hildebrand, *P. cactorum* (Leb. & Cohn) Schröt., and *Phytophthora* sp. (Sneh *et al.*, 1977).

Trichoderma cultures caused rapid disintegration of hyphae of *Pythium* and *Phytophthora* species, probably because of a diffusable toxin (Weindling, 1932), and treatment of beet seeds with *T. viride* and *Penicillium frequentans* Westl. conidia successfully controlled the occurrence of *Pythium ultimum* Trow (Liu and Vaughan, 1965). *Trichoderma harzianum* Rifai was, however, unsuccessful in controlling the disease caused by *Phytophthora cinnamomi* Rands (Kelley, 1966). The coelomycete *Truncatella truncata* (Lév.) Stey. (*Broomella acuta* Shoem. & E. Müll. nom. hol.) shows specific antagonism toward *Pythium ultimum* (Domsch and Gams, 1968) due to ramulosin. For further references regarding fungi antagonistic to *Phytophthora* and *Pythium* see Domsch *et al.* (1980).

D. Zygomycotina

Zygomycetes (113 genera, 610 species) are host to very few fungicolous conidial fungi. Although there are many specialized mucoralean parasites of Mucorales (Barnett, 1964; Benjamin, 1959), very few, if any, other fungi can grow on healthy members of this order. The rapid growth of these fungi, their ephemeral nature, and perhaps also their cell wall composition, may all contribute to this.

The ability of *Aspergillus* and *Penicillium* species to enter sporangiophores and grow intracellularly has been known for over a century (see Weindling, 1938), but whether these are parasitic or merely utilizing nutrient-rich hyphal contents as sporulation of the host is being completed and the composition of the cell walls is changing, is unclear. *Actinomucor elegans* (Eidam) C. Berry and *Rhizopus oryzae* Went. & Prinsen-Geerl. are reported to be attacked by both *Myrothecium verrucaria* (Alb. & Schwein.) Ditm. ex Fr. and *Trichoderma viride,* the latter also parsitizing *Syncephalastrum racemosum* Cohn ex Schröt., *Zygorhynchus moelleri* Vuill., and four other mucoraceous fungi (Durrell, 1968). *Gliocladium roseum* may attack at least *Thamnidium elegans,* and *Fusarium solani* (Mart.) Sacc. parasitizes *Mucor hiemalis* Wehm. and *Phycomyces blakesleeanus* Burg. (Domsch and *et al.*, 1980). *In vitro, Mycogone perniciosa* (Magnus) Delacr., *Verticillium albo-atrum* Reinke & Bert hold, *V. dahliae* Kleb. and *V. psalliotae* Treschow can all grow intracellularly in sporangiophores and spores of *Rhopalomyces elegans* Corda (Barron and Fletcher, 1970, 1972; Dayal and Barron, 1970); *V. psalliotae* attacks six other mucoraceous fungi *in vitro* (Dayal and Barron, 1970). Swart (1975) found that *Mucor hiemalis* and *Phycomyces blakesleeanus*, when attacked by *Fusarium solani*, formed callus-like tubes around the invading hyphae.

The growth of *R. stolonifera* (Ehrenb. ex Link) Lind is inhibited by *Penicillium nigricans* Bain. ex Thom (Saksena and Lilly, 1967) and also by wortmanin from the *Penicillium* anamorph of *Talaromyces wortmanii* (Klöcker) C. R. Benj. (Brian *et al.*, 1957). Products of *Fusarium oxysporum* Schlecht. can inhibit sporangiospore germination in *Cunninghamella elegans* Lendner (Robinson *et al.*, 1968).

Ampelomyces quisqualis Ces., predominantly a parasite of Erysiphales, can also grow intracellularly and severely damage many members of the Mucorales (Linnemann, 1968); a further species of this genus, *A. abramovii* Babayan & Nelen, has been described from the hyphae of an undertermined *Mucor* Mich. ex Fr. *Sepedonium mucorinum* Harz. is associated with *Mortierella polycephala* Cocm., and there are reports of *S. curviseum* Harz and *Gonatobotrys flava* Bonord. from unidentified mucoraceous fungi. The only other conidial fungi reported from Mucorales are *Dendrodochium tenue* Petch (on discharged sporangia of *Pilobolus* Tode ex Fr. sp.) and "*Cephalosporium*" *macrocarpus* Corda (on *Rhizopus stolonifera* and *Thamnidium* Ehrenb. sp.); however, this latter is probably only a species of *Mortierella* (Gams, 1971).

So exceptional is it for any conidial fungus to be confined to a mucoralean host that Matruchot (1903) used this to support his thesis that *Cunninghamella echinulata* (Thaxt.) Thaxt. (*C. africana* Matr.) belonged to the Mucorales rather than the Hyphomycetes. He was right.

The Entomophthorales are almost devoid of conidial fungi. Drechsler (1952), however, discovered that *Dactylella helminthodes* could invade zygospores of *Cochlonema megalosomum* Drechs. in culture. No other conidial fungi seem to have been described from this order, but *Entomophthora* Fres. species commonly become fortuitously intermixed with ubiquitous saprophytic fungi (particularly *Cladosporium* species) on dead insect hosts.

No conidial fungi are known from the remarkable enigmatic group Trichomycetes (30 genera, 100 species) which occurs only attached to the gut linings of living arthropods.

E. Basidiomycotina

1. Teliomycetes

a. Uredinales (126 genera, 5000 species). The rusts support a wide range of conidial fungi, but relatively few belong to genera comprising only uredinicolous species. Among the latter are several monotypic genera: the poorly understood blastomycete *Sporidiobolus johnsonii* Nyland on *Phragmidium rubi-idaei* (DC.) P. Karst., the sphaeropsidaceous *Creonecte biparasitica* Petr. (on *Uromyces costaricensis* Syd.), and the Hyphomycetes *Paraphaeoisaria alabamensis* de Hoog & Morgan-Jones on aecial galls of *Cronartium quercuum* (Berk.) Miyabe & Shirai, *Stenospora uredinicola* Deighton (on *P. kraussiana* Cooke), and *Triposporina uredinicola* Höhn. (on *P. periodica* Racib.; other species referred to this genus are not congeneric, Deighton and Pirozynski, 1972). The formerly monotypic *Redbia* Deight. & Piroz., introduced for *R. pucciniicola* Deight. & Piroz. (on *P. holoserica* Cooke; Fig. 5C), now includes the saprophytic *R. elegans* Piroz. & C. Hodges (on fallen leaves).

Tuberculina Sacc. comprises 10 Hyphomycetes, most of which are exclusively uredinicolous. The commonest species in this genus, *T. persicina* (Ditm.) Sacc. (Fig. 4B) is known from at least 26 rust species in 10 genera and is most frequent on aecial states. *Tuberculina maxima* Rost. (doubtfully distinct from *T. persicina*) causes a "purple mold" on *Cronartium* Fr. species, especially *C. ribicola* J. C. Fischer (blister rust of pines) where it grows over aecial scars (Hubert, 1935). On *C. fusiforme* it sometimes grows away from aecia (Kuhlman and Miller, 1976), and with *C. comandrae* Peck it mainly occurs on the rust-induced canker, nevertheless reducing aeciospore production (Powell, 1971). Attempts to use *T. maxima* in the control of *C. ribicola* have been abandoned because of the slow spread of the mycoparasite (Hiratsuka and Powell, 1976), but *T. persicina* has been reported to prevent the development of *Gymnosporangium fuscum* Hedw. fil. on already infected leaves (Mijuškovíc & Vucinic, 1974).

The most widespread uredinicolous fungus is, however, undoubtedly the coelomycete *Sphaerellopsis filum* (Biv.-Bern. ex Fr.) Sutt. (*Darluca filum* (Biv.-Bern. ex Fr.) Berk.; Fig. 4A), the ubiquitous anamorph of *Eudarluca caricis* (Fr.) O. Erikss., although in its later stages it probably also derives some nutrients from the host plant (Eriksson, 1966). This is now known from at least 226 rusts (13 artificially infected), predominantly those on Cyperaceae and Gramineae, and distributed through 20 genera (Kranz, 1974), but it readily grows *in vitro* (Calpouzos *et al.*, 1957). It is a destructive mycoparasite, growing in the sori, the ascomata developing after the pycnidia and eventually occupying the whole sorus. The urediniospores are penetrated by unspecialized hyphae, presumably enzymatically (Carling *et al.*, 1976). There is evidence for some physiological specialization in the anamorphs (Keener, 1934). *Sphaerellopsis filum* may be potentially useful for the control of *Cronartium strobilinum* (Arth.) Hedgc., but not of *C. fusiforme* Hedgc. & Hunt ex Cumm., because of their different life cycles (Kuhlman and Matthews, 1976); spraying with the mycoparasite prior to rust colonization was considered unlikely to be successful in controlling *Puccinia recondita* Rob. ex Desm. (Swendsrud and Calpouzos, 1972).

Uredinicolous representatives of several more catholic genera are described,

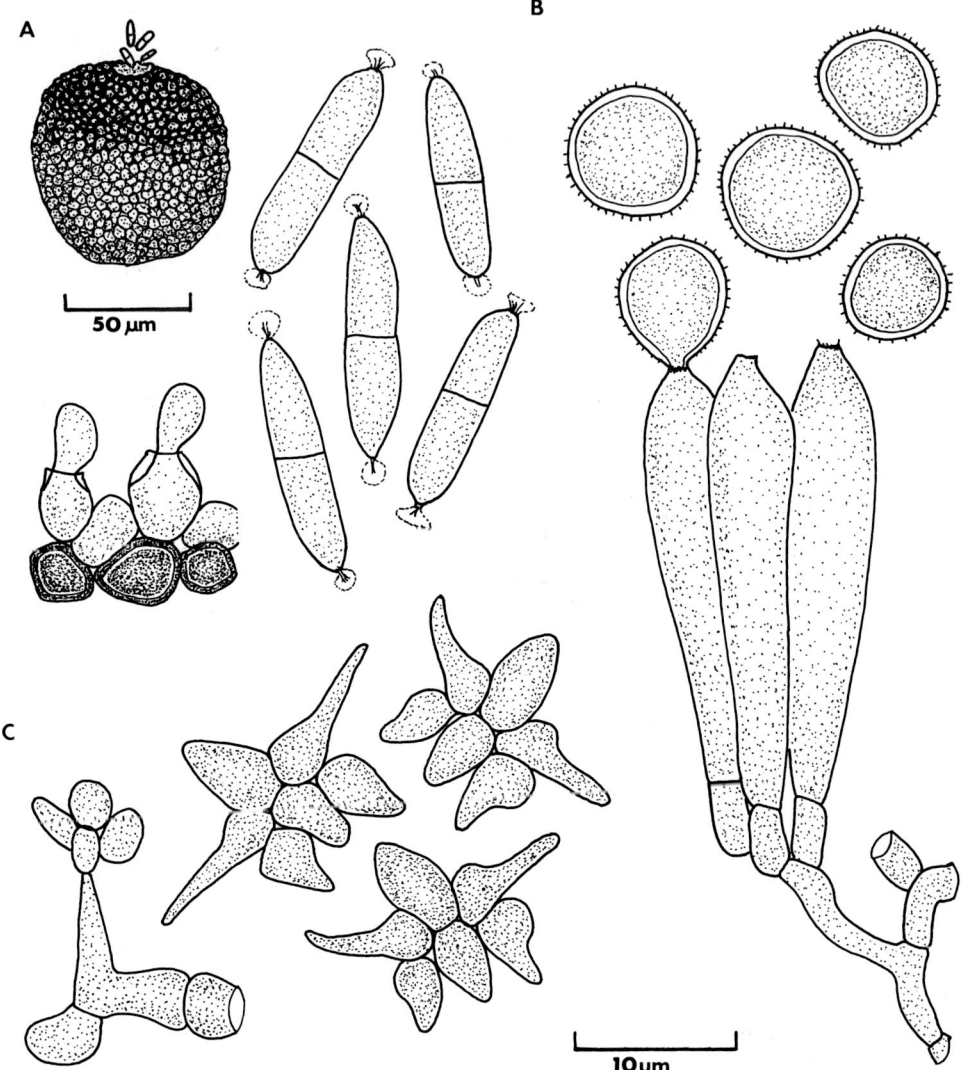

Fig. 4. Conidial fungi occurring on Teliomycetes (Uredinales). (A) *Sphaerellopsis filum*; (B) *Tuberculina persicina*; (C) *Titaea hemileiae*.

for example, *Bactridium gymnosporangii* (Jaap) Wollenw. (*Calonectria gymnosporangii* Jaap nom. hol., on *Gymnosporangium confusum* Plowr.; see Mijušković, 1976), *Cercospora uromycestri* Pollack (on *Uromyces cestri* Mont.), *Cladosporiella uredinicola* Deight. (on *Ravenelia* Berk. and *Uredo* Pers.), *C. uredinis* Deight. (on five *Puccinia* Pers. and three *Uredo* species; Fig. 5A), *Cladosporium aecidiicola* Thüm., *C. gallicola* Sutt. [on *Endocronartium*

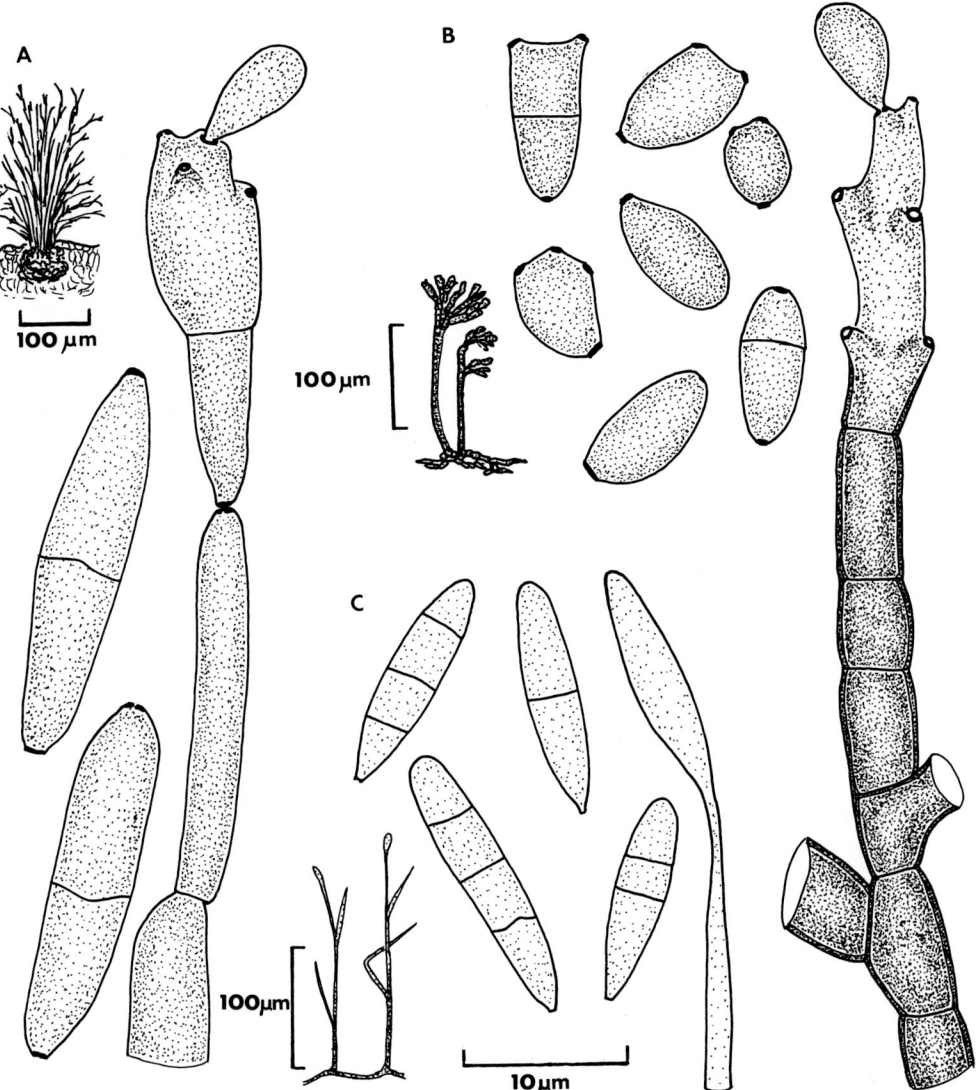

Fig. 5. Conidial fungi occurring on Teliomycetes (Uredinales). (A) *Cladosporiella uredinis*; (B) *Cladosporium gallicola*; (C) *Redbia puccinicola*.

harknessii (J. R. Moore) Hirats. and *Cronartium comandrae;* Fig. 5B], *C. hemileiae* Stey. (on *Hemileia vastatrix*), *Colletotrichum aeciicola* Tehon (on *Puccinia cnici-oleracei* Pers. ex Desm.), *Hendersonia uredinophila* Syd. (on *P. kuehniae* Schwein.), "*Phyllosticta*" *aecidiarum* Petr. (on *P. saccardoi* Ludw.), *Ramularia coleospori* Sacc. (on *Aecidium clematidis* DC.), *Scytalidium uredinicola* Kuhlman et al. (on *Cronartium fusiforme* Hedgc. & Hunt ex Cumm.), and *Titaea* Sacc. species, several of which are not uncommon and include *T. hemileiae* Hansf. (on *Hemileia vastatrix*; Fig. 4C).

The primarily entomogenous but perhaps too widely circumscribed *Verticillium lecanii* (*V. hemileiae* Bouriquet ex Bour. & Bass.; see Locci *et al.*, 1971; Singh, 1974) and the primarily agaricicolous *V. lamellicola* (E. F. Sm.) W. Gams, and *V. psalliotae* Treschow (*Cephalosporium curtipes* var. *uredinicola* Sukap. & Thirum.) are also occasionally found on rusts and may be harmful. Extracts of the myxomyceticolous *Aphanocladium album* cause premature teliospore formation in three *Puccinia* species (Forrer, 1977). *Cladosporium uredinicola* Speg., in contrast, is primarily uredinicolous but also recorded on rust-free leaves and associated with *Peronospora* species. *Scolecobasidium acanthacearum* (Cooke) M. B. Ellis can colonize four rust genera. *Fusarium trichothecioides* Wollenw. (*F. bactridioides* Wollenw.) has been reported to be a particularly destructive parasite of *Cronartium* species (Wollenweber, 1934), and *Sphaeropsis sapinea* (Fr.) Dyko & Sutt. [*Macrophoma sapinea* (Fr.) Pat.] completely killed galls of *Endocronartium harknessii* (Byler *et al.*, 1972).

Numerous fungi can be isolated from rust-infected tissues. Powell (1971), for example, obtained 56 fungi from *Cronartium ribicola*, but most were saprophytes or opportunists. *Alternaria tenuissima* and a *Septoria* Sacc. were, however, found to be active against *Puccinia pelargonii zonalis* Doidge (Mijušković, 1972). The widespread plant parasites *Colletotrichum dematium* (Pers. ex Fr.) Grove (*C. pucciniophilum* Togashi, *C. uredinophilum* Huela) and *C. gloeosporioides* (Penz.) Sacc. [*Gloeosporium roesteliaecola* Bubák & Serebr., *Glomerella cingulata* (Stonem.) Spauld. & Schrenk nom. hol.] are occasionally uredinicolous and may be mycoparasitic in some instances (e.g., Singh, 1975).

b. Ustilaginales (48 genera, 850 species). The smut fungi are extraordinarily free of fungicolous fungi, and only three Hyphomycetes appear to have been described from them. *Acontium ustilaginicola* Dickson [on *Ustilago hordei* (Pers.) Lagerh.] and *"Macrosporium" ustilaginis* Kell. & Swing. [on *Ustilago* (Pers.) Roussel sp.] are both rather doubtful taxa. *Fusarium ustilaginis* Kell. & Swing. [on *U. avenae* (Pers.) Rostr.] is conspecific with the ubiquitous *F. avenaceum* (Corda ex Fr.) Sacc. (Booth, 1971). *Acremonium bactrocephalum* W. Gams was originally isolated from an unidentified smut and may have been parasitic on its mycelium and spores, but this species is also known from other habitats (Gams, 1971).

The primarily uredinicolous *Sphaerellopsis filum* (see above) is also recorded on a few *Ustilago* species, which may also be affected by the contact mycoparasite *Stephanoma phaeospora in vitro* (Rakvidhyastra and Butler, 1973).

It is notable that all the conidial fungi reported from smuts are on a single genus (*Ustilago*) in the Ustilaginaceae. None appear to be known from the other family of the Ustilaginales, the Tilletiaceae, the "conidia" of the basidiomycete *Itersonilia perplexans* Derx on *Entyloma calendulae* (Dud.) de Bary excepted (Brady, 1960).

2. Hymenomycetes

a. Agaricales (200 genera, 5000 species). Decaying large fleshy sporocarps of members of the Agaricales provide a nutrient-rich substrate capable of supporting a wide range of ascomycetous, mucoraceous, and conidial fungi. The number of such fungi able to invade and parasitize maturing and mature fruit bodies is, however, more restricted. As molds on decaying agarics are conspicuous in nature, and fungi attacking commercially grown mushrooms are of economic importance, it was to be expected that these had received more attention than those occurring on some other groups of fungi. No synthesis of these fungi is available, but several regional studies have been prepared, among which those of Arnold (1970), Nicot (1963, 1967), and Tubaki (1955) merit particular mention.

Most exclusively agaricicolous conidial fungi are moniliaceous Hyphomycetes forming conidia holoblastically, although several phialidic species are also known.

One of the most important genera of agaricicolous conidial fungi is *Cladobotryum* Nees, species of which have teleomorphs in *Hypomyces* (Fr.) Tul. (Gams and Hoozemans, 1970). A comprehensive bibliography of the Hypomycetaceae (inclusive of the anamorphs) prepared by Arnold (1976) includes host lists based on the 595 papers abstracted. Most common on decaying agarics are *C. apiculatum* (Tub.) W. Gams & Hooz. [*Blastotrichum puccinioides* Preuss, *H. ochraceus* (Pers. ex Fr.) Tul. nom. hol.], *C. mycophilum* (Oud.) W. Gams & Hooz. (*H. odoratus* G. Arnold nom. hol.), and *C. verticillatum* (Link ex Gray) Hughes. *C. apiculatum* and *C. verticillatum* are particularly frequent on *Lactarius* Pers. ex Gray and *Russula* (Pers. ex Fr.) Gray species. Two further species, *C. dendroides* (Bull. ex Mérat) W. Gams & Hooz. [*Dactylium dendroides* Bull. ex Mérat, *H. rosellus* (Alb. & Schw. ex Fr.) Tul. nom. hol.] and *C. varium* Nees ex Duby [*H. aurantius* (Pers. ex Fr.) Tul. nom. hol.; Fig. 6B], occur on agarics but are more frequent on Aphyllophorales (see Section III,E,2,b). A fungus with an allied teleomorph, *Sepedonium chrysospermum* (Bull. ex Purton) Link [*H. chrysospermus* Tul. nom. hol., syn. *Apiocrea chrysosperma* (Tul.) Syd.; Fig. 7C], is a conspicuous golden-yellow mold, primarily on decaying Boletaceae but also known from some species of the Agaricaceae, Aphyllophorales, and Gasteromycetes. More restricted in its host range is *S. brunneum* Peck (*Peckiella completa* G. Arnold nom. hol.), confined to *Suillus* Gray (*Boletinus* Kalchbr.) species.

Mycogone Link ex Chev., a genus much in need of taxonomic revision, includes several agaricicolous species, some of which are economically important as pathogens of cultivated mushrooms as well as other agarics. *M. perniciosa* (Magnus) Delacr. (Fig. 6D) causes "wet bubble disease" in which the mushroom stipe enlarges, the gills are deformed, and finally the carpophores decompose, exuding brown drops; the invading hyphae generally pass between those of the host and exceptionally just enter the cell walls (Smith, 1934); the

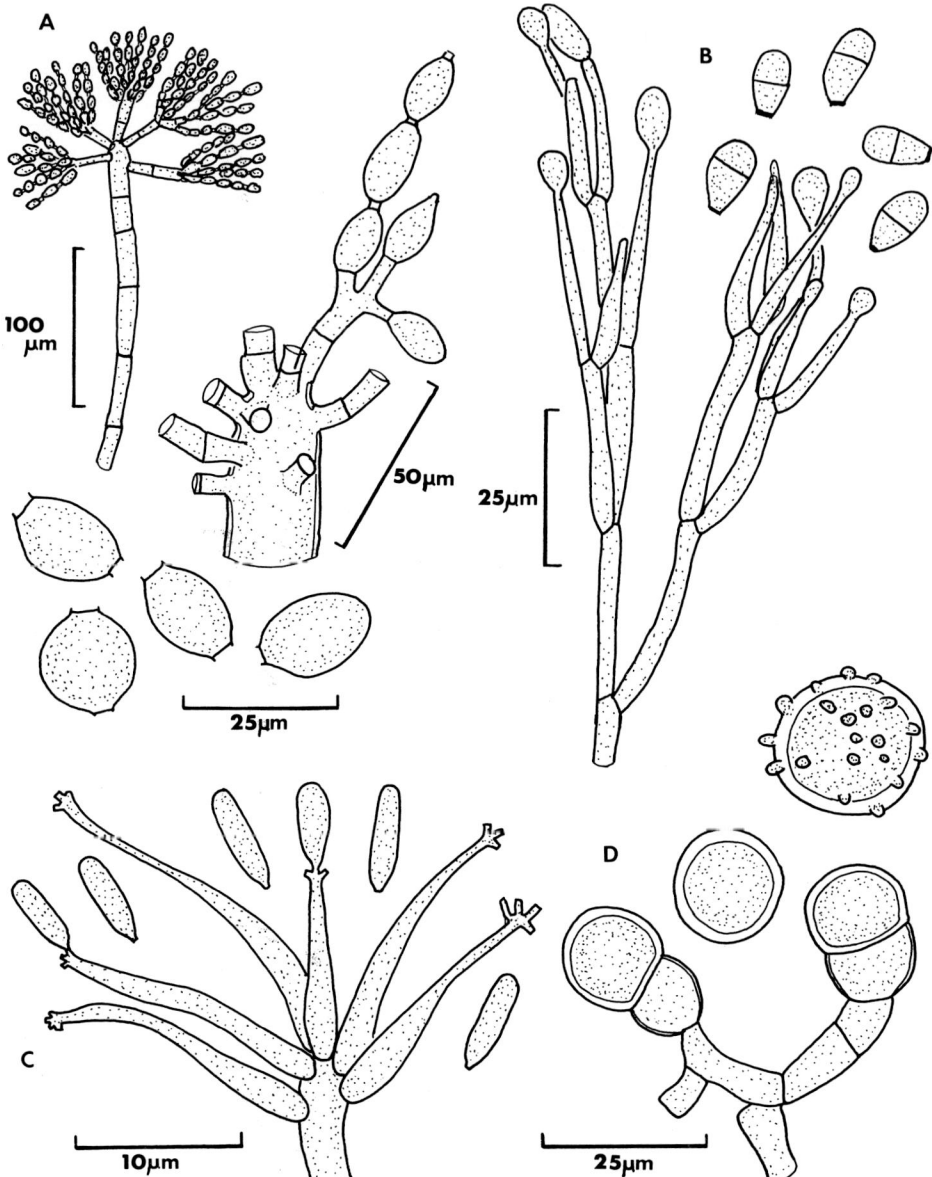

Fig. 6. Conidial fungi occurring on Hymenomycetes (Agaricales). (A) *Amblyosporium spongiosum*; (B) *Cladobotryum varium*; (C) *Calcarisporium arbuscula*; (D) *Mycogone perniciosa*.

considerable literature on this disease is summarized by Brady and Gibson (1976). Much more widespread on wild agarics is *M. rosea* (Link ex Pers.) Link, which leads to brown and often zoned discolorations of the pileus and an underlying yellow staining in cultivated mushrooms (Brady and Gibson, 1976); Arnold (1976) listed this species from nine host genera. Both *M. perniciosa* and *M. rosea* occasionally also produce *Verticillium*-like anamorphs and appear to be primarily transmitted through soil. *Mycogone perniciosa* and *Verticillium psalliotae* grow intracellularly in *Rhopalomyces elegans*, something which might facilitate their dispersal (Barron and Fletcher, 1970, 1972).

Myceliophthora lutea Cost., causal agent of "mat disease" or *vert de gris* in cultivated mushroom compost, forms a mat of yellowish flakes which can lead to crop losses of 50–70% (Manns, 1947) but is perhaps mainly associated with the compost, not the mushrooms themselves. *Myceliophthora* includes two further species (van Oorschot, 1977), one of which, *M. fergusii* (Klopotek) Oorschot [*Corynascus fergusii* (Fergus & Sinden) Klopotek nom. hol.] was also isolated from mushroom compost.

Amblyosporium Fres. comprises four species of which two are common on decaying agarics (Pirozynski, 1969), *A. botrytis* Fres. (also on Aphyllophorales but isolated from a wide range of other substrates), and *A. spongiosum* (Pers.) Hughes (most commonly on *Lactarius* species; Fig. 6A). Nicot and Durand (1965) studied the latter in some detail (incorrectly as "*A. botrytis*"), noting that its sclerotia might be important for its survival in soil and that it might have an inoperculate discomycete as its teleomorph.

A few monotypic agaricicolous moniliaceous hyphomycete genera are known. *Asterophora lycoperdoides* Ditm. ex Gray is the chlamydosporic phase of the obligately agaricicolous agaric *Nyctalis asterophora* Fr.; this species and *N. parasitica* (Bull. ex Fr.) Fr., which has a smooth-walled chlamydosporic phase, were investigated in depth by Buller (1924). *Calcarisporium arbuscula* Preuss (on numerous agarics, but also recorded from polypores, stromatic ascomycetes, and wood; Fig. 6C) generally grows endophytically (Watson, 1955; Nicot, 1968), giving rise to few external symptoms; but when the equilibrium is disrupted, for example by other microorganisms (including other fungi), *C. arbuscula* increases and contributes to the death of the host (Nicot, 1968). "*Harziella*" *capitata* Cost. & Matr. is only recorded on *Lepista nuda* (Bull. ex Fr.) Cooke (Fontana, 1960), *Sibirina fungicola* G. Arnold on *Lentinus* Fr. sp. (Fig. 7B), and *Tilachlidium brachiatum* (Batsch ex Fr.) Petch (syn. *Corethropsis epimyces* Massee, *Pseudonectria tilachlidii* W. Gams nom. hol.) predominantly occurs on *Mycena* Pers. ex Gray and *Xeromphalina* Kühner & Maire species. However, *Gabarnaudia tholispora* Arnaud ex Sams. & W. Gams (Fig. 10B), originally described from *Russula nigricans* (Bull. ex Mérat) Fr., really appears to be a wide-ranging mycophilous saprophyte (Samson, 1974), and the name *Verticilliopsis infestans* Cost. (reported to cause "false plaster of Paris" in

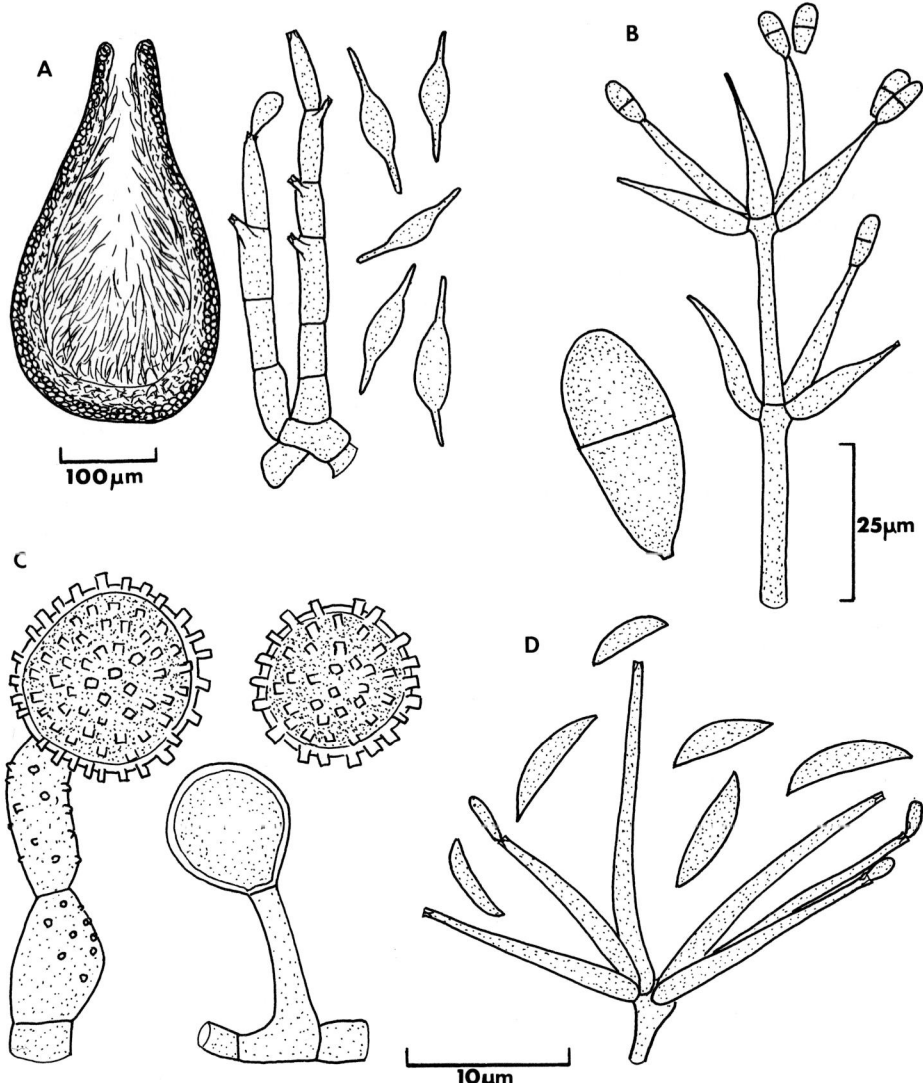

Fig. 7. Conidial fungi occurring on Hymenomycetes (Agaricales). (A) *Eleutheromyces subulatus*; (B) *Sibirina fungicola*; (C) *Sepedonium chrysospermum*; (D) *Verticillium psalliotae*.

cultivated mushrooms) is of uncertain application (Kendrick and Carmichael, 1973).

Several ubiquitous hyphomycete genera include species described from agarics, for example, *Acremonium hypholomatis* (Boed.) D. Hawksw. [on *Hypholoma fasciculare* (Huds. ex Fr.) Kummer; also a saprophyte], *A. incrus-*

tatum W. Gams [on *Armillaria mellea* (Vahl ex Fr.) Kummer], *A. tulasnei* G. Arnold [on *Lactarius deliciosus* (L. ex Fr.) Gray], *Cladosporium fuligineum* Bon. (on several genera), *"Geotrichum cyphellae"* Arnaud [on *Lachnella alboviolascens* (Alb. & Schw. ex Fr.) Fr.], *Gliocladium agaricinum* Cooke & Massee (on *Agaricus campestris* L. ex Fr.), *"Isaria" fruticosa* Demelius [on *Tricholomopsis rutilans* (Schaeff. ex Fr.) Sing.], *Myrothecium inundatum* Tode ex Gray [especially common on *Russula adusta* (Pers. ex Fr.) Fr. and *R. nigricans*], *Polyscytalum fungorum* Sacc. (on *Nyctalis parasitica*), *Verticillium fungicola* (Preuss) Hassebr. (primarily on *Agaricus* species), *V. lamellicola* (F. E. V. Sm.) W. Gams (on *A. arvensis* Schaeff. ex Secr. and *A. campestris*), and *V. psalliotae* Treschow (primarily on *Agaricus* but also known from other groups, see Sections III,D; E,1,a; F,4,b Fig. 7D). *Verticillium fungicola* causes the important "dry bubble" disease of cultivated mushrooms in which the pileus becomes brown-spotted and the host becomes distorted, swollen, and leathery; similar effects can be caused by *V. psalliotae*, and information on these species is summarized by Brady and Waller (1976) and Brady and Gibson (1976), respectively.

Several species of *Sporotrichum* Link ex Fr. and *Stilbum* auct. are described from agarics, but are all in need of critical study. *Fusarium agaricorum* Sarraz. (on *A. campestris*) does not appear to belong to this genus (Booth, 1971), and the identity of *Stachybotrys klebahnii* Burchard (on *A. campestris*), one of the few dematiaceous Hyphomycetes reported, needs reexamination.

Many of the primarily agaricicolous Hyphomycetes are occasionally isolated from woodland leaf litter or soil, but it is uncertain whether many can grow in nature in this situation in the absence of decaying agaric carpophores. Some of these reports at least must be based on resting propagules or sites where agarics have very recently decayed.

Of the ubiquitous mycoparasites, *Trichoderma viride* Pers. ex Gray *sensu lato* is particularly destructive: it can cause up to 85% losses in cultivated *Agaricus* species (Treschow, 1942) and is lethal to cultivated *Lentinus edodes* (Berk.) Singer (Komatsu and Hashioka, 1964). It also attacks *Armillaria mellea* (Aytoun, 1953), and heavy inoculations can be used to control this species (Garrett, 1958); most species of this genus, apart from *T. hamatum* (Bonord.) Bain., are very antagonistic toward *A. mellea*. *Acremonium crotocinigenum* (Schol-Schwarz) W. Gams and *Scytalidium aurantiacum* Klingström & Beyer (Morquer and Touvet, 1974) are also actively antagonistic toward this species.

Not surprisingly, many ubiquitous saprophytic Hyphomycetes have been found on decaying Agaricales (see, e.g., Nicot, 1963). It would be superfluous to enumerate these fully, but the following may be mentioned as examples: *Acremonium furcatum* F. & R. Moreau ex W. Gams, *A. strictum*, *Aspergillus candidus* Link ex Fr., *A. glaucus sensu lato*, *A. niger* van Tieghem, *Cladosporium cladosporioides*, *C. herbarum*, *C. macrocarpum* Preuss, *Epicoccum purpurascens, Fusarium oxysporum* (syn. *F. mycophilum* Sacc.), *F. sporotrichioides* Sherb. (particularly on *Hygrophorus* Fr. and *Omphalina* species), *Gliocladium viride* Matr., *Mariannaea elegans* (Corda) Sams., *Paecilomyces marquandii* (Massee) Hughes (with a preference for *Hygrophorus* species), numerous *Penicillium* species (*P. brevicompactum* is particularly common), and *Trichoderma harzianum* Rifai.

Coelomycetes are relatively scarce on Agaricales, but two monotypic genera, studied in detail by Seeler (1943), are known: *Eleutheromyces subulatus* (Tode ex Fr.) Fuckel (on decaying agarics and on *Grifola frondosa* (Dicks. ex Fr.) Gray and *Lenzites* Fr. sp.; Fig. 7A), and *Hyalopycnis blepharistoma* (Berk.) Seeler (on a very wide range of agarics, but also known from *Leotia lubrica* Pers. and some herbaceous plant materials). The latter may interact with *Calcarisporium arbuscula* to cause considerable damage to the host (Nicot, 1968). Two further Coelomycetes described from agarics in ubiquitous genera require reexamination: *Pestalotia duportii* Pat. (on *Boletus* sp.) and *Phoma agaricola* Rostrup (on *Agaricus* sp.).

Putrifying agarics provide a substrate suited to numerous yeasts. Anderson and Skinner (1947) found yeasts (mainly ascomycetous) on 21 species, and Ramírez Gómez (1957) obtained 117 yeast species from 105 macromycetes (mainly Agaricales). Several blastomycete yeasts have been described as new from agarics, of which *Candida anomala* Ramírez (on Agaricaceae indet.), *C. buffonii* (Ramírez) van Uden & Buckley (on *Boletus edulis* Bull. ex Fr.), *C. obtusa* var. *arabinosa* Montrocher [on *Clitopilus prunulus* (Scop. ex Fr.) Kumm.], and *Torulopsis kruisii* Kocková-Kratochvílová & Ondrušova (on *Boletus purpureus* Pers.), are still known only from agarics; several which have subsequently been discovered on other substrates are omitted here. The cosmopolitan airborne *Sporobolomyces roseus* Kluyver & van Niel (*S. boleticola* Ramírez) has been isolated from *Leccinum aurantiacum* Gray. Also of interest is the anamorph of the yeast *Endomyces decipiens* Rees, *Geotrichum armillariae* v. Arx, discovered on *Armillaria mellea*.

b. Aphyllophorales (375 genera, 1000 species). Only seven genera of conidial fungi are confined to the Aphyllophorales. All are monotypic. They are the moniliaceous Hyphomycetes *Eurasina bondarzewiae* G. Arn. [on *Polyporus brumalis* (Fr.) Fr.; Fig. 8A], *Pseudohansfordia irregularis* G. Arn. [on *Coriolus versicolor* (L. ex Fr.) Quél.], and *Rhinotrichiella globulifera* Arnaud ex de Hoog [on *Ganoderma lucidum* (W. Curtis ex Fr.) P. Karst. and *Inonotus dryadeus* (Pers. ex Fr.) Murr.; see Udagawa and Horie, 1971a]; the dematiaceous Hyphomycetes *Exosporiella fungorum* (Fr.) P. Karst. (on *Corticium* Pers. ex Gray and *Thelephora* Ehrh. ex Fr.; Fig. 8C), *Spondylocladiella botrytioides* Linder (on hymenium of ? *Corticium*), and *Zakatoshia hirschiopori* Sutt. [on *Hirschioporus abietinus* (Pers. ex Fr.) Donk; Fig. 8B]; and the Coelomycetes *Eleutheromycella mycophila* Höhn. and *Scopaphoma corioli* Dearn. & House (both on *Coriolus versicolor*).

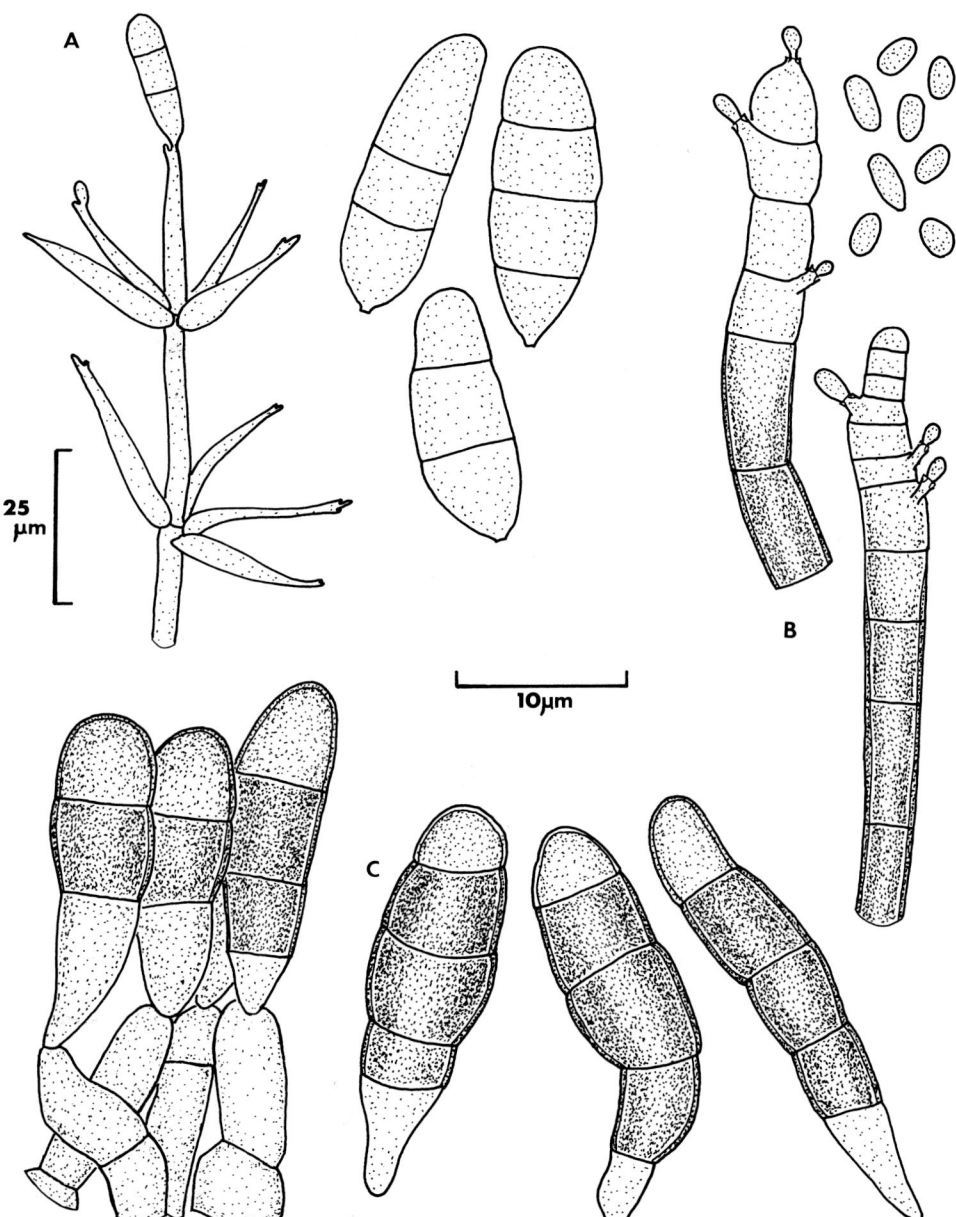

Fig. 8. Conidial fungi occurring on Hymenomycetes (Aphyllophorales). (A) *Eurasina bondarzewiae*; (B) *Zakatoshia hirschiopori*; (C) *Exosporiella fungorum*.

Sympodiophora G. Arn. was originally described for the single species *S. stereicola* G. Arn. [on *Stereum hirsutum* (Willd. ex Fr.) Gray], but the genus now includes a further seven species, mainly restricted to foliicolous ascomycetes (Deighton and Pirozynski, 1972). The sporodochial *Schizotrichella lunata* Morris, originally described from *Cantharellus odoratus* Schwein. ex Fr. and subsequently found on the moss *Mnium cuspidatum* Hedw. (Bowen, 1968), is only *Colletotrichum dematium* (Pers. ex Fr.) Grove (Tulloch, 1972, but see Pirozynski, 1972, regarding *Neottiosporella*). "*Sporocephalum peniophorae*" Arnaud [on *Peniophora longispora* (Pat.) Pat.] can probably be accommodated in *Oedocephalum* Preuss (Kendrick and Carmichael, 1973).

Some primarily agaricicolous genera may also occur on Aphyllophorales: *Amblyosporium botrytis* (particularly on Clavariaceae), *Calcarisporium arbuscula*, the *Sepedonium* anamorph of *Hypomyces chrysospermus*, *Mycogone calospora* (P. Karst.) Höhn. [on *Ramaria decolorans* (P. Karst.) Corner], and *Myrothecium inundatum*. *Cladobotryum* (see Gams and Hoozemans, 1970) includes two species particularly frequent on bracket fungi (*C. dendroides* and *C. varium*, the latter also on Hydnaceae) and a further species, *C. leptosporum* (Sacc.) W. Gams, known only from *Corticium, Peniophora* Cooke, and *Stereum* Pers. ex Gray. *Myceliophthora fusca* Doger [on *Merulius lacrymans* (Wulf. ex Fr.) Schröt.] is, however, only *Ptychogaster rubescens* Boud., an anamorph of a member of the Polyporaceae (von Arx, 1973).

The perennating basidiomata of many members of this order renders them much more available to colonization by other fungi than the shorter-lived Agaricales. Consequently, a considerable number of species have been described in a wide range of genera from bracket fungi, but it is possible that many of these will later be found in other habitats also. The following are examples: *Acremonium crotocinigenum* (on many polypores but also known from agarics and *Ustulina deusta*), *A. domschii* W. Gams [on *Fometopsis* (Fr.) P. Karst. and *Inonotus obliquus* (Pers.) Pilát], *A. exiguum* W. Gams [on *Tubulicium dussii* (Pat.) Oberw.], *Acrodontium hydnicola* (Peck) de Hoog (on ? *Hydnum* L. ex Fr. sp.), *Asterostomella parasitica* Rick (on *Polyporus* Mich. ex Fr.), *Chalara fungorum* (Sacc.) Sacc. (on *Hydnum compactum* Pers. ex Fr.), *C. microspora* (Corda) Hughes (on Hydnaceae indet.), *Coniothyrium epimyces* Cooke & Massee (on *Polyporus portentosa* Berk.), *Didymostilbe obovoidea* Mats. (on *Ganoderma lucidum*), "*Endophragmia*" *dennisii* M. B. Ellis [on *Hyphodontia sambuci* (Pers. ex Pers.) J. Erikss.], *Fusicladium poriicola* Bonar (on *Poria ferrea* Pers.), *Gliocladium penicilloides* Corda (*Nectriopsis aureonitens* (Tul.) Maire nom. hol., on *Merulius, Stereum,* and so on), *Helminthosporium conviva* Malençon & Bertault [on "*Hyphoderma calyciferum*" (Litsch.) Mal. & Bert.], *Humicola asteroidea* Udagawa & Horie (on *Coriolus* sp.), *Melanconium parasiticum* Westend. [on *Ramaria aurea* (Fr.) Quél.], *Phialophora brevicollaris* W. Gams (on *Phellinus* Quél. sp.), *Phoma portentosa* Cooke & Massee (on

Polyporus portentosa), *"Scolecotrichum" clavariarum* (Desm.) Sacc. (on numerous Clavariaceae), *Septocylindrium lindtneri* Kirschst. (on *Ganoderma lucidum*), *Sporothrix fungorum* de Hoog & de Vries [on *Fomes* (Fr.) Fr. sp.], *Stilbum mycetophilum* Ahmad (on *Poria* Pers. ex Gray), *Tritirachium fungicola* Shvartsman (on *Fomitopsis pinicola*), *Verticillium berkeleyanum* P. Karst. [*Nectriopsis berkeleyanum* (Plowr. & Cooke) Maire nom. hol., on *Hymenochaeta rubiginosa* (Schrad.) Lév. and *Stereum hirsutum*], *V. fusisporum* W. Gams [on *Coltricia perennis* (Fr.) Murr. but also from soil], *V. olivaceum* W. Gams [on *Inonotus radiatus* (Sow. ex Fr.) P. Karst.], and *Zythia compressa* Schw. (on *Grifola frondosa*). In addition, the Blastomycetes *Candida santamariae* var. *membranaefaciens* Montr. (on *Hymenochaete rubiginosa*) and *Torulopsis schatavii* Kochová-Kratochílová & Ondrušova (on *Fomitopsis pinicola*) should be mentioned.

Trichoderma species are antagonistic to most members of the Aphyllophorales so far tested. Of particular note is their action against *Ganoderma* P. Karst. species (Varghese *et al.*, 1976), *Pellicularia sasakii* (Hashioka and Fukita, 1969), and *Phaeolus schweinitzii* (Fr.) Pat. (Aytoun, 1953), in the treatment of stumps for the control of *Heterobasidion annosum* (Fr.) Bref. (Negrutskii, 1963), and in preventing the decay of standing poles (Ricard, 1976). Fungi attacking sclerotia of the plant pathogenic *Athelia rolfsii* (Curzi) Tu & Kimbrough have received considerable attention: species of *Alternaria* Nees ex Wallr., *Aspergillus,* and *Geotrichum* Link ex Pers. are all effective (Domsch and Gams, 1980), although *T. harzianum* and *T. viride* proved most useful for its biological control (Wells *et al.*, 1972).

The upper surfaces of bracket fungi, in addition to their own inherent nutrient content, frequently support algae and insect frass and retain moisture. Consequently they provide an ideal habitat for ubiquitous saprophytic fungi, and many have been recorded from them; Udagawa and Horie (1971a,b) provide full descriptions and drawings of some of them. The richness of larger polypores for saprophytic species is evidenced by the occurrence of 12 Hyphomycetes on a single specimen of *Meripileus giganteus* (Pers. ex Fr.) P. Karst. (Nicot, 1962). The species encountered include *Acremoniella atra* (Corda) Sacc., *Acremonium butyri* (van Beyma) W. Gams, *Cladosporium herbarum, C. macrocarpum, Colletotrichum dematium, Doratomyces putredinis* (Corda) Morton & G. Sm., *Epicoccum purpurascens, Fusidium hypophleoides* Corda, *Gliocladium roseum, Penicillium brevicompactum, P. thomii* Maire, *Periconia byssoides* Pers. ex Schw., *Phialophora luteo-viridis* (van Beyma) Schol-Schwarz [*Coniochaeta velutina* (Fuckel) Cooke nom. hol.], *Taeniolella scripta* (P. Karst.) Hughes, *Trichoderma polysporum* (Link ex Pers.) Rifai (*Tolypomyria fungicola* P. Karst.), *Verticillium tenerum* (Nees ex Pers.) Link (*Nectria inventa* Pethybr. nom. hol.), *Virgaria nigra* Link ex Gray, and *Veronaea coprophila* (Subram. & Lodha) von Arx.

c. **Other Orders** (66 genera, 300 species). The remaining orders of the Hymenomycetes have very few conidial fungi reported from them, and most are evidently rare. Only one genus has been described from them, "*Flahaultia hyalina*" Arnaud, which has been compared by Watling and Kendrick (1979) to the anamorph of *Sebacina incrustans* (Pers. ex Fr.) Tul. The primarily agaricicolous *Cladobotryum verticillatum* is recorded on *Auricularia mesenterica* (Dicks. ex Gray) Pers., *Cladosporium fuligineum* appears particularly frequent on *Exobasidium* Woron. species, and *Fusarium larvarum* Fuckel (*Nectria aurantiicola* Berk. & Br. nom. hol.) commonly infects the entomogenous *Septobasidium clelandii* Couch (*Harpographium corynelioides* Cooke & Massee nom. an.) (Coles and Talbot, 1977). *Sympodiophora mycophila* (Tub.) Deight. & Piroz. was described from *Hirneola* Fr., *Bulgaria* Fr., and *Marasmius* Fr. species. Other conidial fungi discovered include *Diplococcium clarkii* M. B. Ellis [on *Cristella confinis* (Bourd. & Galz.) Donk], *Phoma tremellae* Pat. (on *Tremella* Dill. ex Fr. sp.), and two blastomycete yeasts: *Tilletiopsis pallescens* Gokhale (on *Sirobasidium* Lagerh. & Pat. sp.), and *Torulopsis auriculariae* Nakase [on *Auricularia auricula-judae* (Bull. ex St.-Am.) Wettst.]. *Sporobolomyces coralliformis* Tub. (on *Exidia* Fr. sp.) is conspecific with the widespread *S. holseticus* Wind.

3. **Gasteromycetes** (150 genera, 700 species)

Considering that the sporocarps in many gasteromycete genera are tough and persistent, it is remarkable that no genus of conidial fungi is exclusive to them. Furthermore, few Hyphomycetes are described from them, all in need of critical study, for example, *Monilia fungicola* Ell. & Barth. [on *Langermannia gigantea* (Batsch. ex Pers.) Rostk.] and *Sporotrichum phalloidearum* (Corda) Rabenh. (on *Phallus impudicus* L. ex Pers.). Several *Fusarium* species described prove to be only synonyms of widespread saprophytic species, such as *F. avenaceum* (Corda ex Fr.) Sacc. (syn. *F. detonianum* Sacc.), *F. equiseti* (Corda) Sacc. (syn. *F. sclerodermatis* var. *lycoperdonis* Picb.), and *F. sambucinum* Fuckel (syn. *F. sclerodermatis* Oud.) (see Booth, 1971).

Hollós (1906-1907), an experienced student of the Gasteromycetes, made a particular study of the fungi present on them in Hungary. He found Coelomycetes on 10 species, describing 16 as new, for example, *Diplodina geasterina* Hollós (on *Geastrum ambiguum* Mart.), *Phoma gasteropsidis* Hollós (on *Geasteropsis conrathii* Hollós), *Pyrenochaeta geasteris* Hollós [on *Geastrum fornicatum* (Huds. ex Winch) Hook.], *Robillarda geasteris* Hollós (on *G. ambiguum*), and *Stagonospora geastericola* Hollós (on *G. hungaricum* Hollós and *G. minimum* Schw.). In addition, two Coelomycetes have been described on *Tulostoma volvulatum* Borshchov: *Diplodina tylostomatis* Pat. and *Phoma herbarum* var. *tylostomatis* Pat. These data, together with a reference to the occurrence of *Asteroma fugax* Rob. & Desm. (on *Tulostoma brumale* Pers.), suggest that the Gasteromycetes are a rich source of Coelomycetes meriting a special study.

The primarily agaricicolous *Amblyosporium botrytis* is known on several species of *Lycoperdon* Pers., and the *Sepedonium* anamorph of *Hypomyces chrysospermus* is not infrequently reported in temperate woods (e.g., on *Melanogaster variegatis* Tul., *Octaviania asterosperma* Vitt., *Rhizopogon* Fr. & Nordholm sp., *Scleroderma citrinum* Pers., *S. verrucosum* Bull. ex Pers.).

A single blastomycete yeast has been described from the subclass, *Candida boleticola* Nakase [on *Astraeus hygrometricus* (Pers.) Morgan].

F. Ascomycotina

The fungicolous habit reaches its zenith in the Ascomycotina, particularly in those which are foliicolous. Hansford (1946) has reviewed the information on conidial fungi found on foliicolous non-lichenized ascomycetes, and Hawksworth (1979) has prepared a revision of all lichenicolous Hyphomycetes so far described. In the absence of an entirely satisfactory scheme for the classification of all Ascomycotina above the family level, the lichenicolous conidial fungi are treated together here even though the lichen-forming species are distributed through a very wide range of orders and subclasses (Hawksworth, 1978).

1. Hemiascomycetes (60 genera, 325 species)

These, like the entirely blastomycetous yeasts, are devoid of obligately fungicolous conidial fungi. Fortuitous contamination of yeast cultures, for example by species of *Aspergillus, Geotrichum candidum* Link ex Pers., and *Penicillium*, inevitably occurs occasionally and may be detrimental to the yeast colonies.

The Taphrinales are mainly subepidermal or subcuticular and so would not be expected to be easily attacked by other fungi. Only one hyphomycete has been described from this order, *Cladosporium exoasci* Lindau (on *Taphrina pruni* Tul.). Some Coelomycetes referred to ubiquitous genera are also reported, but require a reappraisal, as they may have been associated with the host of the *Taphrina* Fr. rather than the *Taphrina* itself: *Aposphaeria parasitica* Allesch. [on *Taphrina betulae* (Fuckel) Johanson], *Hendersonia taphrinicola* Tracey & Earle (on *Taphrina* sp.), *Pestalotia taphrinicola* Ell. & Ev. [on *T. caerulescens* (Desm.) Tul.], *Phoma parasitica* Ell. & Ev. (on *T. caerulescens*), and *P. parasitica* var. *taphrinae-pruni* Allesch. (on *T. pruni*). No conidial fungi appear to be known from the Protomycetales, members of which occupy a habitat similar to that of the Taphrinales.

2. Plectomycetes (100 genera, 200 species)

With the exclusion of the Erysiphales and Meliolales, now placed in the Pyrenomycetes (Ainsworth *et al.*, 1973a), the Plectomycetes become a group of predominantly saprophytic species mainly known in culture. Apart from cultural contaminants, the Plectomycetes are almost entirely free of conidial fungi and none appear to have been described from them. The only reports located are of unidentified *Acremonium* and *Eriomycopsis* Speg. species on the tropical foliicolous *Dexteria pulchella* Stevens and a few saprophytes on *Onygena equina* Willd. ex Fr. [*Acremonium, Aspergillus*, and *Stephanosporium cerealis* (Thüm.) Swart].

There are some reports of antagonistic behavior against Plectomycetes by conidial fungi. For example, *Pseudeurotium zonatum* van Beyma is inhibited by *Pseudocercosporella herpotrichoides* (Fron) Deight. *in vitro* (Domsch and Gams, 1968).

3. Pyrenomycetes (710 genera, 8000 species)

a. Coronophorales (2 genera, 26 species). This small order, which may be better subsumed under the Sphaeriales (Nannfeldt, 1975), has a single dematiaceous hyphomycete described from it, *Acrodictys obliqua* M. B. Ellis (on *Nitschkia* Otth sp. perithecia), but is probably a saprophyte, since it also occurred directly on wood.

b. Erysiphales (20 genera, 100 species). The most important genus of conidial fungi on Erysiphales is certainly the destructive coelomycete *Ampelomyces quisqualis* (*Cicinnobolus cesatii* de Bary; Fig. 9A) which occurs on both anamorphic and teleomorphic phases of species of *Erysiphe* Hedw. f. ex Fr., *Microsphaera* Lév., *Phyllactinia* Lév., *Sphaerotheca* Lév., and *Uncinula* Lév. About 27 species have been described in *Ampelomyces* (*Cicinnobolus* de Bary, *Cicinnobolus* Ehrenb.) merely on the basis of the host, but it is doubtful if any of these are physiologically or morphologically distinct from *A. quisqualis* (Blumer, 1933). A detailed study of this species (Emmons, 1930) demonstrated that it was at first intracellular but that later some hyphae also lived saprophytically on dead host hyphae; the pycnidia form either within the host perithecia or its mycelium, and their shape may be affected by the structure in which they develop (Clare, 1964). After the death of the host, *Ampelomyces* may persist as a saprophyte in dead leaf tissue. The destructive effect of this species on *Erysiphe trifolii* Grev. has been demonstrated by inoculation experiments (Yarwood, 1932), but it may not be successful in the biological control of powdery mildews because the host and parasite are favored by different climatic conditions (Yarwood, 1957). *Ampelomyces* is almost exclusively restricted to the Erysiphales and their anamorphs; exceptions are afforded only by species able to grow on or which have been described from Mucorales and *Plasmopara*.

Remarkably few other conidial fungi are known on Erysiphaceae. *Acremonium byssoides* W. Gams & Lim was described as a hyperparasite of *Oidium heveae* Steinm. and may perhaps be restricted to it, and the mainly myxomyceticolous *Aphanocladium album* (Fig. 3A) can parasitize *Erysiphe*

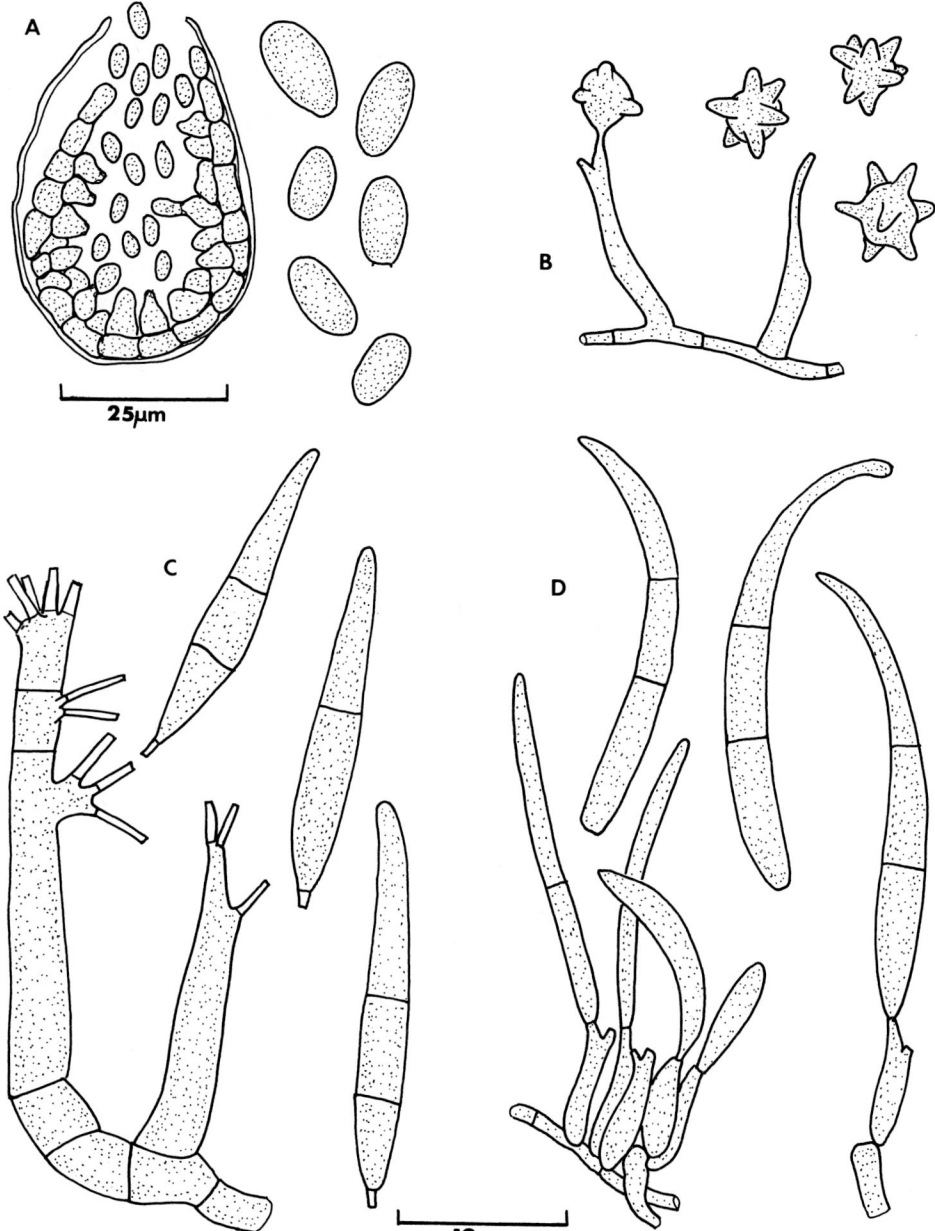

Fig. 9. Conidial fungi occurring on Pyrenomycetes (Erysiphales–Meliolales). (A) *Ampelomyces quisqualis*; (B) *Tuberculispora jamaicensis*; (C) *Trichoconis hibernica*; (D) *Spermosporella pulvinata* Deighton & Piroz.

cichoracearum DC. (Mitov and Ibrahim, 1977). Raghavendra Rao and Pavgi (1978) found that the development and maturation of *Phyllactinia corylea* (Pers.) P. Karst. ascospores were arrested by the usually saprophytic *Cladosporium oxysporum* Berk. and Curt. and further that *Verticillium lecanii* caused premature dislodgement of the conidia in *Oidium tingtanium* Carter. Many other saprophytes can be found intermixed with aged erysiphaceous colonies, for example, *Acremonium strictum, Alternaria longissima* Deight. & MacGarvie, and *Penicillium brevicompactum*, but apparently cause little damage.

Two genera of the Perisporiaceae, perhaps better referred to the Meliolales, support widespread tropical fungicolous Hyphomycetes: *Eriocercospora balladynae* (Hansf.) Deight. [on *Balladynopsis entebbeensis* (Hansf.) Petr.; Fig. 11C] and *Domingoella asterinarum* Petr. & Cif. (on *Linotexis deightonii* Hansf.; Fig. 11E).

c. Meliolales (50 genera, 2000 species). The Meliolales are exceptionally rich in conidial fungi, indeed it may be difficult to collect fungi of this group without other fungi growing on them (Deighton, 1960). Most are dematiaceous Hyphomycetes with annellidic, blastic, or tretic conidiogenesis and may often be biotrophic rather than necrotrophic, obtaining nutrients leaking out from the host. The most luxuriant and largest melioline specimens are often the richest in conidial fungi, which suggests that the latter are not very harmful.

Several genera of Hyphomycetes are known only from Meliolales. These include *Spermosporella* Deight. (two species; Fig. 9D), *Tuberculispora* Deight. & Piroz. [two species; the unnamed anamorph of *Pseudonectria pipericola* Stev. nom. hol. and *T. jamaicensis* Deight. & Piroz., on *Irenopsis aciculosa* (Wint.) Stevens and *I. cryptocarpa* (Ell. & Mart.) Hansf.; Fig. 9B] and the monotypic genera *Monosporiella meliolicola* (Speg.) Speg. (on *Meliola bidentata* Cooke), *Spermatoloncha maticola* Speg. (on *M. clerodendricola* P. Henn.), and *Vermispora grandispora* Deight. & Piroz. (on *I. aciculosa*). *Divinia diatricha* Cif. is of uncertain application, and *Nascimentoa pseudoendogena* Cif. & Bat. is a synonym of *Spiropes* Cif. (Kendrick and Carmichael, 1973). A single monotypic coelomycete genus is known from Meliolales, *Capitorostrum asteridiellae* Cif. & Bat. [on *Asteridiella fraseriana* (Syd.) Hansf.].

In the tropics there is an assemblage of conidial fungi which occur on a wide range of foliicolous ascomycetes, including meliolines, but are not confined to particular hosts. Many are described in detail by Deighton and Pirozynski (1972) and include *Atractilina parasitica* (Wint.) Deight. & Piroz. (exceedingly common on meliolines; Fig. 11A), *Chionomyces meliolicola* (Cif.) Deight. & Piroz. (Fig. 11B), *Cylindrocarpon macrosporum* (Speg.) Deight. & Piroz. (Fig. 11D), *Domingoella asterinarum* (Fig. 11E), *Eriocercospora balladynae* (Fig. 11C), *Eriomycopsis biseptata* Chevaug., *E. bonplandii* Speg. (*Melioliphila melioloides* (Speg.) Piroz. nom. hol., ? *Dactylaria domina-gregum* Cif.), *E. flagellata* (Fig. 12B), *E. minima* Hansf. [? *Calloriopsis gelatinosa*

8. A Survey of the Fungicolous Conidial Fungi

(Ell. & Mart.) H. Syd. & Syd. nom. hol.], *Hansfordiella cupulifera* (Hansf.) Hughes (Fig. 13D), *Heteroconium solanium* (Sacc. and Syd.) M. B. Ellis (Fig. 12A), *Periconia doidgeae* Hansf., *Spiropes dorycarpus* (Mont.) M. B. Ellis, *Trichoconis africana* (Hansf.) Deight. & Piroz. (Fig. 12C), and *T. angustispora* (Hansf.) Deight & Piroz.

States of *Trichothyrium* Speg. species have also been considered conidial fungi, for example, *Hansfordiella meliolae* (Hansf.) Hughes (*T. hansfordii* Hughes nom. hol.), *Spegazzinia meliolae* Zimm. [*T. asterophorum* (Berk. & Br.) Höhn. nom. hol.], and *S. chandleri* Hansf. [*T. reptans* (Berk. & Curt.) Hughes nom. hol.] (see Hughes, 1953). The unnamed *Acremonium*-like anamorph of *Calonectria cephalosporii* Hansf. is also widespread on meliolines (Gams, 1971).

Several genera of widespread tropical fungicolous fungi include species only so far known on meliolines, for example, *Acremoniula suprameliola* Cif. (on *Meliola* sp.), *Chionomyces chorleyi* (Hansf.) Deight. & Piroz. (very common), *C. sclerochitonis* (Hansf.) Deight. & Piroz. (on *Meliola* species), *Eriocercospora olivacea* Piroz. [on *Meliolina mollis* (Berk. & Curt.) Höhn.], *Eriomycopsis bosquieae* Hansf. (on *Meliola soroceae* Speg.), *E. meliolinae* Hansf. [on *Meliolina cladotricha* (Lév.) H. Syd. & Syd.], *E. paraensis* Bat. & Peres (on Meliolales indet.), 16 species of *Spiropes* [of which the commonest are *S. capensis* (Thüm.) M. B. Ellis, *S. guareicola* (Stev.) Cif. and *S. melanoplaca* (Berk. & Curt.) M. B. Ellis], *Trichoconis capitata* Piroz. (on *Meliolina mollis*), *T. hamata* (Hansf.) Deight. & Piroz. (on *Meliola* species), *T. hibernica* Deight. & Piroz. (on *Appendiculella calostroma* (Desm.) Höhn.; Fig. 9C), *T. sigmoidea* Deight. & Piroz. (on *M. kawandensis* Hansf.), *Titaea doidgeae* Hansf. (on *Irene nuxiae* Syd. and *M. deinbolliae* Hansf.), and *T. triradiata* Hansf. (on *M. arundinis* Pat.). Several not exclusively tropical genera of Hyphomycetes also include species known from meliolines, for example, *Aphanocladium meliolae* (Hansf.) W. Gams (on *M.* sp., also known from *Hemileia vastatrix*), *Mycogone meliolarum* Pat., and *Sympodiophora meliolae* (Stev.) Deight. & Piroz. (on *M. paulliniae* Stev.).

Numerous Coelomycetes have been referred to the genera *Asbolisia* Speg., *Ectosticta* Speg., and particularly *Cicinnobella* P. Henn., from tropical foliicolous ascomycetes, including eight described in *Cicinnobella* from meliolines. The name *Cicinnobella* itself was based on an immature ascomycete (Sutton, 1977), and most taxa placed here appear to be anamorphs of Dimeriaceae (Pleosporales) (Hansford, 1946; Hughes, 1976); many taxa in the Dimeriaceae are fungicolous, but their conidial anamorphs are in need of further study.

A large number of conidial fungi are described from meliolines in widespread saprophytic or plant parasitic genera, but many of these require critical study. To illustrate the range of these the following may be cited: *Acremoniella melioliphila* Cif. (on *Meliola swieteniae* Cif.), *Brachysporium minutum* Bat. &

Maia (on *M. sapindacearum* Speg.), *Chloridium meliolae* Hansf. (on *M. carissae* Doidge), *Coniothyrium glabroides* Stev. (on *Meliola* sp.), *Cylindrocarpon ukolayi* Thaung (on *M. tabernaemontanicola* Hansf. & Thirum., *Calonectria ukolayi* Thaung nom. hol.), *Deightoniella leonensis* M. B. Ellis (on *Asteridiella tetracerae* Hansf. & Deight.), *Didymobotryum hymenaearum* Bat. & Peres (on *M. melanochylae* Hansf.), *Dinemasporium meliolicola* Speg. (on *Meliola* sp.), *Epistigme erodens* Bat. & Garnier, *E. teucrii* Bat. & Maia (on *Laeviomeliola cassiae* Bat.), *Fusarium dominicanum* Cif. (on *M. byrsonimae* Stev.), *Haplaria melioliphila* Cif. (on *M. myriopoda* Cif.), *Isaria meliolae* Hansf. (on *Meliola* species), *Monacrosporium melioliphilum* Cif. (on *M. kaduae* Stev.), *Periconiella ellisii* Merny & Hug. ex M. B. Ellis [on *Asteridiella glabra* var. *coffeae* (Roger) Hansf.], *Podosporium ugandense* (Hansf.) Cif. (on *M. ugandensis* Hansf.), *Rhinotrichum alterosum* Viégas (on *Meliola* sp.), *Sepedonium epimeliola* Cif. (on *M. ambigua* Pat. & Gaill.), *Sporothrix setiphila* (Deight. & Piroz.) de Hoog (on *M. clerodendri* Hansf.), and *Sporotrichum meliola* (Stev.) Hansf. (on *Meliola* sp.).

It is interesting to note that "ubiquitous" mycoparasitic and saprophytic fungi are exceptionally rare on meliolines, as they are on most tropical foliicolous ascomycetes.

d. Sphaeriales (500 genera, 5000 species). The Sphaeriales are a vast heterogeneous assemblage and comprise many genera in need of extensive taxonomic revision. Remarkably few genera of conidial fungi are restricted to the Sphaeriales, and most of those that are known are found on members of the largely tropical Polystigmataceae. Coelomycetes described from this family include the monotypic genera *Ciliophora cryptica* Petr. (on *Phyllachora brenesii* Syd.), *Geastrumia polystigmatis* Bat. (on *Polystigma pusillum* Syd.), *Perizomella inquinans* Syd. (on *Phyllachora* Nitschke sp.), and *Sitochora ellipsospora* Upad. (on *P. paspalicola* P. Henn.). Other Coelomycetes described from the family include some in the more widespread genera *Ascochytella* Tassi (one species), *Coniothyrium* Corda (three species), *Davisiella* Petr. (one species), *Microdiplodia* Tassi (one species), *Phomopsis* (Sacc.) Sacc. (one species), *Pleurophoma* Höhn. (one species), *Stagonospora* (Sacc.) Sacc. (six species), and *Zythia* Fr. (one species). *Coniothyrium occultum* Syd. (? *Didymosphaeria winteri* Niessl nom. hol.) is a particularly common mycoparasite of *Phyllachora* species; invaded colonies become lusterless, and a necrotic zone forms around them (Parbery, 1978). Some of the Coelomycetes may be states of the presumed hosts, while others may be only using the *Phyllachora* as a point of entry into the host plant. Parbery (1978) noted that *Cercospora* and other dematiaceous Hyphomycetes tended to attack aging colonies and thus might really be saprophytes; those described from the Polystigmataceae include *Cladosporium phyllachorae* M. B. Ellis (on *P. pseudis* Rehm; Fig. 10E), *Dactylaria dimorpha* Mats. (on

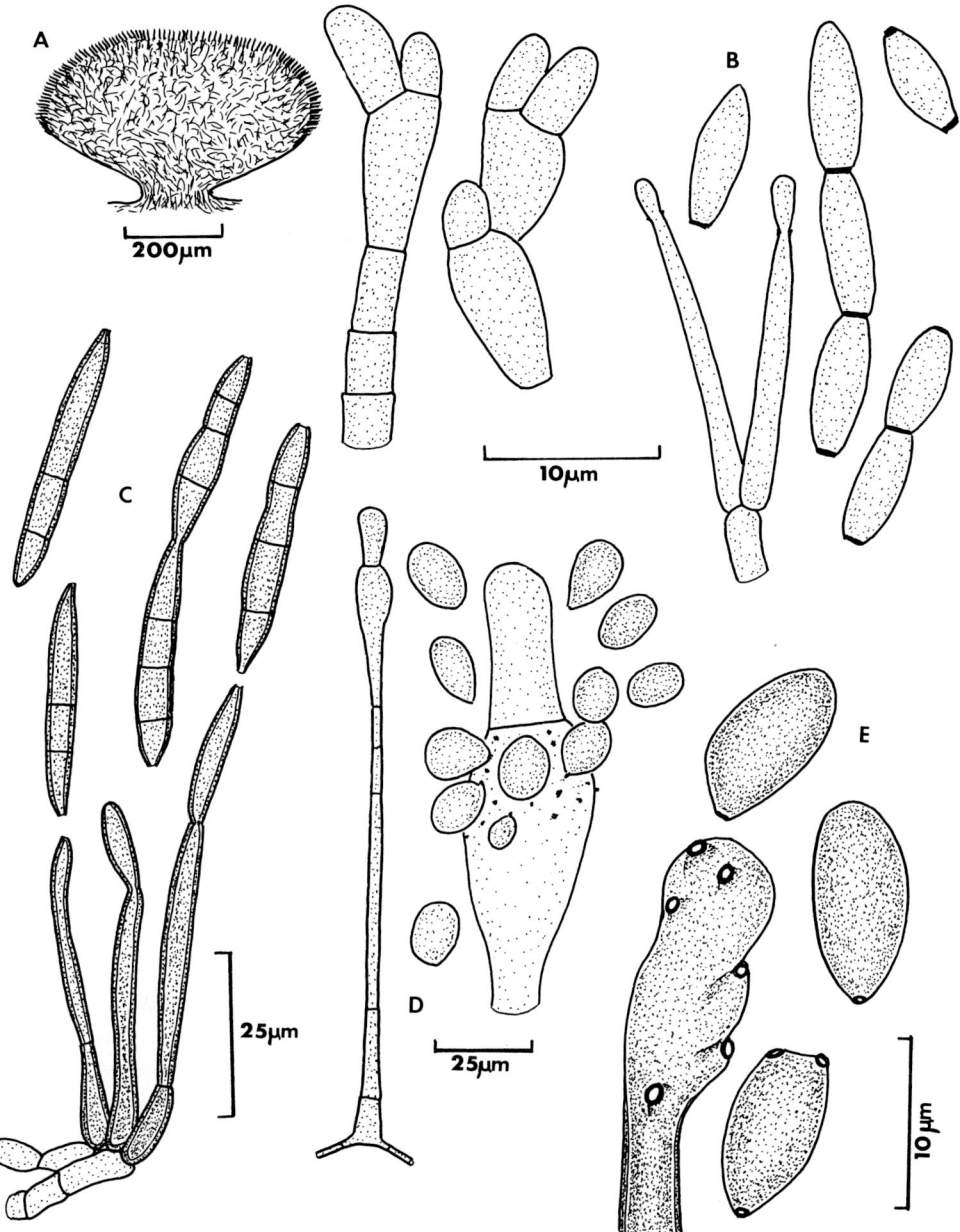

Fig. 10. Conidial fungi occurring on Pyrenomycetes (Sphaeriales). (A) *Annellodochium ramulisporum*; (B) *Gabarnaudia tholispora*; (C) *Heteroconium tetracoilum*; (D) *Nematogonium ferrugineum*; (E) *Cladosporium phyllachorae*.

Glomerella Schrenk & Spauld. sp.), *Didymopsis phyllachorae* Syd. (on *Phyllachora* sp.), *Fusoma telimenellae* Shvarts. [on *Telimenella gangraena* (Fr.) Petr.], *Graphium irradians* Petr. (on *Coccodiella* Hara sp.), *Pseudocercospora heveae* (Vinc.) Deight. (on *Phyllachora* and also directly on *Hevea*), and *Tuberculina phyllachoricola* Bat. & Bez. (on *P. whetzelii* Chard.). Many more widespread saprophytic species are also reported from Polystigmataceae.

Outside this family, only a single coelomycete genus is described from the Sphaeriales, the monotypic *Scolecozythia valsivora* Curzi (on *Valsa ceratophora* Tul.). Six hyphomycete genera are described from other families in the Sphaeriales, but three are not validly published and dubious: *Annellodochium ramulisporum* Deight. (on *Diatrype* Fr. sp.; Fig. 10A), *Dennisographium episphaeria* Rifai [on *Ustulina deusta* (Hoffm. ex Fr.) Lind], "*Jacobia conspicua*" Arnaud [on *Lasiosphaeria ovina* (Fr.) Ces. & De Not.], *Lindquistia indica* Subram. & Chandr. [on *Podosordaria leporina* (Ell. & Ev.) Dennis], and "*Peyronelina glomerulata*" Arnaud (on *Lasiosphaeria* Ces. & De Not. sp.). "*Acrostaphylus hypoxylii*" Arnaud is the *Nodulisporium* Preuss anamorph of a *Hypoxylon* Bull. ex Fr. sp. (Barron, 1968).

About 80 conidial fungi are described in widespread plant-parasitic or saprophytic genera from the remaining families of Sphaeriales. A considerable number of these occur on effete stromata or perithecia, which persist long after the ascospores have been released, and may really be widespread saprophytes awaiting discovery on other substrates. It would be superfluous to enumerate all of them here, but the following list will indicate the range of genera involved: *Acremonium arxii* W. Gams (on *Hypoxylon* sp.), *Cladosporium stromatum* Preuss [on *Eutypa leioplaca* (Fr.) Cooke], *Coniothyrium insuetum* Syd. (on *Nectria prodigiosa* Syd.), *Cytoplea parasitica* Ahmad [on *H. rubiginosum* (Pers. ex Fr.) Fr.], *Dicranidion inaequalis* Tub. & Tokoy. (on ? *Diaporthe* Nits. sp.), *Endophragmiella canadensis* (Ell. & Ev.) M. B. Ellis [on *Valsa ambiens* (Pers. ex Fr.) Fr.], *E. eboracensis* Sutt. [on *Diatrype stigma* (Hoffm. ex Fr.) Fr.], *Fusarium epistromum* (Höhn.) Booth [on *Diatrypella favacea* (Fr.) Sacc. and *D. quercina* (Pers. ex Fr.) Cooke; *Nectria magnusiana* Rehm nom. hol.], *Gliocladium caespitosum* Petch (on *Nectria mammoidea* Phill. & Plowr.), *Heteroconium tetracoilum* (Corda) M. B. Ellis [commonest on *Diatrype stigma* but also on *Anthostoma turgidum* (Pers. ex Fr.) Nits., *Eutypa flavovirens* (Pers. ex Fr.) Tul., and *Peroneutypa heteracantha* (Sacc.) Berl.; Fig. 10C], *Hormiactis nectriae* P. Karst. [on *Nectria coccinea* (Pers. ex Fr.) Fr.], *Microdiplodia mycophaga* Petch (on *Physalospora montana* Sacc.), *Oncopodiella hyperparasitica* D. Hawksw. [on *Lasiosphaeria spermoides* (Hoffm. ex Fr.) Ces. & De Not.], *Phoma hypocrellae* Saw. (on *Hypocrella aleyrodis* Saw.), *Rhinocladiella anceps* (Sacc. & Ell.) Hughes (on *Diatrype* sp.), *R. epichloes* (Ell. & Dearn.) Henneb. [on *Epichloe typhina* (Pers. ex Fr.) Tul.], *Sporotrichum isarioides* Petch (on *Cordyceps dipterigena* Berk. & Br.), *Trichosporium saccardoi* Lindau [on *Eutypa velutina* (Wallr.) Sacc.], and *Trinacrium mycogonis* Tassi (on *Nectria* Fr. sp.).

No conidial fungi are described from the Coryneliaceae, Halosphaeriaceae, and Ophiostomataceae, and only two particularly doubtful ones are reported from each of the Chaetomiaceae (Melanosporaceae) and Sordariaceae (excluding Lasiosphaeriaceae).

The widespread mycophilic *Gabarnaudia tholispora* (*Fusidium parasiticum* auct.; Fig. 10B), is known on *Nectria cinnabarina* (Tode ex Fr.) Fr. and *Xylaria oxyacanthae* Tul.; it has been found to grow particularly well on a decoction agar of the latter, in which it causes the stromata to shrink and disintegrate and the ascospores to abort (Backus and Stowell, 1953). (*G. tholispora* is also recorded from *Elaphomyces* Nees ex Fr. sp., *Russula nigricans*, and *Scleroderma citrinum*.) A few largely agaricicolous fungi occur occasionally on persistent Sphaeriales, for example, *Cladosporium fuligineum* and *Calcarisporium arbuscula* (on several genera of the Xylariaceae). The often lignicolous contact mycoparasite *Nematogonium ferrugineum* (*Gonatorrhodiella highlei*; Fig. 10D) has been isolated from aged polypores and found to be particularly destructive to *Nectria coccinea* (Ayres, 1941; Perrin, 1977) and *N. cinnabarina* (Blyth, 1949), growing prolifically on fungal extracts of the latter. The allied, rarer *N. parasiticum* (Thaxt.) Hughes is known from species of *Hypocrea* Fr., *Hypomyces* (Fr.) Tul., and *Nectria*. *Gliocladium roseum* can parasitize both *Ceratocystis fimbriata* (Barnett and Lilly, 1962) and *C. fagacearum* (Bretz) Hunt (Shigo, 1958), and *Ceratocystis* Ell. & Halst. species are also susceptible to the contact mycoparasite *Gonatobotryum fuscum* (Shigo, 1960); *Penicillium implicatum* Biourge can inhibit both *C. ips* (Rubm.) C. Moreau and *C. minor* (Hedgc.) Hunt *in vitro* (Barras, 1969).

Species antagonistic to the plant-pathogenic *Gaeumannomyces graminis* (Sacc.) von Arx & Oliver have received particular attention. *Acremonium murorum* (Corda) W. Gams is highly antagonistic (Mangan, 1967), as is *A. rutilum* W. Gams (Domsch and Gams, 1968), and it has been suggested that *Phialophora radicicola* Cain could be used in its control (Deacon, 1976). Its hyphae can be parasitized by *Trichoderma* (Slagg and Fellows, 1947) and *Fusarium sambucinum* Fuckel [*Gibberella pulicaris* (Fr.) Sacc. nom. hol.] (Domsch, 1960); several species of *Penicillium* (e.g., *P. citrinum* Thom, *P. janthinellum* Biourge) can also inhibit it (Domsch *et al.*, 1980). *Fusarium lateritium* Nees ex Link [*G. baccata* (Wallr.) Sacc. nom. hol.] is antagonistic toward the pathogen *Eutypa armeniacae* Hansf. & Carter and has potential for its biological control (Carter and Price, 1974), and both *F. oxysporum* (Robinson *et al.*, 1968) and *Memnoniella echinata* (Riv.) Gall. (Quaiser Abbas and Ghaffar, 1973) inhibit *Neocosmospora vasinfecta* E. F. Sm.

Chemical interactions may be important in the coprophilous Sphaeriales, to judge from the observation that metabolites of *Stilbella erythrocephala* are somewhat inhibitory to *Sordaria fimicola* (Rob.) Ces. & De Not. (Singh and Webster, 1973).

Common ubiquitous saprophytic fungi regularly occur on effete stromata and

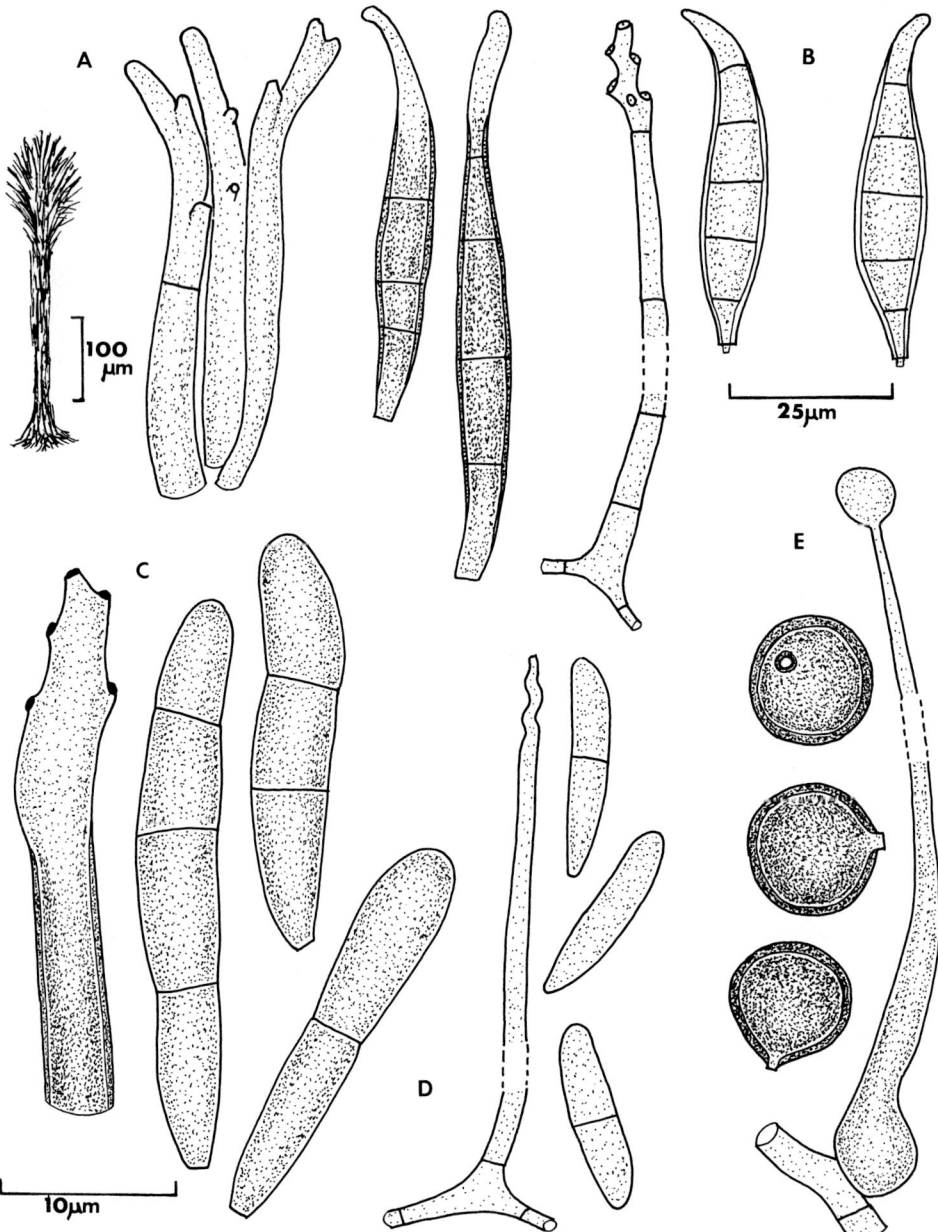

Fig. 11. Conidial fungi widespread on tropical foliicolous ascomycetes. (A) *Atractilina parasitica*; (B) *Chionomyces meliolicola*; (C) *Eriocercospora balladynae*; (D) *Cylindrocarpon macrosporum*; (E) *Domingoella asterinarum*.

perithecia in the temperate zone. Examples are *Acremoniella atra, Acremonium strictum, Fusarium avenaceum, Penicillium stoloniferum* Thom, *Sporidesmium inflatum* (Berk. & Rav.) M. B. Ellis, and *Stachybotrys atra* Corda.

4. Discomycetes (425 genera, 3000 species)

No conidial fungi appear to have been described from the Medeolariales (1 genus, 1 species) or Ostropales (6 genera, 200 species; excluding lichen-forming species). Santesson (1945) listed those reported from the Cyttariales (1 genus, 10 species), but all are described in ubiquitous genera and are of dubious status (*Anthracoderma hookeri* Speg., *A. selenospermum* Speg., *Chalara cyttariae* Bomm. & Rouss., *Cladosporium cyttariicola* Speg., *Coniothyrium hookeri* Speg., and *Torula darwinii* Speg.); *Coniothyrium cyttariae* Bomm. & Rouss. is anamorphic *Cyttaria darwinii* Berk.

a. Helotiales (225 genera, 1750 species). Very few conidial fungi are known from this order. These include the monotypic Hyphomycetes *"Pleurocatena acicularis"* Arnaud [on *Phialea albida* (Rob. & Desm.) Gill.] and *Pseudographiella variiseptata* Morris (on ? *Geoglossum glabrum* Pers. ex Fr.). Several species have been described in ubiquitous hyphomycete genera, but all are of doubtful status; these include *Chalara minima* Höhn. [on *P. sordida* (Fuckel) Sacc.], *Papulospora candida* Sacc. (on *G. glabrum*), and *Stephanoma tetracoccum* Zinderen-Bakker (on *G. glabrum*).

Coniothyrium minitans Campbell is a destructive parasite almost entirely confined to the sclerotia of *Botryotinia fuckeliana* (de Bary) Whetzel, *Sclerotinia sclerotiorum* (Lib.) de Bary, *S. trifoliorum* Erikss., and *Sclerotinia cepivorum* Berk. It can drastically reduce the percentage of viable sclerotia (Hoes and Huang, 1975; Huang and Hoes, 1976), most effectively parasitizing and killing sclerotia formed on the root surface (Huang, 1977); pycnidial dust in soil and as a seed dressing may be useful for the control of *S. cepivorum* (Ahmed and Tribe, 1977). *Gliocladium roseum, Harzia acremonioides* (Harz) Cost., *Microsphaeropsis centaureae* Morgan-Jones (Watson and Miltimore, 1975), *Penicillium spinulosum* Thom, *Sporidesmium sclerotivorum* Uecker *et al.*, and particularly *Trichoderma* species, are among other conidial fungi invading sclerotia of *Sclerotinia sclerotiorum*.

Few other Coelomycetes are known only from the Helotiales, for example, *Dothiorellina quickii* Bonar [on *Godronia abieticola* (Zell. & Godd.) Seav.], *"Microdiplodia cenangicola"* Newod. (on *Cenangium ferruginosum* Fr. ex Fr.), and *Phoma sclerotivora* (Bref.) Sacc. (on *S. libertiana* Fuckel).

A few primarily agaricicolous fungi are occasionally found on Helotiales, especially the more fleshy species (e.g., *Calcarisporium arbuscula* on *Dasyscyphus* Gray species, *Cladobotryum verticillatum* on *Geoglossum* Pers. ex Fr.). With generally less persistent ascomata, members of this order rarely support the more ubiquitous saprophytic Hyphomycetes.

b. Pezizales (90 genera, 600 species). In view of the open, nutrient-rich apothecial discs seen in many genera of the Pezizales, it is remarkable that so few conidial fungi are adapted to this habitat. It is conceivable that some of the chemical products in the apothecia, which may be in high concentrations, are inhibitory, but it must be remembered that numerous lichen-forming conidial fungi are restricted to lichens with high levels of secondary metabolites; this observation may have some evolutionary significance.

No genus of conidial fungi appears to be confined to the Pezizales; *Coleomyces rufus* Moreau [on *Phaeobulgaria inquinans* (Fr.) Nannf.] is merely a *Cylindrocarpon* (Booth, 1966).

A few Hyphomycetes in more widespread genera are, however, known only from this order; for example, *Didymopsis helvellae* (Corda) Sacc. & Marchal (common on *Helvella* L. ex St.-Am., also on *Acetabula calyx* Sacc.), *D. spicata* Rich. (on *Geopyxis ciborium* Vahl), *Diplosporium morchellae* Beeli (on *Morchella esculenta* Pers. ex St.-Am.), *Septocylindrium morchellae* Oud. (on *M. esculenta*), *Gabarnaudia fimicola* Sams. & W. Gams (on *Ascophanus* Boud. sp. and isolated from dung, *Sphaeronaemella fimicola* Marchal nom. hol.), and *Stephanoma strigosum* Wallr. (on various genera, *Hypomyces* nom. hol.).

Mycogone species [particularly *M. cervina* Ditm., *M. peziza* (Rich.) Sacc., and *M. rosea*] are recorded from most larger genera of the Pezizales. The predominantly agaricicolous *Calcarisporium arbuscula*, *Cladosporium fuligineum*, and *Tilachlidium brachiatum* are also encountered occasionally on fleshy members of the order. The mainly lichenicolous *Acremonium lichenicola* W. Gams is also known from *Phaeobulgaria inquinans*, and this host can further support the saprophyte "*Alysidium*" *fuscum* Bon.; *Dactylaria mycophila* Tubaki was described from a *Bulgaria* Fr. species. *Sporotrichum fungicola* (Corda) Sacc., described from Clavariaceae, is also recorded on *Morchella*.

The inhibition of *Ascobolus crenulatus* by *Stilbella erythrocephala* has already been alluded to. Widespread saprophytes are occasionally reported from decaying Pezizales, for example, *Allescheriella crocea* (Mont.) Hughes, *Cladosporium herbarum,* and *Penicillium* species.

c. Phacidiales (64 genera, 250 species). Two genera of conidial fungi have been described from the Phacidiales, but neither proves to be entirely restricted to the Discomycetes. *Cornutispora limaciformis* Piroz., a remarkable coelomycete with triradiate conidia found in the hymenium of *Therrya fuckelii* (Rehm) Kujala, was first described as a monotypic genus, but a second species was subsequently found on lichen-forming ascomycetes. *Columnophora rhytismatis* Bubák (on *Rhytisma* Fr.) has been found to be synonymous with the foliicolous genus *Stigmina* Sacc. (Kendrick and Carmichael, 1973). Most other conidial fungi described from Phacidiales have proved on reexamination to be synonyms of widespread taxa, but a few remain to be critically investigated, for example, *Cladosporium lophodermii* Georgescu & Tutunaru [on *Lophodermium pinastri*

(Schrad. ex Hook.) Grev.; interestingly, *C. herbarum* is evidently not uncommon on old apothecia of this species, according to Mitchell *et al.,* (1978)]; *Phoma colpomatis* Rich. [on *Colpoma quercinum* (Pers. ex Fr.) Wallr.], *Phyllosticta gallicola* Ell. & Ev. (on *Rhytisma solidaginis* Schw.), and *Tuberculina davisiana* Sacc. & Trav. (on *Rhytisma* sp.).

Attention is also drawn here to the occurrence of secondary fungi on conifer needles primarily attacked by Hypodermataceae; about 30 species are known only from this niche (not all conidial fungi) and are apparently associated with joint occurrences of Hypodermataceae and conifer needles (Darker, 1967).

d. Tuberales (35 genera, 150 species). With a subterranean habit it might be expected that no fungi would be able to adapt to grow exclusively on members of this order. Some Pyrenomycetes belonging to *Microthecium* Corda have, however, achieved this (Hawksworth and Udagawa, 1977), and a few conidial fungi in widespread genera are described from Tuberales (although most require reinvestigation): *Colletotrichum umemurai* Imai (on *Elaphomyces japonicus* Lloyd), *Nodulisporium tuberum* Font. & Bonf. ex de Hoog (on *Tuber* Mich. ex Fr. sp.), *Oospora placentiformis* (Corda) Sacc. & Vogl. [on *T. gulonum* (Corda) Paol.], and *O. tuberum* (Corda) Sacc. & Vogl. (on Tuberaceae indet.). There are also reports of the more widespread fungicolous species *Verticillium epimyces* Berk. & Br. (on *Elaphomyces* Nees ex Fr. sp. and *Hydnotrya tulasnei* Berk. & Br.) and *V. psalliotae* (on *Tuber* sp.). Decaying ascomata can be colonized by *Aspergillus* (e.g., *A. cervinus* Massee group) and *Penicillium* species.

5. Loculoascomycetes (430 genera, 2000 species)

Two different detailed schemes for the classification of the Loculoascomycetes have been produced in recent years, those of Luttrell (in Ainsworth *et al.,* 1973a) and von Arx and Müller (1975). The latter authors recognized only a single order, the Dothideales, with 34 families. A fundamentally different approach to the group was outlined by Barr (1976), but a comprehensive account of her system is not yet available. In view of these difficulties I have not tried to discuss the conidial fungi on them by order, and the subclass is treated as a whole. Nevertheless, many of the families constitute cohesive units, and it is notable that most of the conidial fungi confined to the Loculoascomycetes occur on members of three families which are mainly tropical: the Asterinaceae, Englerulaceae, and Parodiellinaceae. A lesser number occur on the heterogeneous Pleosporaceae, and it is surprising that rather few are confined to the tropical Capnodiaceae.

Five monotypic hyphomycete genera are known on Loculoascomycetes: *Acrostaurus turneri* Deight. & Piroz. [on *Asterina* Lév. sp. and *Circosia manaosensis* (P. Henn.) Arnaud; Fig. 13A], *Irpicomyces schiffnerulae* Deight. (on *Schiffnerula solani* Hansf.), *Isariella auerswaldiae* P. Henn. [on *Coccostroma puttemansii* (P. Henn.) Theiss. & Syd.], *Paratrichoconis chinensis* (Hansf.) Deight. & Piroz. (on *A. linderae* Hansf.; Fig. 13C), *Ramalia veronicae* Bat. [on

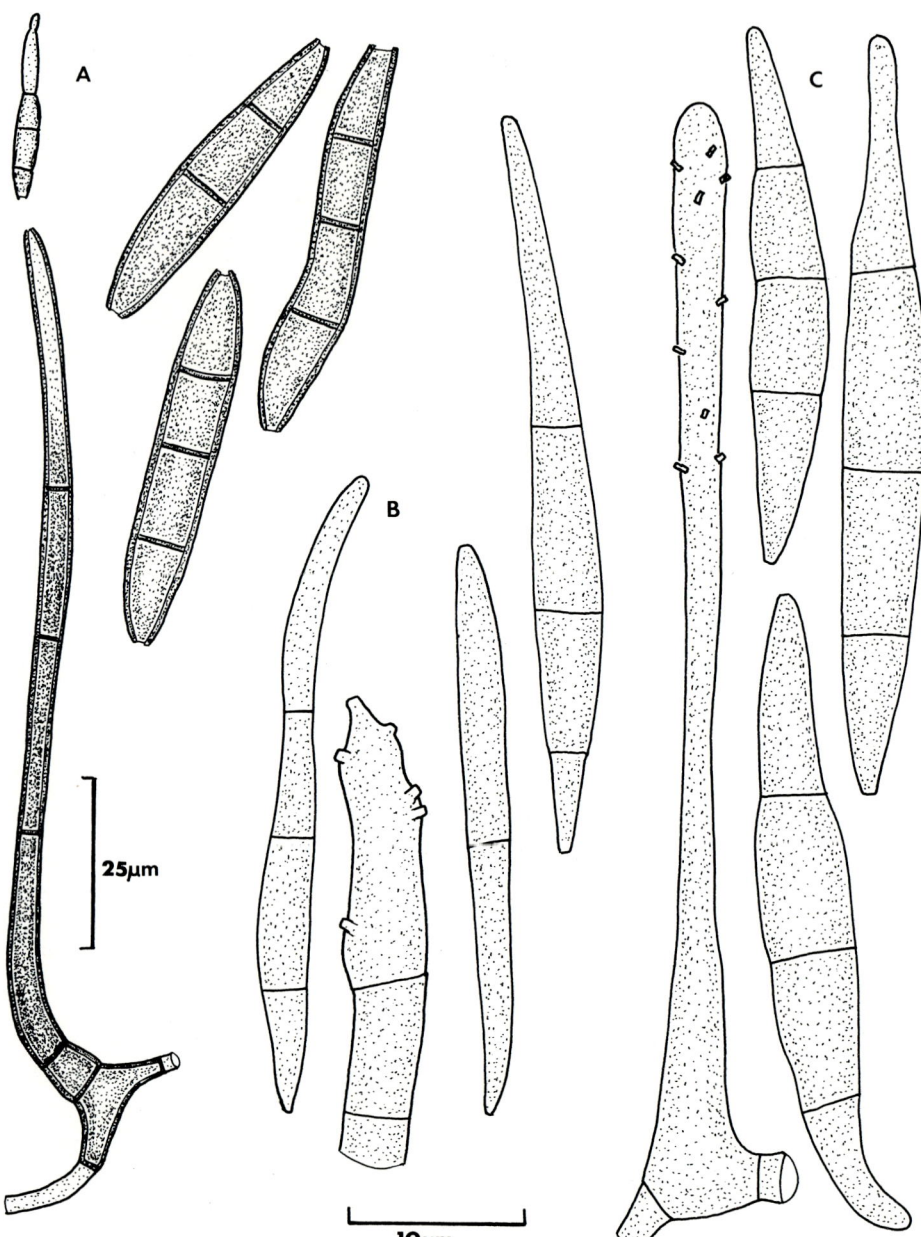

Fig. 12. Conidial fungi widespread on tropical foliicolous ascomycetes. (A) *Heteroconium solanium*; (B) *Eriomycopsis flagellata*; (C) *Trichoconis africana*.

Asterina veronicae (Lib.) Cooke, also on "*Asteromella veronicae*" Bat.], and "*Titaeella capnophila*" Arnaud (on *Capnodium meridionale* Arnaud in France).

In comparison with other groups of fungi, a large number of coelomycete genera appear to be restricted to Loculoascomycetes. These include *Hendersonula* Speg. which is now understood to comprise only *H. australis* Speg. (on *Dothidella australis* Speg.) and *H. symplocii* (Berk. & Br.) Dyko & Sutt. (on *Dermatodothis* Racib. sp. and *D. zeylanica* Syd.), and *Protostegiomyces* Bat. & Vital which also includes two species, *P. asterinarum* Barr & Farr (on *Asterina colliculosa* Speg.) and *P. lembosiae* Bat. & Vital (on *Lembosia byrsonimae* P. Henn.). In addition five monotypic genera of Coelomycetes are known: *Chiastospora parasitica* Riess [on *Cucurbitula berberidis* (Pers. ex Fr.) Gray, *Massaria pupula* (Fr.) Tul., and *M. pyxidata* Riess], *Dialaceniopsis landolphiae* Bat. (on *Asterina* sp.), *Endozythia moravica* Petr. [on *Leptosphaeria derasa* (Berk. & Br.) Auersw.], *Lysotheca suprastromatica* Cif. [on *Balladynella amazonica* (Höhn.) Theiss. & Syd.], and *Mycopara shawii* Bat. & Bez. (on "*Dimerosporium macrocarpum*" Bat. & Bez.).

Some described genera have been found to be unacceptable. *Chaetophomella asterinarum* (Speg.) Speg. is based on discordant elements, and the identity of other taxa referred to it (e.g., *C. parasitica* Petr. on *Asterina dallasica* Petr.) requires reinvestigation. *Chondropodiola falcispora* Petr. & Cif. is of uncertain application; *Cryptogene parodiellae* Syd. and *Cryptogenella parodiellae* Syd. are not fungicolous and are probably close to *Ascochytopsis* P. Henn.; and *Metabotryon connatum* Syd. and *Parabotryon comatum* Syd. may merely be synonyms of the anamorph of *Eudarluca caricis* (Sutton, 1977).

Many widespread fungi present on other tropical foliicolous ascomycetes (particularly meliolines) may also occur in some cases predominantly on Loculoascomycetes (particularly Asterinaceae and Englerulaceae). Many of these are illustrated by Deighton and Pirozynski (1972). Examples are *Acremoniula deightonii* Cif., *A. sarcinellae* (Pat. & Har.) Arnaud ex Deight., *Atractilina parasitica* (Fig. 11A), *Cylindrocarpon macrosporum* (Fig. 11D), *Domingoella asterinarum* (Fig. 11E), *Eriocercospora balladynae* (Fig. 11C), *Eriomycopsis flagellata* (Fig. 12B), *E. minima*, *Hansfordia pulvinata* (Berk. & Curt.) Hughes, *Hansfordiella cupulifera* (Fig. 13D), *Heteroconium solanium* (extremely common on Asterinaceae; Fig. 12A), *Periconia doidgeae*, *Spiropes dorycarpus*, *Titaea ugandae* Hansf., *Trichoconis africana* (Fig. 12C), *T. angustispora*, *T. schiffnerulae* (Hansf.) Deight. & Piroz. and *T. viridula* Deight. & Piroz. Other species described in widespread tropical fungicolous genera confined to particular groups of Loculoascomycetes include *Atractilina asterinae* (Hansf.) Deight. & Piroz. (on Asterineae), *Domingoella pycnopeltarum* Bat. (on *Pycnopeltis conspicua* Bat.), several *Eriomycopsis* species (e.g., *E. englerulae* Hansf., *E. schiffnerulae* Hansf.), *Hansfordiella asterinarum* Hughes (on Asterineae), five

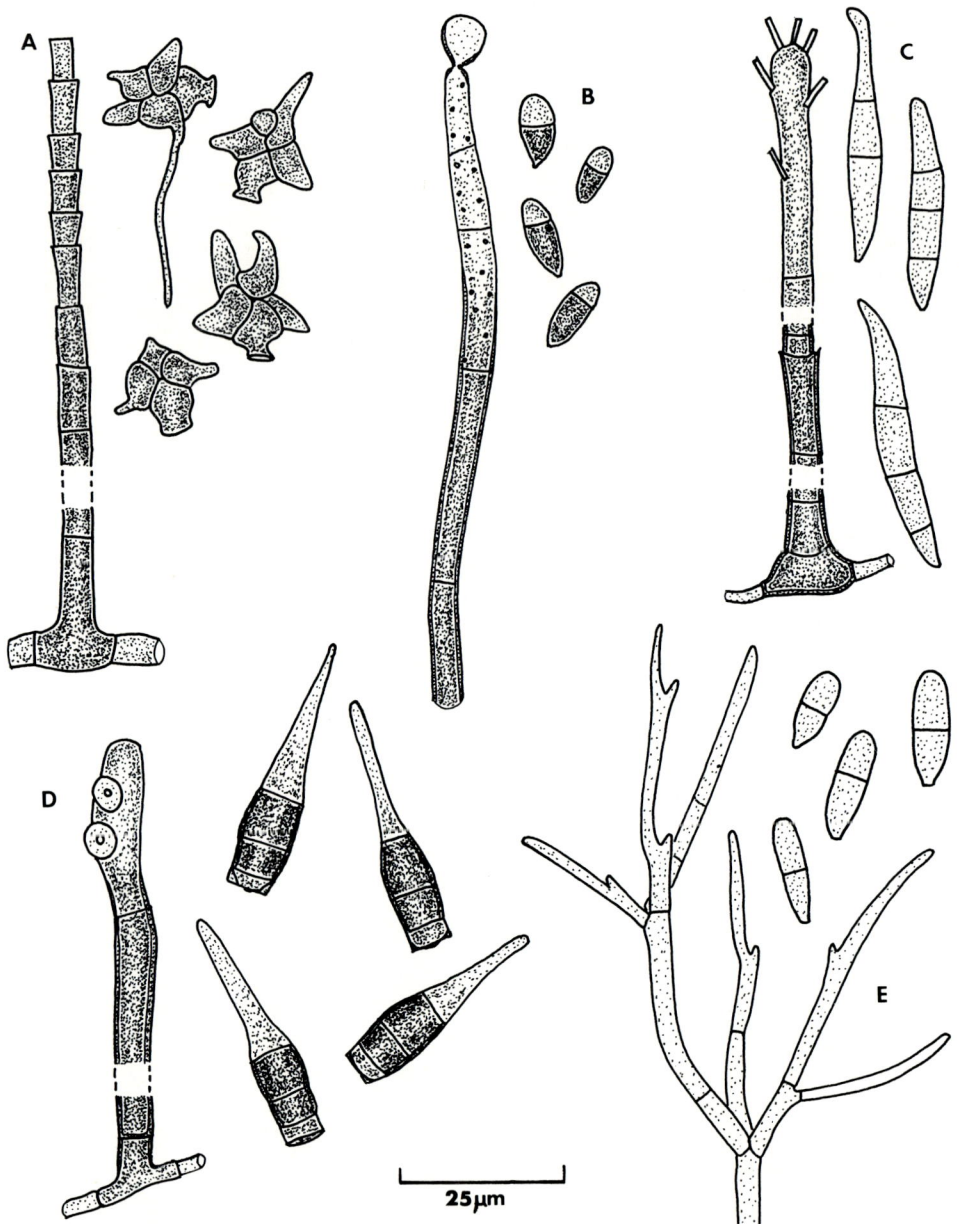

Fig. 13. Conidial fungi on Loculoascomycetes. (A) *Acrostaurus turneri*; (B) *Spiropes asterinae*; (C) *Paratrichoconis chinensis*; (D) *Hansfordiella cupulifera*; (E) *Sympodiophora didyma*.

Spiropes species on different Asterineae [e.g., *S. asterinae* M. B. Ellis (Fig. 13B) and *S. scopiformis* (Berk.) M. B. Ellis], *Tetraposporium asterinearum* Hughes (on *Asterina* species), *Titaea callispora* Sacc. (on Asterineae), and several species of *Trichoconis* [e.g., *T. appendiculata* Deight. & Piroz., *T. englerulae* (Hansf.) Deight. & Piroz., and *T. trichiliae* (Hansf.) Deight. & Piroz.]. About eight species have been placed in the dubious coelomycete genus *Cicinnobella*, for example, *C. megastoma* Syd. and *C. tetracericola* Bat. & Vital (both on Asterinaceae), *C. domingensis* Petr. & Cif. (on *Schiffnerula* Höhn. sp.), and *C. sydowii* Arnaud (on *Cleistosphaeria macrostegia* Syd.).

About 50 conidial fungi have been described in more widespread genera from Loculoascomycetes. Among the Hyphomycetes are *Acrodictys balladynae* M. B. Ellis (on *Balladyna* Racib. species but also known from other fungi), *Cercospora chandleri* Hansf. (on *Asterina* sp.), *Cladosporium asterinae* Deight. (on *Asterina contigua* Syd.), *C. balladynae* Deight. [on *Balladyna magnifica* (Syd.) Hansf.], *C. elsinoes* H. C. Greene (on *Elsinoe wisconsinensis* H. C. Greene), *Cylindrocarpon luteoviride* Deight. & Piroz. (on ? *Microthelia* Körb. sp.), *Fusarium sphaeriae* Fuckel [on *Leptosphaeria acuta* (Hoffm. ex Fr.) P. Karst. and *L. dolium* (Pers. ex Fr.) Ces. & De Not.; *Nectria leptosphaeriae* Niessl nom. hol.], *Oospora dothideae* P. Henn. [on *Pseudothis coccodes* (Lév.) Theiss. & Syd.], *Ramularia episphaeriae* (Desm.) Gunn. [on *Sphaerella isariphora* (Desm.) De Not.], *Septonema trichomeriicola* Cif. et al. (on *Trichomerium jambosae* Bat. & Cif.), *Sporidesmium biseptatum* M. B. Ellis (on *Echidnodes* Theiss. & Syd. sp.), *Sporotrichum hospicida* Schulz. & Sacc. (on *Melogramma vagans* De Not.), *Trichothecium sublutescens* (Peck) Sacc. [on *Teichospora obducens* (Fr.) Fuckel], *Triposporium ledermannii* Hansf. [on *Balladynopsis ledermannii* (Syd.) Hansf.], and *Veronaea filicina* Dingley (on *Rhagadolobium bakerianum* Sacc.). The Coelomycetes described in similarly widespread genera are mostly in need of critical study, particularly as some may merely represent anamorphs of the presumed "hosts." The following examples indicate the range of genera involved: *Ascochytella stegasphaeriae* Petr. (on *Stegasphaeria pavonia* Syd.), *Coniothyrium dolium* Dorog. (on *Leptosphaeria dolium*), *C. massariae* Petr. [on *Massaria conspurcata* (Wallr.) Sacc.], *Ectosticta popowiae* Bat. (on *Asterina* sp.), *Epistigme parmulariicola* Bat. et al. [on *Lembosiodothis parmularioides* (P. Henn.) Bat. et al.], *Hendersonia roblediae* Petr. [on *Botryostoma eupatorii* (Stev.) E. Müll. & von Arx], "*Hendersonula*" *monochaetiella* Tomm. & Langdon (on *Hypnotheca graminis* Tomm.), *Phoma dothideicola* Naumov (on *Dothidea ribesia* Fr.), *P. pyrenophoricola* Bat. et al. (on *Pyrenophora* Fr. sp.), *Rhabdospora mycophaga* Zil. [on *Leptosphaeria macrospora* (Fuckel) Thüm.], *Septogloeum robustum* (Davis) Wollenw. [on *Apiosporina collinsii* (Schw.) Höhn.], *Septoria leptosphaericola* Bat. et al. (on *Leptosphaeria ? wegeliniana* Sacc.), and *Zythia stromaticola* P. Henn. & Shiv. (on Microthyriaceae indet.).

Two species of the broad-spectrum fungicolous genus *Sympodiophora* are known on *Dothidella derridis* (P. Henn.) Theiss. [*S. didyma* Deight. & Piroz. (Fig. 13E) and *S. pulchella* Deight. & Piroz.]. The mycophilic *Zygosporium mycophilum* (Vuill.) Sacc., which can also occur on conidial fungi, has been reported from an unnamed *Capnodium* Mont. The contact mycoparasite *Hansfordia parasitica* occurs on some species of *Physalospora* Niessl. (*Botryosphaeria* Ces. & De Not.?) and allied taxa, including some anamorphs (Barnett, 1964), but cannot attack all species of this genus (Barnett and Lilly, 1958). A further species of *Hansfordia* Hughes, *H. ugandense* (Hansf.) Hughes, is known from *Cycloschizon macarangae* (Hansf.) von Arx. *Epicoccum purpurascens, Phoma humicola* Gilm. & Abb., *Trichoderma viride* and, to a lesser extent, *Myrothecium verrucaria* (Alb. & Schw.) Ditm. ex Steudel inhibit the pathogenicity of the conidial anamorph of *Cochliobolus sativus* by parasitizing its mycelium (Campbell, 1956); this same species is also inhibited by *Alternaria alternata* (Csuti *et al.*, 1965). *Aspergillus fumigatus* Fres. has been reported to be inhibitory to *Venturia inaequalis* (Cooke) Wint.

A few ubiquitous saprophytic fungi have inevitably been discovered on effete ascomata and stromata of Loculoascomycetes, for example, *Acremonium hyalinulum* (Sacc.) W. Gams, *Alternaria tenuissima* (Kunze ex Pers.) Wilts. (on *Cucurbitaria* Gray ex Grev. sp.), *Chalara fusidioides* (Corda) Rabenh. (on effete *Mycosphaerella* Johanson ascomata), and *Fusarium avenaceum* [on *Cucurbitula elongata* (Fr.) Grev.].

6. Laboulbeniomycetes (130 genera, 1500 species)

This isolated group of entomogenous fungi appears to have had only a single conidial fungus described from it, *Fusarium laboulbeniae* Cépède (on *Laboulbenia blanchardi* Cépède), but this proves to be only *F. larvarum* Fuckel (*Nectria aurantiicola* Berk. & Br. nom. hol.), primarily found on scale insects (Booth, 1971).

7. Lichen-Forming Ascomycotina (500 genera, 20,000 species)

The majority of ascomycetes are lichen-forming but as no entirely comprehensive system of the Ascomycotina exists they are treated collectively here. The Hyphomycetes on lichen-forming fungi have been revised by Hawksworth (1979) who accepted 44 species in 23 genera as obligately or primarily lichenicolous; 60 taxa were excluded for a variety of reasons, something which illustrates the extent of revisionary work required in the fungicolous fungi generally.

Ten hyphomycete genera are obligately lichenicolous: *Ampullifera* Deighton (six species, all on foliicolous lichens; Fig. 14A), *Dictyophrynella* Bat. & Caval. (one species; Fig. 14C), *Hansfordiellopsis* Deight. (five species, mainly on setose foliicolous lichens; Fig. 14E), *Illosporium* Mart. ex Ficin. & Schub. (two species; the numerous reported nonlichenicolous species do not appear to be congeneric), *Leightoniomyces* D. Hawksw. & Sutt. (one species), *Milospium*

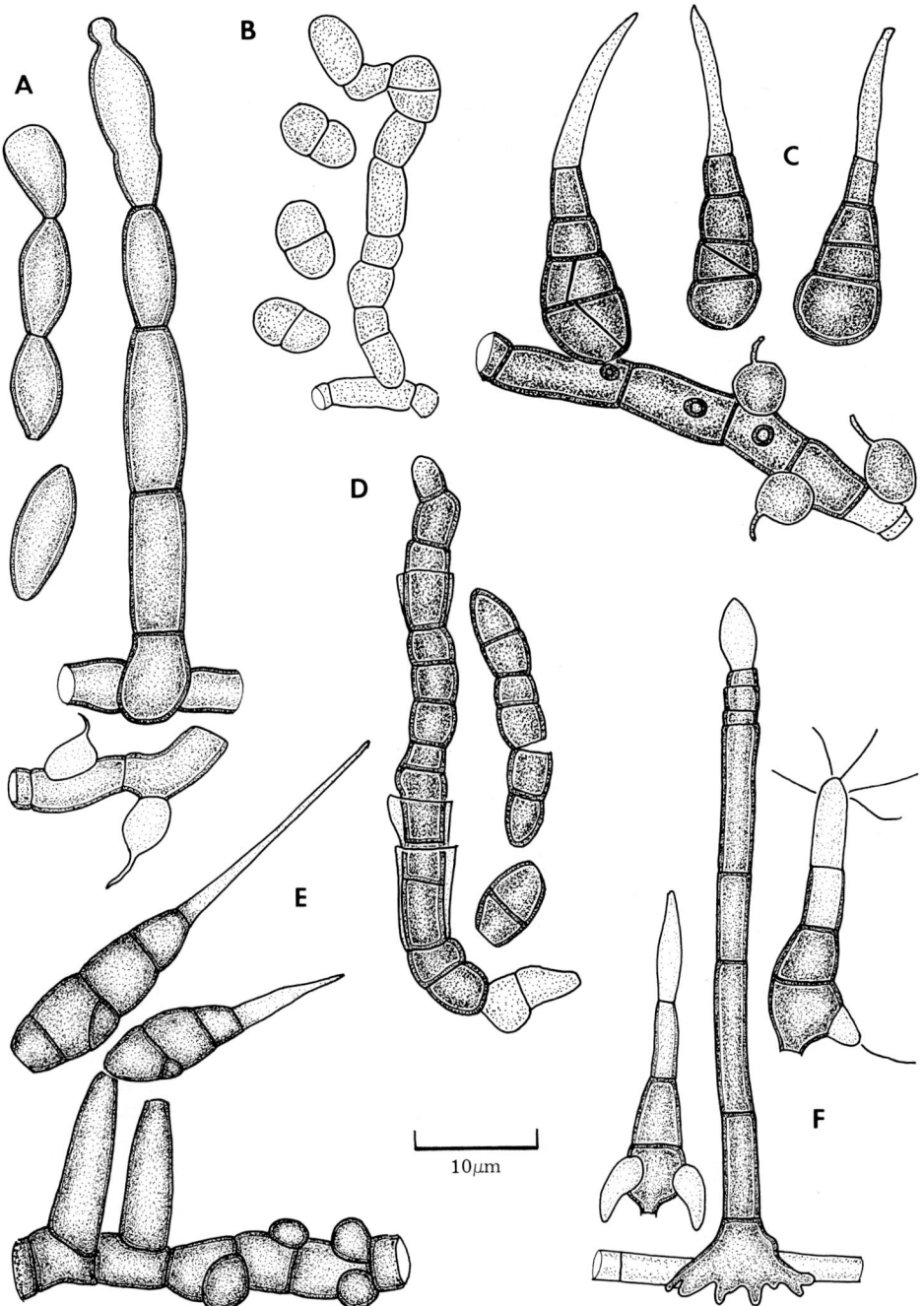

Fig. 14. Conidial fungi on lichen-forming ascomycetes. (A) *Ampullifera foliicola* Deighton; (B) *Bispora christiansenii*; (C) *Dictyophrynella bignoniacearum* Bat. & Caval.; (D) *Taeniolella delicata* M. S. Christ. & D. Hawksw.; (E) *Hansfordiellopsis lichenicola* (Bat. & Maia) Deighton; (F) *Teratosperma anacardii* Hansf.

D. Hawksw. (one species; Fig. 15F), *Refractohilum* D. Hawksw. (three species), *Sclerococcum* Fr. ex Fr. (two species; Fig. 15E), *Sessiliospora* D. Hawksw. (one species), and *Xanthoriicola* D. Hawksw. (one species). The remaining lichenicolous Hyphomycetes are scattered through a wide range of genera: *Acremonium* Link ex Fr. (four species, one also known from other fungi), *Bispora christiansenii* D. Hawksw. (common in Europe; Fig. 14B), *Cladosporium arthoniae* M. S. Christ. & D. Hawksw. [on *Arthonia impolita* (Hoffm.) Borr.], *Dendrodochium subeffusum* Ell. & Galw. (on *Physcia millegrana* Degel.), *Endophragmiella hughesii* D. Hawksw. [on *Lobaria pulmonaria* (L.) Hoffm.], *Fusarium peltigerae* Westend. [on *Peltigera rufescens* (Weis) Humb.], the *Monocillium* Saks. anamorph of *Niesslia cladoniicola* D. Hawksw. & W. Gams (on *Cladonia rangiformis* Hoffm.), *Monodictys* Hughes (two species, one persistent), *Psammina stipitata* D. Hawksw. [on *Schismatomma decolorans* (Turn. & Borr. ex Sm.) Clauz. & Vězda], *Pseudocercospora lichenum* (Keissl.) D. Hawksw. (on *Haematomma cismonicum* Beltr.), *Taeniolella* Hughes (four species, *T. delicata* M. S. Christ. & D. Hawksw. being particularly common in Europe; Fig. 14D), *Teratosperma* Syd. [two species, on *Strigula elegans* (Fée) Müll. Arg.; Fig. 14F], and *Trimmatostroma lichenicola* M. S. Christ. & D. Hawksw. [on *Candelariella vitellina* (Hoffm.) Müll. Arg.].

Various saprophytic Hyphomycetes are recorded (see Hawksworth, 1979), but these are rather rare, and it is conceivable that the lichen products inhibit many potential colonists. It is of interest that none of the widespread tropical fungicolous fungi so common on melioliines and asterines are known to occur on tropical foliicolous lichens.

The extent of host specificity and pathogenicity varies greatly. In some cases a species may be common on only a single host species [e.g., *Sclerococcum sphaerale* (Ach. ex Ficin. & Schub.) Fr. on *Pertusaria corallina* (L.) Arnold (Fig. 15E); *Xanthoriicola physciae* (Kalchbr.) D. Hawksw. on *Xanthoria parietina* (L.) Th. Fr.]. The types of host specificity in the lichenicolous conidial fungi may, however, be better illustrated by reference to the coelomycete genus *Lichenoconium* Petr. & Syd. (Fig. 15C). As revised by Hawksworth (1977), this genus comprised 10 species (all but one, which is probably not really congeneric, obligately lichenicolous) attacking 58 (plus 11 unconfirmed) hosts. When the species were defined morphologically without concern for the host, some in fact proved to be host-restricted [e.g., *L. pyxidatae* (Oud.) Petr. & Syd. on several *Cladonia* Wigg. species, *L. echinosporum* D. Hawksw. on *Heterodea muelleri* (Hampe) Nyl.], and others much more widespread [e.g., *L. usneae* (Anzi) D. Hawks. on 21 species in 9 genera]. In several cases different species differed in the extent to which they were pathogenic on the same host species and gave rise to distinct symptoms [e.g., *L. erodens* M. S. Christ. & D. Hawksw. and *L. lecanorae* (Jaap) D. Hawksw. on *Parmelia saxatilis* (L.) Ach. thalli].

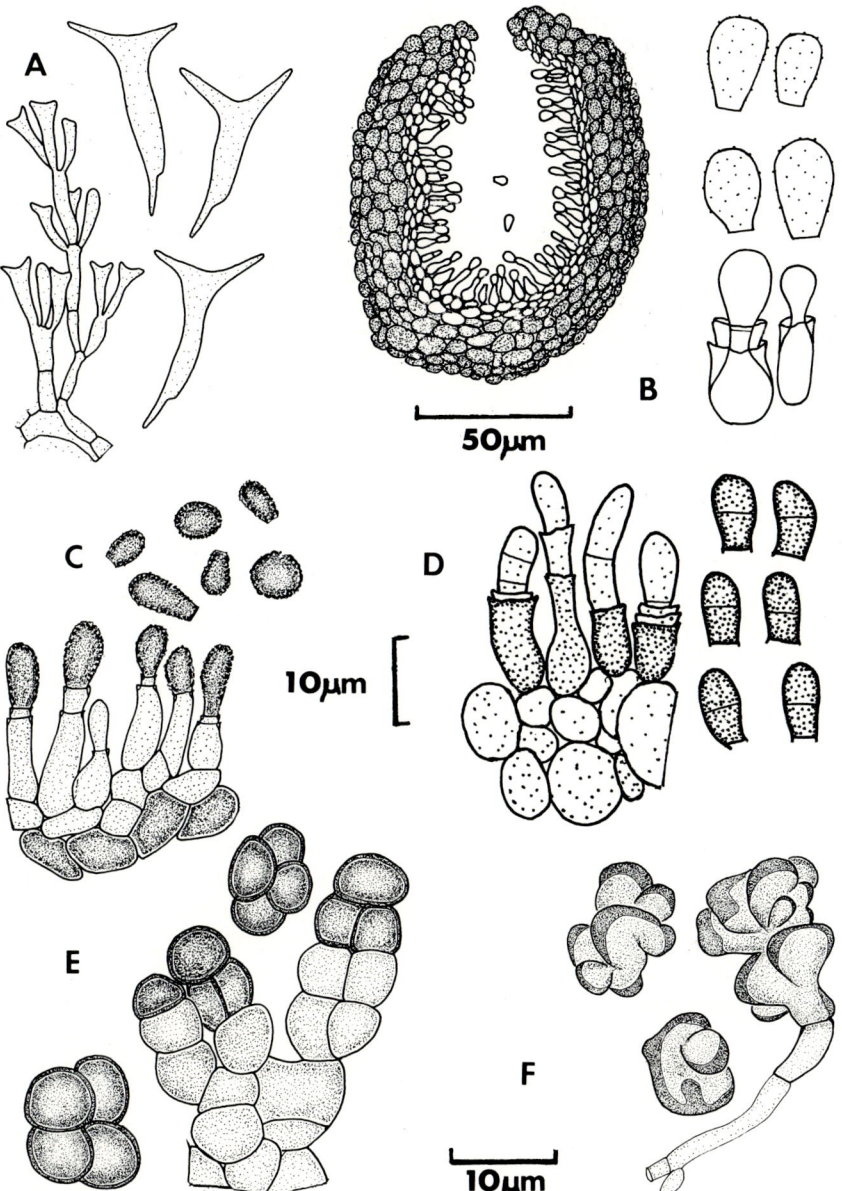

Fig. 15. Conidial fungi on lichen-forming ascomycetes. (A) *Cornutispora lichenicola*; (B) *Vouauxiomyces truncatus*; (C) *Lichenoconium lichenicola* (P. Karst.) Petr. & H. Syd.; (D) *Lichenodiplis lecanorae* (Vouaux) Dyko & D. Hawksw.; (E) *Sclerococcum sphaerale*; (F) *Milospium graphideorum* (Nyl.) D. Hawksw.

Other obligately lichenicolous coelomycete genera are *Lichenodiplis* Dyko & D. Hawksw. (two species; Fig. 15D), anamorphs of *Microcalicium* Vain. (two species, on Caliciales), *Pyrenotrichum* Mont. (nine species, with different host ranges on foliicolous lichens; see Santesson, 1952), *Vouauxiella* Petr. & Syd. (three species), and further the following monotypic genera: *Libertiella peltigerae* (Lib.) Keissl. [on *Peltigera polydactyla* (Neck.) Hoffm.], *Lichenosticta alcicornaria* (Linds.) D. Hawksw. (on several *Cladonia* species), *Verrucaster lichenicola* Tobler (on *Cladonia*), and *Vouauxiomyces truncatus* (B. de Lesd.) Dyko & D. Hawksw. [on *Parmelia caperata* (L.) Ach., perhaps parasymbiotic; Fig. 15B]. *Lichenophoma* Keissl. (two species) and *Sphaeromma mazosiae* Upad. and *Sporhaplus rondiensis* Upad. [both on *Mazosia phyllosema* (Nyl.) Zahlbr.] need further study; *Phaeoantenariella lichenicola* Caval. is based on sterile mycelium, and *Pleurosticta lichenicola* Petr. on pycnidia of the "host." I know of at least four further monotypic genera of lichenicolous Coelomycetes awaiting description.

A second species of *Cornutispora*, *C. lichenicola* D. Hawksw. & Sutt. (Fig. 15A), is now known from five lichenized genera. Further lichenicolous Coelomycetes have been referred to diverse genera and include the following: *Ascochyta lichenoides* (A. L. Sm.) D. Hawksw. (host indeterminate), *Phoma cytospora* (Vouaux) D. Hawksw. (on several *Parmelia* species), *P. physciicola* Keissl. [on *Physcia aipolia* (Ehrh. ex Humb.) Hampe], *P. ramalinae* Nord in [*Abrothallus suecicus* (Kirschst.) Nord. nom. hol., on several *Ramalina* Ach. species], *Sphaeronema lichenophilum* Dur. & Mont. [on ? *Diploicia canescens* (Dicks.) Manal or *Dirina ceratoniae* Fr.], *Stagonopsis peltigerae* P. Karst. [on *Peltigera canina* (L.) Willd.], *Stagonospora sandsteadeana* Keissl. [on *Cladonia furcata* (Huds.) Schrad.], and *Pyrenochaeta collematis* Vouaux [on *Collema tenax* (Sw.) Ach.]. Lichenicolous fungi referred to *Rhabdospora* (Sacc.) Sacc. so far studied (e.g., *R. lecanorae* Vouaux, *R. thallicola* Tassi) have proved to be merely based on the pycnidia of their supposed hosts (see Chapter 9). No fortuitously lichenicolous Coelomycetes are currently known to me.

A blastomycete yeast, *Candida fermenticarens* van der Walt, has recently been described from an unnamed corticolous lichen.

G. Deuteromycotina

1. Blastomycetes (20 genera, 200 species)

No conidial fungi appear to be described from the anamorphic yeasts, though, as with the Endomycetales, cultures may occasionally be contaminated by ubiquitous saprophytic Hyphomycetes.

2. Coelomycetes (870 genera, 7000 species)

Only one genus of conidial fungi is confined to the Coelomycetes, the monotypic hyphomycete *Engelhardtiella alba* Funk (on *Cytospora abietis* Sacc.

and *Pestalotia funera* Desm.). *Ramalia veronicae* (on "*Asteromella veronicae*") also occurs on *Parasterina veronicae*.

A considerable number of conidial fungi are described from Coelomycetes but how many are strictly confined to them is less certain, because most have not been studied critically. Some may be fortuitously present on effete stromata and pycnidia or simply using the existing coelomycete as a point of entry into the host. These include the Coelomycetes *Chaetophoma stromaticola* Speg. [on *Botryodiplodia lecanidion* (Speg.) Petr. & Syd.], *Dothiorella parasitica* Bubák (on *Cytospora* Ehrenb. ex Fr. sp.), *Hendersonia leptostromatis* Petr. (on *Leptostroma ahmadii* Petr.), "*Hendersonula*" *monochaetiella* (on *Monochaetiella themeda* Kandasw. and its teleomorph), *Microdiplodia anthurii* Trinch. (on *Phyllosticta cavarae* Trinch.), *Mycosticta cytosporicola* Frolow (on *Cytospora prunorum* Sacc. & Syd.), *Phoma consocians* Naumov (on *Septoria didyma* Fuckel), *Phyllosticta consors* Sacc. [on *Phleospora mori* (Lév.) Sacc.], *Sirosperma floridana* West (on *Aschersonia* Mont. sp.), *Sphaerographium microperae* (Cooke) Sacc. (on *Micropera* Lév. sp.), and *Vermicularia oligochaeta* Sacc. (on *Septoria antirrhinonum* Thorp). "*Asbolisia*" *indica* Agarwal & Sharma (on *Microxyphium alangii* Agarwal & Sharma) and *Coniothyrium crepinianum* Sacc. & Roum. (on *Heteropatella lacera* Fuckel) were probably fortuitously fungicolous.

The Hyphomycetes described from Coelomycetes are more numerous and include *Acladium ellipticum* Bat. (on *Colletotrichum* Corda sp.), *Alternaria olivacea* (Ell. & Ev.) van Hook (on *Sphaeropsis asiminae* Ell. & Ev.), *Cladosporium tuberculatum* Fr. [on *Cytospora leucosperma* (Pers.) Fr. ostioles], *Clasterosporium parasiticum* (Cooke) Sacc. [on *Phleosporella maculans* (Sandri) Höhn.], *Coccospora parasitica* Bomm. *et al.* [on *Coryneum brachyurum* Link, *Pseudovalsa laviciformis* (Fr.) Ces. & De Not. nom. hol., also originally mentioned as on *Eutypella stellulata* and *Pleomassaria siparia* (Berk. & Br.) Sacc.], *Cylindrocolla fugax* Sacc. (on *Diplodia castaneae* Sacc. old pycnidia), *Dendrodochium parasiticum* Chevaug. (on *Botryodiplodia theobromae* Pat.), *Didymaria acervulicola* Bat. & Nasc. (on *Gloeotrochila anthuriicola* Bat. & Nasc.), *Endophragmiella pallescens* Sutt. [on *Cytospora* cf. *chrysosperma* (Pers.) Fr.; Fig. 16B], *Graphium hendersonulae* Chevaug. (on "*Hendersonula*" *toruloidea* Nattr.), *Oidium fungicola* Bat. & Vital (on *Stromatopycnis rosetum* Vital), *Sirodesmium rosae* Bubák (on *Phoma pusilla* Schulz. & Sacc. old pycnidia), *Sporotrichum fallax* (Schulz.) Sacc. & Trev. (with *Diplodia cydoniae* Sacc.), and *Trinacrium subtile* Riess (on *Stilbospora* Pers. ex Mérat sp.). *Hymenula socia* Sacc. [on *Dothiorella stratosa* Sacc., *Phaeobotryosphaeria plicatula* (Berk. & Br.) Petch nom. hol.] is probably not fungicolous but an anamorph of a *Nectria* Fr. species (Hughes, 1953).

Some characteristically fungicolous genera include species on Coelomycetes also, for example, *Davisiella botryodiplodiae* Ahmad (on *Botryodiplodia* Sacc. sp.), *Ectosticta insignis* Petr. & Cif. [on *Aschersonia turbinata* Berk., also on

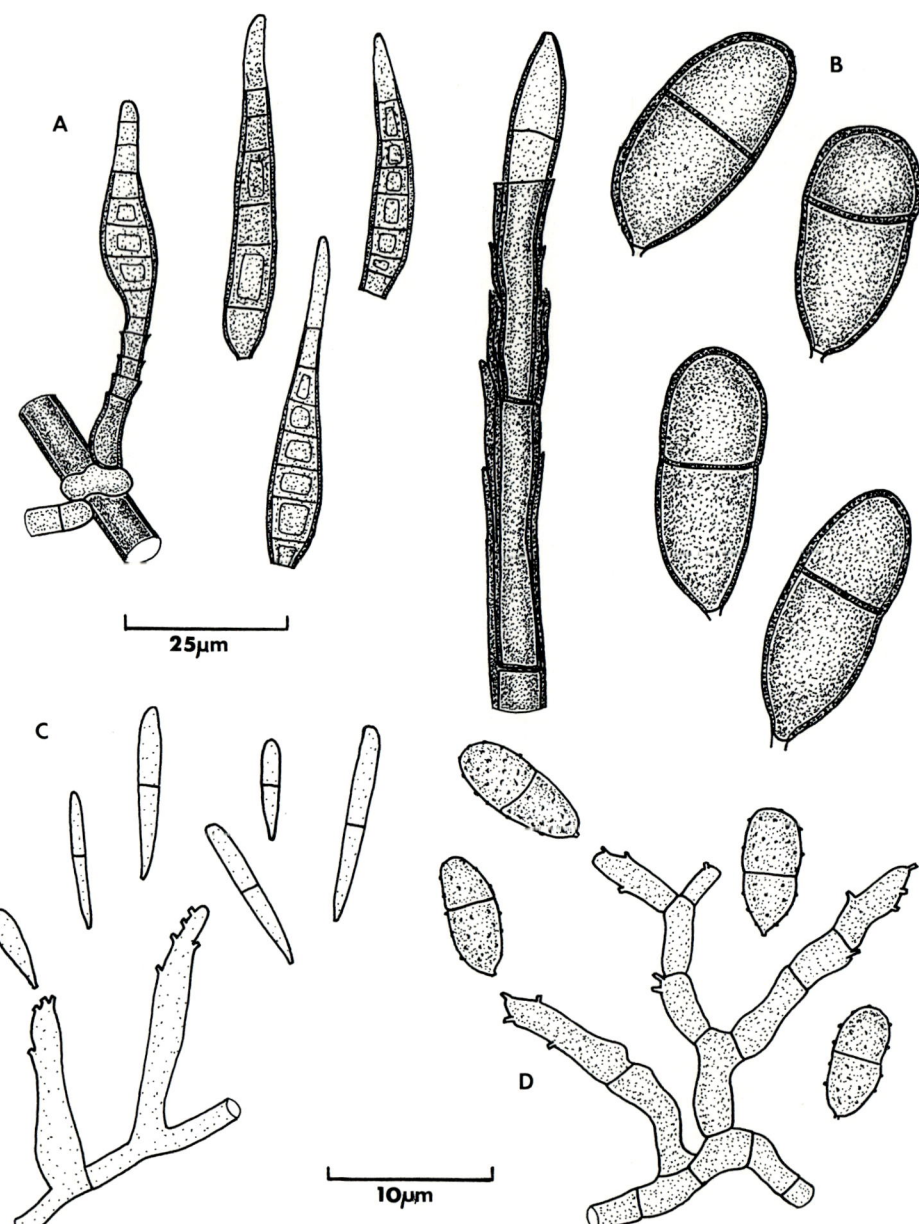

Fig. 16. Conidial fungi on Deuteromycotina. (A) *Annellophora dendrographii*; (B) *Endophragmiella pallescens*; (C) *Eriomycopsis minuta*; (D) *Scolecobasidium dendroides*.

Hypocrella turbinata (Berk.) Seav.], *Hansfordia alba* J. A. Meyer (on *Botryodiplodia theobromae*), *Hansfordiella diedickeae* Deight. (on *Diedickea piptadeniae* Deight.), *Nematogonium niveum* W. Gams (on the agaricicolous *Eleutheromyces subulatus*) the relationship of which with its host is discussed on the basis of cultures by Gams (1976), *Sympodiophora venezuelensis* Deight. & Piroz. (overgrowing *Colletotrichum* sp.), *Titaea callispora* Desm. (on *Ascochyta graminicola* Sacc. and *Septoria graminis* Desm. but commonest on Asterineae), and *T. submutica* Sacc. (on *S. forskahleana* Sacc.). The contact mycoparasite *Hansfordia parasitica* can occur on *Botryosphaerostoma quercina* Petr. and some *Physalospora* anamorphs. *Trichoderma viride sensu lato* can inhibit *Phoma lingam* (Tode ex Fr.) Desm. [*Leptosphaeria maculans* (Desm.) Ces. & De Not. nom. hol.] *in vitro* (Upitis, 1956) and partially suppressed both *Colletotrichum lini* (Westerd.) Toch. (Fedorinchik and Vanderflass, 1954) and *Macrophomina phaseolina* (Tassi) Goid. in soil (Norton, 1954). *Aspergillus aculeatus* Iizuka applied to seeds or soil may prevent damping off by *M. phaseolina* (Dhingra and Khare, 1973). Metabolites from *Stilbella erythrocephala* are inhibitory to *Phoma medicaginis* var. *pinodella* (L. K. Jones) Boerema (Singh and Webster, 1973).

An isolate close to *Acremonium sordidulum* W. Gams & D. Hawksw., denominated as f. *colletotrichum-dematii* Singh *et al.*, smothered and reduced the growth and conidial production of *Colletotrichum dematium* f. *truncatum* (Schw.) von Arx *in vitro*, while in nature it formed lesions in the colonies of the host on *Albizzia lebbek* Benth. pods (Singh *et al.*, 1978).

Numerous Hyphomycetes occur fortuitously on Coelomycetes, particularly stromatic species, for example, *Acarocybella jasminicola* (Hansf.) M. B. Ellis (on *Peltaster* H. & P. Syd. sp.), *Acremonium hyalinulum* [on *Colletotrichum capsici* (Syd.) Butler & Bisby], *Aureobasidium pullulans* (on *Sphaceloma ampelinum* de Bary), *Fusarium aquaeductuum* (*F. volutella* Ell. & Ev., on ? *Cytospora* sp.), *Helicoma stigmateum* (Riess) Linder (*Helicomyces niveus* Bres. & Jaap, on *Diplodia celtidigena* Ell. & Ev.), *Helicosporium vegetum* Nees [*Tubeufia cerea* (Berk. & Curt.) Booth nom. hol., on *Colletotrichum dematium* (Pers. ex Fr.) Grove], and *Kmetia exigua* Bres. & Sacc. (on Leptostromataceae sp.).

3. Hyphomycetes (930 genera, 7500 species)

There are a great many types of associations and interactions between Hyphomycetes and other conidial fungi. At the generic level, however, very few are exclusive to the group, and all of these are monotypic Hyphomycetes: *Basididyma perexigua* Cif. [on *Sirosporium gliricidiae* (Syd.) Deight.], *Gibellula pulchra* (Sacc.) Sacc. (on *Isaria* Pers. ex Fr. sp.), *Pseudofusidium hansfordii* Deight. (on *Mycovellosiella* Rangel sp.), *Sphaerulomyces coralloides* Marvanová [on the aquatic *Anguillospora crassa* Ingold, *A. longissima* (Sacc. & Syd.) Ingold, *Tricladium splendens* Ingold, and *Varicosporium elodeae* Kegel], and *Tympanosporium parasiticum* W. Gams [*Oospora candicula* auct., on *Tubercularia vulgaris* Tode ex Fr., *Nectria cinnabarina* (Tode ex Fr.) Fr. nom. hol.]. *Cercosporiella* Deight., originally described for *C. cercosporicola* Deight. (on *Cercospora koepkei* Krüger), is now known also to include some uredinicolous species. *Hemispora stellata* Vuill. [a culture contaminant of *Aspergillus repens* (Corda) Sacc.] is *Sporendonema epizoum* (Corda) Cif. & Red. and not a mycophilous species.

Approximately 80 Hyphomycetes have been described in ubiquitous plant parasitic or saprophytic genera. The following selection will demonstrate the range of genera involved: *Alternaria dendritica* (Souza da Camara) Joly [on *Passalora dendritica* (Wallr.) Sacc.], *Annellophora dendrographii* M. B. Ellis (on *Dendrographium atrum* Massee conidiophores; Fig. 16A), *A. sydowii* M. B. Ellis [on *Sporidesmium baccharidis* (Syd.) M. B. Ellis], *Bactridium parasiticum* Petr.(on *Tubercularia* Tode ex Fr. species), *Botrytis cercosporicola* Hara (on *Cercospora kakivora* Hara), *B. yuae* Muntanola [on *C. unamunoi* (Unam.) E. Castel., which Matta (1962) found it partly controlled, and also *Fulvia fulva* (Cooke) Cif.], *Cladosporium penicillioides* Preuss (on *Tubercularia* species), *Coniosporium helminthosporii* Corda [on *Pseudospiropes simplex* (Kunze ex Pers.) M. B. Ellis], *Cylindrocephalum stellatum* (Harz) Sacc. (on *Stilbum* species), six species referred to *Fusidium* Link ex Fr. (e.g., *F. hormiscii* Corda on *Taeniolella hormiscii* Corda), *Helicoon elegans* (Corda) Arnaud [on *Bispora antennata* (Pers.) Arnaud], *Helicosporium brunneolum* Berk. & Curt. (on *Helminthosporium* Link ex Fr. sp.), *Oedocephalum crystallinum* Ces. (on *Sporidesmium* Link ex Fr. sp.), *"Pediliospora" ramularioides* Bubák [on *Bispora betulina* (Corda) Hughes], *Rhinotrichum gossypinum* Speg. (on *Cercospora carioae* Speg.), *Riessia minima* Sacc. (on *Spiropes dorycarpus*), *Scolecobasidium dendroides* Piroz. & Hodges [on *Idriella fertilis* (Piroz. & Hodges) Mats.; Fig. 16D], *S. pusillum* Deight. & Piroz. [on *Exosporium stilbaceum* (Moreau) M. B. Ellis], *Speira densa* Viala (on *Dematophora necatrix* Hartig, *Rosellinia necatrix* Prill. nom. hol.), *Spermosporella elegans* Sutt. [on *Haplographium delicatum* Berk. & Br., *Hyaloscypha dematiicola* (Berk. & Br.) Nannf. nom. hol.], *Spicaria valdiviensis* Speg. (on *Heterosporium tupae* Speg.), *Sporotrichum biparasiticum* Bubák (on *Fusarium sphaeriae*), *Stigmina septoidii* Hansf. [on *Septoidium lateritium* (Syd.) Arnaud], *Trichothecium helminthosporii* (Thüm.) Sacc. [on *Drechslera ravenelii* (Curt.) Subram.], *Tritirachium isariae* (Petch) de Hoog [mixed with *Paecilomyces farinosus* (Dicks. ex Fr.) A. Brown & G. Sm.], and *Verticillium epiphytum* Hansf. [on *Pseudocercospora triumfettigena* (Yen & Gilles) Deight.].

In addition a few Hyphomycetes known only from other Hyphomycetes are described in predominantly or entirely fungicolous genera: *Cladobotryum australe* Viégas (on *Septoidium didymopanacis* Viégas), *Eriomycopsis minuta* Deight. & Piroz. [on *Melanographium citri* (Gonz., Frag. & Cif.) M. B. Ellis, intrahyphal and around mycelium and conidiophores; Fig. 16C], *Mycogone sporotrichi* (Corda) Sacc. & Trav. (on *Sporotrichum* sp.), and *Sympodiophora varanasiensis* Deight. & Piroz. (among *Cercospora vestita* (Ramakr.). The tropical *Trichoconis caudata* Deight. & Piroz. also occurs on *Periconia byssoides* Pers. ex Mérat. The contact mycoparasite *Stephanoma phaeospora* was originally described from *Fusarium* sp., which it needs for sporulation (Butler and McCain, 1968; Hoch, 1978), but it also attacked 14 of a wide range of other fungi it was tested against (Rakvidhyasastra and Butler, 1973). The widespread tropical mycophilic *Hansfordia pulvinata* (Berk. & Curt.) Hughes, which can also occur on leaves and lignum, is exceptionally common on *Cercospora* but is now known from approximately 70 Hyphomycetes distributed through 15 genera. A further species of *Hansfordia*, *H. triumfettae* (Hansf.) Hughes, occurs on *Pseudocercospora triumfettigena*. The common tropical *Acrodictys balladynae* is also frequent on foliicolous Hyphomycetes.

In contradistinction to the situation with the Hyphomycetes, as might be expected, relatively few Coelomycetes are described from Hyphomycetes. Almost all of those traced require reinvestigation and comprise "*Cicinnobella*" *heterothea* Syd. (on *Ovulariopsis* Pat. & Har. sp.), "*Cicinnobolus*" *sporophagus* Golovin (on *Helminthosporium delphinii* Golovin conidia), *Coniothyrium tuberculariae* Pass. (on *Tubercularia* sp.), *Didymosporium conglutinatum* Corda (on *Cladosporium herbarum*), *Microsphaeropsis sarcinellae* (Sahni) Morgan-Jones [on *Sarcinella palawanensis* (Syd.) Sahni], *Naemosphaerella chalaroides* Keissl. (on *Sporotrichum* sp.), *N. epimyces* Petch (on *Aegerita weberi* Fawcett), *Patellina epimyces* Petch (on *Hirsutella entomophila* Pat. and *H. versicolor* Petch), *Phoma stemphylii* Gonz. Frag. (on *Stemphylium anomalum* Gonz. Frag.), *Phyllosticta pivensis* Bubák [on *Ramularia geraniphaei* (C. Massal.) Magnus], *Pyrenochaeta mitteriellae* Sahni (on *Mitteriella zizyphina* Syd.) and *Sirosperma sparsum* Petch (on *Acremonium* sp.). *Phyllosticta bauhinicola* Rangel (with *Cladosporium* sp.) and *Rhabdospora elettariae* Penz. & Sacc. (with ? *Helminthosporium* sp.) at least were associated with Hyphomycetes but perhaps only fortuitously.

Considerable experimental work has been carried out *in vitro* on the antagonism and inhibition of numerous plant-pathogenic and soil-inhabiting Hyphomycetes by other conidial fungi. Information regarding the commoner soil fungi is compiled in Domsch *et al.* (1980), and the following examples illustrate the types of reactions found. *Gliocladium roseum* is particularly antagonistic, and among the species strongly inhibited by it are *Botrytis allii* Munn, *Cylindrocarpon destructans* (Zinssm.) Scholten (*Nectria radicicola* Gerlach & L. Nilsson nom. hol.), *Trichothecium roseum*, and *Verticillium dahliae*. This species is not, however, active against all other Hyphomycetes; *G. roseum* was less antagonistic toward *Helminthosporium turcicum* Pass. than several other of 15 microfungi tested, especially *Trichoderma harzianum* which inhibited its growth by coiling (Mickala-Doukaga *et al.*, 1978). Differential activity of species against the anamorph of *Cochliobolus sativus* has already been mentioned (Section III,F,5). *Trichoderma* species include numerous Hyphomycetes among their potential hosts, and some pathogens are sensitive to it, for example, *Fusarium solani* (Mart.) Sacc., the occurrence of which in soil is negatively correlated with that of *T. viride* (Joffe, 1966), and *F. culmorum* (W. G. Sm.) Sacc. hyphae are penetrated by *Epicoccum purpurascens*, *Trichoderma aureoviride* Rifai, and *T. harzianum* (Wu, 1977). *Trichoderma* cultures themselves may exceptionally be attacked, as by *Nematogonium parasiticum* (Davidson, 1935), and *Alternaria* and *Cladosporium* species are hosts for the contact mycoparasite *Gonatobotrys simplex* (Whaley and Barnett, 1963). *Memnoniella echinata* (Riv.) Gall. has been reported to inhibit a wide range of Hyphomycetes *in vitro* including *Alternaria alternata*, *Fusarium moniliforme* Sheld., *Myrothecium cinctum* (Corda) Sacc., *Stachybotrys atra*, *Trichoderma pseudokoningii* Rifai, and *Trichothecium roseum* (Qaiser Abbas and Ghaffar, 1973).

Acremonium strictum is parasitic on *Drechslera teres* (Sacc.) Shoem. and *D. poae* (Baudys) Shoem., invading the conidia, conidiophores, and hyphae (Kenneth and Isaac, 1964). *Fusarium semitectum* parasitizes the conidia and conidiophores of *Pseudocercospora mori* (Hara) Deight. and other *Cercosporae* both in nature and in culture, where coiling and colony overgrowth occurs (Rathaiah and Pavgi, 1973). *Paecilomyces carneus* (Duché & Heim) A. Brown & G. Sm. shows specific antagonism toward *Pseudocercosporella herpotrichoides*, which is also strongly inhibited by *Penicillium granulatum* Bain. and *P. spinulosum* (Domsch and Gams, 1968).

Culture filtrates of *Aspergillus* species are particularly rich in antifungal metabolites (Domsch *et al.*, 1980). Chemical antagonism is also shown by some Coelomycetes, although they are less well studied; *Phoma exigua* Desm. produces an antibiotic active against *Fusarium solani* var. *coeruleum* (Sacc.) C. Booth but not against *Aspergillus niger* or *Penicillium expansum* (Logan and O'Neill, 1970).

Numerous saprophytic Hyphomycetes may be found growing intermixed with one another, particularly dematiaceous species on dead herbaceous stems and lignum, but most such associations are perhaps merely fortuitous. However, in some instances overgrowth or facultative parasitism may be exhibited, as with *Acremonium acutatum* W. Gams (on *Cercospora atromarginalis* Atk.), *Alternaria tenuis* (on *Cladosporium*, *Helminthosporium*, *Torula herbarum*, and so on), *Cheiromycella microscopica* (P. Karst.) Hughes [over *Alysidium resinae* (Fr.) M. B. Ellis, and so on], *Chloridium botryoideum* (Corda) Hughes (on various dematiaceous species on wood), *Monotospora megalospora* (Berk. & Br.) Massee [*Farlowiella carmichaeliana* (Berk.) Sacc. nom. hol., on *Graphium flexuosum* (Mass.) Sacc.], *Sepedonium monosporum* Peyronel (on *Cladosporium herbarum*, and so on), *Sporidesmium socium* M. B. Ellis (lignicolous with a predilection for *Helminthosporium velutinum* Link ex Ficin. & Schub.), and *Zygosporium mycophilum* (Vuill.) Sacc. (culture contaminant of *Gonatobotryum fuscum*, also overgrowing other fungi).

4. Mycelia Sterilia (28 genera, 200 species)

The Mycelia sterilia (Agonomycetales) include several anamorph genera for perennial resting structures which are consequently available for colonization by conidial fungi. Most important of these are sclerotia developed in diverse groups of fungi, and conidial fungi associated with some of these have already been mentioned. *Gliocladium roseum* and *Tichoderma* species are particularly widespread on sclerotia and may be pathogenic to them; these fungi do not attack previously killed sclerotia readily, suggesting that they receive some stimulus from the host (Weindling, 1938).

The sclerotial phase *Rhizoctonia solani* (*Thanatephorus cucumeris* nom. hol.) has received particular attention in view of its importance as a plant pathogen. These sclerotia are directly and severely parasitized by several conidial fungi, including *Fusarium oxysporum*, *Papulaspora stoveri* Warren, *Penicillium vermiculatum* Dang., and *Trichoderma viride sensu lato* (Boosalis, 1956), but *R. solani* is also acted on by numerous other fungi, for example, *Acremonium cerealis* (P. Karst.) W. Gams, *Humicola grisea* Traaen, *Paecilomyces marquandii*, *Penicillium brevicompactum*, and *P. janthinellum* Biourge (Domsch et al., 1980).

Most conidial fungi found on sclerotia are ubiquitous species, but a few have been described from them, most in need of critical study: *Acrostalagmus charceus* (Corda) Sacc. (on *Sclerotium* Tode ex Fr. sp.), the pathogen *Coniothyrium minitans* (see Section III,F,4,a), *Fusarium heterosporioides* Fautr. (perhaps a yeast; Booth, 1971), *F. heterosporum* Nees ex Fr. (a widespread species), *Hymenula sclerotii* Crouan and *Phyllosticta sclerotialis* Cocc. (all on *S. clavus* DC.), and *Phoma scleroticola* Pat. (on *S. patouillardii* Sacc. & Syd.). Other fortuitously or facultatively occurring Hyphomycetes include *Acremoniella atra*, *Acremonium kiliense* Grütz, *Fusarium solani*, *Geotrichum candidum* (*G. bipunctatum* Roll.), *Monotospora priceana* Sacc., and *Papulaspora dodgei* Conners.

A few species have also been described from rhizomorphs (all on *Rhizomorpha subcorticalis* Grev.): *Acrogenospora setiformis* (Wallr.) M. B. Ellis, *Cladotrichum opacum* Schulz. & Sacc., *Harpographium rhizomorphum* (Mart.) Sacc., and *Periola hirsuta* (Schum.) Fr. The mycophilous *Amblyosporium botrytis* is also reported from rhizomorphs (*R. subterranea* Pers.).

IV. DISCUSSION

The main conclusion of this survey must necessarily be that a great deal of taxonomic work remains to be carried out on numerous conidial fungi originally described from other fungi; perhaps as many as 70% have not been restudied since their first description. There is also no physiological, ultrastructural, or experimental work available for all but a minority of species (under 1%). This lack of basic information is unfortunate, because studies on mycoparasites may, as stressed by Barnett and Binder (1973), help in the elucidation of some fundamental principles of fungal parasitism of vascular plants, since their use saves time and space and also allows environmental and nutritional parameters to be controlled. However, one is forced to concur with Barnett (1964) that "... with the almost unlimited supply of test fungi, it will be many years before our information will permit us to establish clearly the basic principles of mycoparasitism."

A. Evolutionary Considerations

The evolution of obligate parasites must necessarily follow, or be contemporaneous with, that of their hosts. The extent of phenetic isolation (divergence) of conidial fungicolous fungi varies considerably from group to group, and some interesting points emerge, related not only to the fungi themselves but also to their hosts.

Almost all the host ranges of obligately fungicolous conidial fungi given above are based on direct observations of what has been found in nature. There have been very few thorough field or *in vitro* trials to assess the potential host ranges of species, compared with the numbers of taxa involved. More investigations along the lines of those of Butler (1957) and Rakvidhyastu and Butler (1973), who tested mycoparasitic species against 59 and 97 other fungi, respectively, are needed to assess this. In the course of random sampling of the world's mycoflora, it is inevitable that some taxa originally described as fungicolous will later prove to be more catholic as further material becomes available, and some instances of this are mentioned in the preceding sections. However, bearing in mind these difficulties and problems of reliably estimating the numbers of potential host taxa, some generalizations might be justified in view of the size of the sample.

First, it is interesting to note that in the conidial fungi the numbers of species known only from the major divisions is, proportional to the numbers of potential hosts, remarkably similar in the Basidiomycotina (1:85) and Ascomycotina (1:79), but in the Deuteromycotina (1:96), Mastigomycotina (1:103), and especially the Zygomycotina (1:236), considerably less (Table I, see p.181). If genera are considered, ratios for the Basidiomycotina (1:676) and Ascomycotina (1:636) are even closer, and other groups more divergent, for example, the Deuteromycotina (1:2500).

The scarcity of conidial fungi confined to the Mastigomycotina and Zygomycotina may be due to (1) differences in cell wall composition

(Bartnicki-Garcia, 1970) or other biochemical features, and (2) the fact that the fungicolous ecological niche on these groups had already been occupied prior to the evolution of the conidial fungi. The early origin of these, as opposed to the other fungal divisions, has already been postulated on other grounds (Bartnicki-Garcia, 1970; Pirozynski and Malloch, 1975). The comparable proportions of conidial genera and species on Ascomycotina and Basidiomycotina may imply that these divisions have been available for colonization by conidial fungi for approximately the same length of evolutionary time. Anamorphs may be presumed to be continually arising from ascomycete teleomorphs and so could not have originated earlier than the Ascomycotina. It is also interesting that the at first apparently somewhat surprising figures for the Myxomycota (genera 1:282, species 1:51) and also the Basidiomycotina (especially if the Teliomycetes are excluded from consideration here) mainly have moniliaceous Hyphomycetes with teleomorphs in the Sphaeriales (Hypomycetaceae and Nectriaceae) confined to them, an order that may have expanded relatively recently (see below).

When this concept is applied within the taxa on Ascomycotina, the Discomycetes (species: hosts 1:100) might be considered more recent than the Loculoascomycetes (1:21) and Pyrenomycetes (1:36). If the generic frequencies are interpreted as an indication of divergence, the proportion in the Loculoascomycetes (1:166) suggests that these may be more ancient than the Pyrenomycetes (1:421), and especially the Discomycetes (1:1500). The conidial taxa on lichen-forming fungi present a special problem here because, (1) they are perhaps now better understood than those on other groups of ascomycetes (with many published names now excluded), and (2) they are relatively rarely collected and reported. Both these considerations lead to the numbers considered here being lower than in the other groups. However, despite these considerations, more *genera* of conidial fungi, 22, are actually confined to lichenized hosts than any other group of fungi distinguished in Table I, most are dematiaceous (perhaps implying loculoascomycete links), and furthermore many are monotypic or small and have developed parasymbiotic relations with their hosts or adapted to narrow host ranges (in several cases single host species). These associations may reasonably be assumed to have taken an extremely long time to evolve and consequently support the concept that the lichen habit is a particularly ancient one.

As "lichen" is most satisfactorily viewed as a continually evolving and devolving nutritional state (Hawksworth, 1978), extant lichen-forming taxa necessarily include those derived from recent and not only ancient ascomycetes. In view of these considerations the information on conidial fungi suggests that descendants of the most ancient ascomycetes should be sought among the extant Loculoascomycetes and lichen-forming taxa. It would be interesting to see how these conclusions relate to those derived from a survey of other fungicolous fungi.

The Deuteromycotina present a special case in that the basis of genetic variation in the hosts as well as the conidial fungi must mainly be in extinct or extant teleomorphs, host and conidial anamorphs continually being formed and exploiting new microhabitats as they arise. This evolutionarily more fluid situation may account for the proportionately fewer conidial fungi which have been able to adapt to growth on other members of their division.

Some conidial fungi appear to be confined to a restricted number of hosts, while others can utilize a wide range of other fungi. This situation is seen at both generic and specific levels, though, as would be expected, all types of intermediates occur. In addition to ubiquitous fungal species, several genera include individual species which are only known from particular groups of fungi (Table II, p. 182). This specialization may be seen in both catholic (e.g., *Acremonium, Fusarium*) and exclusively fungicolous (e.g., *Sympodiophora*) genera, and the restriction may be to single host genera, species, or higher categories.

A very high proportion of the conidial fungi described are currently known only from single collections, hence on single host species. To what extent such restrictions are real or due merely to a lack of material is necessarily unclear.

Allied obligately fungicolous fungi might be expected to need similar nutrient or other compounds, and so in certain cases it is conceivable that they may serve as taxonomic aids in the separation of allied host fungi. Perhaps one of the best examples of the integration of host and fungus systematics is afforded by the nine species of *Pyrenotrichum*, all of which have different but clearly defined ranges over a few genera of foliicolous lichenized ascomycetes (Santesson, 1952). In carrying out taxonomic work on any group of fungi it may always be worth ascertaining how the fungicolous fungi tie in with any remodeling being considered.

B. Practical Considerations

It will be evident from the information reviewed here that there is now a considerable body of *in vitro* experimental work on the inhibition and antagonism between economically important and other fungi. In comparison to laboratory reports, ones on the use of such interactions in the biological control of fungal pathogens are relatively sparse. References to the use of conidial mycoparasitic fungi in the control of pathogens belonging to diverse groups of fungi have already been made above and will not be repeated here. The degree of success obtained in field trials varies greatly, and major problems arise both in effectively disseminating the fungi and encouraging their spread. Fungicolous fungi with narrow host ranges may tend to form biotrophic rather than necrotrophic relationships with their hosts, and the most effective mycoparasites in control tend to be the more ubiquitous. For example, dense populations of *Aureobasidium pullulans* prevent or reduce damage by *Alternaria zinniae* Pope on beans (van den Heuvel, 1969), *A. porri* (Ellis) Cif. on onion (Fokkema and

Lorbeer, 1974), and *Cochliobolus sativus* on rye (Fokkema *et al.*, 1975). Caution is also required, as a species may reduce the population of one pathogen but at the same time promote another. *Aureobasidium pullulans,* for example, can stimulate rhizomorph development in *Armillaria mellea* (Pentland, 1965). Such interactions are not always easy to foresee, as in the use of benomyl against a *Cercospora* Fres. leaf spot which enhanced the development of *Athelia rolfsii* as a result of the supression of *Trichoderma* and other fungi normally antagonistic to the *Athelia* (Backman *et al.*, 1975).

Trichoderma species inoculated into soil have long been known to influence the virulence of plant pathogens in soil (Porter and Carter, 1938), and their action has frequently been mentioned in the preceding sections of this survey. Spreading these fungi has been tried by various methods, including coating of seeds with conidia, but the promotion of natural populations may have much to commend it, for example, by stimulation with carbon disulfide (Darley and Wilbur, 1954), crop rotation, or soil amendments (Boosalis, 1964).

In cases where chemical products are inhibitory, it may be safer to synthesize and distribute these, rather than the fungi themselves, but this is inevitably much more costly.

The greatest barrier to the increasingly effective use of fungicolous fungi in the biological control of plant-pathogenic fungi is undoubtedly our lack of basic knowledge of the ecology of microfungi and the factors underlying the responses of one species to another. As already stressed by Boosalis (1964), if the tide of the interfungal wars can be directed in favor of the fungicolous mycoparasites "man will reap the spoils of hyperparasitic victories."

ACKNOWLEDGMENTS

I am very grateful to Professor Dr. K. H. Domsch and Dr. W. Gams for access to the manuscript for their forthcoming compendium of information on selected soil fungi; to Dr. K. A. Pirozynski for most valuable discussions and reading the manuscript; to my colleagues at the Commonwealth Mycological Institute for their continuing assistance during the preparation of this contribution; and to Professor J. Webster, Dr. C. Dennis, and the Cambridge University Press and for permission to reproduce Figs. 1–2.

Mrs. P. Townson assisted in the compilation of the card index which greatly facilitated this work, and Mrs. M. Rainbow typed the final copy.

REFERENCES

Ahmed, A. H. M., and Tribe, H. T. (1977). Biological control of white rot of onion (*Sclerotium cepivorum*) by *Coniothyrium minitans*. *Plant Pathol.* **26**, 75–78.

Ainsworth, G. C. (1971). "Ainsworth & Bisby's Dictionary of the Fungi," 6th ed. Commonw. Mycol. Inst., Kew, Surrey, England.

Ainsworth, G. C., Sparrow, F. K., and Sussman, A. S., eds. (1973a). "The Fungi. An Advanced Treatise," Vol. 4A. Academic Press, New York.
Ainsworth, G. C., Sparrow, F. K., and Sussman, A. S., eds. (1973b). "The Fungi: An Advanced Treatise," Vol. 4B. Academic Press, New York.
Anderson, K. W., and Skinner, C. E. (1947). Yeasts in decomposing fleshy fungi. *Mycologia* **39**, 165-170.
Arnold, G. R. W. (1970). Predvarītel'nȳī obzor mikofil'nȳkh gribov SSR. *Nov. Sist. Niz. Rast.* **7**, 108-120.
Arnold, G. R. W. (1971). O nekotorȳkh gribakh, redkikh ili novȳky diya SSR i nauki. *Nov. Sist. Niz. Rast.* **8**, 130-138.
Arnold, G. R. W. (1976). Internationale Bibliographie der Hypomycetaceae (Mycophyta, Ascomycotina). *Bibliogr. Mitt. Univ.-bibliotek Jena* **25**, i-iv, 1,129.
Ayres, T. T. (1941). The distribution and association of *Gonatorrhodiella highlei* with *Nectria coccinea* in the United States. *Mycologia* **33**, 178-187.
Aytoun, R. S. C. (1953). The genus *Trichoderma*: Its relationship with *Armillaria mellea* (Vahl ex Fries) Quél. and *Polyporus schweinitzii* Fr., together with preliminary observations on its ecology in woodland soils. *Trans. Proc. Bot. Soc. Edinburgh* **36**, 99-114.
Backman, P. A., Rodriguez-Kabana, R., and Williams, J. C. (1975). The effect of peanut leafspot fungicides on the nontarget pathogen *Sclerotium rolfsii*. *Phytopathology* **65**, 773-776.
Backus, M. P., and Stowell, E. A. (1953). A *Fusidium* disease of *Xylaria* in Wisconsin. *Mycologia* **45**, 836-847.
Barnett, H. L. (1963). The nature of mycoparasitism by fungi. *Annu. Rev. Microbiol.* **17**, 1-44.
Barnett, H. L. (1964). Mycoparasitism. *Mycologia* **56**, 1-19.
Barnett, H. L., and Binder, F. L. (1973). The fungal host-parasite relationship. *Annu. Rev. Phytopathol.* **11**, 273-292.
Barnett, H. L., and Lilly, V. G. (1958). Parasitism of *Calcarisporium parasiticum* on species of *Physalospora* and related fungi. *W. Va., Agric. Exp. Stn. Bull.* **420**, 1-37.
Barnett, H. L., and Lilly, V. G. (1962). A destructive mycoparasite: *Gliocladium roseum*. *Mycologia* **54**, 72-77.
Barr, M. E. (1976). Perspectives in the Ascomycotina. *Mem. N.Y. Bot. Gard.* **28**, 1-8.
Barras, S. J. (1969). *Penicillium implicatum* antagonistic to *Ceratocystis minor* and *C. ips*. *Phytopathology* **59**, 520.
Barron, G. L. (1968). "The Genera of Hyphomycetes from Soil." Williams & Wilkins, Baltimore, Maryland.
Barron, G. L., and Fletcher, J. T. (1970). *Verticillium albo-atrum* and *V. dahliae* as mycoparasites. *Can. J. Bot.* **48**, 1137-1139.
Barron, G. L. and Fletcher, J. T. (1972). *Rhopalomyces elegans* Corda, a host of *Mycogone perniciosa*. *Mushroom Sci.* **8**, 383-386.
Bartnicki-Garcia, S. (1970). Cell wall composition and other biochemical markers in fungal phylogeny. *In* "Phytochemical Phylogeny" (J. B. Harborne, ed.), pp. 81-103. Academic Press, New York.
Benjamin, R. K. (1959). The merosporangiferous Mucorales. *Aliso* **4**, 321-433.
Blumer, S. (1933). Die Erysiphaceen Mitteleuropas mit besonderer Berücksichtigung der Schweiz. *Beitr. Kryptogamenflora Schweiz* **7**(1), 1-483.
Blyth, W. (1949). Studies on *Gonatorrhodiella highlei* A. L. Sm. *Trans. Proc. Bot. Soc. Edinburgh* **35**, 157-179.
Boller, R. A., and Schroeder, H. W. (1972). Self-parasitism in *Aspergillus flavus*. *Mycologia* **64**, 433-437.
Boosalis, M. G. (1956). Effect of soil temperature and green-manure amendment of unsterilized soil on parasitism of *Rhizoctonia solani* by *Penicillium vermiculatum* and *Trichoderma* sp. *Phytopathology* **46**, 473-478.

Boosalis, M. G. (1964). Hyperparasitism. *Annu. Rev. Phytopathol.* **2**, 363-375.
Booth, C. (1966). The genus *Cylindrocarpon. Mycol. Pap.* **104**, 1-56.
Booth, C. (1971). "The Genus *Fusarium*." Commonw. Mycol. Inst. Kew, Surrey, England.
Bowen, W. R. (1968). The imperfect fungus *Schizotrichella lunata* on the moss *Mnium cuspidatum. Bryologist* **71**, 124-126.
Brady, B. L. [K.] (1960). Occurrence of *Itersonilia* and *Tilletiopsis* on lesions caused by *Entyloma. Trans. Br. Mycol. Soc.* **43**, 31-50.
Brady, B. L. K., and Gibson, I. A. S. (1976). "CMI Descriptions of Plant Pathogenic Fungi and Bacteria," Nos. 498-500. Commonw. Mycol. Inst. Kew, Surrey, England.
Brady, B. L. K., and Waller, J. M. (1976). "CMI Descriptions of Plant Pathogenic Fungi and Bacteria," No. 497. Commonw. Mycol. Inst. Kew, Surrey, England.
Brian, P. W. (1960). Antagonistic and competitive mechanisms limiting survival and activity of fungi in soil. *In* "The Ecology of Soil Fungi" (D. Parkinson and J. S. Ward, eds.), pp. 115-129. Liverpool Univ. Press, Liverpool.
Brian, P. W., Curtis, P. J., Hemming, H. G., and Norris, G. L. F. (1957). Wortmannin, an antibiotic produced by *Penicillium wortmanni. Trans. Br. Mycol. Soc.* **40**, 365-368.
Buller, A. H. R. (1924). "Researches on Fungi," Vol. 3. Longmans, Green, New York.
Butler, E. E. (1957). *Rhizoctonia solani* as a parasite of other fungi. *Mycologia* **49**, 354-373.
Butler, E. E., and McCain, A. H. (1968). A new species of *Stephanoma. Mycologia* **60**, 955-959.
Butt, Z. L., and Ghaffar, A. (1972). Inhibition of fungi, actinomycetes and bacteria by *Stachybotrys atra. Mycopathol. Mycol. Appl.* **47**, 241-252.
Byler, J. W., Cobb, F. W., and Parmeter, J. R. (1972). Occurrence and significance of fungi inhabiting galls caused by *Peridermium harknessii. Can. J. Bot.* **50**, 1275-1282.
Calpouzos, L., Thies, T., and Batlle, C. M. R. (1957). Culture of the rust parasite, *Darluca filum. Phytopathology* **47**, 108-109.
Campbell, W. P. (1956). The influence of associated microorganisms on the pathogenicity of *Helminthosporium sativum. Can. J. Bot.* **34**, 865-874.
Carling, D. E., Brown, M. F., and Millikan, D. F. (1976). Ultrastructural examination of the *Puccinia graminis—Darluca filum* host-parasite relationship. *Phytopathology* **66**, 419-422.
Carter, M. V., and Price, T. V. (1974). Biological control of *Eutypa armeniacae*. II. Studies of the interaction between *E. armeniacae* and *Fusarium lateritium*, and their relative sensitivities to benzimidazole chemicals. *Aust. J. Bot.* **25**, 105-119.
Clare, B. G. (1964). *Ampelomyces quisqualis (Cicinnobolus cesatii)* on Queensland Erysiphaceae. *Univ. Queensl. Pap., Dep. Bot.* **4**, 147-149.
Coles, R. B., and Talbot, P. H. B. (1977). *Septobasidium clelandii* and its conidial state *Harpographium cornelioides. Kew Bull.* **31**, 481-488.
Cooke, R. (1977). "The Biology of Symbiotic Fungi." Wiley, New York.
Csuti, É., Lemaire, J. M., Ponchet, J., and Rapilly, F. (1965). Exemples d'interactions fongiques au niveau de plantules de blé contaminées par l'*Helminthosporium sativum. Ann. Epiphyt.* **16**, 37-44.
Darker, G. D. (1967). A revision of the Hypodermataceae. *Can. J. Bot.* **45**, 1399-1444.
Darley, E. F., and Wilbur, W. D. (1954). Some relationships of carbon disulphide and *Trichoderma viride* in the control of *Armillaria mellea. Phytopathology* **44**, 485.
Davidson, R. W. (1935). Forest pathology notes. *Plant Dis. Rep.* **19**, 95-97.
Day, P. R. (1974). "Genetics of Host-Parasite Interactions." Freeman, San Francisco, California.
Dayal, R., and Barron, G. L. (1970). *Verticillium psalliotae* as a parasite of *Rhopalomyces. Mycologia* **62**, 826-830.
Deacon, J. W. (1976). Biological control of the take-all fungus, *Gaeumannomyces graminis*, by *Phialophora radicicola* and similar fungi. *Soil Biol. & Biochem.* **8**, 275-283.
de Hoog, G. S. (1974). The genera *Blastobotrys, Sporothrix, Calcarisporium* and *Calcarisporiella* gen. nov. *Stud. Mycol.* **7**, 1-84.

Deighton, F. C. (1960). Collecting fungi in the tropics. *In* "Herb. I.M.I. Handbook," pp. 78-83. Commonw. Mycol. Inst. Kew, Surrey, England.
Deighton, F. C. (1969). Microfungi. IV. Some hyperparasitic Hyphomycetes, and a note on *Cercosporella uredinophila* Sacc. *Mycol. Pap.* **118**, 1-41.
Deighton, F. C., and Pirozynski, K. A. (1972). Microfungi. V. More hyperparasitic Hyphomycetes. *Mycol. Pap.* **128**, 1-110.
Dennis, C., and Webster, J. (1971a). Antagonistic properties of species-groups of *Trichoderma*. I. Production of non-volatile antibiotics. *Trans. Br. Mycol. Soc.* **57**, 25-39.
Dennis, C., and Webster, J. (1971b). Antagonistic properties of species-groups of *Trichoderma*. II. Production of volatile antibiotics. *Trans. Br. Mycol. Soc.* **57**, 41-48.
Dennis, C., and Webster, J. (1971c). Antagonistic properties of species-groups of *Trichoderma*. III. Hyphal interaction. *Trans. Br. Mycol. Soc.* **57**, 363-369.
De Vay, J. E. (1956). Mutual relationships in fungi. *Annu. Rev. Microbiol.* **10**, 115-140.
Dhingra, O. D., and Khare, N. M. (1973). Biological control of *Rhizoctonia bataticola* on urid bean. *Phytopathol. Z.* **76**, 23-29.
Domsch, K. H. (1960). Das Pilzspecktrum einer Bodenprobe. III. Nachweis der Einzelpilze. *Arch. Mikrobiol.* **35**, 310-339.
Domsch, K. H., and Gams, W. (1968). Die Bedeutung vorfruchtabhängiger Verschiebungen in der Bodenmikroflora. II. Antagonistische Einflüsse auf pathogene Bodenpilze. *Phytopathol. Z.* **63**, 165-176.
Domsch, K. H., Gams, W., and Anderson, T. H. (1980). "A Compendium of Selected Soil Fungi," 2 vols. Academic Press, New York (in press).
Drechsler, C. (1937). Some Hyphomycetes that prey on free-living terricolous nematodes. *Mycologia* **29**, 447-552.
Drechsler, C. (1938). Two Hyphomycetes parasitic on oospores of root-rotting Oomycetes. *Phytopathology* **28**, 81-103.
Drechsler, C. (1943). Another hyphomycetous fungus parasitic on *Pythium* oospores. *Phytopathology* **33**, 227-331.
Dreschsler, C. (1952). Another nematode-strangulating *Dactylella* and some related Hyphomycetes. *Mycologia* **44**, 533-556.
Drechsler, C. (1962). Two additional species of *Dactylella* parasitic on *Pythium* oospores. *Sydowia* **15**, 92-97.
Drechsler, C. (1963). A slender-spored *Dactylella* parasitic on *Pythium* oospores. *Phytopathology* **53**, 1050-1053.
Durell, L. W. (1968). Hyphal invasion by *Trichoderma viride*. *Mycopath. Mycol. Appl.* **35**, 138-144.
Emmett, R. W., and Parbery, D. G. (1975). Appressoria. *Annu. Rev. Phytopathol.* **13**, 146-167.
Emmons, C. W. (1930). *Cicinnobolus cesatii*: A study in host-parasite relationships. *Bull. Torrey Bot. Club* **57**, 421-441.
Eriksson, O. (1966). On *Eudarluca caricis* (Fr.) O. Erikss., comb. nov., a cosmopolitan uredinicolous pyrenomycete. *Bot. Not.* **119**, 33-69.
Fedorinchik, N. S. (1961). Die Verwendung des Pilzes *Trichoderma lignorum* zur Bekaempfung phytopathogener Pilze im Boden. *Tagungsber. Dtsch. Akad. Landwirtschaftswiss. Berlin* **41**, 109-118 (not seen).
Fedorinchik, N. S., and Vanderflass, L. K. (1954). [Effect of the antagonistic activity of the soil fungus *Trichoderma lignorum* Harz on increases in yields of agricultural crops.] *Tr. Vses. Nauchno-Issled. Inst. Zashch. Rast.* **5**, 17-37 (not seen).
Fokkema, N. J., and Lorbeer, J. W. (1974). Interactions between *Alternaria porri* and the saprophytic mycoflora of onion leaves. *Phytopathology* **64**, 1128-1133.
Fokkema, N. J., van de Laar, J. A. J., Nelis-Blomberg, A. L., and Schippers, B. (1975). The

buffering capacity of the natural mycoflora of rye leaves to infection by *Cochliobolus sativus*, and its susceptibility to benomyl. *Neth. J. Plant Pathol.* **81**, 176-186.

Fontana, A. (1960). Sopra un parassita di *Rhodopaxillus nudus* (Fr.) Maire nuovo per l'Italia: *Harziella capitata* Cost. et Matr. *Allionia* **6**, 35-41.

Forrer, H. R. (1977). Der Einfluss von Stoffwechselprodukten des Mycoparasiten *Aphanocladium album* auf die Teleutosporenbildung von Rostpilzen. *Phytopathol. Z.* **88**, 306-311.

Gain, R. E., and Barnett, H. L. (1970). Parasitism and axenic growth of the mycoparasite *Gonatorrhodiella highlei*. *Mycologia* **62**, 1122-1129.

Gams, W. (1971). "*Cephalosporium*-artige Schimmelpilze (Hyphomycetes)." Fischer, Stuttgart.

Gams, W. (1976). *Nematogonium niveum*, a new hyperparasitic species of Hyphomycetes. *Rev. Mycol.* **39**, 273-278.

Gams, W., and Hoozemans, A. C. M. (1970). *Cladobotryum*-konidienformen von *Hypomyces*-Arten. *Persoonia* **6**, 95-110.

Garrett, S. D. (1958). Inoculum potential as a factor limiting lethal action by *Trichoderma viride* Fr. on *Armillaria mellea* (Fr.) Quél. *Trans. Br. Mycol. Soc.* **41**, 157-164.

Gilman, J. C., and Tiffany, L. H. (1952). Fungicolous fungi from Iowa. *Proc. Iowa Acad. Sci.* **59**, 99-110.

Hansford, C. G. (1946). The foliicolous ascomycetes, their parasites and associated fungi, especially as illustrated by Uganda specimens. *Mycol. Pap.* **15**, 1-240.

Hashioka, Y. (1973a). Mycoparasitism in relation to phytopathogens. *Forsch. Geb. Pflanzenkrankh.* **8**, 179-190.

Hashioka, Y. (1973b). Scanning electron microscopy on the mycoparasites, *Trichoderma*, *Gliocladium* and *Acremonium*. *Rep. Tottori Mycol. Inst.* **10**, 473-484.

Hashioka, Y., and Fukita, T. (1969). Ultrastructural observations on mycoparasitism of *Trichoderma*, *Gliocladium* and *Acremonium* to phytopathogenic fungi. *Rep. Tottori Mycol. Inst.* **7**, 8-18.

Hashioka, Y., Ishikawa, H., Komatsu, M., and Arita, I. (1961). *Trichoderma viride* as an antagonist of the wood-inhabiting Hymenomycetes. II. A metabolic product of *Trichoderma* fungistatic to the Hymenomycetes. *Rep. Tottori Mycol. Inst.* **1**, 10-15.

Hawksworth, D. L. (1974). "Mycologist's Handbook." Commonw. Mycol. Inst., Kew, Surrey, England.

Hawksworth, D. L. (1977). Taxonomic and biological observations on the genus *Lichenoconium* (Sphaeropsidales). *Persoonia* **9**, 159-198.

Hawksworth, D. L. (1978). The taxonomy of lichen-forming fungi: Reflections on some fundamental problems. *In* "Essays in Plant Taxonomy" (H. E. Street, ed.), pp. 211-243. Academic Press, New York.

Hawksworth, D. L. (1979). The lichenicolous Hyphomycetes. *Bull. Br. Mus. (Nat. Hist.), Bot.* **6**, 183-300.

Hawksworth, D. L., and Udagawa, S. (1977). Contributions to a monograph of *Microthecium*. *Trans. Mycol. Soc. Jpn.* **18**, 143-154.

Hiratsuka, Y., and Powell, J. M. (1976). "Pine Stem Rusts of Canada," For. Tech. Rep. No. 4. Department of the Environment, Ottawa.

Hoch, H. C. (1977a). Mycoparasitic relationships: *Gonatobotrys simplex* parasitic on *Alternaria tenuis*. *Phytopathology* **67**, 309-314.

Hoch, H. C. (1977b). Mycoparasitic relationships. III. Parasitism of *Physalospora obtusa* by *Calcarisporium parasiticum*. *Can. J. Bot.* **55**, 198-203.

Hoch, H. C. (1978). Mycoparasitic relationships. IV. *Stephanoma phaeospora* parasitic on a species of *Fusarium*. *Mycologia* **70**, 370-379.

Hoes, J. A., and Huang, H. C. (1975). *Sclerotinia sclerotiorum*: Viability and separation of sclerotia from soil. *Phytopathology* **65**, 1431-1432.

Hollós, L. (1906-07). Pöffetegeken termö új Gombak. (Fungi novi in Gasteromycetis habitantes.) *Ann. Hist.-Nat. Mus. Natl. Hung.* **4**, 532-536; **5**, 278-284.

Huang, H. C. (1977). Importance of *Coniothyrium minitans* in survival of sclerotia of *Sclerotinia sclerotiorum* in wilted sunflower. *Can. J. Bot.* **55**, 289-295.

Huang, H. C., and Hoes, H. J. (1976). Penetration and infection of *Sclerotinia sclerotiorum* by *Coniothyrium minitans*. *Can. J. Bot.* **54**, 406-410.

Hubert, E. E. (1935). Observations on *Tuberculina maxima*, a parasite of *Cronartium ribicola*. *Phytopathology* **25**, 253-261.

Hughes, S. J. (1951). Studies on micro-fungi. IX. *Calcarisporium, Verticicladium,* and *Hansfordia* (gen. nov.). *Mycol. Pap.* **43**, 1-25.

Hughes, S. J. (1953). Fungi from the Gold Coast. II. *Mycol. Pap.* **50**, 1-104.

Hughes, S. J. (1976). Sooty moulds. *Mycologia* **68**, 693-820.

Ikediugwu, F. E. O., and Webster, J. (1970a). Antagonism between *Coprinus heptemerus* and other coprophilous fungi. *Trans. Br. Mycol. Soc.* **54**, 181-204.

Ikediugwu, F. E. O., and Webster, J. (1970b). Hyphal interference in a range of coprophilous fungi. *Trans. Br. Mycol. Soc.* **54**, 205-210.

Ing, B. (1974). Mouldy Myxomycetes. *Bull. Br. Mycol. Soc.* **8**, 25-30.

Joffe, A. Z. (1966). Quantitative relations between some species of *Fusarium* and *Trichoderma* in a citrus grove in Israel. *Soil Sci.* **102**, 240-243.

Karling, J. S. (1942). Parasitism among the chytrids. *Am. J. Bot.* **29**, 24-35.

Karling, J. S. (1960). Parasitism among chytrids. II. *Bull. Torrey Bot. Club* **87**, 326-336.

Keener, P. D. (1934). Biological specialization in *Darluca filum*. *Bull. Torrey Bot. Club* **81**, 475-490.

Kelley, W. D. (1966). Evaluation of *Trichoderma harzianum*-impregnated clay granules as a biocontrol for *Phytophthora cinnamomi* causing damping-off of pine seedlings. *Phytopathology* **66**, 1023-1027.

Kendrick, W. B., and Carmichael, J. W. (1973). Hyphomycetes. *In* "The Fungi: An Advanced Treatise" (G. C. Ainsworth, F. K. Sparrow, and A. S. Sussman, eds.), Vol. 4A, pp. 323-509. Academic Press, New York.

Kenneth, R., and Isaac, P. K. (1964). *Cephalosporium* species parasitic on *Helminthosporium* (*sensu lato*). *Can. J. Plant Sci.* **44**, 182-187.

Komatsu, M., and Hashioka, Y. (1964). *Trichoderma viride*, as an antagonist of the wood-inhabiting Hymenomycetes. V. Lethal effects of the different *Trichoderma* forms on *Lentinus edodes* inside log-woods. *Rep. Tottori Mycol. Inst.* **4**, 11-18.

Kranz, J. (1974). A host list of the rust parasite *Eudarluca caricis* (Fr.) O. Eriks. *Nova Hedwigia* **24**, 169-180.

Kuhlman, E. G., and Matthews, F. R. (1976). Occurrence of *Darluca filum* on *Cronartium strobilinum* and *C. fusiforme* infecting oak. *Phytopathology* **66**, 1195-1197.

Kuhlman, E. G., and Miller, T. (1976). Occurrence of *Tuberculina maxima* on fusiform rust galls in the southeastern United States. *Plant Dis. Rep.* **60**, 627-629.

Linnemann, G. (1968). *Ampelomyces quisqualis* Ces., ein Parasit auf Mucorineen. *Arch. Mikrobiol.* **60**, 59-75.

Liu, S. -Y., and Vaughan, E. K. (1965). Control of *Pythium* infection in table beet seedlings by antagonistic microorganisms. *Phytopathology* **55**, 986-989.

Locci, R., Ferrante, G. M., and Rodrigues, C. J. (1971). Studies by transmission and scanning electron microscopy on the *Hemileia vastatrix—Verticillium hemileiae* association. *Riv. Patol. Veg.* [4] **7**, 127-140.

Logan, C., and O'Neill, R. (1970). Production of an antibiotic by *Phoma exigua*. *Trans. Br. Mycol. Soc.* **55**, 67-75.

Madelin, M. F. (1968). Fungi parasitic on other fungi and lichens. *In* "The Fungi: An Advanced

8. A Survey of the Fungicolous Conidial Fungi

Treatise'' (G. C. Ainsworth and A. S. Sussman, eds.), Vol. 3, pp. 253-269. Academic Press, New York.

Mangan, A. (1967). Studies on wheat rhizosphere soil fungi. *Ir. J. Agric. Res.* **6**, 9-14.

Manns, T. F. (1947). The vert-de-gris disease of the cultivated mushroom occurring in the United States. *Plant Dis. Rep.* **31**, 417-418.

Manocha, M. S., and Lee, K. Y. (1971). Host-parasite relations in mycoparasites. I. Fine structure of host, parasite, and their interface. *Can. J. Bot.* **49**, 1677-1681.

Marvanová, L. (1977). A contact biotrophic mycoparasite on aquatic hyphomycete conidia. *Trans. Br. Mycol. Soc.* **68**, 485-488.

Matruchot, L. (1903). Une Mucorinée purement conidienne, *Cunninghamella africana*: Étude ethologique et morphologique. *Ann. Mycol.* **1**, 45-60.

Matta, A. (1962). Segnalazione della *Cercospora unamunoi* (Unam.) E. Castel. e di un suo parassita su peperone in sena. *Ann. Fac. Sci. Agrar. Univ. Torino* **1**, 307-312.

Mickala-Doukaga, E., Albertini, L., and Petitprez, M. (1978). Action *in vitro* d'antagonistes fongiques sur la croissance mycélienne de l'*Helminthosporium turcicum* Pass. parasite du mais: Note préliminaire. *Bull. Trimest. Soc. Mycol. Fr.* **94**, 33-47.

Mijuškovíc, M. (1976). *Calonectria gymnosporangii* Jaap [*Bactridium gymnosporangium* (Jaap) Wr.], kao superparazit na *Gymnosporangium confusum* Plowr. u Crnoj Gori. *Zast. Bilja* **27**, 137-138.

Mijuškovíc, M., and Vucinic, Z. (1974). Nove pojave superparazita gliva u Crnoj Gori. *Zast. Bilja* **25**, 128-129.

Mitchell, C. P., Millar, C. S., and Minter, D. W. (1978). Studies on decomposition of scots pine needles. *Trans. Br. Mycol. Soc.* **71**, 343-348.

Mitov, N., and Ibrahim, I. (1977). Edna noya khiperparazitna g'ba po prichinitelya na brashnestata mana po bamyata (*Erysiphe cichoracearum* DC.). *Gradinar. Lozar. Nauka* **14**, 93-98.

Moreau, F. (1939). Sur deux champignons parasites des Saprolégniacées. *Bull. Trimest. Soc. Mycol. Fr.* **55**, 95-98.

Morquer, R., and Touvet, A. (1974). Action comparée de champignons antagonistes sur divers Hymenomycetes parasites des arbres résineux. *C. R. Hebd. Séances Acad. Sci., Ser. D* **278**, 709-713.

Moser-Rohrhofer, M. (1975). "Physiologische und vergleichende Anatomie der Flechtenpilze," Vol. 1. Akad. Druck- und Verlagsanstalt, Graz.

Müller, F. W. (1941). Zur Wirkstoffphysiologie des Bodenpilzes *Mucor ramanianus*. *Ber. Schweiz. Bot. Ges.* **51**, 165-256.

Nannfeldt, J. A. (1975). Stray studies in the Coronophorales (Pyrenomycetes) 1-8. *Svensk Bot. Tidskr.* **69**, 49-66, 289-335.

Negrutskiĭ, S. F. (1963). Ob ispol'zowanii gribov-antagonistov dlya bor'by s gribom *Fomitopsis annosa* (Fr.) Karst. *Mikrobiologiya* **32**, 632-635.

Nicot, J. (1962). En marge du salon du champignons: La flore des moisissures d'un polypore. *Rev. Mycol.* **27**, 87-92.

Nicot, J. (1963). Les moisissures des champignons supérieurs: Liste préliminaire des espèces récoltées en 1959, 1960, 1961. *Bull. Trimest. Soc. Mycol. Fr.* **78**, 221-238.

Nicot, J. (1967). Clé pur la détermination des espèces banales de champignons fongicoles. *Rev. mycol.* **31**, 393-399.

Nicot, J. (1968). Sur le mycoparasitisme de *Calcarisporium arbuscula* Preuss. *Bull. Trimest. Soc. Mycol. Fr.* **84**, 87-92.

Nicot, J., and Durand, F. (1965). Remarques sur la moisissure fongicole *Amblyosporium botrytis* Fres. *Bull. Trimest. Soc. Mycol. Fr.* **81**, 623-649.

Norton, D. C. (1954). Antagonism in soil between *Macrophomina phaseoli* and selected soil inhabiting organisms. *Phytopathology* **44**, 522-524.

Oudemans, C. A. J. A. (1918-24). "Enumeratio Systematica Fungorum," 5 vols. Nijhoff, The Hague.
Parbery, D. G. (1978). *Phyllachora, Linochora* and hyperparasites. *In* "Taxonomy of Fungi" (C. V. Subramanian, ed.), Vol. 1, pp. 263-277. University of Madras, Madras.
Pentland, G. D. (1965). Stimulation of rhizomorph development of *Armillaria mellea* by *Aureobasidium pullulans* in artificial culture. *Can. J. Microbiol.* **11**, 345-350.
Perrin, R. (1977). *Gonatorhodiella highlei* A. L. Smith, hyperparasite de *Nectria coccinea* Pers. ex Fries, un des agents de la maladie du hetre. *C. R. Séances Acad. Agric. Fr.* **63**, 67-70.
Pirozynski, K. A. (1966). The genus *Xenosporium. Mycol. Pap.* **105**, 21-35.
Pirozynski, K. A. (1969). Reassessment of the genus *Amblyosporium. Can. J. Bot.* **47**, 325-334.
Pirozynski, K. A. (1972). Microfungi of Tanzania. *Mycol. Pap.* **129**, 40-64.
Pirozynski, K. A., and Malloch, D. W. (1975). The origin of land plants: A matter of mycotrophism. *BioSystems* **6**, 153-164.
Poelt, J. (1977). Types of symbiosis with lichens. *In* "Abstracts, Second International Mycological Congress" (H. E. Bigelow and E. G. Simmons, eds.), Vol. M-Z, p. 526. Second International Mycological Congress, Tampa, Florida.
Porter, C. C., and Carter, J. C. (1938). Competition among fungi. *Bot. Rev.* **4**, 165-182.
Powell, J. M. (1971). Fungi and bacteria associated with *Cronartium comandrae* on lodgepole pine in Alberta. *Phytoprotection* **52**, 45-51.
Qaiser Abbas, S., and Ghaffar, A. (1973). Inhibition of certain fungi by *Memnoniella echinata. Pak. J. Bot.* **5**, 169
Raghavendra Rao, N. N., and Pavgi, M. S. (1977). A mycoparasite on *Sclerospora graminicola. Can. J. Bot.* **54**, 220-223.
Raghavendra Rao, N. N., and Pavgi, M. S. (1978). Two mycoparasites on powdery mildews. *Sydowia* **30**, 145-147.
Rakvidhyasastra, V., and Butler, E. E. (1973). Mycoparasitism by *Stephanoma phaeospora. Mycologia* **65**, 580-593.
Ramírez Gómez, C. (1957). Contribucion al estudio de la ecologia de las levaduras I. Estudio de levaduras aisladas de hongos carnosos. *Microbiol. Espan.* **10**, 215-247.
Rathaiah, Y., and Pavgi, M. S. (1973). *Fusarium semitectum* mycoparasitic on *Cercosporae. Phytopathol. Z.* **77**, 278-281.
Reiss, J. (1972). Toxicity of rubratoxin B to fungi. *J. Gen. Microbiol.* **71**, 167-172.
Ricard, J. (1976). Biological control of decay in standing creosote-treated poles. *J. Inst. Wood Sci.* **7**, 6-9.
Richardson, D. H. S. (1974). Photosynthesis and carbohydrate movement. *In* "The Lichens" (V. Ahmadjian and M. E. Hale, eds.), pp. 249-288. Academic Press, New York.
Robinson, P. M., Park, D., and Garrett, M. K. (1968). Sporostatic products of fungi. *Trans. Br. Mycol. Soc.* **51**, 113-124.
Rudakov, O. L. (1978). Physiological groups in mycophilic fungi. *Mycologia* **70**, 150-159.
Saksena, S. B., and Lilly, K. (1967). Studies on the interaction between soil micro-organisms with special reference to *Penicillium nigricans. J. Indian bot. Soc.* **46**, 185-192.
Samson, R. A. (1974). *Paecilomyces* and allied Hyphomycetes. *Stud. Mycol.* **6**, 1-119.
Samuels, G. J. (1973). The myxomyceticolous species of *Nectria. Mycologia* **65**, 401-420.
Santesson, R. (1945). *Cyttaria*, a genus of inoperculate Discomycetes. *Svensk Bot. Tidskr.* **39**, 319-345.
Santesson, R. (1952). Foliicolous lichens I. A revision of the obligately foliicolous lichenized fungi. *Symb. Bot. Ups.* **12**(1), 1-590.
Seeler, E. V. (1943). Several fungicolous fungi. *Farlowia* **1**, 119-133.
Shigo, A. L. (1958). Fungi isolated from oak-wilt trees and their effects on *Ceratocystis fagacearum. Mycologia* **50**, 757-769.

Shigo, A. L. (1960). Parasitism of *Gonatobotryum fuscum* on species of *Ceratocystis*. *Mycologia* **52**, 584-598.
Singh, N., and Webster, J. (1973). Antagonism between *Stilbella erythrocephala* and other coprophilous fungi. *Trans. Br. Mycol. Soc.* **61**, 487-495.
Singh, U. P. (1974). *Cephalosporium coccorum* Petch. A mycoparasite on *Ravenelia* species. *Sydowia* **26**, 63-66.
Singh, U. P. (1975). *Colletotrichum gloeosporioides* Penzig—a mycoparasite on *Ravenelia sessilis* Berkeley. *Curr. Sci.* **44**, 623-624.
Singh, U. P., Vishwakarma, S. N., and Basuchaudhury, K. C. (1978). *Acremonium sordidulum* mycoparasitic on *Colletotrichum dematium* f. *truncatum* in India. *Mycologia* **70**, 453-455.
Slagg, C. M., and Fellows, H. (1947). Effects of certain soil fungi and their by-products on *Ophiobolus graminis*. *J. Agric. Res.* **75**, 279-293.
Slocum, R. D. (1980). Light and electron microscopic investigations in the Dictyonemataceae (basidiolichens). II. *Dictyonema irpicinum*. *Can. J. Bot.* **58**, 1005-1015.
Smith, F. E. V. (1934). Three diseases of cultivated mushrooms. *Trans. Br. Mycol. Soc.* **10**, 81-97.
Sneh, B., Humble, S. J., and Lockwood, J. L. (1977). Parasitism of oospores of *Phytophthora megasperma* var. *sojae*, *P. cactorum, Pythium* sp., and *Aphanomyces euteiches* in soil by Oomycetes, Chytridiomycetes, Hyphomycetes, Actinomycetes, and bacteria. *Phytopathology* **67**, 622-628.
Sutton, B. C. (1973). Hyphomycetes from Manitoba and Saskatchewan, Canada. *Mycol. Pap.* **132**, 1-143.
Sutton, B. C. (1977). Coelomycetes. VI. Nomenclature of generic names proposed for Coelomycetes. *Mycol. Pap.* **141**, 1-253.
Swart, H. J. (1975). Callosities in fungi. *Trans. Br. Mycol. Soc.* **64**, 511-515.
Swendsrud, D. P., and Calpouzos, L. (1972). Effect of inoculation sequence and humidity on infection of *Puccinia recondita* by the mycoparasite *Darluca filum*. *Phytopathology* **62**, 931-932.
Treschow, C. (1942). Bekæmpelse af *Mycogone perniciosa* i champignonkulturer. *Friesia* **2**, 232-238.
Tubaki, K. (1955). Studies on the Japanese Hyphomycetes. (II) Fungicolous group. *Nagaoa* **5**, 11-40.
Tulloch, M. (1972). The genus *Myrothecium* Tode ex Fr. *Mycol. Pap.* **130**, 1-42.
Udagawa, S., and Horie, Y. (1971a). Taxonomical notes on mycogenous fungi. I. *J. Gen. Appl. Microbiol.* **17**, 141-159.
Udagawa, S., and Horie, Y. (1971b). Isolation of mushroom-inhabiting fungi. *Bull. Natl. Sci. Mus., Tokyo* **14**, 516-544.
Upitis, V. V. (1956). [The importance of soil saprophytic fungi for the control of pathogens of agricultural plants.] *Sb. Tr. Zashch. Rast. Riga*, 181-190 (not seen).
van den Heuvel, J. (1969). Effects of *Aureobasidium pullulans* on numbers of lesions on dwarf bean leaves caused by *Alternaria zinniae*. *Neth. J. Plant Pathol.* **75**, 300-307.
van Oorschot, C.A.N. (1977). The genus *Myceliophthora*. *Persoonia* **9**, 401-408.
Varghese, G., Chew, P. S., and Lim, J. K. (1976). Biologically and chemically assisted biological control of *Ganoderma*. *Proc. Int. Rubber Conf.* **3**, 278-292.
Viégas, A. P. (1961). "Indice de Fungos da América do Sul." Instituto Agronômico, Campinas.
von Arx, J. A. (1973). Further observations on *Sporotrichum* and similar fungi. *Persoonia* **7**, 127-130.
von Arx, J. A., and Müller, E. (1975). A re-evaluation of the bitunicate ascomycetes with keys to families and genera. *Stud. Mycol.* **9**, 1-159.
Watling, R., and Kendrick, W. B. (1979). *Osteomorpha* Arnaud—a validation. *Yorks. Nat.* **104**, 1-4.
Watson, A. K., and Miltimore, J. E. (1975). Parasitism of the sclerotia of *Sclerotinia sclerotiorum* by *Microsphaeropsis centaureae*. *Can. J. Bot.* **53**, 2458-2461.

Watson, R. (1955). *Calcarisporium arbuscula* living as an endophyte on apparently healthy sporophores of *Russula* and *Lactarius*. *Trans. Br. Mycol. Soc.* **38**, 409-414.
Webster, J. (1970). Coprophilous fungi. *Trans. Br. Mycol. Soc.* **54**, 161-180.
Weindling, R. (1932). *Trichoderma lignorum* as a parasite of other soil fungi. *Phytopathology* **22**, 837-845.
Weindling, R. (1938). Association effects of fungi. *Bot. Rev.* **4**, 475-496.
Wells, H. D., Bell, D. K., and Jaworski, C. A. (1972). Efficacy of *Trichoderma harzianum* as a biological control for *Sclerotium rolfsii*. *Phytopathology* **62**, 442-447.
Whaley, J. W., and Barnett, H. L. (1963). Parasitism and nutrition of *Gonatobotrys simplex*. *Mycologia* **55**, 199-210.
Willoughby, L. G. (1956). Studies on soil chytrids. I. *Rhizidium richmondense* sp. nov. and its parasites. *Trans. Br. Mycol. Soc.* **39**, 125-141.
Wollenweber, H. W. (1934). *Fusarium bactridioides* sp. nov., associated with *Cronartium*. *Science* **79**, 572.
Wu, W. -S. (1977). Antibiotic and mycoparasitic effects of several fungi against seed- and soil-borne pathogens associated with wheat and oats. *Bot. Bull. Acad. Sin.* **18**, 25-31.
Yarwood, C. E. (1932). *Ampelomyces quisqualis* on clover mildew. *Phytopathology* **22**, 31.
Yarwood, C. E. (1957). Powdery mildews. *Bot. Rev.* **23**, 235-301.
Zopf, W. (1897). Über Nebensymbiose (Parasymbiose). *Ber. Dtsch. Bot. Ges.* **15**, 90-92.

9

Conidial Lichen-Forming Fungi

G. Vobis and D.L. Hawksworth

I.	Introduction	245
II.	Coelomycetous Anamorphs	246
	A. Pycnidia	246
	B. Conidiophores	255
	C. Conidiogenesis	258
	D. Conidia	264
III.	Hyphomycetous Anamorphs	264
IV.	Parasymbiotic Conidial Fungi	265
V.	Conidial Lichen-Forming Fungi	266
VI.	Sterile Lichen-Forming Fungi	267
VII.	Discussion	270
	References	271

I. INTRODUCTION

The majority of lichen-forming fungi have ascomycetous teleomorphs, and their classification has naturally been based on the features of the teleomorph, supplemented (particularly in the macrolichens) by characters derived from the lichen itself (i.e., the composite fungus–alga association). Numerous lichen-forming fungi, perhaps as many as 8000 species, have conidial anamorphs but, because most occur regularly with teleomorphic sporocarps or are on thalli morphologically and chemically indistinguishable from those with teleomorphic fruits, they have received only cursory attention in lichenological investigations. In addition, there are 41 genera of conidial lichen-forming fungi described for which no teleomorphic phase is known, conidial parasymbiotic fungi, and entirely sterile lichenized taxa.

The lichen-forming ascomycetes are themselves heterogeneous, with taxa distributed through 6 subclasses, 12 orders, and 42 families (Hawksworth, 1978). They may most satisfactorily be interpreted as fungi characterized by a novel method of nutrition which, in different groups, includes examples of different degrees of evolutionary specialization to this habit. The conidial lichen-forming fungi and the anamorphs of lichenized ascomycetes might consequently be expected to be an equally diverse assemblage meriting inclusion in any comprehensive treatise on conidial fungi. Their existence has, however, been almost entirely ignored by mycologists working on Fungi Imperfecti.

This chapter reviews the currently available information on this numerically very important group of conidial fungi in the hope that it will stimulate both mycologists and lichenologists to consider them more adequately in future investigations. A biological group comprising 35% of all known conidial fungi should not continue to be passed over.

II. COELOMYCETOUS ANAMORPHS

Tulasne (1852) was probably the first author to appreciate the range of form of lichen pycnidia, which he termed "spermogonia," and he described in detail and illustrated selected examples. A much wider range of taxa was meticulously documented and illustrated by Lindsay (1859, 1872) whose work remains of outstanding value even though he did not attempt a synthesis. Glück (1899) extended the studies of earlier workers and proposed classifications of the structures he observed; his investigations have remained the basis of the consideration of lichen pycnidia by lichenologists up to the present time. A useful review of the earlier studies on lichen pycnidia was provided by Smith (1921) whose work should be consulted for further historical information. Pycnidia were illustrated for numerous taxa by Galløe (1927–1972), but little critical work has been attempted in recent decades. This section is based almost entirely on our own investigations.

It must be stressed that this account is necessarily biased toward the relatively few species so far studied in detail and that further variants may be expected to occur. Foliicolous and other crustose lichens have been particularly inadequately studied in this respect.

A. Pycnidia

Examination of mature pycnidia can be used to establish a separation into unilocular and multilocular types, but in order to understand their structure properly ontogenetic studies are necessary. In addition to providing information on the origin of the locules, ontogeny also assists in interpretation of the origin of the pycnidial wall and any ostiole that may be present. Since the studies of Glück (1899), the most important investigations of pycnidial ontogeny have concerned individual taxa: *Diploicia canescens* (Dicks.) Massal. (Letrouit-Galinou and Lallemant, 1977), *Collema* Wigg. species (Degelius, 1954), *Cryptolechia carneolutea* (Turn.) Massal. (Letrouit-Galinou, 1973), *Lobaria laetevirens* (Lightf.) Zahlbr. (Letrouit-Galinou, 1972), and *Umbilicaria cinereorufescens* (Schaer.) Frey (Janex-Favre, 1977). [After our survey was completed, Honnegger's (1978) important ultrastructural study of *Rhizocarpon* Ram. ex DC., which includes transmission electron microscope (TEM) investigations of conidiogenesis, came to our attention. Her paper should be consulted in conjunction with this chapter.] These studies have now been supplemented by the ontogenetic examination of taxa from other groups of lichen-forming fungi, and five main types of development have so far been distinguished. These will now be considered in turn.

1. *Lecanactis* Type (Fig. 1; Plate II A)

This type of development results in a cylindrical to cupuliform pycnidium with a single primary locule. Conidiophores of this type comprise elongated conidiogenous cells subtended most frequently by short cells which are sometimes branched; they belong to what we have designated morphological type II (Fig. 6) following the system proposed by Glück (1899). The *Lecanactis* type of development is now known in *Arthonia medusula* (Pers.) Nyl., *Byssoloma subdiscordans* (Nyl.) P. James, *Lecanactis abietina* (Ach.) Körb., and *Lecanactis subabietina* Coppins & P. James.

The first stage consists of a compact ball of interwoven short-celled hyphae (Fig. 1a). The cells in the peripheral parts of the primordium then increase so that the volume also increases and a cavity is initiated in the center where young conidiophores are soon seen (Fig. 1b). The peripheral cells branch, and it is from the new branches that conidiogenous cells grow out into the central cavity. Conidia form acrogenously, and the locule becomes filled with a gelatinous matrix mixed with conidia; under moist conditions the mucilage swells and exerts sufficient pressure to rupture the upper wall of the pycnidium and leave a more or less well-defined ostiole (Fig. 1c). Subsequent development is determined by the multiplication of conidiophores, the pycnidia tending to become wider at the base and taller at the edges (Fig. 1d). After reaching a particular size the pycnidium wall may become invaginated so that in section the pycnidium appears to be secondarily multiloculate.

2. *Roccella* Type (Fig. 2; Plate IB)

In the Roccellaceae (Arthoniales) all species so far investigated [*Combea mollusca* (Ach.) De Not., *Dirina ceratoniae* Fr., *Dolichocarpus chilensis* R. Sant., *Lobodirina cerebriformis* (Mont.) Follm., *Roccella hypomecha* (Ach.)

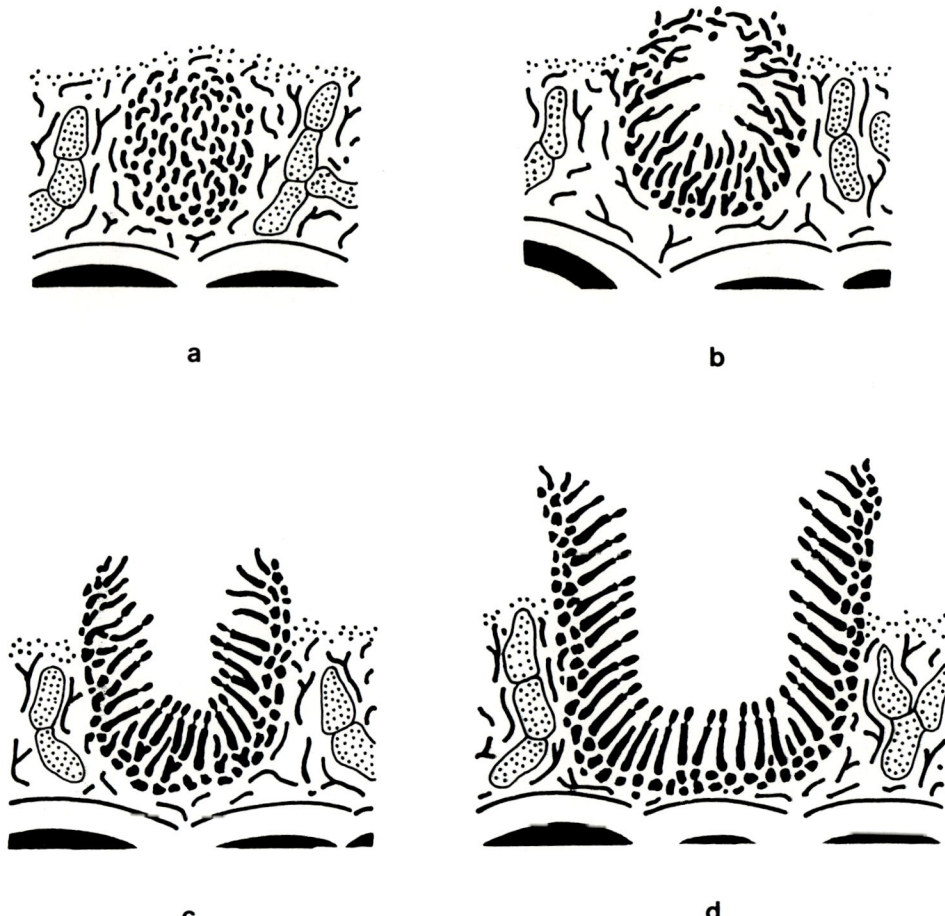

Fig. 1. Pycnidium development of the *Lecanactis* type.

Bory, and *R. portentosa* (Mont.) Darb.] have a common ontogeny. The mature pycnidia are primarily unilocular, frequently immersed, and ovoid, with conidiophores of morphological type III (Fig. 6) and filiform conidia arising acrogenously on elongated conidiogenous cells.

The primordium arises in the algal layer and comprises numerous thin and long-celled hyphae which are irregularly interwoven (Fig. 2a). These then assume a radial arrangement (Fig. 2b) as new hyphae grow from the periphery toward the center to form young conidiophores. As the conidiophores branch repeatedly, the pycnidium enlarges (Fig. 2c), and an ostiole starts to be formed by hyphae which, instead of growing toward the center, are directed vertically;

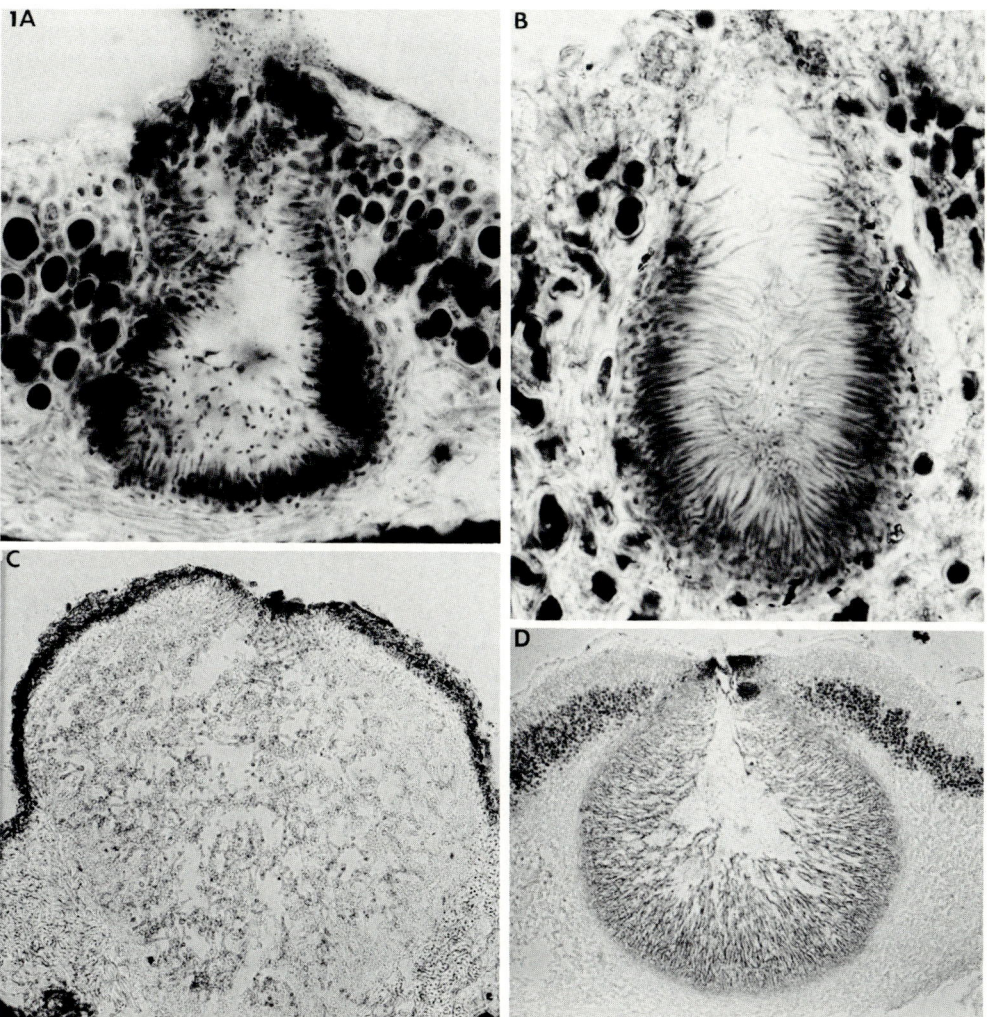

Plate I. Vertical sections of pycnidia of coelomycetous anamorphs of lichen-forming fungi. (A) *Phaeophyscia orbicularis, Umbilicaria*-type ontogeny. ×450. (B) *Dirina ceratoniae, Roccella*-type ontogeny. ×550. (C) *Teloschistes capensis, Xanthoria*-type ontogeny. ×130. (D) *Lobaria amplissima, Lobaria*-type ontogeny. ×65.

the latter are short-celled and frequently become pigmented. Glück (1899) termed them "orifice cells." Under pressure from the mass of conidia forming, the orifice cells are forced apart, and the conidia are liberated through the ostiole (Fig. 2d). The pycnidium wall is formed by hyphae extending around the pycnidium which form conidiophores toward the center.

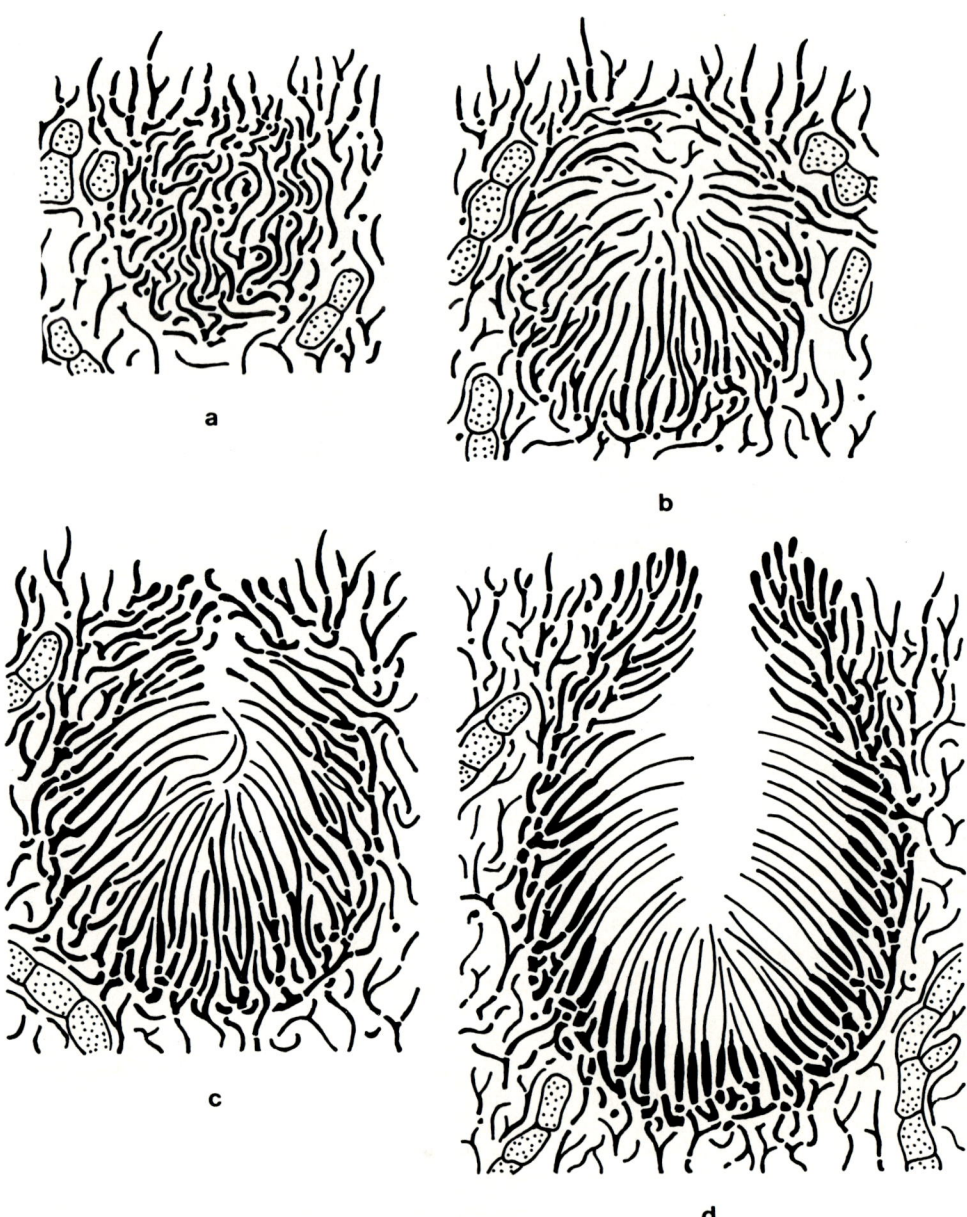

Fig. 2. Pycnidium development of the *Roccella* type.

3. *Umbilicaria* Type (Fig. 3; Plate IA)

The *Umbilicaria* type of development is widespread in the Lecanorales. The mature pycnidia are pyriform or ovoid and possess uni- or multilocular cavities and conidiophores of either type V or VI (Fig. 6). This type has been studied in detail in *Acroscyphus sphaerophoroides* Lév., *Cetraria islandica* (L.) Ach., *Hypogymnia physodes* (L.) Nyl., *Parmelia acetabulum* (Neck.)

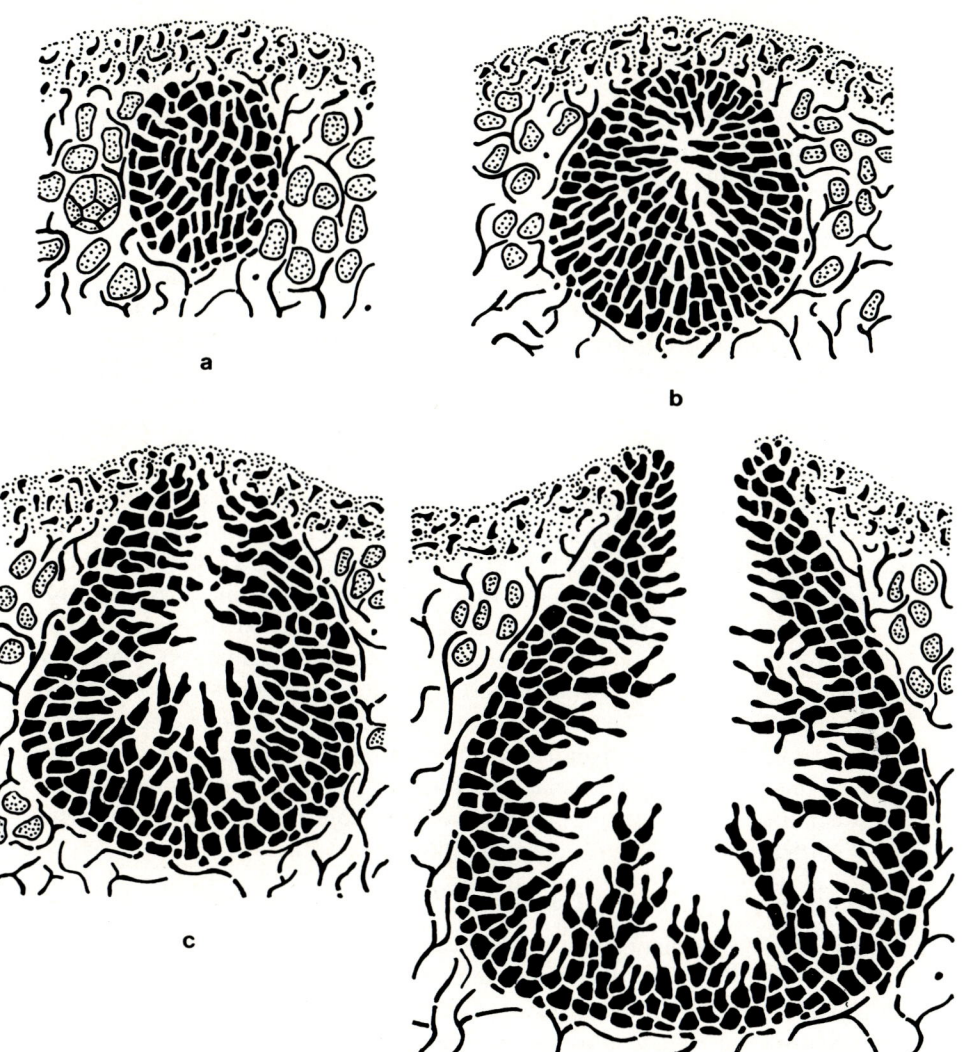

Fig. 3. Pycnidium development of the *Umbilicaria* type.

Duby, *Umbilicaria cinereorufescens* (see Janex-Favre, 1977), and *U. crustulosa* (Ach.) Frey.

The pycnidia originate from a ball of more-or-less isodiametric cells within the algal layer (Fig. 3a). The cells divide toward the outside, while those in the center show little activity, with the result that the whole expands, leaving a small cavity in the center (Fig. 3b). As this process continues, the original inner cells become further drawn to the periphery, so that the cavity enlarges further and young conidiophores are able to develop into the locule (Fig. 3c). Simultaneously the cells nearest the thallus surface grow upward and then part to form a channel opening as an ostiole at the surface (Fig. 3c–d). Within the pycnidium the multicellular conidiophores branch and increase and may divide the original cavity to form a secondarily multilocular pycnidium (Fig. 3d). The basal cells of the conidiophores and the cells of the pycnidial wall arise from the same pseudoparenchymatous tissue.

4. *Lobaria* Type (Fig. 4; Plate ID)

This type was originally described from *Lobaria linita* (Ach.) Rabenh. (Glück, 1899), was studied in detail in *L. laetevirens* (Letrouit-Galinou, 1972), and has subsequently also been found to occur in *L. amplissima* (Scop.) Forss. As no other members of the Stictaceae (Lecanorales) have been examined, it would be premature to regard this type as restricted to *Lobaria* (Schreb.) Hoffm. species. The mature pycnidia in this type are more or less globose, sometimes slightly compressed, and unilocular, with the cavity almost filled by very long, branched conidiophores which are short-celled and belong to type VII (Fig. 6).

The primordium is made up of short-celled and mostly interwoven hyphae and arises in the algal layer (Fig. 4a). The cells of these hyphae elongate, branch, and divide so as to arrange themselves tangentially to the longitudinal axis (Fig. 4b). The pycnidium almost reaches its final size by this method of growth and is visible from above as a papillate swelling with a darkened center, this pigmentation being due to the original cortical cells assuming a dark-brown color and also elongating (Fig. 4b–c). The young conidiophores which fill the pycnidial cavity become more radially arranged and commence conidiogenesis. When the gaps between the conidiophores have become filled with conidia and mucilage, under moist conditions the mucilage swells and the resultant pressure causes the elongated cortical cells to tear, forming an ostiole (Fig. 4d).

5. *Xanthoria* Type (Fig. 5; Plate IC)

This type of pycnidium appears to have a rather strange distribution among the lichen-forming fungi. It is characteristic of *Xanthoria* (Fr.) Th. Fr. (Glück, 1899) but also occurs in other genera of the Teloschistaceae (Lecanorales) and has been studied in *Caloplaca ferruginea* (Huds.) Th. Fr. and *Teloschistes capensis* (L. fil.) Malme. Surprisingly, it was found by Glück (1899) also in the

9. Conidial Lichen-Forming Fungi

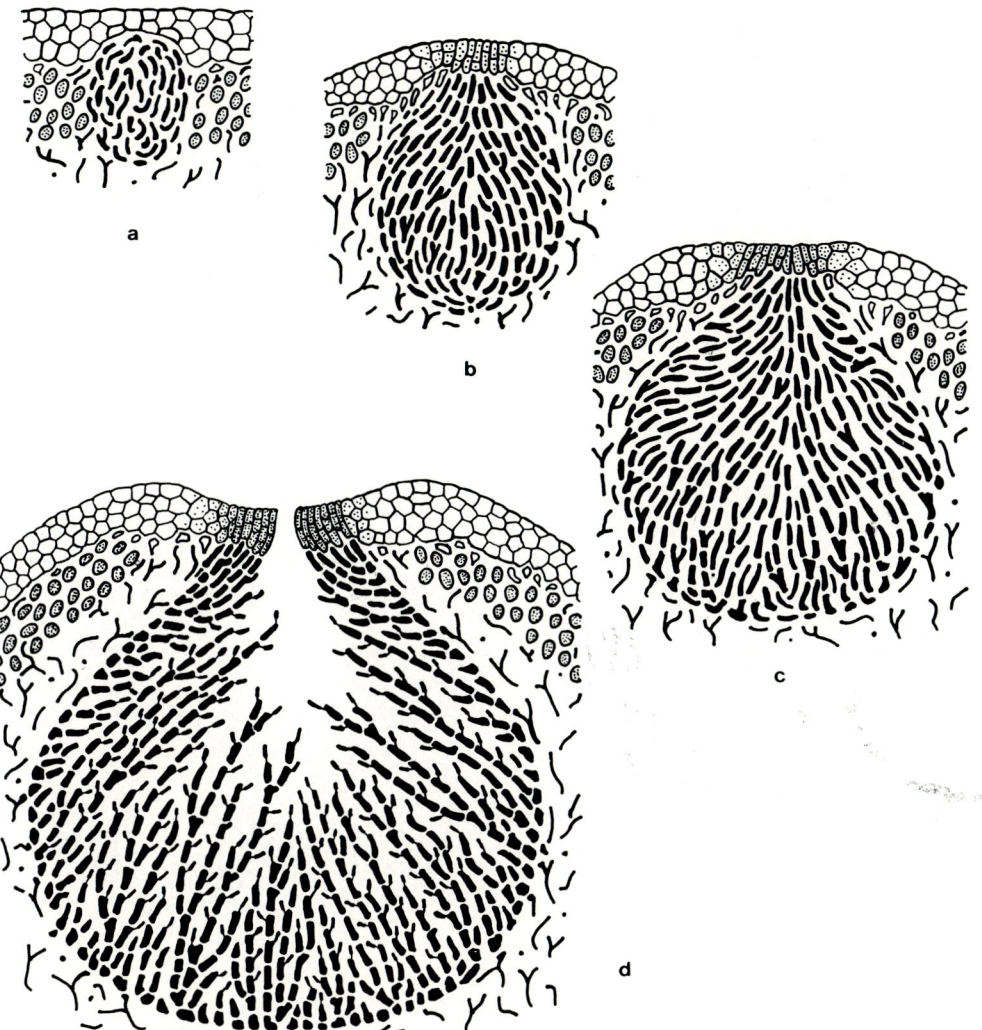

Fig. 4. Pycnidium development of the *Lobaria* type.

genera *Dermatocarpon* Eschw. and *Endocarpon* Hedw., both members of the Verrucariaceae (Verrucariales, Loculoascomycetes). The main feature of this pycnidium type is the complex multilocular nature of the mature pycnidium; the conidiophores belong to type VIII (Fig. 6).

The pycnidia first appear as an ovoid or globose complex of isodiametric cells in the algal and upper part of the medullary layer (Fig. 5a), which resembles that of the *Umbilicaria* type (Fig. 3a). The cells do not become radially arranged,

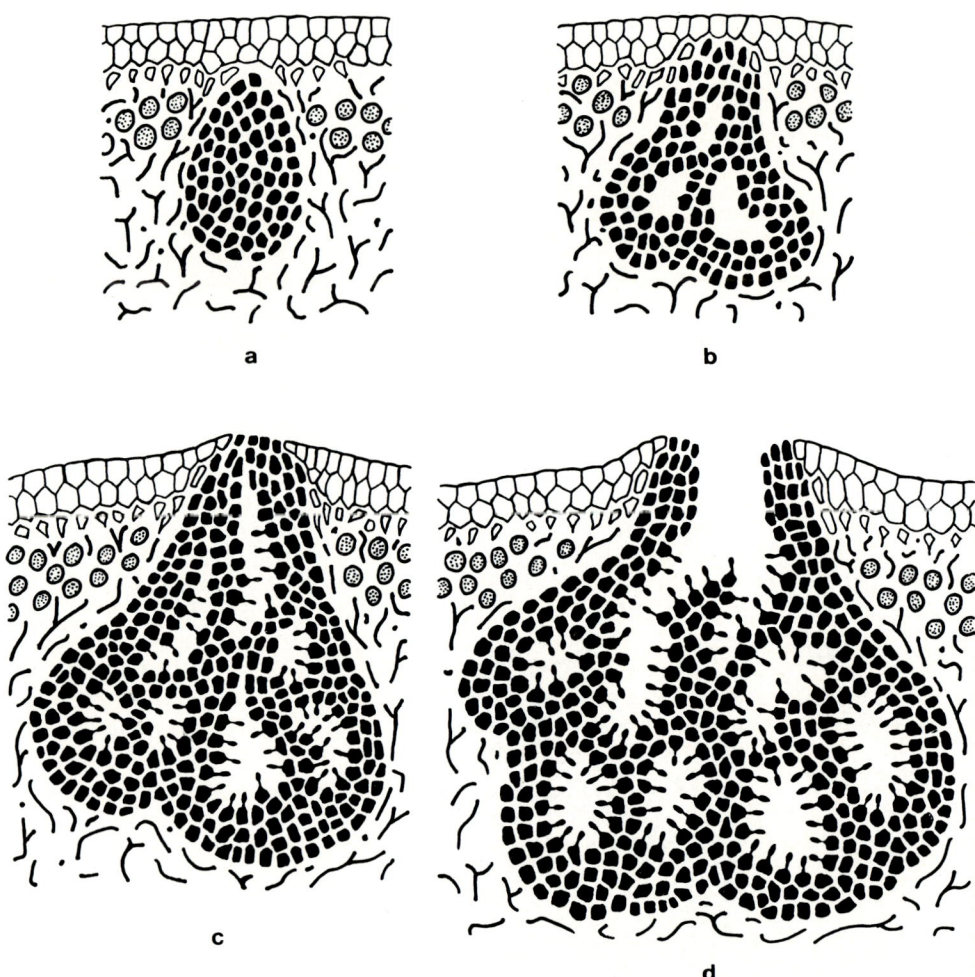

Fig. 5. Pycnidium development of the *Xanthoria* type.

however, but multiply in both the outer and inner parts of the young pycnidium and retain their original isodiametric shape. The first-formed locules originate by tissue rupture (Fig. 5b), so that the pycnidia are primarily multilocular. The cells lining the locules so produced give rise to the conidia (Fig. 5c). Later the pycnidium extends outward into the medulla, forming new chambers as it proceeds, so that the oldest locules are toward the center, not the periphery (Fig. 5d). The ostiole is produced in a way similar to that seen in the *Umbilicaria* type, by the upper cells growing toward the surface and losing their adhesion so that they can be forced apart by the pressure of conidia. The numerous locules may not always be open to one another, and so the upper locules are the first to be able to release their conidia. Liberation of the remaining conidia was considered by Glück (1899) to be due to a disintegration of the upper parts of the pycnidium.

B. Conidiophores

The original division of lichen conidiophores into the "simple" and "articulate sterigmata" of Nylander (1858–1860) was replaced by Glück (1899) with a scheme of eight morphological types ranging from the most simple to the most complex. A great range of types occurs, and the most important of these are summarized in Fig. 6 which is adapted and modified from that proposed by

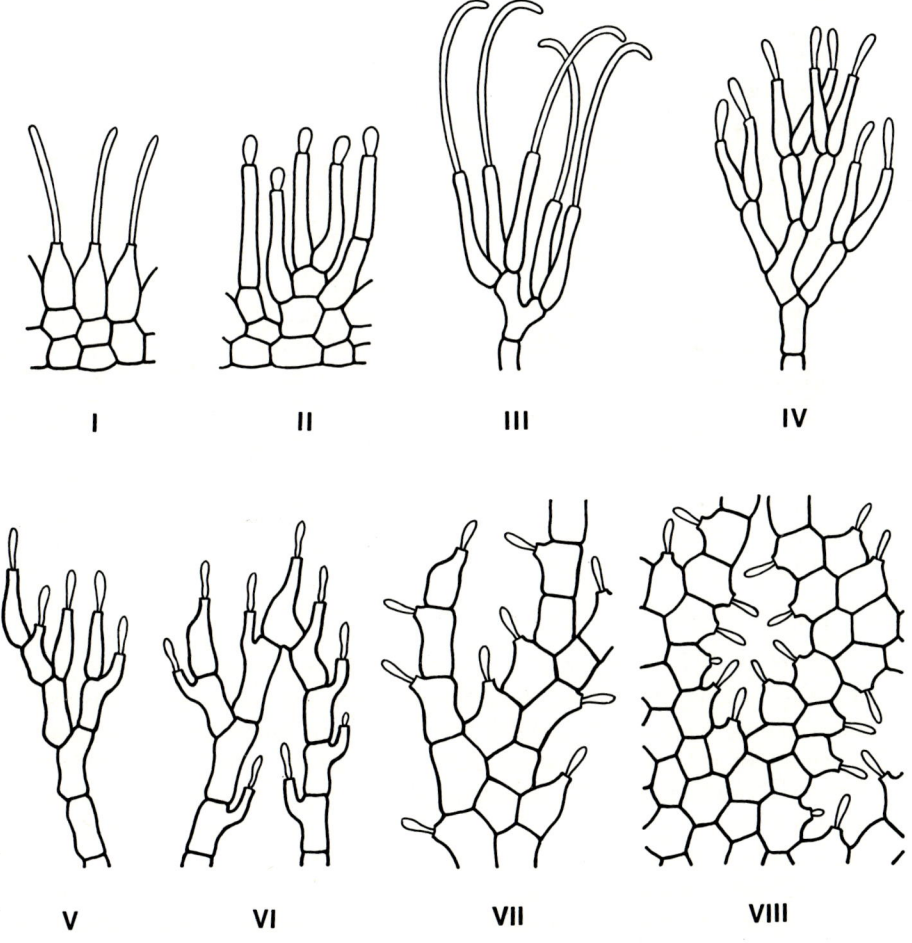

Fig. 6. Types of conidiophores in coelomycetous anamorphs of lichen-forming fungi.

Glück. In this scheme conidiogenesis is deliberately omitted, as is ontogeny, and the arrangement is viewed as noda on a morphological continuum.

Types I–IV are all characterized by terminal conidiogenous cells with the conidia formed apically; these correspond to the "exobasidial fulcra" of Steiner (1901). Types V–VIII, in contrast, represent the "endobasidial fulcra" of Steiner and comprise intercalary conidiogenous cells with laterally formed conidia. These types may sometimes be so modified that it is not easy to refer the conidiophores encountered in a particular taxon being studied to any one of them. This may be due to differences in the shape of the conidiogenous cells, increased numbers of septa in the conidiophores, or branching.

1. Type I

In this type the conidiogenous cells are sometimes hardly differentiated from the inner cells of the pycnidial wall and are doliiform to ampulliform. The isodiametric cells subtending the conidiogenous cells can scarcely be termed conidiophores, as they are also wall cells. This type is very rare in lichen-forming fungi, although it is well known in some lichenicolous Coelomycetes, but is seen for example in *Arthonia galactites* (DC.) Duf. It is possible that *Tholurna dissimilis* Norm. also belongs to this type (Henssen and Jahns, 1973).

2. Type II

The conidiogenous cells of this type are much more clearly differentiated than in type I, as they are elongated. They are, however, also borne on cells integrated with and scarcely separable from the pseudoparenchymatous tissue of the pycnidial wall. If the conidiophores branch, they do so by the production of cells similar to those of the wall. This type, which comprises both the *Peltigera* and *Psora* types of Glück, occurs in at least *Byssoloma subdiscordans*, *Lecanactis subabietina*, *Peltigera tomentosa* Vain. (see Lindahl, 1959), and *Thermutis velutina* (Ach.) Th. Fr. (see Henssen and Jahns, 1973). It is also present in the coelomycetous anamorph of the lichenicolous fungus *Microcalicium subpedicellatum* (Schaerer) Tibell (Tibell, 1978).

3. Type III

This type can be considered to be derived from type II by an increasing elongation and differentiation of the conidiophore into a branched, slightly swollen, but nonseptate cell. This, the *Placodium* type of Glück, is most frequent in the Roccellaceae, where it occurs in *Combea mollusca*, *Dirina ceratoniae*, *D. repanda* Fr., *Dolichocarpus chilensis*, *Lobodirina cerebriformis*, and *Roccella portentosa*.

4. Type IV

Here the conidiophores are more specialized, with the cells bearing conidiogenous cells similar to those in type III but forming part of a more complex branched and septate conidiophore with a main axis and several lateral branches. This type is seen in the Cladoniaceae and Ramalinaceae, for example, *Cladonia chlorophaea* (Flörke) Zopf, *C. turgida* (Ehrh.) Hoffm., and *Ramalina siliquosa* (Huds.) A. L. Sm.; it constitutes the *Cladonia* type of Glück.

5. Type V

The conidiophores are septate and branched as in the preceding type, but the conidiogenous cells are terminal, and intercalary conidiogenous cells also develop; in addition there is only a single axis with a few branches. This previously unrecognized type is found in representatives of several genera of the Parmeliaceae, notably *Alectoria ochroleuca* (Hoffm.) Massal., *Cetraria islandica, Omphalodium pisacommense* Meyer & Flotow, and *Parmelia acetabulum* (Neck.) Duby.

6. Type VI

This type comprises intercalary, erect conidiogenous cells forming complex branching patterns in which the conidiogenous cells have bayonet-like processes near the upper cell septum. Where conidiophores anastomose, a supplementary terminal conidiogenous cell may occur, as first demonstrated by Glück in *Hypogymnia physodes*. Early in the development of this type the conidiophores are very close together and become separated from the base as a result of the increasing size of the pycnidial wall. Conidiophores of this type, the *Parmelia* type of Glück, occur in a wide range of families, for example, *Acroscyphus sphaerophoroides* (Caliciaceae), *Hypogymnia physodes* (Parmeliaceae), *Phaeophyscia orbicularis* (Neck.) Moberg and *Physconia pulverulacea* Moberg (both Physciaceae), and *Umbilicaria crustulosa* (Umbilicariaceae).

7. Type VII

A section of a pycnidium with this type of conidiophore reveals a dense mass of richly branched and elongate, septate conidiophores comprising rows of intercalary conidiogenous cells. Toward the pycnidial wall the conidiophore cells are pressed together. Most conidia arise laterally, but some terminal cells with apical conidia are also present. The *Sticta* and *Physcia* types of Glück are combined here into this type, as intergradations may occur in the same species. This type is known only from the Lecanorales among the lichen-forming fungi but occurs in several families of this order, for example, in *Anaptychia ciliaris* (L.) Körb. (Physciaceae), *Lobaria amplissima* (Stictaceae), *Nephroma laevigatum* Ach. (Peltigeraceae), and *Psoroma hypnorum* (Vahl) Gray (Pannariaceae).

8. Type VIII

This type, seen only in pycnidia of the *Xanthoria* type, was referred to by Glück as the *Endocarpon* type. The pycnidia are almost entirely filled by pseudoparenchymatic tissue, which includes numerous small locules lined by conidiogenous cells similar in shape to those of type I; they differ from type I in that they are internal, not intimately associated with the pycnidial wall. As examples of this type the following species may be mentioned: *Dermatocarpon weberi* (Ach.) Mann (Verrucariaceae), *Teloschistes capensis*, and *Xanthoria parietina* (L.) Th. Fr. (both Teloschistaceae).

C. Conidiogenesis

The last 25 yr has seen considerable research on the precise method of conidiogenesis in the Fungi Imperfecti. Conidiogenesis in the coelomycetous anamorphs of lichen-forming fungi has, however, largely been ignored by mycologists and lichenologists, perhaps mainly because of the minute size of the points at which conidia arise on conidiogenous cells which are often less than 1.5 μm wide. Holoblastic development was reported by Gilenstam (1969) in *Conotrema urceolatum* (Ach.) Tuck. (Ostropales), but in all other lichen-forming fungi examined in recent years conidiogenesis has been interpreted as phialidic: *Bacidia vezdae* Coppins & P. James (Coppins and James, 1978), *Diploicia canescens* (Dicks.) Massal. (Letrouit-Galinou and Lallemant, 1977), *Lecanactis abietina* (Henssen and Jahns, 1973), *L. subabietina, Lecanora quercicola* Coppins & P. James, *Micarea pycnidiophora* Coppins & P. James, *M. stipitata* Coppins & P. James (Coppins and James, 1979), *Microcalicium subpedicellatum* (Tibell, 1978), *Santessonia namibensis* Hale & Vobis (Hale and Vobis, 1978), and *Thermutis velutina* (Henssen and Jahns, 1973).

As part of a survey of the anamorphs of lichen-forming fungi, conidiogenesis has now been studied by TEM in the following 11 species: *Anaptychia ciliaris, Combea mollusca, Dermatiscum thunbergii* (Ach.) Nyl., *Dirina ceratoniae, Heterodermia* Trevis. sp., *Hypogymnia physodes, Lecanactis abietina, Lobaria amplissima, Phaeophyscia orbicularis, Physcia semipinnata* (Gmel.) Moberg, and *Teloschistes capensis*. Freshly collected material retained in moist chambers for 2 days at 15°C was fixed in 4% glutaraldehyde in a 0.05 M sodium cacodylate buffer (pH 7.0) for 2 h and then postfixed in 1% osmium tetroxide in the same buffer overnight. Blocks were dehydrated in ethanol and embedded in Epon (Spurr, 1969). Sections prepared on an LKB ultramicrotome were poststained with uranyl acetate and lead citrate (Reynolds, 1963) prior to examination in a Siemens Elmiskop IA. Full details of these investigations will be presented elsewhere, and only a summary of the results is included here.

In the Arthoniales and Hysteriales anamorphs examined (i.e., *Combea mollusca, Dirina ceratoniae,* and *Lecanactis abietina*) the first formed conidium is produced by an apical extension of the conidiogenous cell (Fig. 7a–b, Plate IID) and a subsequent deposition of wall layers (Fig. 7c) which is extended downward inside the apex of the conidiogenous cell. The outer wall is ruptured (Fig. 7d), none of the outer wall material apparently remaining on the apex of the

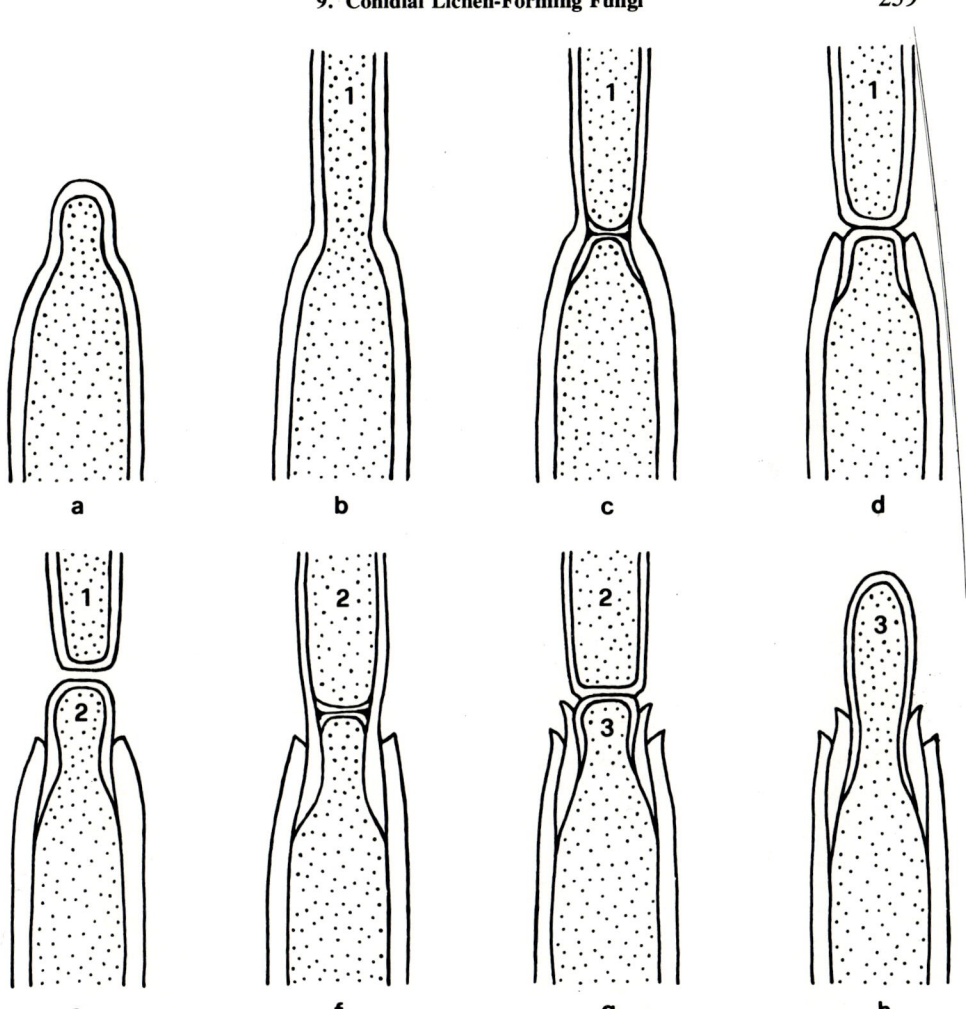

Fig. 7. Conidiogenesis in the species of Arthoniales examined.

newly formed conidium, and the remaining part of the new wall extends (Fig. 7e, Plate IIB and E) to form the next conidium in a similar manner (Fig. 7f–g). Annellations are consequently evident in this form of conidiogenesis (Plate II C–F).

The remaining eight species studied, all members of the Lecanorales, follow a slightly different pattern, which is similar whether the conidia are being produced apically or laterally. The outer cell wall begins to extend, and simultaneously a new inner wall layer appears to be deposited (Fig. 8b). The outer wall ruptures (Fig. 8c, Plate IIIB), and the new inner wall extends to form the new conidium

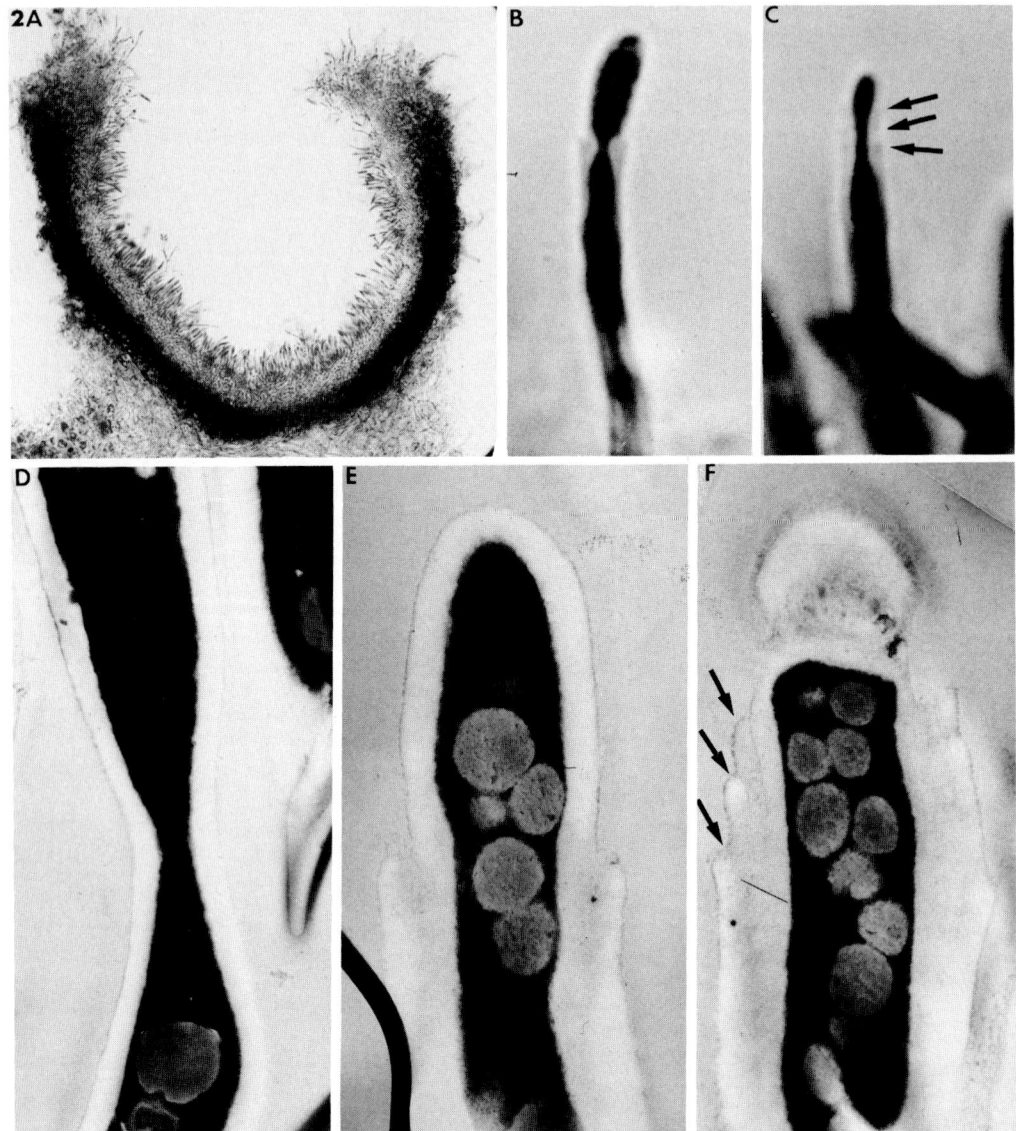

Plate II. Coelomycetous anamorph of *Lecanactis abietina*. (A) Vertical section of pycnidium, *Lecanactis*-type ontogeny. ×130. (B) Conidiogenous cell with no evident annellations. ×2550. (C) Conidiogenous cell with distinct annellations. ×2550. (D) Conidiogenous cell with the first-formed conidium ×19,300. (E) Conidiogenous cell with no annellations. ×20,300. (F) Conidiogenous cell with distinct annellations. ×20,300. Annellations are arrowed.

(Fig. 8d and e′) which is then released by a rupture of the new wall after the formation of a new wall (Fig. 8f′, Plate IIIC). In some cases, in the same pycnidium, the inner wall may cease extension at the stage shown in Fig. 8d, and a further wall be laid down within it (Fig. 8e″); this newer wall then gives rise to a conidium (Fig. 8f″) in the way mentioned above (Fig. 8e′–f′). In the latter case annellations are clearly seen (Plates IIID and IVA), although they are less obvious in the former (Plate IIIA–B). A modification occurs in *Teloschistes capensis*, the only species with *Xanthoria*-type ontogeny and conidiophore type VIII studied, where each new set of wall layers is produced at a lower level than is indicated in Fig. 8a–f′; the neck of the conidiogenous cell consequently becomes thicker with the production of each successive conidium (Plate IVB).

The interpretation of conidiogenesis in these taxa with respect to the situation

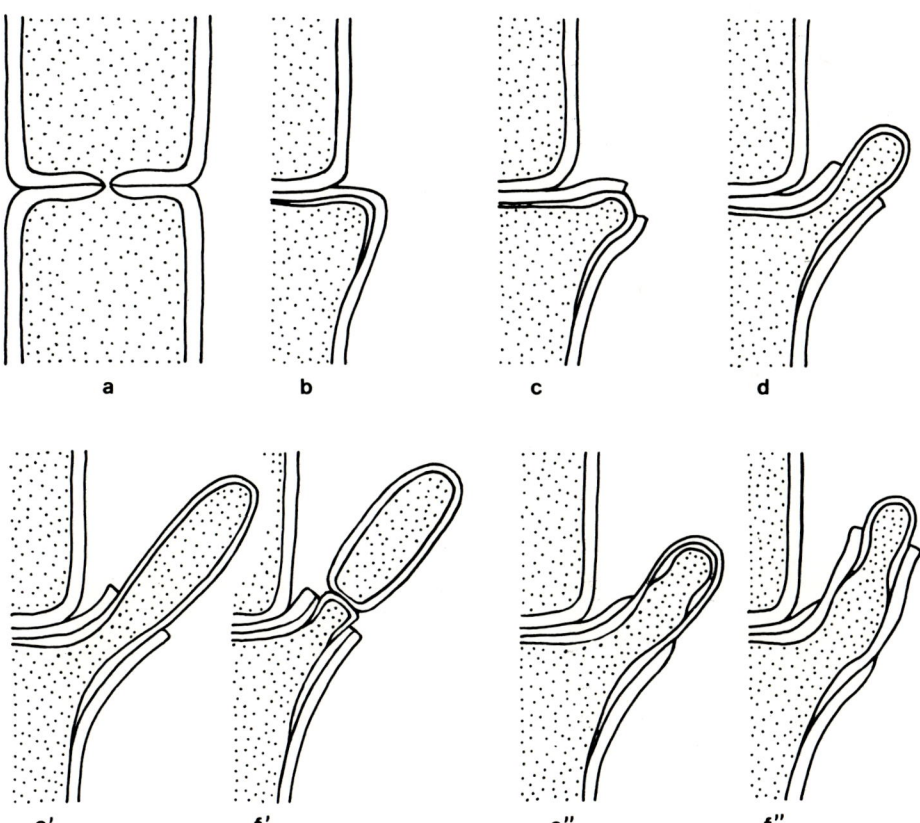

Fig. 8. Conidiogenesis in the species of Lecanorales examined.

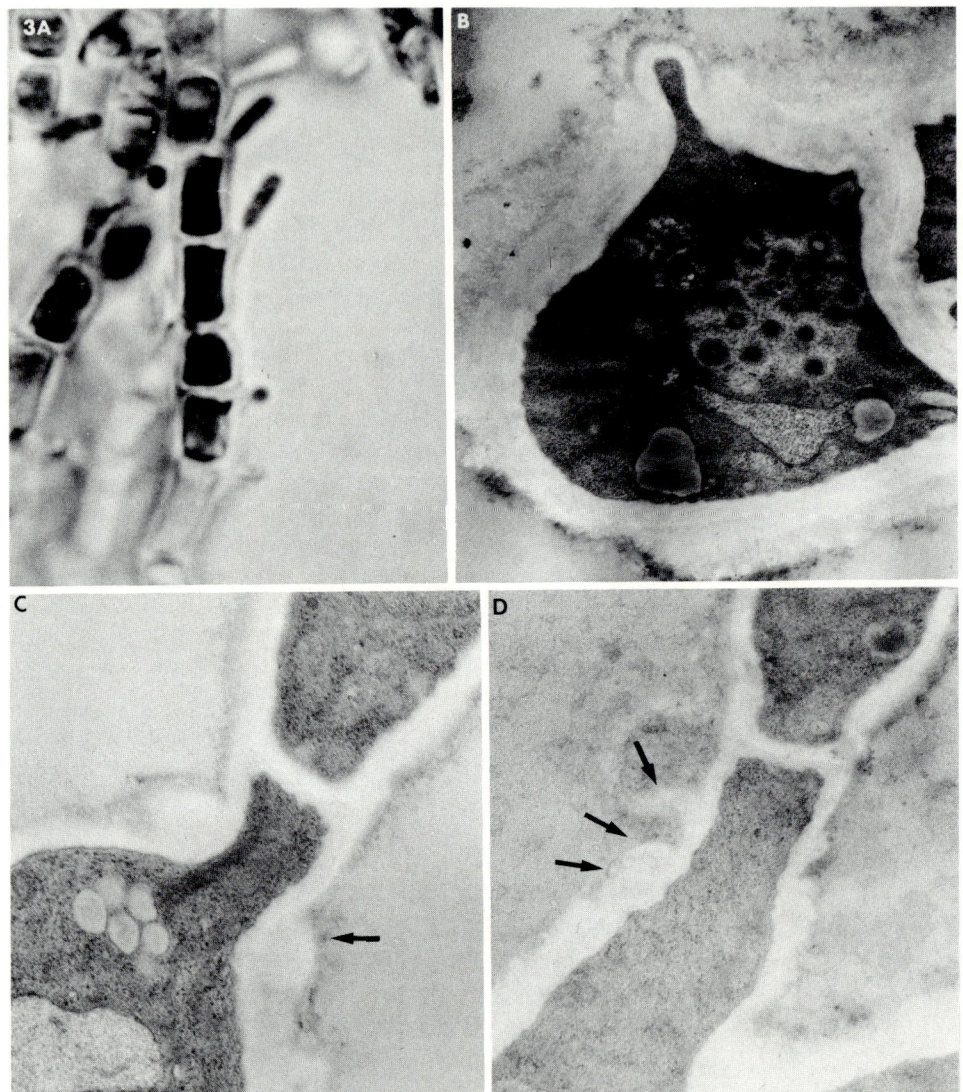

Plate III. Coelomycetous anamorph of *Lobaria amplissima*. (A) Intercalary conidiogenous cells. ×2000. (B) Conidiogenous cell showing conidiogenesis as in Fig. 8e'–f'. ×22,000. (C) As (B) but later stage. ×26,700. (D) Conidiogenous cell showing conidiogenesis as in Fig. 8e"–f". ×26,700. Annellations are arrowed.

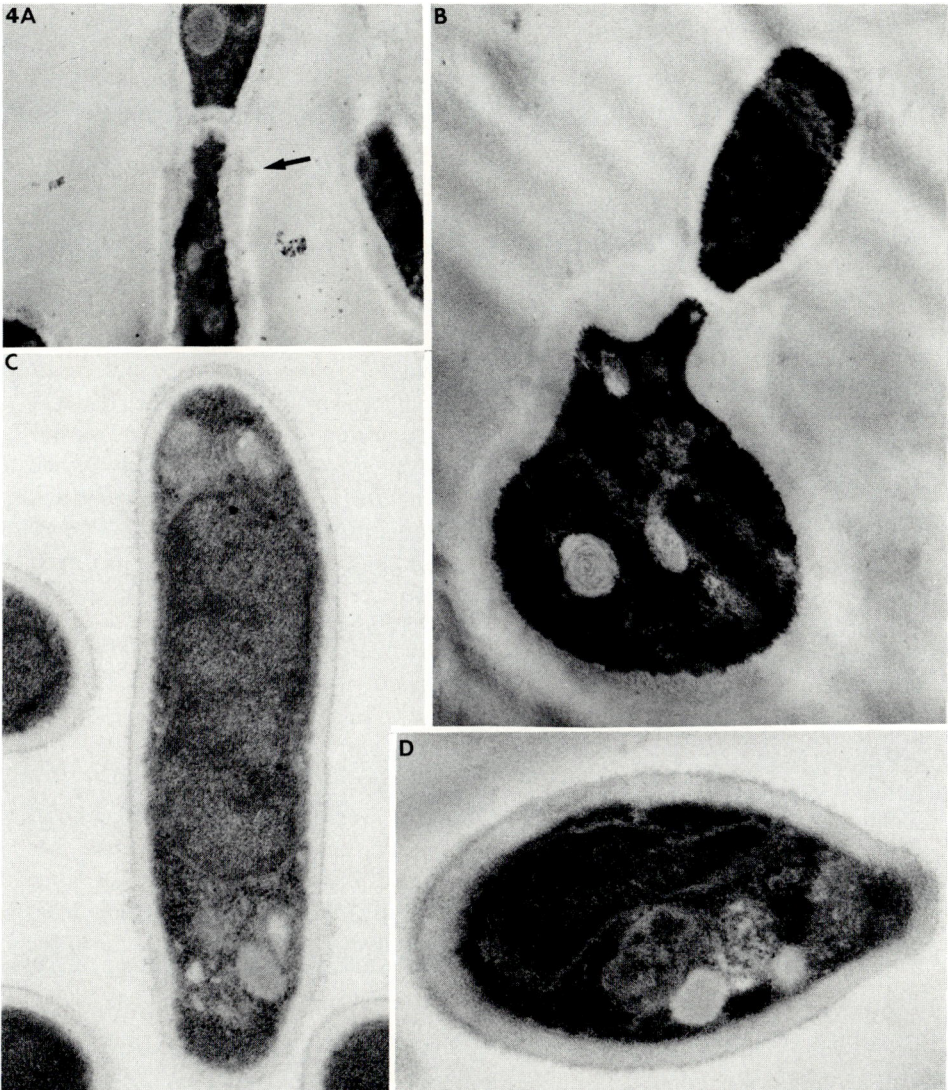

Plate IV. (A) Conidiogenous cell of *Physcia semipinnata* showing a single annellation. ×20,000. (B) Conidiogenous cell of *Teloschistes capensis* where the wall layers delimiting the conidium arise at a lower level than in other such types studied. ×24,000. (C) Conidium in *Heterodermia* sp. showing the large nucleus and basal ribosomes. ×40,000. (D) Conidium in *Phaeophyscia orbicularis* showing the nucleus. ×40,000. Annellations are arrowed.

in other Coelomycetes is difficult, as so few TEM investigations on non-lichen-forming Coelomycetes have been published. In species traditionally regarded as phialidic, conidia form at about the same level, and with each successive conidium the periclinal region of the apex becomes thicker as extra wall tissue is deposited (Boerema and Bollen, 1975; Jones, 1976); this does not add appreciably to the length of the conidiogenous cell. In species generally accepted as annellidic, however, the wall material from each episode of conidiogenesis adds to the length of the conidiogenous cell (Sutton and Sandhu, 1969). This situation is unfortunately made extremely complicated by the ability of some phialides to "proliferate" percurrently, in which case with light microscopy they can be confused with annellides (Morgan-Jones *et al.*, 1972). Whether such a separation between annellides and proliferating phialides is taxonomically significant must await numerous future studies on Coelomycetes with TEM. In the course of routine determinations such fine distinctions requiring the use of TEM are clearly unworkable. As a result, Sutton (1980) adopted a practical approach: When distinct annellations were visible by *light* microscopy the conidiogenesis was termed annellidic; if not, or if a collarette (distinct or indistinct) was evident, conidiogenesis was referred to as phialidic. In any case separation on this basis may not be taxonomically significant, because even in a single genus and species both types can occur (Hawksworth, 1977). This debate consequently need not concern us further here.

D. Conidia

The conidia in all species studied in detail are hyaline, smooth-walled, and uninucleate (Plate IV C and D). Their shape varies a great deal from subglobose, ellipsoidal, or cylindrical to elongate and falcate, depending on the species (see the illustrations of Lindsay, 1859, 1872, and Galløe, 1927-1972). In some species both micro- and macroconidia can be formed (e.g., *Lecanactis abietina*), but they are usually produced in identical pycnidia, or most commonly in the same pycnidium. Santesson (1952) noted the occurrence of two conidium types in numerous foliicolous lichen-forming fungi. Conidia are almost always nonseptate, but exceptions are found in the single-septate conidia of some members of the Porinaceae, for example, *Porina viridiseda* (Nyl.) Zahlbr. (Henssen and Jahns, 1973), and *Strigula elegans* (Fée) Müll. Arg. (Santesson, 1952).

III. HYPHOMYCETOUS ANAMORPHS

In contrast to the situation with pycnidia, the evidence for hyphomycetous anamorphs of lichen-forming fungi is very limited and mainly in need of critical reexamination. Von Istvánffi (1895) described simple orthotropic conidiogenous cells and cylindrical, hyaline conidia (spermatia) in *Buellia punctata* (Hoffm.)

Massal. [*B. punctiformis* (Hoffm.) Massal.], but this observation has not been confirmed. More reliable is the evidence for the occurrence of entirely immersed conidiogenous structures in *Collema* Wigg. These were first discovered by Bachmann (1912) in *C. bachmanianum* (Fink) Degel. and are entirely immersed in the thallus, not associated with a pycnidium-like structure. These "internal conidia" are simple, hyaline, and cylindrical and borne in groups of 2-15, adhering laterally and more rarely terminally on the conidiophores; Bachmann believed that these structures functioned as spermatia, because trichogynes were evidently attracted to them. Bachmann's observations have been confirmed by Degelius (1954), who restudied her material and also discovered similar, although slightly smaller, conidia in *C. multipunctatum* Degel. Riedl (1976) reported that *Bacidia chlorococca* (Stenh.) Lett. had a dematiaceous hyphomycetous anamorph very similar to *Coniothecium toruloides* Corda, but this is difficult to accept (Hawksworth, 1979).

More convincing is the study of Hasenhüttl and Poelt (1978) which demonstrated that the *Brutkörner* of the undersurfaces of lobes and rhizinae of some *Umbilicaria* Hoffm. spp. were conidia formed by the fungal partner. These conidia can adhere to damp surfaces and germinate.

Cultural studies with lichen-forming fungi are still in their infancy, and most of the work so far carried out has been directed at lichen synthesis rather than at studying the behavior of the fungi in pure culture. It has been reported that ascospores of *Buellia stillingiana* Steiner give rise to the dematiaceous hyphomycete *Sporidesmium folliculatum* (Corda) Mason & Hughes (Hale, 1957), but it was most probably a contaminant (Hawksworth, 1979). Of particular interest here are the reports of conidium production in pure cultures of the lichen-forming fungi *Lecidea erratica* Körb. and *Phaeographina fulgurata* (Fée) Müll. Arg. (Ahmadjian, 1963); superficially these fungi do not appear to resemble any other hyphomycete, and they merit further investigation. The production of conidia from ascospores has been found to occur in *Vezdaea aestivalis* (Ohl.) Tsch.-Woess & Poelt (Tschermak-Woess and Poelt, 1976).

Attention is also drawn here to the enigmatic synnematous peltate structures termed "hyphophores" (Vězda, 1973, 1975, 1979; Sérusiaux, 1977) which occur in the foliicolous genera *Echinoplaca* Fée and *Tricharia* Fée. They appear to be an integral part of the lichen rather than an invading lichenicolous or parasymbiotic fungus and might act as spermatia or asexual diaspores. A detailed study of their development and structure has not, however, so far been carried out but is necessary for their interpretation in mycological terms.

IV. PARASYMBIOTIC CONIDIAL FUNGI

Lichenicolous fungi forming apparently harmless associations with already existing lichens, termed by Poelt (1977) "three-membered symbioses" but more familiar to lichenologists as parasymbionts, may arguably be considered as lichenized themselves, since they presumably derive their carbohydrate needs from the algal partner of the lichen, just as the fungus predominant in the association does (Hawksworth, 1978). In view of the practical difficulties in determining whether a lichenicolous fungus, which is not obviously pathogenic to the lichen association it occurs on, is really participating in a three-membered symbiosis or not, it is more expedient to consider the lichenicolous fungi as a whole regardless of the type of biotrophic association developed. The known conidial lichenicolous fungi comprise about 87 species in 35 genera, 22 of the genera being exclusively lichenicolous. As these fungi are considered elsewhere in this volume (Chapter 8) they are not discussed further here.

V. CONIDIAL LICHEN-FORMING FUNGI

As numerous ascomycetes, several genera of basidiomycetes, and at least one zygomycete have evolved methods of nutrition by the formation of lichen associations, it was to be expected that some conidial fungi with no known teleomorphic phase would be lichen-forming. This indeed proves to be the case on the basis of published reports, mainly those of Batista and his co-workers. Between 1961 and 1972 these investigators described 37 genera of lichen-forming conidial fungi, with varying numbers of species, almost all of which were discovered on leaves in Brazil (Table I, Figs. 9 and 10). A further genus from North America was added by Funk (1973), and another has recently been discovered in Ireland (Hawksworth *et al.*, 1980).

These genera have been almost completely ignored by other lichenologists, and in the absence of critical studies on the type material no comprehensive survey of them can be attempted. It is possible that some of these genera are based on pycnidia of lichen-forming fungi with known teleomorphic phases, that some are saprophytic fungi fortuitously intermixed with algae, and that others are parasymbiotic or lichenicolous fungi on undetermined hosts. To judge from the experience with lichenicolous Hyphomycetes described by these workers (Hawksworth, 1979), we would expect a reasonable proportion of the described taxa to be soundly based. The original drawings of several of these conidial lichen-forming fungi are presented in Figs. 9–10 to give some indication of the types of structures described.

Three further genera have been based on coelomycetous anamorphs: *Chaetothyriolum* Speg. (teleomorph indeterminate), *Conicosolen* Schilling (*Psorotheciopsis* Rehm), and *Pleurosticta* Petr. (*Parmelia acetabulum*).

As several ascomycete genera comprise both lichen-forming and non-lichen-forming species, it is interesting that an at least partially lichenized species of *Coniosporium* Link ex Fr., *C. aeroalgicola* Turian, has been described; this

TABLE I

Genera of Conidial Lichen-Forming Fungi So Far Described[a]

Aciesia Bat. & Bez. (1)	*Kilikiostroma* Bat. & Bez. (1)
Acleistomyces Bat. (2)	*Lagenomyces* Caval. & Silva (1)
Actinoteichus Caval. & Poroca (3)	*Lyromma* Bat. & Maia (2)
Aderkomyces Bat. (1)	*Microlychnus* Funk (1)
Alysia Caval. & Silva (1)	*Microxyphiomyces* Bat. et al. (5)
Amazonomyces Bat. (1)	*Microlychnus* Funk (1)
Ameropeltomyces Bat. & Maia (1)	*Oncosporomyces* Bat. (1)
Amoebomyces Bat. & Maia (1)	*Phallomyces* Bat. & Valle (1)
Anconomyces Caval. & Silva (1)	*Podoxyphiomyces* Bat. et al. (1)
Arthrobotryomyces Bat. & Bez. (1)	*Psathyromyces* Bat. & Peres (1)
Asbolisiomyces Bat. & Maia (1)	*Pycnociliospora* Bat. (3)
Astrabomyces Bat. (1)	*Pyriomyces* Bat. & Maia (1)
Blarneya D. Hawksw. et al. (1)	*Pyripnomyces* Caval. (1)
?*Blodgettia* Wright (1)	*Septoriomyces* Caval. & Silva (1)
Caprettia Bat. & Maia (1)	*Setomyces*[b] Bat. & Peres (4)
Chaetomonodorus Bat. & Maia (1)	*Spinomyces*[b] Bat. & Peres (1)
Crocicreomyces Bat. & Peres (1)	*Sporocybomyces* Maia (1)
Cyrta Bat. & Maia (1)	*Stephosia* Bat. & Maia (1)
Didymaster Bat. & Maia (1)	*Tauromyces* Caval. & Silva (1)
Dothiomyces Bat. & Bez. (1)	*Tegoa*[b] Bat. & Fonseca (1)
Ephelidium Dodge & Rudolph (non Speg.) (1)	

[a] The numbers in parentheses is the number of species described in the genus.
[b] Generic names which have not been validly published. The genera *Setomyces* and *Spinomyces* may be sterile, in which case they would be appropriately listed in Table II.

appears to be widespread, as it is known from both Switzerland and the United States (Turian, 1977). We feel it is probable that other similar associations occur but are generally overlooked by both lichenologists and mycologists.

VI. STERILE LICHEN-FORMING FUNGI

Numerous lichen-forming fungi form sporocarps rarely or not at all. For example, Brodo and Hawksworth (1977) found that only 0.4% of specimens of *Bryoria capillaris* (Ach.) Brodo & D. Hawksw. examined had apothecia, and described several new species in this and allied genera in which ascomata are unknown. Individual specimens and species, particularly of macrolichens, can usually be confidently referred to species or genera, respectively, on the basis of the structure of the composite thallus and/or its chemical products. Although more hazardous with the crustose lichen-forming fungi, this practice is also often adopted with them to avoid the separation of what will probably prove to be

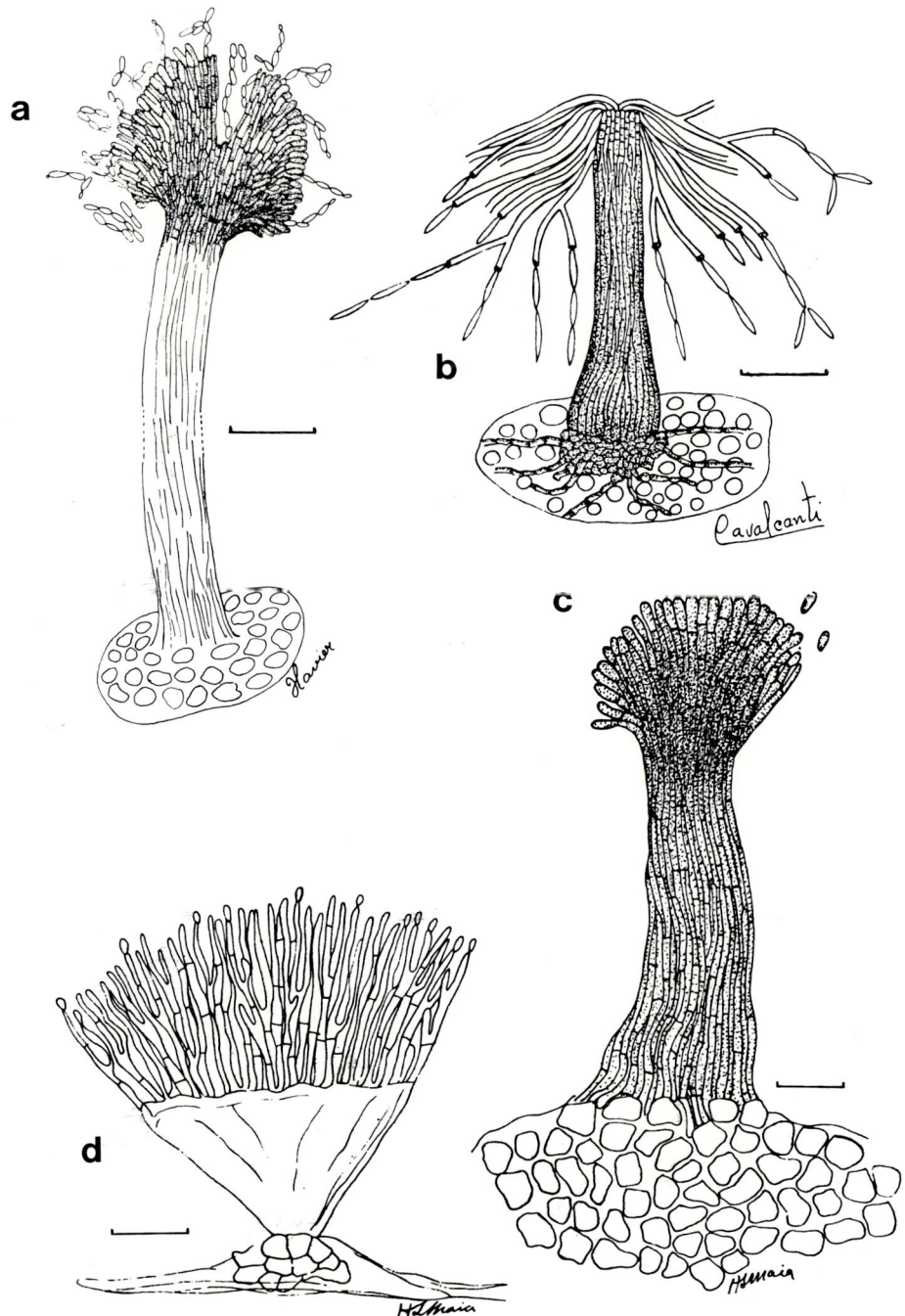

Fig. 9. Some lichen-forming Hyphomycetes described by Batista and co-workers. (a) *Phallomyces palmae* Bat. & Valle (from Batista *et al.*, 1961); (b) *Microxyphiomyces astrocaryifolii* Bat.,

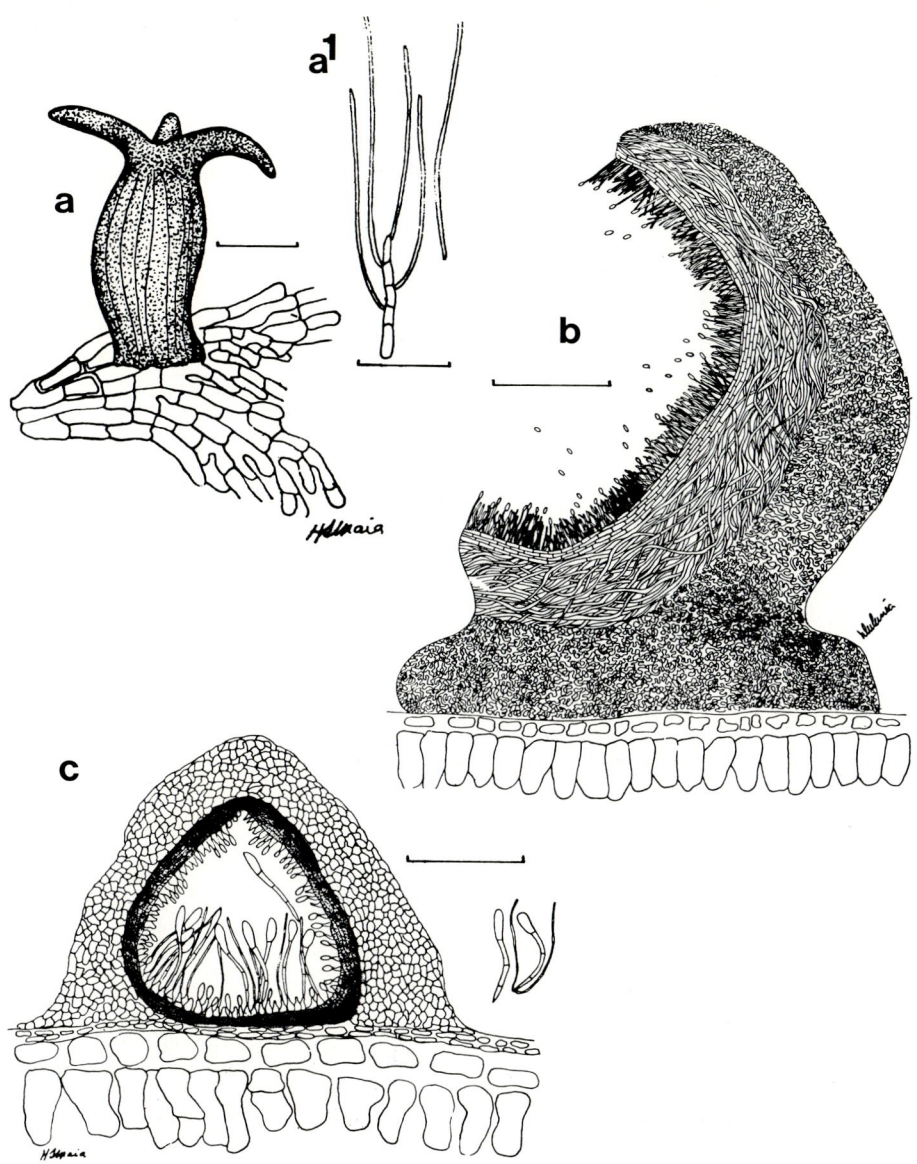

Fig. 10. Some lichen-forming Coelomycetes described by Batista and co-workers. (a) *Lyromma nectandrae* Bat. & Maia (from Batista and Maia, 1965); (b) *Acleistomyces rionegrensis* Bat. & Maia (from Batista, 1961); (c) *Cyrta licaniae* Bat. & Maia (from Batista and Maia, 1961). Scales: (a) 50 μm; (a¹) 20 μm; (b) 100 μm; (c) 100 μm.

Fig. 9 (*continued*).

Bez. & Caval. (from Batista *et al.*, 1961); (c) *Sporocybomyces pulcher* Maia (from Batista and Maia, 1967); (d) *Stephosia protii* Bat. & Maia (from Batista and Maia, 1967). Scales: (a and b) 20 μm; (c) 50 μm; (d) 30 μm.

TABLE II

Genera of Sterile Lichen-Forming Fungi Currently Accepted[a,b]

Byssophytum Mont. (2)	*Lichenothrix* Henss. (1)
Cystocoleus Thwaites (1)	*Phyllophiale* R. Sant. (1)
Endocena Cromb. (1)	*Racodium* Pers. (1) (syn.
Lepraria Ach. (ca. 30)	*Racodiopsis* Donk)
Leprocaulon Nyl. (8)	*Siphula* Fr. (ca. 25)
Leproplaca (Nyl.) Hue (2)	*Thamnolia* Ach. ex Schaer. (2)

[a]The number in parentheses is the number of species described in the genus.
[b]The genera *Setomyces* and *Spinomyces* may belong in this table (see Table I).

congeneric taxa when sporulating material is eventually found.

At least eleven genera which are completely sterile are, however, known and accepted by contemporary lichenologists (Table II). These are delimited on the basis of the anatomy and morphology of the composite thallus, in some cases supplemented by chemical characters. Several of these genera are exceptionally widely distributed and are presumed to be dispersed by thallus fragments or other propagules which incorporate both fungal and algal partners. It seems probable that most of these genera were originally able to form teleomorphic structures but have lost this ability during their evolutionary history.

VII. DISCUSSION

Too few conidial anamorphs of lichen-forming ascomycetes have been studied in detail to enable generalizations to be made about the types found in the various ascomycete groups involved. From the ontogenetic standpoint, the pycnidia of the *Lecanactis* and *Roccella* types appear to be very closely related, both having conidiophores with terminal conidiogenous cells and becoming multilocular only if the pycnidial wall invaginates; the conidiophores are not able to grow into the pycnidial cavity as in the *Umbilicaria* type. These two ontogenetic types are associated with the method of conidiogenesis shown in Fig. 7.

The *Lobaria* and *Xanthoria* types of ontogeny could theoretically be derived from the *Umbilicaria* type. The primordia of all three types are very similar. In the *Lobaria* type, the young conidiophores fill the whole of the pycnidial cavity and can isolate themselves almost completely. In contrast, the conidiophore cells of the *Xanthoria* type continue to adhere in a pseudoparenchymatic tissue, forming numerous small locules schizogenously. It is interesting that these three types are all associated with the method of conidiogenesis shown in Fig. 8.

These observations suggest that investigations of the pycnidia of lichen-forming ascomycetes may be of interest in the consideration of separations at higher taxonomic levels. It is further conceivable that they may shed some light on the discussions of the ancestry of the ascomycetes as a whole by emphasizing affinities with non-lichen-forming groups (see also Chapter 8, this volume).

Pycnidial characters have been used occasionally by lichenologists since they were emphasized by Nylander (1858-1860) in certain separations. For example, conidial differences have been used to substantiate generic segregations in the Physciaceae (Moberg, 1977) and also in species separations, as in some *Opegrapha* Ach. species (Duncan, 1970) and more recently in *Heterodea* Nyl. (Filson, 1978). In general, characters derived from anamorphs of lichen-forming fungi have been insufficiently considered by lichenologists, and more attention should be directed to them in the future.

The role of conidia produced by lichen-forming fungi has long been a matter of debate (Smith, 1921). They may function as spermatia, asexual propagules, or perhaps both. Attempts to germinate these conidia have met with varying degrees of success, but it is clear that some can grow in pure culture to form mycelia (Vobis, 1977). The demonstration of abundant trichogynes in other species (Jahns, 1973) can leave little doubt that some function as spermatia. Some crustose lichens commonly form pycnidial conidiomata but so rarely produce ascomata that their conidia must be a major method of dispersal (e.g., *Lecanactis abietina*, *L. subabietina*). In a considerable number of lichen-forming fungi, if the results of Batista and his co-workers are to be accepted, conidia may be the only propagules known.

ACKNOWLEDGMENTS

We are very grateful to Dr. B. C. Sutton for discussions of the methods of conidiogenesis in the coelomycetous anamorphs of lichen-forming fungi and also to Professor E. Silva and the Centro de Ciências Biológicas for permission to reproduce illustrations combined in Figs. 9 and 10.

REFERENCES

Ahmadjian, V. (1963). The fungi of lichens. *Sci. Am.* **208**, 122-132.
Bachmann, F. M. (1912). A new type of spermogonium and fertilization in *Collema*. *Ann. Bot. (London)* **26**, 747-760.
Batista, A. C. (1961). Um pugilo de gêneros novos de liquens imperfeitos. *Publ. Inst. Micol. Recife* **320**, 1-31.
Batista, A. C., and Maia, H. da S. (1961). *Asbolisiomyces, Cyrta* e *Chaetomonodorus,* novos gêneros de liquens imperfeitos. *Publ. Inst. Micol. Recife* **322**, 1-19.
Batista, A. C., and Maia, H. da S. (1965). Alguns novos gêneros de liquens imperfeitos assinalados no IMUR. *Atas Inst. Mic. Recife* **2**, 351-373.
Batista, A. C., and Maia, H. da S. (1967). Novos liquens imperfeitos do Amazonas e de Pernambuco. *Atas Inst. Mic. Recife* **5**, 55-71.

Batista, A. C., de Valle, R. C., Cavalcanti, W. C., Peres, G. E. P., and Bezerra, J. L. (1961). Tres novos gêneros de liquens imperfeitos, do Amazonas. *Publ. Inst. Micol. Recife* **319**, 1-42.
Boerema, G. H., and Bollen, G. J. (1975). Conidiogenesis and conidial septation as differentiating criteria between *Phoma* and *Ascochyta*. *Persoonia* **8**, 111-144.
Brodo, I. M., and Hawksworth, D. L. (1977). *Alectoria* and allied genera in North America. *Op. Bot. Soc. Bot. Lund* **42**, 1-164.
Coppins, B. J., and James, P. W. (1978). New or interesting British lichens II. *Lichenologist* **10**, 179-207.
Coppins, B. J., and James, P. W. (1979). New or interesting British lichens IV. *Lichenologist* **11**, 139-179.
Degelius, G. (1954). The lichen genus *Collema* in Europe. *Symb. Bot. Ups.* **13**(2), 1-499.
Duncan, U. K. (1970). "Introduction to British Lichens." T. Buncle, Arbroath, Scotland.
Filson, R. B. (1978). A revision of the genus *Heterodea* Nyl. *Lichenologist* **10**, 13-25.
Funk, A. (1973). *Microlychnus* gen. nov., a lichenized hyphomycete from western conifers. *Can. J. Bot.* **51**, 1249-1250.
Galløe, O. (1927-1972). "Natural History of the Danish Lichens." 10 vols. Aschehoug, Copenhagen.
Gilenstam, G. (1969). Studies in the lichen genus *Conotrema*. *Ark. Bot.* [2] **7**, 149-179.
Glück, H. (1899). Entwurf zu einer vergleichenden Morphologie der Flechten-Spermogonien. *Verh. Naturhist. Med. Ver. Heidelberg* [N.S.] **6**(2), 81-216.
Hale, M. E. (1957). Conidial stage of the lichen fungus *Buellia stillingiana* and its relation to *Sporidesmium folliculatum*. *Mycologia* **49**, 417-419.
Hale, M. E., and Vobis, G. (1978). *Santessonia*, a new lichen genus from southwest Africa. *Bot. Not.* **131**, 1-5.
Hasenhüttl, G., and Poelt, J. (1978). Über die Brutkörner bei der Flechtengattung *Umbilicaria*. *Ber. Dtsch. Bot. Ges.* **91**, 275-296.
Hawksworth, D. L. (1977). Taxonomic and biological observations on the genus *Lichenoconium* (Sphaeropsidales). *Persoonia* **9**, 159-198.
Hawksworth, D. L. (1978). The taxonomy of lichen-forming fungi: Reflections on some fundamental problems. *In* "Essays in Plant Taxonomy" (H. E. Street, ed.), pp. 211-243. Academic Press, New York.
Hawksworth, D. L. (1979). The lichenicolous Hyphomycetes. *Bull. Br. Mus. (Natl. Hist.), Bot.* **6**, 183-300.
Hawksworth, D. L., Coppins, B. J., and James, P. W. (1980). *Blarneya*, a lichenized hyphomycete from southern Ireland. *Bot. J. Linn. Soc.* **79**, 357-367.
Henssen, A., and Jahns, H. M. (1973). "Lichenes. Eine Einführung in die Flechtenkunde." Thieme, Stuttgart.
Honegger, R. (1978). Ascocarpontogenie, Ascusstruktur und Funktion bei Vertretern der Gattung *Rhizocarpon*. *Ber. Dtsch. Bot. Ges.* **91**, 579-594.
Istvánffi, G. von (1895). Ueber die Rolle der Zellkerne bei der Entwickeung der Pilze. *Ber. Dtsch. Bot. Ges.* **13**, 452-467.
Jahns, H. M. (1973). The trichogynes of *Pilophorus strumaticus*. *Bryologist* **76**, 414-418.
Janex-Favre, M. C. (1977). Le développement et la structure des pycnides de l'*Umbilicaria cinereorufescens*. *Rev. Bryol. Lichénol.* **43**, 1-18.
Jones, J. P. (1976). Ultrastructure of conidium ontogeny in *Phoma pomorum*, *Microsphaeropsis olivaceum*, and *Coniothyrium fuckelii*. *Can. J. Bot.* **54**, 831-851.
Letrouit-Galinou, M.-A. (1972). Études sur le *Lobaria laetevirens* (Lght.) Zahlbr. (Discolichen, Stictaceé). II. Le développement des pycnides. *Bull. Soc. Bot. Fr.* **119**, 477-486.
Letrouit-Galinou, M.-A. (1973). Les pycnospores et les pycnides du *Gyalecta carneolutea* (Turn.) Oliv. *Bull. Soc. Bot. Fr.* **120**, 373-384.

Letrouit-Galinou, M.-A., and Lallemant, R. (1977). Le développement des pycnides du discolichen *Buellia canescens* (Dicks.) D.N. *Ann. Sci. Nat., Bot.* [12] **18**, 119-134.
Lindahl, P. O. (1959). On the occurrence of pycnidia in the lichen genus *Peltigera. Svensk Bot. Tidskr.* **53**, 475-478.
Lindsay, W. L. (1859). Memoir on the spermogones and pycnides of filamentous, fruticulose, and foliaceous lichens. *Trans. R. Soc. Edinburgh* **22**, 101-303.
Lindsay, W. L. (1872). Memoir on the spermogones and pycnides of crustaceous lichens. *Trans. Linn. Soc. London* **28**, 189-318.
Moberg, R. (1977). The lichen genus *Physcia* and allied genera in Fennoscandia. *Symb. Bot. Ups.* **22**(1), 1-108.
Morgan-Jones, G., Nag Raj, T. R., and Kendrick, B. (1972). Conidium ontogeny in Coelomycetes. IV. Percurrently proliferating phialides. *Can. J. Bot.* **50**, 2009-2014.
Nylander, W. (1858-1860). "Synopsis Methodica Lichenum." Martinet, Paris.
Poelt, J. (1977). Types of symbiosis with lichens. *In* "2nd Internat. Mycol. Congr. Abstr. Volume M-Z" (H. E. Bigelow and E. G. Simmons, eds.), p. 526. 2nd Internat. Mycol. Congr., Tampa, Florida.
Reynolds, E. S. (1963). The use of lead citrate at high pH as an electron-opaque stain in electron microscopy. *J. Cell Biol.* **17**, 208-212.
Riedl, H. (1976). Die Flechte *Bacidia chlorococca* (Stenh.) Lett. und ihre Beziehungen zu Formgattungen der Fungi imperfecti. *Phyton (Horn)* **17**, 337-347.
Santesson, R. (1952). Foliicolous lichens I. A revision of the taxonomy of the obligately foliicolous, lichenized fungi. *Symb. Bot. Ups.* **12**(1), 1-590.
Sérusiaux, E. (1977). Quelques lichens foliicoles recoltes à la Réunion (Afrique, Ocean Indien). *Bull. Soc. R. Bot. Belg.* **110**, 39-41.
Smith, A. L. (1921). "Lichens." Cambridge Univ. Press, London and New York.
Spurr, A. R. (1969). A low-viscosity epoxy resin embedding medium for electron microscopy. *J. Ultrastruct. Res.* **26**, 31-43.
Steiner, J. (1901). Über die Function und den systematischen Wert der Pycnoconidien der Flechten. *In* "Fetschrift zur Feier des zweihundert-jährigen Bestandes des k. k. Staats-gymnasiums im VIII. Bezirke Wiens," pp. 119-154. Vienna.
Sutton, B. C. (1980). "The Coelomycetes." Commonw. Mycol. Inst., Kew, Surrey, England.
Sutton, B. C., and Sandhu, D. K. (1969). Electron microscopy of conidium development and secession in *Cryptosporiopsis* sp., *Phoma fumosa, Melanconium bicolor,* and *M. apiocarpum. Can. J. Bot.* **47**, 745-749.
Tibell, L. (1978). The genus *Microcalicium. Bot. Not.* **131**, 229-246.
Tschermak-Woess, E., and Poelt, J. (1976). *Vezdaea*, a peculiar lichen genus, and its phycobiont. *In* "Lichenology: Progress and Problems" (D. H. Brown, D. L. Hawksworth, and R. H. Bailey, eds.), pp. 89-105. Academic Press, New York.
Tulasne, L. R. (1852). Mémoire pour servir à l'histoire organographique et phisiologique des lichens. *Ann. Sci. Nat., Bot.* [3] **17**, 5-249.
Turian, G. (1977). *Coniosporium aeroalgicolum* sp. nov., moisissure Dématiée semi-lichenisante. *Ber. Schweiz. Bot. Ges.* **87**, 19-24.
Vězda, A. (1973). Foliicole Flechten aus der Republik Guinea (W. -Afrika). I. *Čas. Slezskeho Muz., Silesiae, Ser. A* **22**, 67-90.
Vězda, A. (1975). Foliicole Flechten aus Tanzania (Ost. Afrika). *Folia Geobot. Phytotaxon.* **10**, 383-432.
Vězda, A. (1979). Flechtensystematische Studien. XI. Beiträge zur Kenntnis der Familie Asterothyriaceae (Discolichens). *Folia Geobot. Phytotaxon.* **14**, 43-94.
Vobis, G. (1977). Studies on the germination of lichen conidia. *Lichenologist* **9**, 131-136.

III

DISTRIBUTION AND ECOLOGY

10

Ecology of Soil Fungi

Dennis Parkinson

I.	Introduction	277
II.	Development of Ecological Concepts of Soil Fungi	278
III.	Methods for Studying Soil Fungi	282
IV.	Growth and Growth Forms of Soil Fungi	287
V.	Interaction of Soil Fungi and Soil Microarthropods	289
VI.	Conclusion	290
	References	291

I. INTRODUCTION

The fungi, being heterotrophic organisms, are consigned to a saprophytic or to a parasitic or symbiotic existence. In the soil they play crucial roles as agents of organic matter decomposition, as agents of root disease, and as participants in mycorrhizal associations. Therefore they have vital roles in the functioning of terrestrial ecosystems.

Because of the breadth of the topic no attempt will be made here to consider parasitic and mycorrhizal fungi, and indeed the treatment of general soil fungi will be found at best to be idiosyncratic. To balance this presentation it is recommended that the reader consult the excellent reviews of fungal ecology in soil by Griffin (1972) and Warcup (1967) and of soil-borne pathogens by Garrett (1970) and Baker and Cook (1974), and the symposia on mycorrhizal associations edited by Marks and Koslowski (1973) and Saunders *et al.* (1975).

Early in the development of studies on soil fungi the question was raised as to whether the soil contained (and supported) an actively growing mycoflora (i.e., fungal hyphae) or whether it merely contained fungal spores which had fallen from the atmosphere. Workers such as Waksman (1916) and Conn (1922) demonstrated that fungal hyphae were present and active in the soil, and with this demonstration came the acceptance of saprophytic fungi as natural and active members of the soil biota. Nevertheless, as pointed out by Warcup (1967), the soil is also a reservoir of a wide range of organisms from other habitats.

II. DEVELOPMENT OF ECOLOGICAL CONCEPTS OF SOIL FUNGI

While Adametz in the late 1880s is generally considered to have begun the serious study of general soil fungi through his attempts to isolate and identify fungi in the upper layers of soil, Jensen (1912) was the first to attempt a synthesis of data on soil mycoflora. He grouped soil fungi into obligate saprophytes and facultative parasites (the former being the more abundant). From the data he gave it is apparent that, in this early phase of soil mycology, the soil was known to contain a wide variety of fungi, the commonest species belonging to such genera as *Aspergillus, Botrytis, Cladosporium, Mucor, Penicillium,* and *Trichoderma.*

Shortly after this, Waksman began to stamp his mark on the subject. In 1916 he observed, from perusal of a range of isloation data for soils from different parts of the world obtained by numerous workers including himself, that species of *Aspergillus, Mucor, Penicillium*, and *Trichoderma* were present in all the soils investigated. From his studies he found that the genera of soil fungi could be listed in the following decreasing order of importance (i.e., frequency of occurrence): *Penicillium, Mucor, Aspergillus, Trichoderma, Cladosporium, Fusarium, Cephalosporium, Rhizopus, Zygorhynchus, Acrostalagmus, Alternaria, Verticillium.* Waksman (1917) concluded that there was indeed a characteristic fungal flora of the soil, which was essentially as stated above with the addition of *Scopulariopsis,* sterile mycelia, and one or two yeasts. This conclusion was subsequently supported by Bisby *et al.* (1933, 1935) who, in their work on a number of Manitoban soils, confirmed the occurrence of a characteristic soil mycoflora in which, in addition to the genera already mentioned, *Mortierella, Cylindrocarpon,* and *Monotospora* were common. They reiterated the statement by Brierley (1923) that basidiomycetes were rarely, if ever, isolated from soils which were known, from the presence of numerous fruit bodies, to contain such fungi. Thus was developed the concept of a basic cosmopolitan mycoflora capable of growth in soil under normal, appropriate conditions and regularly isolable from soil. Together with such fungi would be found organisms with specialized physiological properties, which therefore had a more complex distribution and were not regularly isolated.

These ideas provided the basis for the concept of soil inhabitants (i.e., regularly isolated species) and soil invaders (i.e., "exotic" species), which is attributed to Waksman and which was successfully applied by Reinking and Manns

(1933) in studies on *Fusarium* spp. in a range of tropical American soils. They showed that certain of these species were common in all the 14 soils studies (i.e., were soil inhabitants), while others were only found locally (i.e., were soil invaders). The extension of these ideas by Garrett (1938, 1950) will be discussed later.

In 1928 Thom recognized four functional classes of soil fungi: species not actively growing but existing in the soil as spores, sclerotia, or other resistant forms; species locally or occasionally active and significant; species which are plant pathogens capable of prolonged, active saprophytic existence in soil, and endemic soil fungi. Later, Thom and Morrow (1937) suggested that soil fungi could be divided into two broad groups: organisms concerned with the primary decomposition of organic residues, and organisms capable of utilizing the soluble products of organic matter decomposition and organic matter suspended in the soil solution. As Thom and Morrow pointed out, this division is somewhat academic, because individual species could well belong to both groups.

Burges (1939) suggested a provisional grouping of fungi present in soil, which involved root parasites, casual parasites and mycorrhizal fungi, facultative parasites and primary saprophytes, and the true soil fungi (comprising fungi of the sugar type and those of the humus type).

Thus through this early phase of soil mycology there was a gradual change in emphasis, from an initial stage dominated by assessment of the number (frequency of occurrence) and nature of taxa present in soil to a later stage where a real interest in the substrate relationships of these fungi developed together with initial concepts of successional patterns of fungal colonization of organic matter in soil (Burges, 1939).

During this period (1916–1939) soil mycoflora was found to vary with soil type, soil treatment, vegetative cover, and soil depth. However, it seemed, from the methods used for fungal isolation and the mode of data presentation, that many workers considered mycoflora to be uniformly distributed in soil. Harley (1948) has pointed out that soil is an extremely heterogeneous environment in terms of both physical and chemical characteristics and that this heterogeneity has great effects on the distribution and activity of individual taxa; i.e., soil fungi are not uniformly distributed in soil, individual taxa varying in frequency of occurrence and metabolic activity with local variations in soil properties. Thus soil is considered a mosaic of microhabitats for microbial development, the amount and chemical quality of the organic matter fraction being the crucial characters determining this development.

The problems of ecological groups of soil fungi, their substrate relationships, and the successional sequences of fungi on fresh plant residues were discussed in detail by Garrett (1951). To him an ecological group of fungi was a group of species possessing some peculiar advantage(s) for the pioneer colonization of a particular substrate. The advantage(s) acting as an ecological determinant would be either the possession of some biochemical or physiological ability (abilities) peculiar to that ecological group, or extreme development of a physiological property common to all fungi. Five groups of soil fungi were given as examples of ecological groups: saprophytic sugar fungi, root-inhabiting fungi, lignin-decomposing fungi, coprophilous fungi, and predaceous fungi.

As mentioned earlier, Garrett (1938) extended the idea of soil inhabitants and soil invaders in distinguishing between two types of behavior in root-infecting fungi. Unspecialized parasites where parasitism was incidental to saprophytism were grouped as soil inhabitants, whereas soil invaders were highly specialized parasites with low (if any) competitive saprophytic ability. In 1950, Garrett proposed that the term "root-inhabiting fungi" replace "soil invaders"; thus his group of root-inhabiting fungi contained mycorrhizal fungi and specialized root parasites, while the group of soil-inhabiting fungi included unspecialized parasites and obligate soil saprophytes.

Garrett's work, particularly in the period 1948-1951, laid a basis for the more recent phase of active ecological work on soil fungi. Certainly it is no coincidence that in the 1950s a great deal of effort was expended on the critical evaluation of existing methods for studying soil fungi, the development of new methods (see later), and synecological studies of fungi in a range of soil types (e.g., Warcup, 1951, 1957; Thornton, 1956; Sewell, 1959). From the accumulating synecological data, attempts were made to characterize fungal associations in soils. For example, Peyronel (1956), following the ideas of phytosociology, proposed the grouping of communities on the basis of frequency of representation of conveniently chosen species (using a compass diagram method for representing these communities). Sappa (1956) also suggested that soil fungal communities could be characterized by species combinations. Thornton (1956) proposed the term "soil fungal pattern" for species isolated from a soil and arranged in decreasing order of percentage frequency of occurrence. Such patterns would be expected to vary from one soil to another and also in the same soil under different conditions (with the vegetation type having a profound influence). In a comparison of fungi in a brown forest soil and in a podzol he found the lists of species from each soil to be very similar, but preparation of soil fungal patterns revealed differences between the soils, e.g., brown forest soil: *Trichoderma viride* tribe II (20%), *T. viride* tribe I (13%), *Mortierella humilis* (10%); and podzol: *Trichoderma viride* tribe I (23%), *Mucor ramannianus* (22%), *Penicillium frequentans* (17%), *Mortierella gracilis* (14%). None of these concepts has been used extensively, although the concept of a typical mycoflora in soil under specific vegetation types has been clearly demonstrated.

In the early 1960s detailed work on the succession of fungi on freshly fallen plant debris (particularly tree leaves) began to gain momentum. The work of Kendrick and Burges (1962) on fungi associated with decomposing needles of *Pinus sylvestris* and that of Hering (1965) on deciduous leaves represent two detailed studies which have served as basic models for much subsequent work. Garrett (1963) has discussed successional patterns of fungi colonizing root tissue, and a generalized pattern of succession of fungi on plant debris was presented by Hudson (1968), which has been reevaluated (and supported) by Hayes (1979). This pattern of succession involves three major stages in which the initial colonization is of living leaves (i.e., phylloplane organisms including saprophytes and weak parasites). The second stage in the successional pattern is effected by common primary saprophytes (e.g., *Cladosporium* spp., *Alternaria tenuis, Epicoccum nigrum, Aureobasidium pullulans,* and *Botrytis cinerea*) which are most important during leaf senescence. Some fungal species are restricted in their occurrence to specific hosts and are termed restricted primary saprophytes. The third major stage occurs on dead leaf litter and involves a range of cellulose- and lignin-decomposing Ascomycetes, Fungi Imperfecti, and Basidiomycetes—termed secondary saprophytes. At this stage secondary sugar fungi (e.g., soil-inhabiting Mucorales) will be found in association with the cellulolytic and ligninolytic species.

Following the mid-1960s, with the inception of the International Biological Program (IBP), particularly the section of this program involving productivity of terrestrial ecosystems, considerable emphasis was placed on the role of fungi as integral parts of decomposer and nutrient cycles. Consequently, emphasis on detailed fungal species diversity in soil and litter tended to give way to studies on fungal standing crop and biomass in different soils. With the urgent requirement for such quantitative data for the studies on energy and nutrient cycling came the realization that the available methodology was imperfect (Parkinson *et al.,* 1971). Nevertheless, studies in the IBP emphasized to ecologists, in general, facts which had been apparent for several decades to agriculturalists and foresters; i.e., the decomposition of organic material with attendant nutrient cycling is a crucial feature of ecosystem functioning and, while the process is a complex multiorganismic phenomenon, the fungi, because of their preeminent role in the initial stage of litter decomposition, play an extremely important role in the energetic and nutrient economy of soil systems.

Interbiome synthesis of data on fungal contributions to decomposition processes is still awaited, but types of data obtained (by the author and colleagues) on fungal standing crops in various terrestrial locations are exemplified below (from such data calculations of energy and nutrient relationships can be made):

   a. Plant litter (figures represent milligrams dry weight of mycelium per gram dry weight of plant tissue)
H layer, *Populus tremuloides* forest (Kananaskis, Alberta)   49.9
Lying gray *Agropyron* litter (Matador, Saskatchewan)   20.3
Gray-black *Dryas* leaves (Devon Island, Northwest Territory)   1.2
   b. Soil (figures represent grams dry weight mycelium per square meter)

0- to 10-cm-depth grassland soil (Matador, Saskatchewan)   60.0
0- to 5-cm-depth tundra mesic meadow (Devon Island, Northwest Territory) 30.4

Unfortunately methodological problems have prevented the attainment of data on fungal turnover in terrestrial locations; therefore detailed energy and nutrient budgets for soil fungi in natural terrestrial locations have not as yet been possible.

Recently interest has focused on several areas of the ecology of soil fungi:

1. Methods for rapidly and accurately estimating fungal biomass and productivity (which will be considered later).

2. An attempt to assimilate data obtained on the ecology of soil fungi into mainstream developments in general ecology, i.e., niche theory and ideas on diversity–stability. This is a contentious issue among many mycologists, who feel that soil fungi are not amenable to such assimilation. It is argued that it is extremely difficult (or impossible) to define an individual fungus at any microsite in soil and that it is equally difficult (or impossible) to isolate all the fungi present at each site; hence application of the developing ideas in population and community ecology must be fruitless. Nevertheless a thorough consideration of these ideas must be made, because at worst they expand the argument on soil fungal ecology and at best they provoke the posing of better questions for experimental attack and allow the soil ecologist to escape the trap of habitual, questionless, mass isolation of fungi from soil samples. Swift (1976), in a stimulating review of terrestrial microbial communities, has attempted to view the ecology of soil fungi against a background of more general ecological ideas.

3. An attempt to study the effects of natural and artificial perturbations of soil systems on the role and nature of soil mycofloras. This is a rapidly expanding topic which has been superficially reviewed by Parkinson (1978) but which should soon be open for more critical evaluation.

In the foregoing brief account of the development of ecological ideas on soil fungi, while mention was made of the importance of the microhabitat concept in soil ecology, no reference was made to the rhizosphere and root surface as microhabitats for fungal development. Of course root development in soil, because of the chemical and physical changes in the root zone effected by living roots and because of input of dead root material, has profound effects on fungal growth. The root regions of higher plants represent an extremely important group of microhabitats in the soil. Detailed reviews of the rhizosphere have been given by such workers as Rovira (1965), Parkinson (1967), and Baker and Cook (1974).

III. METHODS FOR STUDYING SOIL FUNGI

As stated in Section II, the 1950s saw the beginning of a period of reevaluation of methods for the qualitative and quantitative study of soil fungi. In view of the published, detailed treatments of such methods (Parkinson et al., 1971; Johnson and Curl, 1972; Parkinson, 1980), the following comments will be only general.

In the early phase of soil mycological investigations the soil dilution plate method was used almost exclusively for both the isolation of fungi from soil and the (so-called) enumeration of the fungal community in soil samples. More recently, Warcup (1950) has developed the soil plate technique, which is in reality a simple and rapidly applied variant of the soil dilution plate method, and because of this it has been used for various extensive studies on soil fungi (e.g., Warcup, 1951). However, it was clearly demonstrated by Warcup (1955) that the large majority of fungi isolated on both soil dilution plates and soil plates originated from spores or other propagules in the soil samples. Therefore the application of these methods gave little information on fungi present as hyphae (and therefore probably in an active metabolic state) in the soil at the time of isolation. Also, use of the soil dilution plate method does not allow assessment of the microhabitat relationships of the fungi isolated; i.e., no information is provided on the spatial arrangement and substrate relationships of the fungi. It will be appreciated that many of the concepts presented in the previous section must be considered against the background of these criticisms, since the soil dilution method was the method which provided data on which the concepts were based.

Application of the soil dilution plate method in quantitative studies is also open to serious question. The expression of quantitative data on soil fungi in terms of "numbers of fungi" per unit weight of soil is surely unreasonable given the hyphal growth form of the majority of such fungi. Given the comments regarding the selectivity of this method for isolating fungi present as spores in soil samples, counting of fungal colonies developing on dilution plates allows assessment of the spore content of soils. It cannot be used for total fungal biomass determinations.

Realization of the defects of the soil dilution plate method for qualitative studies on soil mycofloras led investigators to attempt to design methods which would allow isolation of fungi present in the soil as active hyphae. Chester (1940) introduced the immersion tube method which subsequently had numerous variants (e.g., Thornton, 1952; Parkinson, 1957; Mueller and Durrell, 1957; Wood and Wilcoxson, 1960; Anderson and Huber, 1965; Luttrell, 1967). All these methods have the same principle; i.e., a chosen agar isolation medium, held in perforated tubes or plates, is placed in the soil, the isolation medium exposed at the perforations being separated from the soil by an air gap. Therefore any fungi eventually isolated from the agar medium must have grown actively across the air gaps. Following retrieval from the soil (after an appropriate predetermined time of burial), portions of the isolation medium are plated, and fungi developing therefrom are subcultured and identified.

These methods suffer from several defects. The placement of immersion apparatus in soil causes changes in moisture and aeration conditions in the immediate vicinity, changes which may stimulate germination of fungal spores;

i.e., fungi entering the isolation medium may not represent species actually active in soil prior to introduction of the immersion apparatus. Soil animals may "track" spores into the isolation medium. There may be competition for entry into and colonization of the isolation medium, fungi with a high competitive ability (i.e., a high growth rate and antagonistic potential) being favored.

Warcup (1955) introduced a direct hyphal isolation method. This involves picking hyphal fragments from soil-water suspensions, attempting to clean such fragments by drawing them through semisolid agar, and then plating the fragments on an appropriate nutrient agar medium. This is apparently an ideal method; however, because it is so time-consuming, it can only be used in restricted studies; also, it is open to considerable observational error and bias (Parkinson et al., 1971). Nevertheless, data provided by Warcup (1957) on fungi isolated from a wheatfield soil by direct hyphal isolation and by soil dilution plating showed great differences in the taxa isolated by the two methods, i.e., the most abundant fungi isolated: (1) By soil dilution plating: species of *Penicillium, Mucor, Cladosporium, Fusarium,* and *Rhizopus.* (2) By direct hyphal isolation: sterile forms, several basidiomycetes, *Rhizoctonia solani, Gaumannomyces graminis, Pythium, Rhizopus, Mortierella,* and *Fusarium.*

More recently, soil and organic matter washing methods have become more widely used. These methods, which involve serial washing of small soil samples, have been shown to be capable of removing large numbers of fungal spores from such samples (the efficiency of the washing varying with soil type), so that the chance of isolating fungi present as hyphae on plated, washed particles is increased (Williams *et al.,* 1965). Soil-washing methods, particularly if automatic, multichambered washing machines (Hering, 1965; Gams and Domsch, 1967; Bissett and Widden, 1972) are used, allow the handling of numerous soil samples in a relatively short time. The incorporation of appropriately sized sieves in the washing apparatus allows separation of the soil samples into different particle sizes and into mineral and organic particles; therefore microhabitat studies are facilitated.

Washing techniques have been applied in the isolation of active fungi from various types of plant debris in or on soil. Harley and Waid (1955) demonstrated the value of serial washing in the isolation of fungi from root surfaces. Subsequently variants of this method have been used in studying fungi associated with decomposing plant debris, and its use has provided the data on which ideas of successional patterns of fungi on organic debris (discussed earlier) have been based.

In all methods for the isolation of fungi from soil, choosing the appropriate isolating medium is of paramount importance. At one time, the aim of investigators involved in synecological studies on soil fungi appeared to be development of a medium which would allow the isolation of any fungus present in soil (i.e., a nonselective medium). Thus a medium such as tap water agar was used as the primary isolation medium. The hope of developing a nonselective medium has been apparently abandoned, and media are used which allow isolation of the maximum number of species from soil (accepting the fact that such media are selective to one extent or another). The choice of an appropriate isolating medium should be made after a properly designed comparative study of a range of media.

By application of the isolation methods described above a large number of species have been isolated from a variety of soils throughout the world. With the use of more diverse types of isolating media the list of fungi in soils continues to increase.

The foregoing discussion refers to methods for synecological studies on soil fungi—studies from which the concepts discussed earlier have been developed. The importance of the application of appropriate methods in soil fungal studies cannot be overemphasized.

For autecological studies a large number of selective isolation methods and media have been developed. Johnson and Curl (1972) have given numerous examples of such methods, therefore no comment will be made here.

In recent years a great deal has been written about methods for quantifying soil fungal communities (Parkinson et al., 1971; Parkinson, 1973, 1980; Frankland et al., 1978). In view of the current interest in the role of fungi in conservation and release of significant quantities of nutrients and energy, the ability to assess accurately (and, preferably, rapidly) the biomass and standing crop of soil fungi becomes of more than academic interest.

The soil dilution plate method, i.e., counting fungal colonies developing on dilution plates and subsequently expressing data as numbers of fungi per unit weight of soil, was exclusively used for the quantification of soil fungi until the 1950s. It must be apparent from previous comments that application of this method in reality can only provide data on the number of fungal spores in soil.

Although direct observation of soil smears and of glass slides (which had been buried in soil for appropriate times) was used to demonstrate the presence of hyphae in soil, these methods were not used to provide data on absolute amounts of hyphae per unit weight of soil. Nicholas and Parkinson (1967) discussed the pros and cons of several types of direct observation methods for quantitative studies on soil mycofloras: modified impression slides, soil sections, and soil-agar films. They concluded that the soil-agar film method was the best available method. This method, devised by Jones and Mollison (1948) and modified by Thomas et al. (1965), involves the preparation of soil–agar suspensions (containing a known weight of soil) and the preparation of soil–agar films of accurately known thickness (using a hemocytometer slide). The slides are observed under the microscope, and hyphal lengths measured in each microscope field. From these measurements and knowing the diameter of the microscope field, thickness of the film, and dilution of the soil suspension, absolute measurements of hyphal length per unit weight of soil can be calculated. With the use of an average value for hyphal diameter, specific gravity, and moisture content, the length measurements can be converted to dry weight values. The method has been criticized on the grounds that it does not easily allow a distinction between live and dead hyphae (the use of phase-contrast microscopy does, at least partially, eliminate this criticism), that it is tedious and time-consuming in application (a criticism that is eliminated with experience), and that the calculations of hyphal weights can only be approximations because of the average values, particularly of specific gravity, which must be used.

To allow the efficient application of fluorescent stains in a more rapid way, various methods involving membrane filtration have been suggested (Hansen *et al.*, 1974; Paul and Johnson, 1977; Sundman and Silelä, 1978). As yet no detailed comparisons of the relative efficiency of such methods versus soil–agar films have been provided.

In an attempt to distinguish between live and dead hyphae, Söderström (1977) used a vital staining method employing fluorescein diacetate (FDA). Membrane filter preparations of staining soil must be observed within 30 s, otherwise the fluorescent stain fades. In applying this method and the conventional soil–agar film method to the A_{01}/A_{02} horizons of a Swedish pine forest soil the following data were obtained: (1) Fungal standing crop (live plus dead hyphae) using soil–agar films: 42,000 mg dry wt hyphae/m². (2) Fungal biomass (live hyphae only) using FDA staining: 680 mg dry wt hyphae m².

While attempts to employ chemical methods (particularly on ATP assay) in estimating total microbial biomass (i.e., bacteria plus fungi) have become more frequent in the last decade, the application of such methods in fungal biomass studies has been much more restricted. Swift (1973a,b) and Frankland *et al.* (1978) have discussed the possibility of using chitin determinations in soil or litter fungal biomass studies. They have concluded that this method could be valuable in studies on the growth of individual fungal species on a specific substrate, but that for normal soil studies where many species are present together with soil invertebrates (many of which possess chitinous exoskeletons) such a method is less preferable than direct observation methods.

Anderson and Domsch (1978) showed that optimally glucose-stimulated soil respiration (carbon dioxide evolution) could be used in the determination of total microbial biomass in soil samples. When this method is coupled with the selective inhibition method discussed by the same authors in 1974, then calculation of fungal biomass in soil samples is both feasible and rapid. This method has not yet been used sufficiently to allow detailed evaluation (Domsch *et al.*, 1979), however, indications of the type of data obtained from its use are given here (data from Domsch *et al.*, 1979). (All figures represent micrograms of carbon per gram dry weight of soil.

Use of the above-mentioned methods allows assessment of total fungal stand-

Soil type	Total microbial biomass	Fungal biomass
Acid brown soil (spruce)	2042	1532
Podzolizing acid brown soil (spruce)	552	386
Sandy loam (permanent grassland)	2107	1485
Lowland bog	1832	1374
Chernozem	898	629
Parabrown soil with pseudogleyization	128	102

ing crop (and perhaps biomass) in soil. They rarely allow detailed study of a particular species or group of related species. With direct observation methods, such specific studies are only possible if the species concerned possesses peculiar morphological features which allow their hyphae to be measured selectively. While this may be feasible for some fungi (e.g., clamp connection-bearing fungi, *Mycelium radicis atrovirens*), in the main the lack of peculiar, distinguishing features makes this approach impractical. The use of chemical and physiological methods cannot be adapted for studies on individual taxa in natural communities.

Development of the fluorescent antibody method (Schmidt and Bankole, 1963; Bohlool and Schmidt, 1968) has provided a potentially powerful tool for autecological investigations. This method was used in combination with a buried-slide technique to study the presence of *Aspergillus flavus* (using antisera against hyphae of this fungus conjugated with fluorescein isothiocyanate to stain the slides which had been buried in soil). Obviously, in applying this method, careful preliminary examination is necessary to ensure the specificity of the conjugate prepared. With the elimination of such problems, this method will allow a much greater understanding of the growth, distribution, and biomass of individual fungi in the complex soil environment.

IV. GROWTH AND GROWTH FORMS OF SOIL FUNGI

If one is genuinely interested in the role of fungi in organic matter decomposition and nutrient cycling in soil systems, then one must address the problems of measuring growth and death rates of fungi and of determining rates of hyphal degradation.

Attempts to measure growth rates of single species of soil fungi in sterile, partially sterilized, and normal soil have been made. Such methods as soil recolonization tubes (Evans, 1955) and replica plating (Stotzky, 1965) have allowed such studies.

Garrett (1963) has summarized the substantial amount of work done to determine the competitive saprophytic ability of various soil-borne pathogens (e.g.,

Curvularia ramosa, Fusarium culmorum, Helminthosporium sativum). The methods described by Garrett (1963) aim to assess the abilities of these (and other) fungi to colonize a given natural substrate in the face of competition from other soil microorganisms. While this method has been used in studying the saprophytic behavior of unspecialized plant-pathogenic fungi, it can be easily adapted for studying the behavior of saprophytic soil fungi. It can allow measurement of rates of colonization of organic debris by individual saprophytes in the face of competition.

Needless to say, a great many data have accumulated on growth rates of pure cultures of individual soil fungi grown under different nutritional and other environmental conditions in agar or liquid culture (or in other forms of growth medium, e.g., glass microbeads, sand). The knowledge gained about physiological properties of individual species by these methods, while important, should be used with care. The activities of an organism in pure culture may differ greatly from those exhibited in soil in a mixed microbial community.

Despite attempts to use autoradiography and other isotope techniques (see contributions in Rosswall, 1973), we still have little real information on the productivity of soil fungal communities (or of individual components thereof). Investigators have tended to fall back on the regular (frequent) use of direct observation methods (e.g., soil–agar films) to achieve information on changes in fungal standing crop over short time periods. The availability and rapidity of execution of the Anderson and Domsch (1978) method makes the study of short-term changes in fungal biomass much more feasible (Parkinson *et al.*, 1978).

The growth form of fungi in soil has been considered by various workers. Burges (1960), basing his conclusions on a detailed study of soil sections, recognized five growth patterns:

1. The *Penicillium* type, where small fragments of organic substrates are densely colonized by the fungus but where no mycelial growth occurs into surrounding soil.

2. The *Mucor ramannianus* type, where, following colonization and exploitation of substrate, the fungus spreads into the surrounding soil where it forms numerous chlamydospores.

3. The *Zygorrhynchus* type, where there is growth of isolated hyphae through soil with no obvious association with organic fragments (presumably such fungi utilize organic substances in the soil solution).

4. The basidiomycete type, where growth from organic substrates is effected by means of mycelial strands or rhizomorphs.

5. The fairy ring type, where the fungus forms a well-defined (dense) mycelial zone which has no apparent connection with special substrates but has a considerable chemical and microbiological effect on the soil.

V. INTERACTION OF SOIL FUNGI AND SOIL MICROARTHROPODS

The interactions of microflora and fauna in soil are numerous and complex. The phenomena involved include predation, competition for food substrates, parasitism, commensalism, and symbiosis. In this section only the effects of the grazing on fungi by microarthropods (Collembola and mites) will be considered, because they have been shown to have potentially important effects on organic matter decomposition and nutrient cycling.

It is generally accepted that decomposition of organic debris on the soil surface involves both abiotic (leaching) and biotic processes. The biological processes in decomposition are effected by the activities of both microflora and soil and litter fauna. The bacteria and fungi play by far the major role in oxidation of the organic constituents of litter—in fact Satchell (1971), discussing data from a deciduous woodland, described the fungi as "the most important decomposer organisms." Data presented by Satchell (1971) on estimated biomass of different components of the soil and litter biota in this deciduous woodland are interesting in this regard:

Bacteria and actinomycetes	2 kg dry wt/ha
Fungi	454 kg dry wt/ha
Soil and litter fauna	36 kg dry wt/ha

Although it has been shown that some soil microarthropods can act as agents of primary decomposition of cellulose (i.e., they possess cellulolytic gut enzymes), the major roles of these members of the soil fauna are considered to be: (1) the fragmentation of organic matter, thus increasing the surface area available for fungal colonization and development, (2) changing the chemical nature of organic matter during passage of this material through the animal gut—these changes may enhance microbial activity or retard it (as is seen in the low decomposition rate of some types of faecal pellets), (3) causing stimulatory or inhibitory effects on the mycoflora as a result of grazing on this mycoflora, (4) transmission of fungi in the organic matter, i.e., "tracking" of spores and other propagules.

Recently attention has been drawn to the effects of the consumption of fungi by microarthropods in natural environments. While attempts to calculate the consumption of fungi are relatively few, Mitchell (1974) calculated that in the organic layer of an aspen forest soil oribatid mites consumed 6 g hyphae/m^2 per year (about 2% of the fungal standing crop). Coleman and McGinnis (1970) found that only 0.2% of test fungus placed in a field location was consumed by soil fauna.

It should be noted here that fungi, during organic matter decomposition, tend to accumulate micro- and macronutrient elements in their thalli. It has also been suggested that grazing on fungi by animals tends to concentrate nutrients further, and also, through the incorporation of these nutrients into longer-lived animal tissue, to restrict fluctuations in nutrient leaching and loss (i.e., animal grazing on fungi acts as a stabilizing phenomenon in nutrient cycling).

From these figures on the amount of consumption of fungi by soil microarthropods it might appear that, at least for the animal taxa considered in temperate biomes, the effects on fungi (and thus on decomposition processes) are small. However, it has also been demonstrated by numerous workers (e.g., Mitchell and Parkinson, 1976; Visser and Whittaker, 1977) that various collembolan species and oribatid mites exhibit preferences in their grazing on fungi; i.e., it has been shown that fungi with dark-pigmented hyphae, e.g., sterile dark fungi, *Cladosporium* spp., are preferentially consumed over basidiomycete hyphae.

This phenomenon of selective grazing adds another dimension to the interaction of microarthropods and fungi in the decomposition of organic matter, i.e., it may affect the pattern of fungal colonization of freshly fallen plant debris. Parkinson *et al.* (1979) showed that, at least in simplified experimental systems in the laboratory, selective grazing by one collembolan species (*Onychiurus subtenuis*) on sterile dark fungi allowed a more effective colonization of aspen poplar leaf material by a basidiomycete. The basidiomycete used in these experiments was avoided by the Collembola and in fact appeared to produce a volatile substance(s) which was lethal to the animals. The importance of this selective grazing was considered even more significant when it was observed that the basidiomycete studied was an active cellulose decomposer, whereas the sterile dark forms were not. Therefore the effects of grazing by microarthropods on litter fungi are a more complex, and probably more important, phenomenon than would be apparent from a mere reference to total amounts of fungal tissue consumed.

A number of attendant issues require detailed study. As stated earlier, it has been suggested that animal grazing enhances fungal growth and fungal metabolic activity by the removal of senescent hyphae and a "pruning effect." This idea requires critical experimental evaluation. Another area which requires careful study is the quantitative assessment of the effects of grazing on fungi on the nutrient dynamics of the litter system.

This section has dealt with one small segment of the problem of invertebrate-fungus interactions, and the reader is reminded of such spectacular examples as the predaceous fungi (and their potential effects on such animal groups as the nematodes and the Enchytraeideae) and the termites.

VI. CONCLUSION

This chapter has dealt with only a small number of general aspects of the ecology of soil fungi which are applicable to the conidial fungi. No reference has been made to the problems of survival, propagation, physiology, or genetics of soil fungi, nor has any consideration been given to the numerous taxonomic problems encountered by the soil mycologist. Nevertheless, it is hoped that the vital and complex role of fungi in the decomposition of organic matter and in nutrient cycling has been emphasized. It must be apparent that, as has been stated many times, our knowledge of such important facets as fungal productivity and the actual roles of individual species in soil or decaying litter can only increase after the development of new methods of investigation and the application of an integrated biological approach.

REFERENCES

Anderson, J. P. E., and Domsch, K. H., (1974). Use of selective inhibitors in the study of respiratory activities and shifts in bacterial and fungal populations in soil. *Ann. Micro* **24**, 189-194.

Anderson, J. P. E., and Domsch, K. H. (1978). A physiological method for the quantitative measurement of microbial biomass in soils. *Soil Biol. & Biochem.* **10**, 215-221.

Anderson, A. L., and Huber, D. M. (1965). The plate profile technique for isolating soil fungi and studying their activity in the vicinity of roots. *Phytopathology* **55**, 592-594.

Baker, K. F., and Cook, R. J. (1974). "Biological Control of Plant Pathogens." Freeman, San Francisco, California.

Bisby, G. R., James, N., and Timonin, M. I. (1933). Fungi isolated from Manitoba soil by the plate method. *Can. J. Res., Sect. C* **8**, 253-275.

Bisby, G. R., James, N., and Timonin, M. I. (1935). Fungi isolated from soil profiles in Manitoba. *Can. J. Res., Sect. C* **13**, 47-65.

Bissett, J., and Widden, P. (1972). An automatic, multi-chamber soil-washing apparatus for removing fungal spores from soil. *Can. J. Microbiol.* **18**, 1399-1409.

Bohlool, B. B., and Schmidt, E. L. (1968). Non-specific staining: Its control in immunofluorescence examination of soil. *Science* **162**, 1012-1014.

Brierley, W. B. (1923). *In* "The Microorganisms of the Soil" (Sir J. Russell, ed.), pp. 115-129, Longmans, Green, New York.

Burges, N. A. (1939). Soil fungi and root infection. *Broteria* **8**, 64-81.

Burges, N. A. (1960). Dynamic equilibria in the soil. *In* "The Ecology of Soil Fungi" (D. Parkinson and J. S. Waid, eds.), pp. 185-191, Liverpool Univ. Press, Liverpool.

Chesters, C. G. C. (1940). A method for isolating soil fungi. *Trans. Br. Mycol. Soc.* **24**, 352-355.

Coleman, D. C., and McGinnis, J. T. (1970). Quantification of fungus-small arthropod food chains. *Oikos* **21**, 134-137.

Conn, H. J. (1922). A microscopic method for demonstrating fungi and actinomycetes in soil. *Soil Sci.* **14**, 149-151.

Domsch, K. H.. Beck, T., Anderson, J. P. E., Söderström, B., Parkinson, D., and Trolldenier, G. (1979). A comparison of methods for soil microbial population and biomass studies. *Z. Pflan zenernaehr. Bodenkd.* **142**, 520-533.

Evans, E. (1955). Survival and recolonization by fungi in soil treated with formalin or carbon disulphide. *Trans. Br. Mycol. Soc.* **38**, 335-346.

Frankland, J. C., Lindley, D. K., and Swift, M. J. (1978). A comparison of two methods for the estimation of mycelial biomass in leaf litter.*Soil Biol. & Biochem.* **10**, 323-333.

Gams, W., and Domsch, K. H. (1967). Beitrage zur Anwendung der Bodenwaschtechik für die Isolierung von Bodenpilzen. *Arch. Microbiol.* **58**, 134-144.

Garrett, S. D. (1938). Soil conditions and the root-infecting fungi. *Biol. Rev. Cambridge Philos. Soc.* **13**, 159-185.
Garrett, S. D. (1950). Ecology of root-inhabiting fungi. *Biol. Rev. Cambridge Philos. Soc.* **25**, 220-254.
Garrett, S. D. (1951). Ecological groups of soil fungi: A survey of substrate relationships. *New Phytol.* **50**, 149-166.
Garrett, S. D. (1963). "Soil Fungi and Soil Fertility." Pergamon, Oxford.
Garrett, S. D. (1970). "Pathogenic Root-Infecting Fungi." Cambridge Univ. Press, London and New York.
Griffin, D. M. (1972). "Ecology of Soil Fungi." Chapman & Hall, London.
Hansen, J. F., Thingstad, T. F., and Groksøyr, J. (1974). Evaluations of hyphal lengths and fungal biomass in soil by a membrane filtration technique. *Oikos* **25**, 102-107.
Harley, J. L. (1948). Mycorrhiza and soil ecology. *Biol. Rev. Cambridge Philos. Soc.* **23**, 127-158.
Harley, J. L., and Waid, J. S. (1955). A method for studying active mycelia on living roots and other surfaces in the soil. *Trans. Br. Mycol. Soc.* **38**, 104-118.
Hayes, A. J. (1979). The microbiology of plant litter decomposition. *Sci. Prog. (Oxford)* **66**, 25-42.
Hering, T. F. (1965). An automatic soil-washing apparatus for fungal isolation. *Plant Soil* **25**, 195-200.
Hudson, H. J. (1968). The ecology of fungi on plant remains above the soil. *New Phytol.* **67**, 837-874.
Jensen, C. N. (1912). Fungus flora of the soil. *N.Y., Agric. Exp. Stn., Ithaca, Bull.* **315**, 414-501.
Johnson, L. F., and Curl, E. A. (1972). "Methods for Research on the Ecology of Soil-Borne Plant Pathogens." Burgess, Minneapolis, Minnesota.
Jones, P. C. T., and Mollison, J. E. (1948). A technique for the quantitative estimation of soil microorganisms. *J. Gen. Microbiol.* **2**, 54-69.
Kendrick, W. B., and Burges, N. A. (1962). Biological aspects of the decay of *Pinus sylvestris* leaf litter. *Nova Hedwigia* **4**, 313-342.
Luttrell, E. S. (1967). A strip bait for studying the growth of fungi in soil and aerial habitats. *Phytopathology* **57**, 1266-1267.
Marks, G. C., and Koslowski, T. T., eds. (1973). "Ectomycorrhizae." Academic Press, New York.
Mitchell, M. J. (1974). Ecology of oribatid mites in an aspen woodland soil. Ph.D. Thesis. University of Calgary, Calgary, Alberta, Canada.
Mitchell, M. J., and Parkinson, D. (1976). Fungal feeding of oribatid mites (Acari: Cryptostigmata) in an aspen woodland soil. *Ecology* **57**, 302-312.
Mueller, K. E., and Durrell, L. W. (1957). Sampling tubes for soil fungi. *Phytopathology* **47**, 243.
Nicholas, D. J., and Parkinson, D. (1967). A comparison of methods for assessing the amount of fungal mycelium in soil samples. *Pedobiologia* **7**, 23-41.
Parkinson, D. (1957). New methods for qualitative and quantitative study of fungi in the rhizosphere. *Pedol. Gand.* **7**, Spec. No., 146-154.
Parkinson, D. (1967). Soil microorganisms and plant roots. *In* "Soil Biology" (N. A. Burges and F. Raw, eds.), pp. 449-478, Academic Press, New York.
Parkinson, D. (1973). Techniques for the study of soil fungi. *Bull. Ecol. Res. Commun.* **17**, 29-36.
Parkinson, D. (1978). The restoration of soil productivity. *In* "The Breakdown and Restoration of Ecosystems" (M. W. Holdgate and M. J. Woodman, eds.), pp. 213-229. Plenum, New York.
Parkinson, D. (1980). Filamentous fungi. *In* "Methods of Soil Analysis," ASA Monogr.
Parkinson, D., Gray, T. R. G., and Williams, S. T. (1971). "Methods for Studying the Ecology of Soil Microorganisms," IBP Handb. 19. Blackwell, Oxford.
Parkinson, D., Domsch, K. H., and Anderson, J. P. E. (1978). Die entwicklung-mikrobieller Biomassen in organichien Horizont eines Fichtenstandortes. *Oecol. Plant.* **13**, 355-366.

Parkinson, D., Visser, S., and Whittaker, J. B. (1979). Effects of collembolan grazing on fungal colonization of leaf litter. *Soil Biol. & Biochem.* **11**, 529–535.
Paul, E. A., and Johnson, R. L. (1977). Microscopic counting and adenosine 5'-triphosphate measurement in determining microbial growth in soils. *Appl. Environ. Microbiol.* **34**, 263–269.
Peyronel, B. (1956). Caractérisation des mycocoénoses de climate et de milieux divers, et nouvelle méthode pour les represente graphiquement. *Rep. Int. Congr. Soil Sci., 6th, 1956* C.45–49.
Reinking, O. A., and Manns, M. M. (1933). Parasitic and other fusaria counted in tropical soils. *Z. Parasitenkd.* **6**, 23–75.
Rosswall, T., ed. (1973). "Modern Methods in the Study of Microbial Ecology," Bull. Ecol. Res. Commun. 17.
Rovira, A. D. (1965). Plant root exudates and their influences on soil micro-organisms. *In* "Ecology of Soil-Borne Pathogens" (K. F. Baker and W. C. Snyder, eds.), pp. 170–186. Univ. of California Press, Berkeley.
Sappa, F. (1956). La mycoflore du sol comme élément structurel des communautes végétales. *Rep. Int. Congr. Soil Sci., 6th, 1956* C.57–61.
Satchell, J. E. (1971). Feasibility study of an energy budget for Meathop Wood. *In* "Productivity of Forest Ecosystems" (P. Duvigneaud, ed.), pp. 619–630. UNESCO, Paris.
Saunders, F. E., Mosse, B., and Tinker, P. B., eds. (1975). "Endomycorrhizas." Academic Press, New York.
Schmidt, E. A., and Bankole, R. O. (1963). The use of fluorescent antibody with the buried slide technique. *In* "Soil Organisms" (J. Doeksen and J. van der Drift, eds.), pp. 197–203. North-Holland Publ., Amsterdam.
Sewell, G. W. F. (1959). Studies of fungi in a *Calluna*-heathland soil. *Trans. Br. Mycol. Soc.* **42**, 343–353.
Söderström, B. E. (1977). Vital staining of fungi in pure cultures and in soil with fluorescein diacetate. *Soil Biol. & Biochem.* **9**, 59–63.
Stotzky, G. (1965). Replica plating techniques for studying microbial interactions in soil. *Can. J. Microbiol.* **11**, 629–636.
Sundman, V., and Silelä, S. (1978). A comment on the membrane filter technique for estimation of length of fungal hyphae in soil. *Soil Biol. & Biochem.* **10**, 399–401.
Swift, M. J. (1973a). Estimation of mycelial growth during decomposition of plant litter. *Bull. Ecol. Res. Commun.* **17**, 323–328.
Swift, M. J. (1973b). The estimation of mycelial biomass by determination of hexosamine content of wood tissue decayed by fungi. *Soil Biol. & Biochem.* **5**, 321–332.
Swift, M. J. (1976). Species diversity and structure of microbial communities in terrestrial habitats. *In* "The Role of Terrestrial and Aquatic Organisms in Decomposition Processes" (J. M. Anderson and A. Macfadyen, eds.), pp. 185–222. Blackwell, Oxford.
Thomas, A., Nicholas, D. P., and Parkinson, D. (1965). Modifications to the agar film technique for assaying lengths of mycelium in soil. *Nature (London)* **205**, 105.
Thornton, R. H. (1952). The screened immersion plate: A method for isolating soil microorganisms. *Research (London)* **5**, 190–191.
Thornton, R. H. (1956). Fungi occurring in mixed oakwood and heath soil profiles. *Trans. Br. Mycol. Soc.* **39**, 485–494.
Visser, S., and Whittaker, J. B. (1977). Feeding preferences for certain litter fungi by *Onychiurus subtenuis*. *Oikos* **29**, 320–325.
Waksman, S. A. (1916). Soil fungi and their activities. *Soil Sci.* **2**, 103–156.
Waksman, S. A. (1917). Is there any fungus flora of the soil? *Soil Sci.* **3**, 565–589.
Warcup, J. H. (1950). The soil plate method for isolation of fungi from soil. *Nature (London)* **166**, 117–118.
Warcup, J. H. (1951). The ecology of soil fungi. *Trans. Br. Mycol. Soc.* **34**, 376–399.

Warcup, J. H. (1955). Isolation of fungi from hyphae present in soil. *Nature (London)* **175**, 953–954.
Warcup, J. H. (1957). Studies on the occurrence and activity of fungi in a wheat field soil. *Trans. Br. Mycol. Soc.* **40**, 237–262.
Warcup, J. H. (1967). Fungi in soil. *In* "Soil Biology" (N. A. Burges and F. Raw, eds.), pp. 51–110. Academic Press, New York.
Williams, S. T., Parkinson, D., and Burges, N. A. (1965). An examination of the soil washing technique by its application to several soils. *Plant Soil* **22**, 167–186.
Wood, F. A., and Wilcoxson, R. D. (1960). Another screened immersion plate for isolating fungi from soil. *Plant Dis. Rep.* **44**, 594.

11

Morphology, Distribution, and Ecology of Conidial Fungi in Freshwater Habitats

J. Webster and E. Descals

I.	Introduction	295
II.	Ingoldian Conidial Fungi	296
	A. Techniques for Study	296
	B. The Classification of Ingoldian Fungi	307
	C. Possible Significance of the Tetraradiate Shape	324
	D. Possible Significance of the Sigmoid Spore Shape	327
	E. Ecology	329
III.	Aeroaquatic Conidial Fungi	335
	A. Techniques for Study	336
	B. Common Anamorph-Genera	337
	C. Affinities of Aeroaquatic Fungi	343
	D. Ecology	343
	References	348

I. INTRODUCTION

There are two main distinct groups of conidial fungi in freshwater habitats, which we shall distinguish as Ingoldian and aeroaquatic. The biology of the two groups is different. Ingoldian fungi abound in babbling brooks and well-aerated lakes, growing on leaves and twigs and forming conidia which are released in water and are readily trapped in foam. Aeroaquatic fungi are more usually found in stagnant ponds, ditches, or slow-running streams and are capable of vegetative growth on submerged leaves or woody substrates. They sporulate only when the substrate is exposed to air, when they form buoyant propagules capable of dispersal when the substrate is again submerged. In addition to these two main groups there are numerous common conidial fungi on reed swamp vegetation and some which grow on submerged shoots of aquatic macrophytes such as *Phragmites, Carex,* and *Schoenoplectus.* These hosts provide substrates for many pycnidial fungi which are not dealt with in this chapter.

II. INGOLDIAN CONIDIAL FUNGI

Ingold (1942) initiated the systematic study of Ingoldian conidial fungi and he has produced an illustrated guide to the common species (Ingold, 1975a). Over 150 species are now known, but more await description. Traditionally the group has been known as aquatic hyphomycetes but, as shown below, they are not exclusively aquatic and an increasing number have been shown to be anamorphs of diverse genera of ascomycetes and basidiomycetes. Some of the teleomorphs fruit on twigs or branches out of water, and it may be more appropriate to call some of them amphibious fungi. In many cases the conidia are large (often spanning more than 50 μm). Two shapes predominate: branched and commonly tetraradiate (see Figs. 1–9), or sigmoid (Figs. 3 and 4). Spores of other shapes are also found.

A. Techniques for Study

1. Fungi on Leaves

A rapidly flowing, nonpolluted stream overhung by deciduous trees provides an ideal place to begin the study of Ingoldian fungi. About 10 days after falling into a stream in temperate climates, leaves of deciduous trees such as *Alnus, Quercus, Acer, Salix,* and *Betula* are well colonized by these fungi. Collect a handful of leaves in a polythene bag and, after rinsing them, incubate them singly in petri dishes containing water. The incubation temperature should preferably be fairly low, say 10°–15°C. It is a mistake to attempt to incubate several leaves in one dish because sporulation is inhibited and heavy bacterial contamination may prevent successful isolation into pure culture. Conidia develop within 1–2 days, and the conidiophores can be seen if the dish containing the colonized leaf is mounted on the stage of a dissecting microscope and the edge of the leaf scanned by transmitted light at a magnification of 50× or 100×. Detached conidia may also be seen floating in the water, held by the meniscus or lying on the bottom of the dish. Since development of the conidium is an important taxonomic criterion, especially when new taxa are described, it may be necessary to remove a small piece of leaf, bearing conidiophores, to a well or hanging-drop slide and to follow conidium development over a period of several hours. The sequence of conidium development can also be followed when small pieces of culture are placed in hanging drops or, better, in a specially designed microscope slide chamber through which sterile water slowly flows. This technique is especially useful when time-lapse cine films of conidial development are made (Descals *et al.*, 1976).

2. Culture Techniques

The isolation of most species and growth in pure culture present no great difficulty. Conidia can be picked up by hand under a dissecting microscope, using a fine capillary pipette, a mounted needle, or a hair, and streaked onto low-nutrient media (e.g., 0.1% malt extract agar) incorporating antibiotic. It is advisable to incubate at 10°–15°C (Descals *et al.*, 1977). Transfers to common laboratory media such as 2% malt extract agar will usually result in colonies capable of sporulation, but in some cases special media may be necessary (Goos, 1970; Miura and Kudo, 1970). After incubation for 10–15 days at 15°–20°C, cultures are generally ready to sporulate. While some species such as *Heliscus lugdunensis* can sporulate on dry agar, most do so only if culture strips are placed in water. Forced aeration in water, which simulates the turbulence of the natural habitat, can markedly stimulate sporulation (Webster and Towfik, 1972; Webster, 1975).

Prolonged incubation of cultures at moderately low temperatures (10°–12°C) under conditions where evaporation loss is minimized can induce the formation of teleomorphs, e.g., apothecia, perithecia, or pseudothecia (see below). Illumination of the cultures with near-ultraviolet light can enhance the formation of teleomorphs. An alternative approach in relating teleomorphs and anamorphs is to make isolations from ascospores or basidiospores collected on twigs and branches near water courses and to induce conidial development in aerated suspensions.

3. Spores in Foam

Foam, which collects around small barriers or in backwaters near turbulent runs in many streams, provides a very effective trap for the conidia of Ingoldian fungi. Experimental estimates of spore loss from aerated suspensions and comparisons of relative spore concentrations in river foam and river water show that the tetraradiate type of propagule is more readily removed from suspension than the sigmoid type. This result leads to the conclusion that foam generally includes a sample of spores biased in favor of branched types. In forms such as *Varicosporium elodeae*, in which conidial morphology can be affected by manipulation of culture conditions, it has been found that spores with more complex branching are removed from suspension more readily than simpler spores (Iqbal and Webster, 1973a). There are numerous descriptive accounts of spores in river foam (for references, see Table I and Miura, 1974; Ingold, 1975a,b, 1976; Gönczöl, 1976b).

TABLE I[a]

No.	Fungus	Source or substrate[b]	Habitat[c]	Climate[d]	Distribution[e]	Figure
	Actinospora					
1	*A. megalospora* Ingold 1952*	W, DL, F	R	TC	GB, Eu, NA	7A
	Alatospora					
2	*A. acuminata* Ingold 1942*	DL, H, M, F	R, L, S, T, D, Po	TWC	GB, W?	2B
3	*A. constricta* Dyko 1978	DL	R	T	NA, GB	
4	*A. crassipes* Marvanová 1977b	DL	R	T	Eu	
5	*A. pulchella* Marvanová 1977b	Fil	R	T	Eu	
	Anguillospora					
6	*A. crassa* Ingold 1958	DL, W, F	R, D	TCW	GB, W?	
7	*A. curvula* Iqbal 1972b	M	R (moors)	T	GB	
8	*A. filiformis* Greathead 1961	DL	R	T	S.Af, NA	
9	*A. furtiva* sp. ined. Descals 1978	F	R	T	GB	
10	*A. gigantea* Ranzoni 1953	DL, W	R, D?	T	NA	
11	*A. longissima* (Sacc. & Syd.) Ingold 1942*	DL, W, H, M, F	R, S, D, Po?	T (C? W?)	GB, W?	7B, 9G
12	*A. pseudolongissima* Ranzoni 1953	DL, F?	R, L	T	NA, Eu?	
13	*A. pulchella* Wolfe 1976	F	R	T	NA	
14	*A. rosea* sp. ined. Descals 1978	F	R	T	GB	
15	*A. virginiana* Wolfe 1976	F	S	T	NA	
	Angulospora					
16	*A. aquatica* Nilsson 1962a*	DL?	L, R?	W	SA	3D
	Articulospora					
17	*A. angulata* Tubaki 1957	DL, F	R, L	T	Pac, NA	
18	*A. atra* sp. ined. Descals 1978	F	R	T	GB	
19	*A. moniliforma* Ranzoni 1953	DL	R (Po)	T	NA, Aus	
20	*A. tetracladia* Ingold 1942*	DL, H, M, F	R, S, L, Po	T, W	GB, W?	4A
	Brachiosphaera					
21	*B. jamaicensis* (Crane & Dumont) Nawawi (Descals *et al.*, 1976)	W	R?	W	Jamaica	

298

#	Species						
22	*B. tropicalis* Nawawi* (Descals et al., 1976)	F		R	W	Pac, Af, CA	7C
	Calcarispora						
23	*C. hiemalis* Marvanová & Marvan 1963*	DL, F?		R	T	Eu	4B
	Campylospora						
24	*C. chaetocladia* Ranzoni 1953*	DL, F, W		R, D	TW	NA, SA, Aus Af, Pac	4C
25	*C. filicladia* Nawawi 1974c	DL, W		R	W	As	
26	*C. parvula* Kuzuha 1973	DL, F?		R	T	Pac, GB?	
	Clavariana						
27	*C. aquatica* Nawawi* (Descals et al., 1976)	F		R	W	As, Pac? Af?	7D
	Clavariopsis						
28	*C. aquatica* de Wildeman 1895*	DL, W, M, F		S, D, R	T, CW?	Eu, W?	4F, 9C
29	*C. brachycladia* Tubaki 1957	DL, H, M		R	T	Pac, Eu	
30	*C. bulbosa* Anastasiou 1962	W		Na	W, T?	NA, Pac	
	Clavatospora						
31	*C. filiformis* Nawawi 1973a	DL		R	W	As, Af	
32	*C. flagellata* Gönczöl 1976	DL, F		R	T	Eu	
33	*C. longibrachiata* (Ingold) Nils. ex Marv. & Nils. 1971*	DL, F		R, S, Na, D, S	T, C	GB, W?	2A
34	*C. stellata* (Ingold & Cox) Nils. ex Marv. & Nils. 1971	DL, F		D, S, Na, R	T, C	GB, Eu, NA	
35	*C. stellatacula* Kirk ex Marv. & Nils. 1971	C		Na	T	NA	
36	*C. tentacula* (Umphlett) Nilsson 1964	DL, F		R, Po	T, W	NA, W?	
	Condylospora						
37	*C. spumigena* Nawawi 1976a*	F		R	W	As	1B
	Culicidospora						
38	*C. aquatica* Petersen 1960*	W, C? F		R, S?	T, C	NA, GB, Eu	4G
39	*C. gravida* Petersen 1963a	DL, F		R, S	T, C	NA, GB, Eu	
	Cylindrotrichum Bonorden 1851 (Type sp. *C. oligospernum*)						
40	*C. helisciforme* Marvanová 1979	F, debris		R	T	Eu	

(continued)

TABLE I (continued)

No.	Fungus	Source or substrate	Habitat	Climate	Distribution	Figure
	Dactylella Grove 1884 (Type sp. *D. minuta*)					
41	*D. microaquatica* Tubaki 1957	DL (Shiia)	R*, S	T, C	Pac. NA, Eu	2E
	Dendrospora					
42	*D. erecta* Ingold 1943a*	DL, W, M, F	R, S (moors)	T, C	GB, Eu, NA	7E
43	*D. fastuosa* Descals & Webster 1980	C, F	R	T	GB	
44	*D. fusca* Descals & Webster 1980	C, F	R	T, C?	GB, NA?	
45	*D. juncicola* Iqbal 1972b	M	R (moors)	T	GB	
46	*D. nana* Descals & Webster 1980	F	R (moors)	T	GB	
47	*D. tenella* Descals & Webster 1980	W	R	T, C?	GB, NA?	
48	*D. torulosa* Descals & Webster 1980	W, DL	R (moors)	T	GB	
	Dendrosporomyces					
49	*D. prolifer* Nawawi* (Nawawi et al., 1977b)	F	R	W	As	7F
50	*D. splendens* (Nawawi) Nawawi nom. nov. ined.	DL	R	W	As	
	Dimorphospora					
51	*D. foliicola* Tubaki 1958*	DL	S, R	T	Pac, GB	3F
	Filosporella					
52	*F. annelidica* (Shearer & Crane) Crane & Shearer	DL	R	T	NA, GB	6A
53	*F. aquatica* Nawawi 1976b*	M	S	W	As	
	Flabellospora					
54	*F. acuminata* sp. ined. Descals 1978	F, W	R	T	GB, As	
55	*F. crassa* Alasoadura 1968a*	DL, M	R	W	Af	6B
56	*F. multiradiata* Nawawi 1976c	F	R	V	As	
57	*F. tetracladia* Nawawi 1973c	F	R	W	As	
58	*F. verticillata* Alasoadura 1968b	DL, M	R	W	Af	

#	Species					
	Flagellospora					
59	*F. curvula* Ingold 1942*	DL, H, M, F	R, Po, S, T, L	T, C, W?	GB, W?	4D
60	*F. fusarioides* Iqbal 1974a	DL	R (moors)	T	GB	
61	*F. penicillioides* Ingold 1944	DL, C, W, P?	R, T	T, C, W?	GB, W?	
62	*F. stricta* Nilsson 1962b	DL	R	C	Eu	
	Fontanospora					
63	*F. alternibrachiata* Dyko 1978	DL	R	T, W?	NA, CA?	
64	*F. eccentrica* (Petersen) Dyko 1978*	DL, F?	R	T	NA, GB?	4H
	Geniculospora					
65	*G. grandis* (Greathead) Nilsson ex Nolan 1972*	DL	R	T	S.Af.	6F
66	*G. inflata* (Ingold) Nils. ex Marv. & Nils. 1971	DL, W, F	R, D, S, Po	T, C, W?	GB, W?	
	Gyoerffyella Kol 1928					
67	*G. biappendiculata* (Arnold) Ingold 1975a	C, F	R? T?	T	Eu, GB	
68	*G. rotula* (v. Höhn.) Marvanová* (Marvanová et al., 1967)	DL	S, R	T, C	Eu, NA	1E
	Heliscus					
69	*G. speciosa* (Miura) Dudka 1974	F	R (moors)	T	Pac, GB, Eu	
70	*G. tricapillata* (Ingold) Marvanová et al., 1967	DL	S, R, T	T, C	GB, NA?	
	Heliscus					
71	*H. lugdunensis* Sacc. & Thérry 1880*	C, W, D, L, F	R, S, Po	T, C	Eu, GB, NA, Aus.	2G, 9A, B
72	*H. submersus* Hudson 1961	DL	R	W, T	CA, SA	
	Ingoldiella					
73	*I. fibulata* Nawawi 1973b	F	R	W	As.	
74	*I. hamata* Shaw 1972*	DL	R	W	Aus, As, Af.	8C, D
	Isthmotricladia					
75	*I. britannica* sp. ined. Descals 1978	C, F	R	T	GB	
76	*I. gombakiensis* Nawawi 1975b	F	R	W	As	
77	*I. laeensis* Matsushima 1971[a]*	M	L? R	W	Pac, Af?	1A
	Jaculispora					
78	*J. submersa* Hudson & Ingold 1960*	DL, F	R, S	W, T	CA, NA, Eu, GB	2C
	Laridospora					
79	*L. appendiculata* (Anastasiou) Nawawi 1976d*	L, W, F	R	W, T	Pac, Af, As, GB	4J

(continued)

TABLE I (*continued*)

No.	Fungus	Source or substrate	Habitat	Climate	Distribution	Figure
	Lateriramulosa (Type sp. *L. uniinflata* Matsushima 1971)					
80	*L. quadriradiata* Miura & Okano 1979	DL	R	T	Pac	
	Lemonniera					
81	*L. alabamensis* Sinclair & Morgan-Jones 1979b	W	R	T	NA	
82	*L. aquatica* de Wildeman 1894*	DL, C, W, H, F	R, S, Na?	T, C	Eu, W?	6D, 9D
83	*L. centrosphaera* Marvanová 1968	DL, F	R	T	Eu, Pac, GB	
84	*L. cornuta* Ranzoni 1953	DL, H, F	R	T	NA, GB, Eu, Pac	
85	*L. filiformis* Petersen ex Dyko (Descals *et al.*, 1977)	DL, F	R	T	NA, GB, Eu	
86	*L. pseudofloscula* Dyko (Descals *et al.*, 1977)	L	R	T	NA	
87	*L. terrestris* Tubaki 1958	W, F	L, R, S	T, C	Pac, W?	
	Lunulospora					
88	*L. curvula* Ingold 1942*	DL, H, W, C, F	R, S	T, W, C	GB, W?	5A, 9F
89	*L. cymbiformis* Miura 1972	DL	R	T	Pac, Aus?	
	Margaritispora					
90	*M. aquatica* Ingold 1942*	DL, F	R, S	T	GB, NA, Eu, Pac?	5B
91	*M. monticola* Dyko 1978	DL	R?	T	NA	
	Monotosporella Hughes 1958 (Type sp. *M. setosa*)					
92	*M. tuberculata* Gönczöl 1976a	DL	R?	T	Eu	3B
	Mycocentrospora					
93	*M. acerina* (Hartig) Deighton 1972*	W, P, Humans? DL, M, C, F, H	T, R, S	T, C	Eu, NA, GB	4E
94	*M. angulata* (Petersen) Iqbal 1974a	DL	R	T	NA,	
95	*M. aquatica* (Iqbal) Iqbal 1974a	W, DL?	R (moors)	T	GB, NA?	
96	*M. clavata* Iqbal 1974a	W, M, DL	R (moors)	T	GB	
97	*M. varians* Sinclair & Morgan-Jones 1979b	L	R	T	NA	
	Obstipispora (=*Condylospora*?)					
98	*O. chewaclensis* Sinclair & Morgan-Jones 1979a*	L	R	T	NA, Pac	see 1B

#	Species	Substrate	Habitat	Region	Fig.
	Orbimyces				
99	*O. spectabilis* Linder* (Barghoorn and Linder, 1944)	W	Na	NA	6E
	Pleuropedium				
100	*P. tricladioides* Marvanová & Iqbal 1973*	M, F	R (moors)	T	Eu, GB	6C
	Polycladium				
101	*P. equiseti* Ingold 1959*	*Equisetum*	S, R?	T, C?	GB, Eu?	8B
	Porocladium				
102	*P. aquaticum* Descals et al. 1976*	F	R	T	GB	6G
	Pseudoanguillospora				
103	*P. gracilis* Sinclair & Morgan-Jones 1979b	DL	R	T	NA	
104	*P. prolifera* Iqbal 1974b	M	R (moors)	T	GB	
105	*P. stricta* Iqbal 1974b*	M	R (moors)	T	GB	5C
	Pyramidospora				
106	*P. casuarinae* Nilsson 1962a*	DL, Fil	R	W	SA, CA	2H
107	*P. constricta* Singh 1972	DL	R	W	Af	
108	*P. densa* Alasoadura 1968c	DL	R	W	Af	
109	*P. fluminea* Miura & Kudo 1971	DL	R	T	Pac	
110	*P. herculiformis* Singh 1976	DL	R	W	Af	
111	*P. ramificata* Miura (Miura and Kudo 1971)	DL	R	T	Pac	
112	*P. stellata* Sinclair & Morgan-Jones 1979b	DL, F	R	T	NA	
	Pyricularia Sacc. 1880 (Type sp. *P. grisea*)				
113	*P. aquatica* Ingold 1943a	DL, W, F	R, S	T, C	GB, Eu, NA, SA?	2F, 9H
114	*P. submersa* Ingold 1944	DL	R, S, D	T, C	GB	
	Ramocercospora gen. nov. ined.				
115	*R. flagelliformis* sp. ined. Descals 1978	DL, F	R, T?	T	GB	
116	*R. foliosa* sp. ined. Descals 1978*	F	R	T	GB	1C
	Scorpiosporium				
117	*S. angulatum* (Ingold) Iqbal 1974c	DL, F	R, S	T, C, W?	GB, W?	
118	*S. anomalum* (Ingold) Iqbal 1974c	M, DL?	R? S, D?	T, C? W?	GB, Eu? Pac?	
119	*S. gracile* (Ingold) Iqbal 1974c	DL, F	R, S	T, C	GB, W?	
120	*S. gracile* var. *oxyphilum* Nimura & Suzuki 1962	?	L	T?	Pac	

(continued)

TABLE I (continued)

No.	Fungus	Source or substrate	Habitat	Climate	Distribution	Figure
121	*S. minutum* Iqbal 1974c*	M	R (moors)	T	GB	5D
122	*S. rangiferinum* sp. ined. Descals 1978	F	R	T	GB	
	Sigmoidea					
123	*S. aurantia* sp. ined. Descals 1978	W	R	T	GB	
124	*S. prolifera* (Petersen) Crane 1968*	Debris	R?	T	NA	3E
	Speiropsis					
125	*S. irregularis* Petersen 1963a	Debris	R?	T, C	NA	
126	*S. pedatospora* Tubaki 1958*	DL	?	T, W	Pac, As	5E
	Sympodiocladium gen. nov. ined.					
127	*S. frondosum* sp. ined. Descals 1978*	F	R	T	GB	5G
	Taeniospora					
128	*T. gracilis* Marvanová 1977a*	DL, W? F	R	T	Eu, GB	1F
	Tetrabrunneospora					
129	*T. ellisii* Dyko 1978*	DL	R	T	NA	
	Tetrachaetum					
130	*T. elegans* Ingold 1942*	DL, F	R, S	T, C, W	GB, W?	8E
	Tetracladium					
131	*T. marchalianum* de Wildeman 1893*	Fil, H, F, P? W, DL, M	S, T, Po, R, D	T, C, W	Eu, W	5H, 9E
132	*T. maxilliforme* (Rostrup) Ingold 1942	P, W, F	T, R	T, C	Eu, NA, GB	
133	*T. setigerum* (Grove) Ingold 1942	W, DL, H, P, F	T, R, L, S	T, C	GB, W?	
	Tricellula van Beverwijk 1954 (Type sp. *T. inaequalis*)					
134	*T. aquatica* Webster 1959c	DL, M	R	T, C	GB, Eu, NA	3A
135	*T. botryosa* sp. ined. Descals 1978	F	R	T	GB	
	Tricladium					
136	*T. attenuatum* Iqbal 1971	DL, F	R	T	GB	

137	T. brunneum Nawawi 1974b	F	R	W	As	
138	T. castaneicola Sutton 1975	W, DL	D, L, S	T	GB	
139	T. caudatum Kuzuha 1973	F	R	T	Pac.	
140	T. chaetocladium Ingold 1974	DL, F	R	T	GB	
141	T. giganteum Iqbal 1971	DL, M	R (moors)	T	GB	
142	T. malaysianum Nawawi 1974b	F	R	W	As.	
143	T. marylandicum Crane 1968	F, DL?	R	T	NA	
144	T. patulum Marvanová & Marvan 1963	DL	R	T	Eu, GB	
145	T. sorghicolum Gupta & Gandhi 1979	P	T only	T	As	
146	T. splendens Ingold 1942*	DL, H, Fil, M, F	R, L, S, D	T, C, W?	GB, W	6H
147	T. terrestre Park 1974a	DL, F	T, R	T	GB	
148	T. varium Jones & Stewart 1972	C	R	T	GB	
149	T. st. of Hymenoscyphus varicosporoides Tubaki 1966b	W	L	T?	Pac	
	Tripospermum Spegazzini 1918 (Type sp. T. acerinum (Sydow) Speg.)					
150	T. camelopardus Ingold et al. 1968a	F	R	T	GB	
151	T. prolongatum Sinclair & Morgan-Jones 1979b	L, F	R	T	NA, Eu?	5F
	Triposporina von Höhnel 1912 (Type sp. T. uredinicola v. Höhn).					
152	T. ceranoica sp. ined. Descals 1978	W, F	R	T, W?	GB, Af?, Eu.	2D
	Triscelophorus					
153	T. acuminatus Nawawi 1975a	DL, W, F	R	W, T?	As, GB?	
154	T. magnificus Petersen 1962	DL	R	T	NA	
155	T. monosporus Ingold 1943b*	L, M, C, F	R, S	T, W, C?	GB, W?	3C
156	T. septatus Wolfe 1976	DL	R	T	NA	
	Varicosporina					
157	V. ramulosa Meyers & Kohlmeyer 1965*	H	Na	T	NA	1D
	Varicosporium					
158	V. aquaticum Vischnevskaja 1955	?	?	C	Eu	
159	V. delicatum Iqbal 1971	DL, F?	R, moors	T	GB	

(continued)

305

TABLE I (*continued*)

No.	Fungus	Source or substrate	Habitat	Climate	Distribution	Figure
160	*V. elodeae* Kegel 1906*	DL, M, H, F	R? S, D, T, L	T, C, W?	Eu, W?	
161	*V. giganteum* Crane 1968	F, DL	R, S, L?	T, C?	NA, GB?	6J
162	*V. helicosporum* Nawawi 1974a	F	R	W	As.	
163	*V. macrosporum* Nawawi 1974a	F	R	W	As.	
164	*V. trimosum* Wolfe 1976	F	R	T	NA	
	Volucrispora Haskins 1958 (Type sp. *V. aurantiaca*)					
165	*V. graminea* Ingold *et al.* 1968b	M, F	L, R, T	T, C	GB, Eu, NA	

[a] The first record for all entries represents data from the type description. Authors of anamorph genera are those of the type species (indicated with *) unless otherwise stated. ?, Data doubtful; ( ), not consistent. Synonymy has been omitted. The following species have not been recorded from wet or submerged habitats: *Articulospora foliicola* Marvanová 1975, *A. ozeensis* Matsushima 1975, *Gyoerffiella entomobryoides* (Boerema & v. Arx) Marvanová 1967, *G. gemellipara* Marvanová 1975, *G. oxalidis* Vanev 1976, *Speiropsis hyalospora* Subr. & Lodha 1964, *S. simplex* Matsushima 1971a, *Tricellula curvatis* Haskins 1958. *T. inaequalis* v. Beverwijk 1954, *Tricladium minimum* Matsushima 1975. *Trisulcosporium acerinum* Hudson & Sutton 1964. *Volucrispora aurantiaca* Haskins 1958, *V. ornithomorpha* (Trotter) Haskins 1958, and various species in *Dactylella, Monotosporella, Mycocentrospora, Pyricularia, Tripospermum* and *Triposporina*. All species are saprophytic unless otherwise stated. Data from publications in cyrillic script or Japanese may not be up-to-date.

[b] DL, Dicot leaves; W, angiosperm wood; C, conifer wood: H, herbaceous dicots; M, monocots; P, also plant parasites; Fil, ferns; Alg., algae; F, foam.

[c] R, Running waters (rivers, streams, falls, springs, and so on); S, static waters (lakes, ponds, pools, and so on); T, "terrestrial" (isolated from sites away from streams, or from soil, including farmland); L, forest litter; D, temporary water courses (ditches, torrents, and so on); Po, slightly polluted waters; Na, brackish or seawater.

[d] C, Cold: T, temperate; W, warm (ignoring the possible climatic influence of altitude, unknown in most cases).

[e] W, Widespread; Af, Africa; A, America; As, Asia; Aus, Australia; Eu, Europe; GB, Great Britain; Pac, Pacific Islands, mainly Japan; N, north; C, central; S, south.

It is possible to preserve foam samples conveniently by using a fixative such as Formalin–acetic alcohol. Care is needed when identifying species from detached spores, because convergent evolution has resulted in the development by different fungi of spores which are remarkably similar in morphology, and it is possible, especially with sigmoid spores, to confuse taxonomically distinct forms (see below). Spores may remain suspended in foam without germinating for at least 1 month at 13°C, retaining a high percentage of viability. It is therefore possible to make isolations from spores in foam using techniques similar to those outlined above. Spores can be carried in foam in a Thermos flask or in a jar surrounded by an ice pack so that isolations can be made several hours after collection. Alternatively, it is possible to smear foam over a previously poured agar plate containing antibiotic and to make isolations from well-separated germinating spores up to a day or more later. Because there are many undescribed species known so far only as spores from foam, a profitable line for future taxonomic work would be to isolate unidentified spores from foam and to induce spore development in the resulting cultures. We have found a mobile laboratory constructed from a converted trailer particularly useful for such studies.

4. Filtration Methods for Estimating Spore Concentrations in Streams

Since the spores of many Ingoldian fungi are so characteristic, it is possible to estimate their concentration when water samples are filtered. Millipore filters of 8 μm pore size are effective because quick filtration is possible. In rivers with a high spore concentration a 250-ml sample is adequate and gives good replication. In less productive streams it may be necessary to filter as much as 5 liters of water to obtain reasonable spore counts. The Millipore filters can be fixed and stained in the field by the addition of a few drops of cotton blue in lactic acid, and the filters can be rendered transparent by heating at 60°C for 1 h (Iqbal and Webster, 1973b, 1977).

B. The Classification of Ingoldian Fungi

It is important to stress that the classification of aquatic Hyphomycetes is based on anamorph-genera which include species having spores with similar morphology and development. However, studies connecting anamorphs to teleomorphs have confirmed that many of the anamorph-genera are heterogeneous, i.e., include taxa of diverse relationships (Table II).

The number of described Ingoldian fungi is increasing at a rapid rate, and many more await isolation, especially in the tropics. A key to their identification is beyond the scope of this chapter. It seems more sensible to discuss various important aspects of this group which may be used in their identification and classification. A list of described species (Table 1) and illustrations of representative aquatic species of all anamorph-genera (Figs. 1–9) are also provided. Impor-

TABLE II
Teleomorphs of Ingoldian fungi

Anamorph	Teleomorph	Classification	Authors of associations
Sigmoid conidia			
Flagellospora penicillioides	*Nectria penicillioides* Ranzoni 1956	Pyrenomycete	(Loc. cit.)
Filosporella anneliidica?	"*Mollisia*" sp.	Helotiales	Webster and Descals (1979)
Anguillospora furtiva	*Pezoloma* sp.	Helotiales	Webster and Descals (1979)
A. rosea	*Orbilia* sp.	Helotiales	Webster and Descals (1979)
A. crassa	"*Mollisia*" sp. Webster 1961	Helotiales	(Loc. cit.)
A. longissima	*Massarina* sp.	Loculoascomycetes	Willoughby and Archer (1973)
Branched conidia			
Pyricularia aquatica	*Massarina aquatica* Webster 1965	Loculoascomycetes	(Loc. cit.)
Clavariopsis aquatica	*Massarina* sp.	Loculoascomycetes	Webster and Descals (1979)
C. bulbosa	*Corollospora pulchella* Kohlmeyer et al. 1967	Marine pyrenomycete	Anastasiou (1962)
Tricladium sp.	*Hymenoscyphus varicosporoides* Tubaki 1966	Helotiales	(Loc. cit.)
Tricladium splendens	*Hymenoscyphus splendens* Abdullah et al. (1981)	Helotiales	Abdullah et al. (1981)
Heliscus lugdunensis	*Nectria lugdunensis* Webster 1959b	Pyrenomycete	(Loc. cit.)
Actinospora megalospora	*Miladina lechithina* (Cooke) Svrček 1972	Pezizales	Descals and Webster (1978)
Clamp connections			
Ingoldiella hamata	*Sistotrema* sp.	Basidiomycete	A. Nawawi (unpublished)
Taeniospora gracilis	*Leptosporomyces galzinii* (Bourd.) Jülich 1972	Basidiomycete	Nawawi et al. (1977a)

tant taxonomic treatments have been provided by Petersen, 1962, 1963a,b; Nilsson, 1964; Dudka, 1974; Ingold, 1975a.

1. Teleomorphs

The majority of Ingoldian fungi are known only from their anamorphs. Thirteen species, however, have been associated with Ascomycetes and two with resupinate Basidiomycetes (Table II). Others are now being studied. A few species have either dikaryotic cells, clamps, or dolipore septa, but most have uninucleate cells (A. Nawawi, unpublished). Members of the Entomophthorales are discussed separately. There are also some tetraradiate forms among the Zoopagaceae, but these are not included here. A more detailed account of the teleomorph–anamorph associations can be found in Webster and Descals (1979). Nevertheless, it is evident that various phylogenetic lines have converged in aquatic habitats toward the formation of sigmoid or tetraradiate conidial shapes.

2. Substrates

Table I presents a range of substrates colonized by these fungi. Attention has concentrated on decaying dicotyledonous leaves, but some species are consis-

---

Fig. 1. (A) *Isthmotricladia laeensis;* (B) *Condylospora spumigena;* (C) *Ramocercospora foliosa;* (D) *Varicosporina ramulosa;* (E) *Gyoerffiella rotula,* primary and secondary conidia; (F) *Taeniospora st.* of *Leptosporomyces galzinii.* (Tracings from the originals.)

Fig. 2. (A) *Clavatospora longibrachiata;* (B) *Alatospora acuminata;* (C) *Jaculispora submersa;* (D) *Triposporina ceranoica* (development of six-horned conidium); (E) *Dactylella microaquatica,* percurrent and sympodial proliferation; (F) *Pyricularia aquatica;* (G) *Heliscus lugdunensis* (aerial conidium without knobs); (H) *Pyramidospora casuarinae.* (Tracings from the originals, except G.)

Fig. 3. (A) *Tricellula aquatica;* (B) *Monotosporella tuberculata;* (C) *Triscelophorus monosporus;* (D) *Angulospora aquatica;* (E) *Sigmoidea prolifera;* (F) *Dimorphospora foliicola.* (Tracings from the originals.)

Fig. 4. (A) *Articulospora tetracladia* (penta- and tetraradiate forms). (B) *Calcarispora hiemalis;* (C) *Campylospora chaetocladia;* (D) *Flagellospora curvula;* (E) *Mycocentrospora acerina;* (F) *Clavariopsis aquatica;* (G) *Culicidospora aquatica;* (H) *Fontanospora eccentrica;* (J) *Laridospora appendiculata.* (Tracings from the originals.)

Fig. 5. (A) *Lunulospora curvula;* (B) *Margaritispora aquatica* (aerial form, phragmosporous); (C) *Pseudoanguillospora stricta;* (D) *Scorpiosporium minutum;* (E) *Speiropsis pedatospora;* (F) *Tripospermum camelopardus;* (G) *Sympodiocladium frondosum;* (H) *Tetracladium marchalianum;* (Tracings from the originals.)

Fig. 6. (A) *Filosporella annelidica* (note annellations); (B) *Flabellospora crassa;* (C) *Pleuropedium tricladioides;* (D) *Lemonniera aquatica;* (E) *Orbimyces spectabilis;* (F) *Geniculospora grandis;* (G) *Porocladium aquaticum;* (H) *Tricladium splendens;* (J) *Varicosporium elodeae.* (Tracings from the originals, except D and H.)

Fig. 7. (A) *Actinospora megalospora;* (B) *Anguillospora longissima;* (C) *Brachiosphaera tropicalis;* (D) *Clavariana aquatica;* (E) *Dendrospora erecta;* (F) *Dendrosporomyces prolifer.* (Tracings from the originals, except A and B.)

Fig. 8. (A) *Casaresia sphagnorum;* (B) *Polycladium equiseti;* (C and D) *Ingoldiella hamata;* (C) aquatic form; (D) aerial form; (E) *Tetrachaetum elegans.* (Tracings from the originals, except A.)

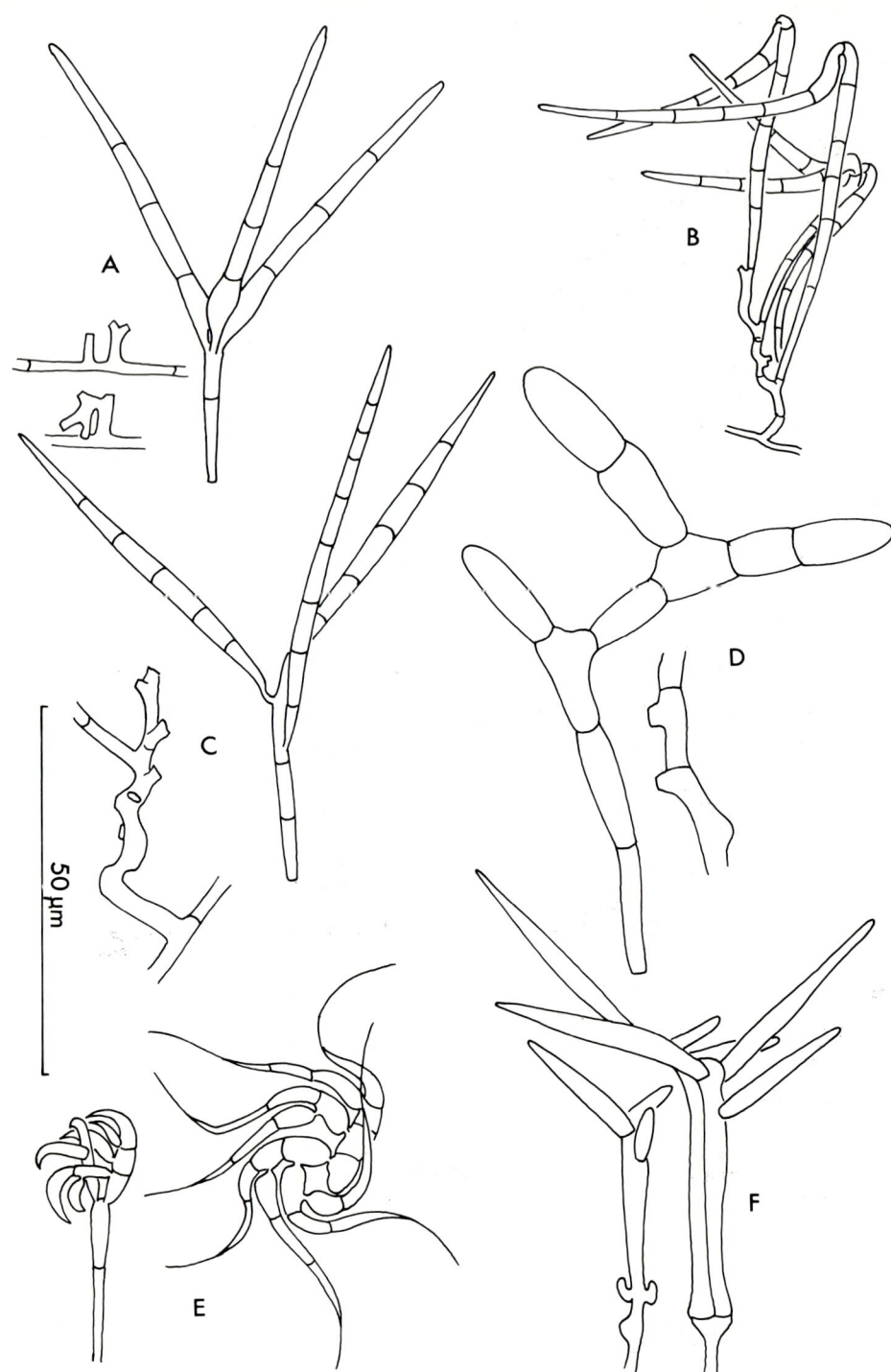

Fig. 1. Caption on page 309.

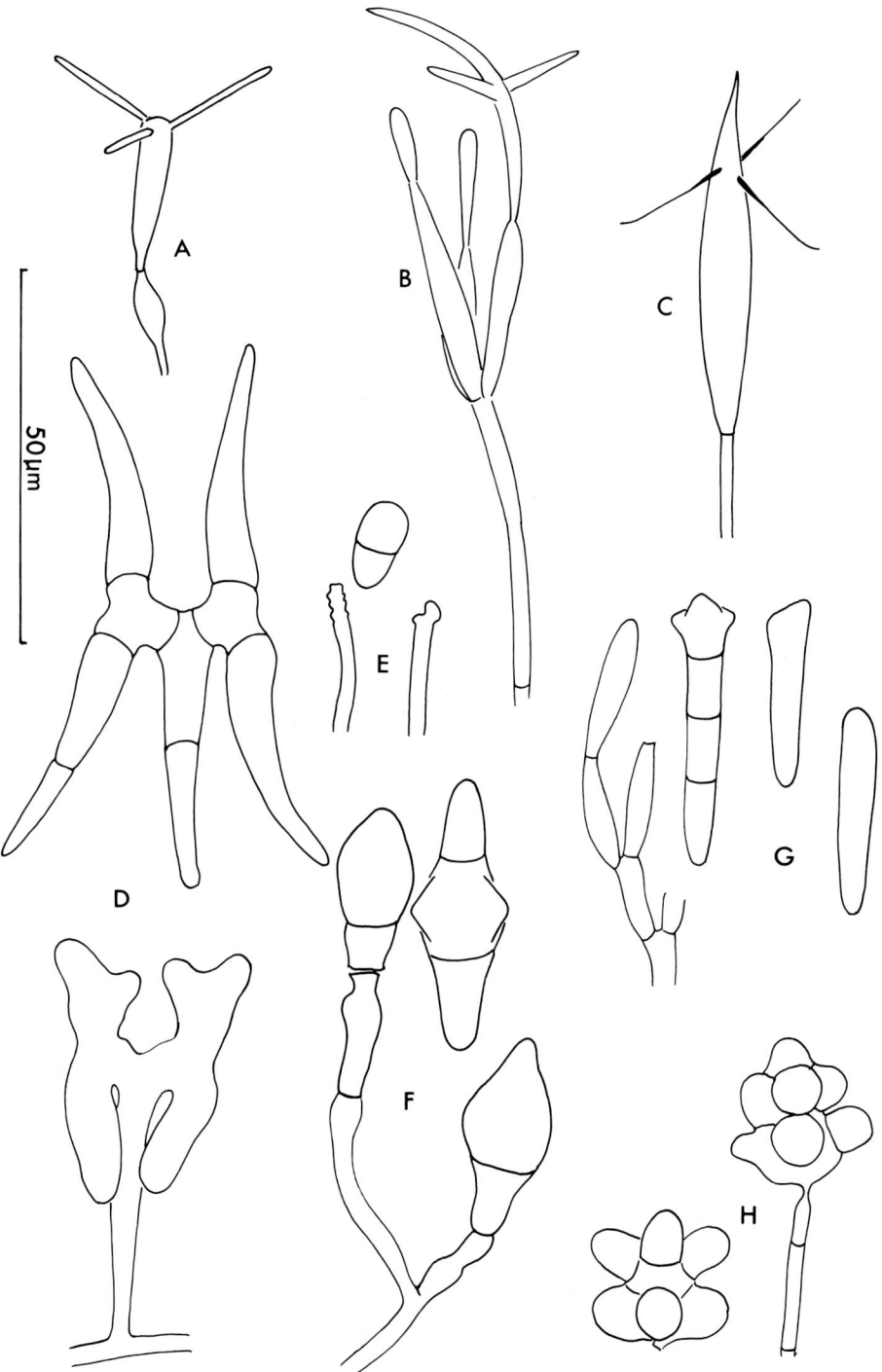

Fig. 2. Caption on page 309.

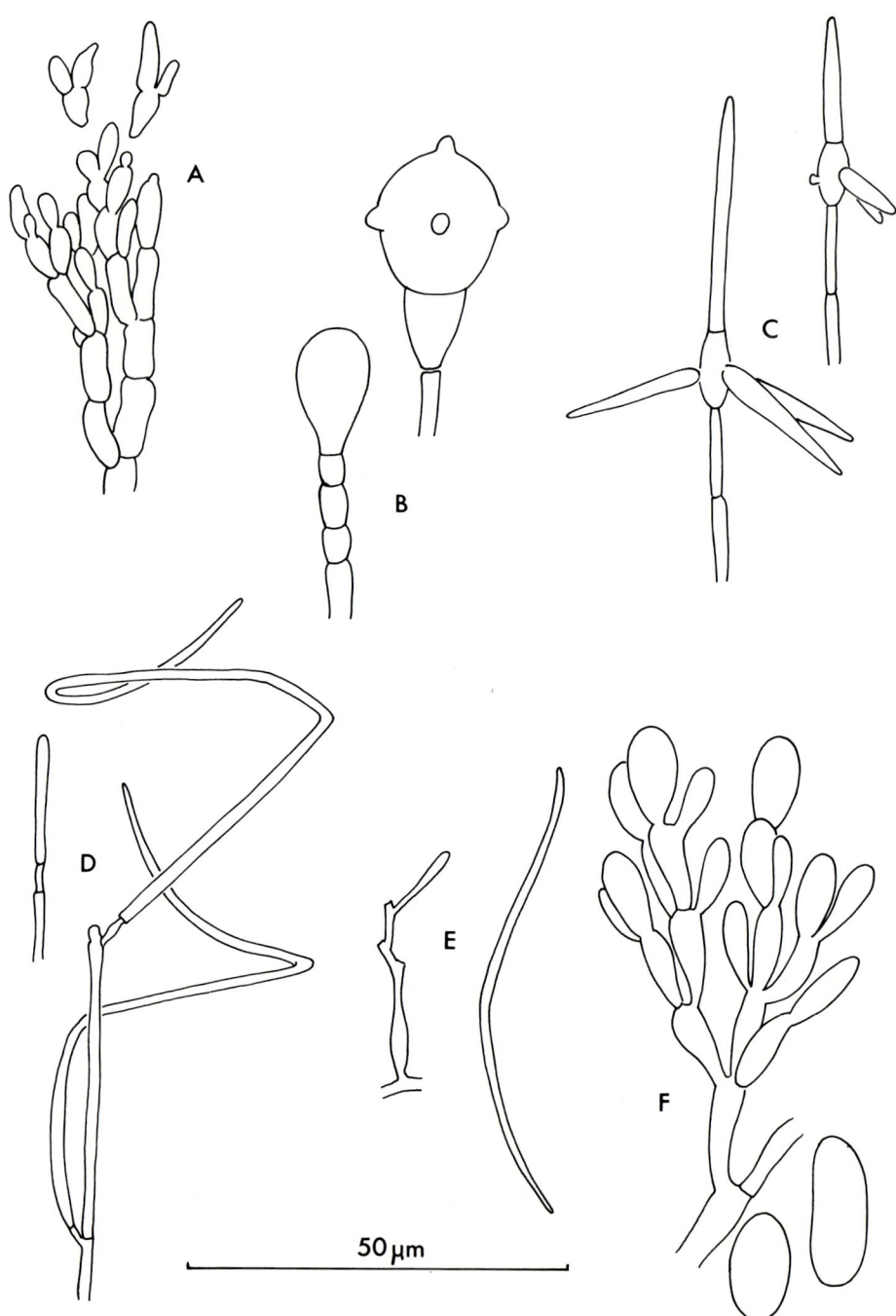

Fig. 3. Caption on page 309.

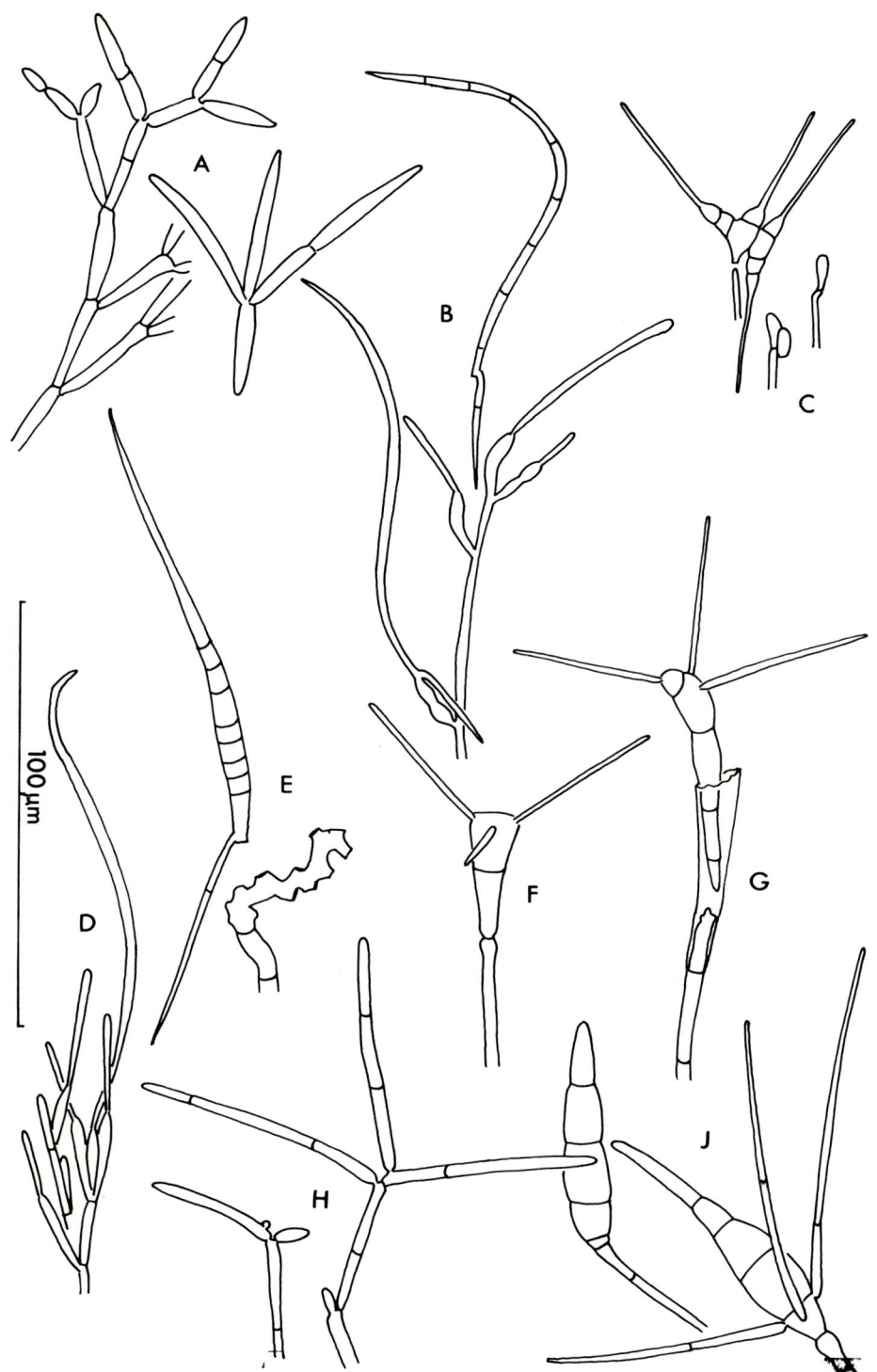

Fig. 4. Caption on page 309.

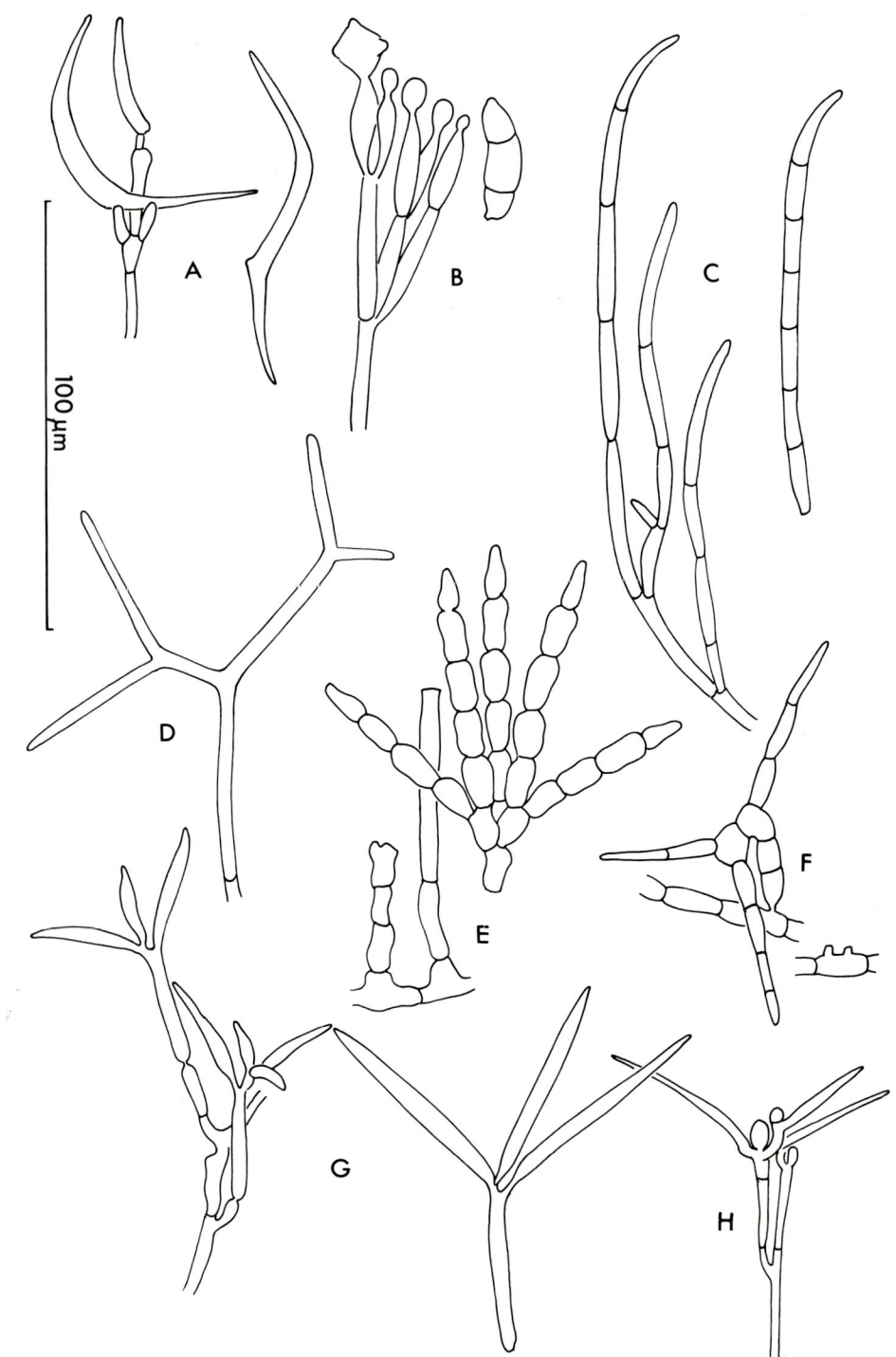

Fig. 5. Caption on page 309.

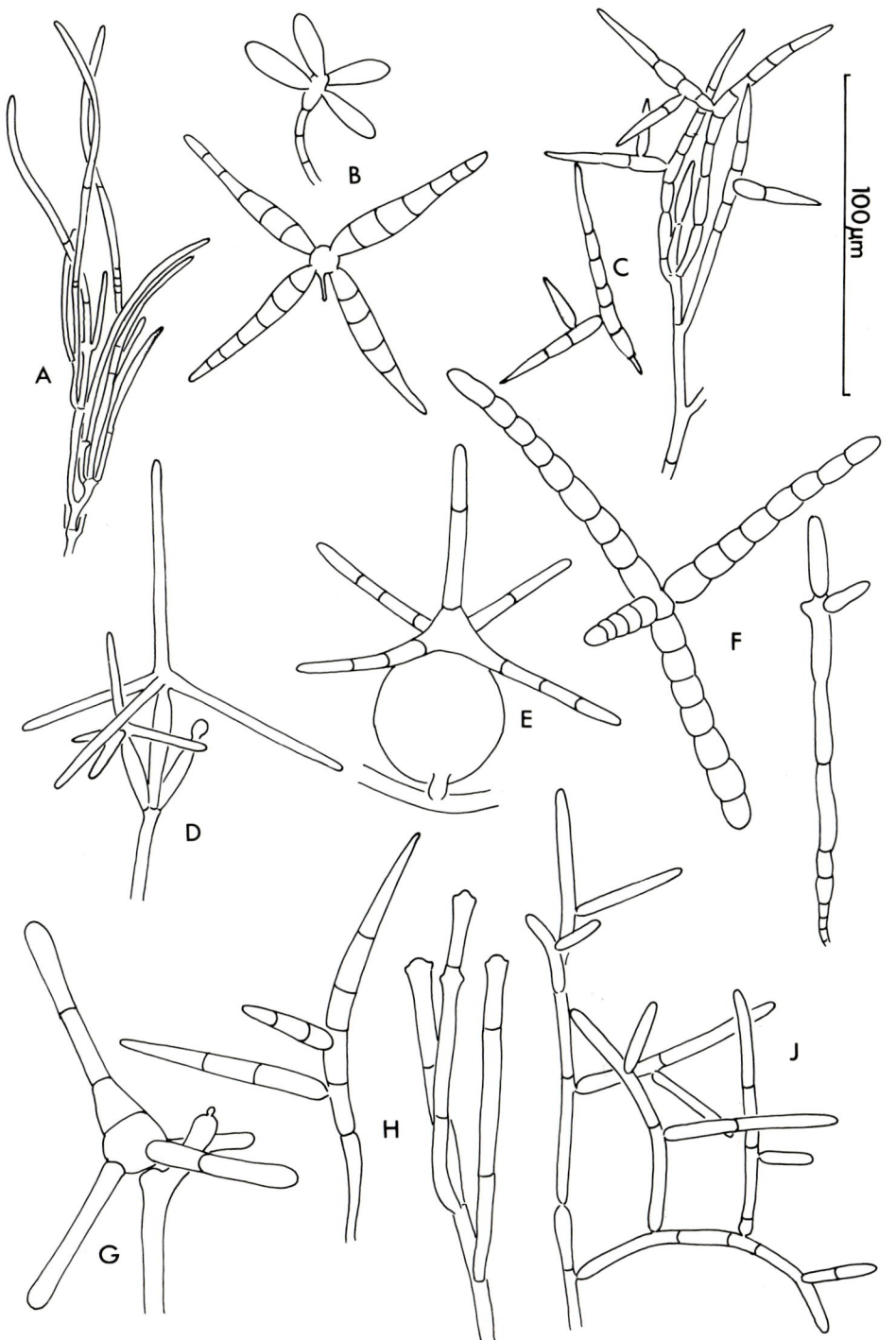

Fig. 6. Caption on page 309.

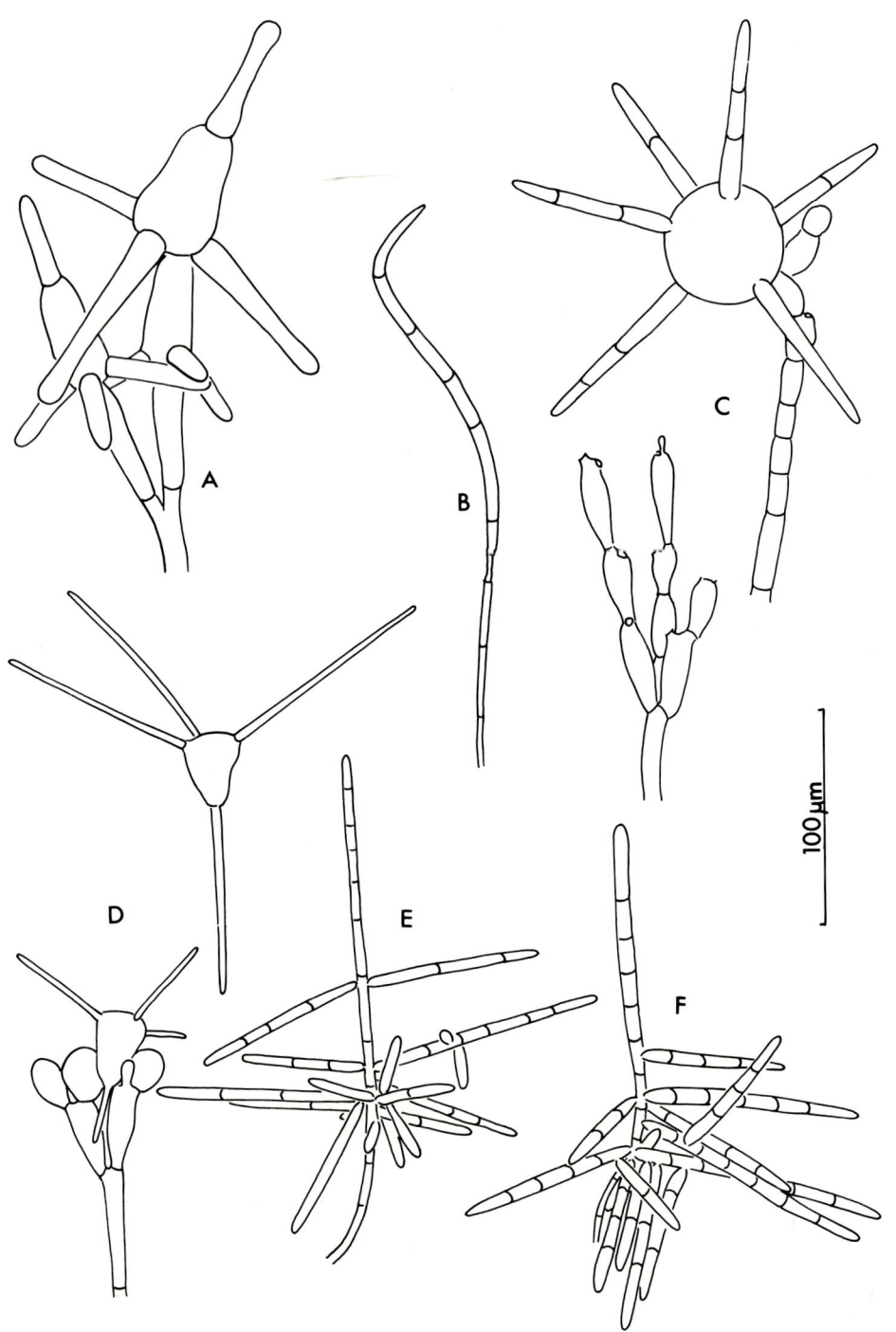

Fig. 7. Caption on page 309.

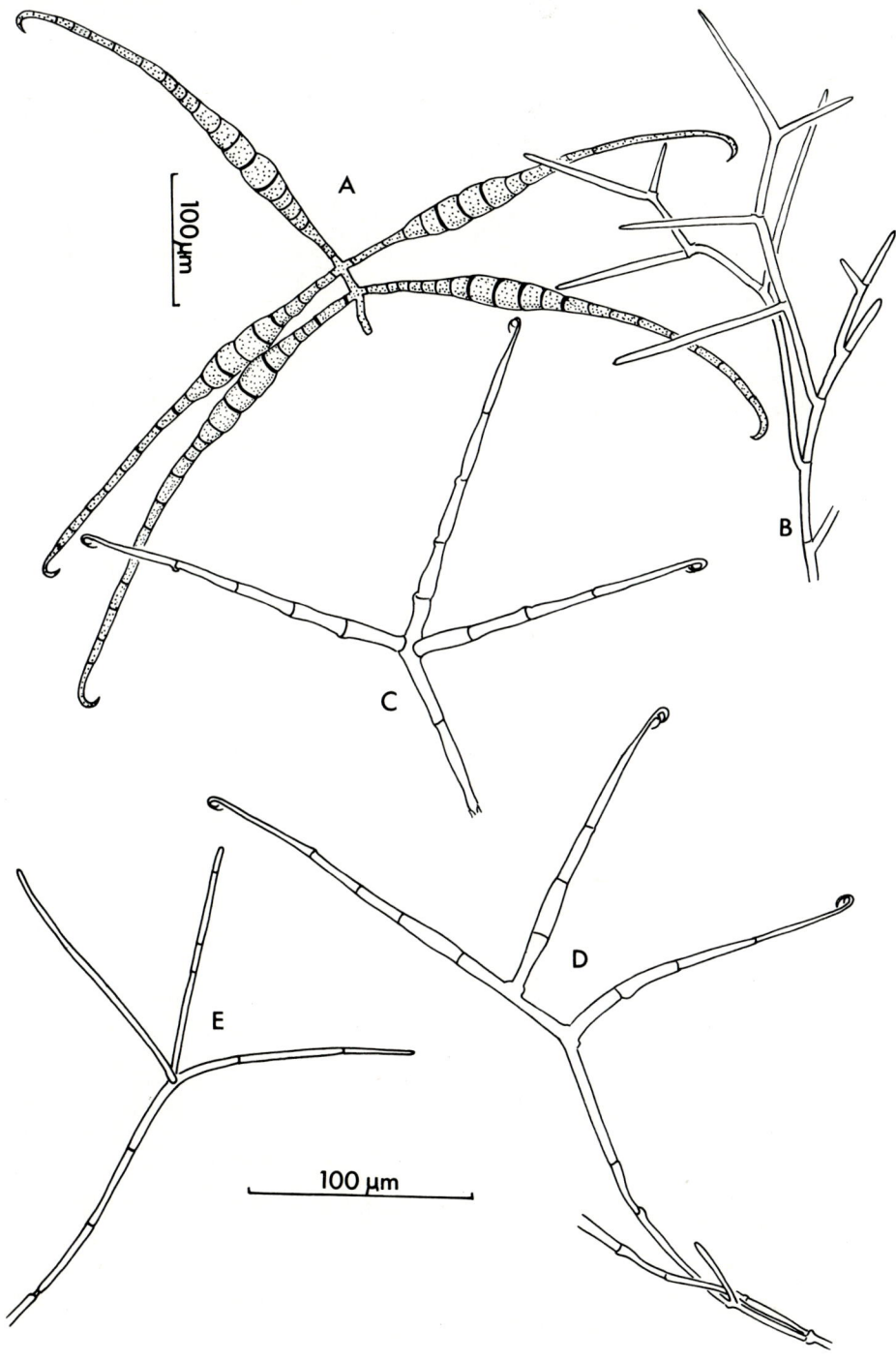

Fig. 8. Caption on page 309.

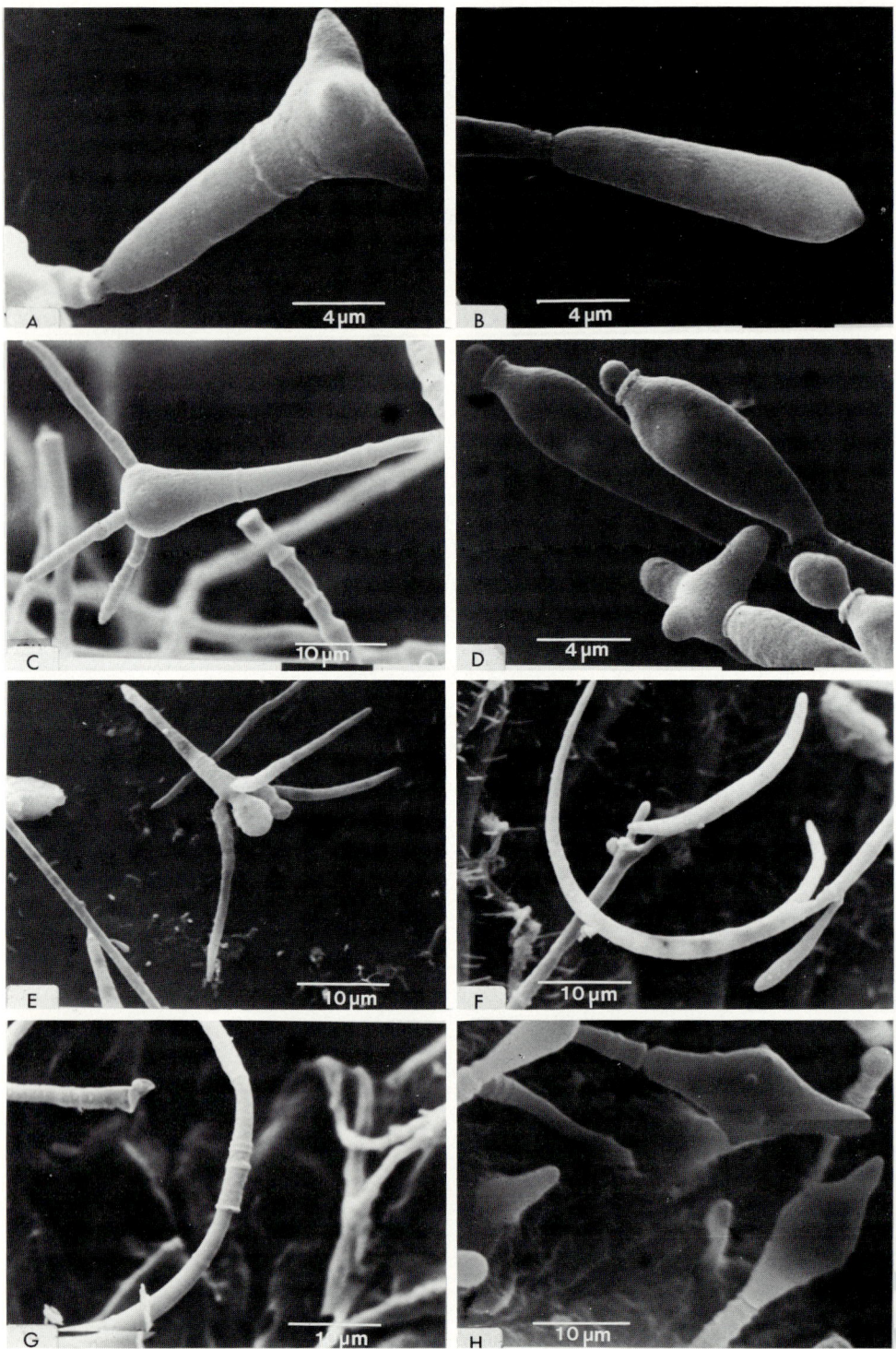

Fig. 9.

tently found on wood and others on grasses and sedges. Aquatic macrophytes and gymnosperms are only now being studied.

3. Habitat

A significant proportion of Ingoldian fungi are known only from conidia in foam (Table I). Their frequent presence in this natural trap and their fruiting response to submersion in pure culture allows us to assume a wet habitat for them. Very few records exist from even slightly polluted waters. Lack of aeration could be an important reason, but other possible causes (toxicity or antibiosis) should not be ignored. A well-defined group appears to inhabit moorland (acidic) streams and lakes. Here we also find *Candida aquatica* (Jones and Sloof, 1965), an anamorphic yeast forming tetraradiate colonies in free water (Webster and Davey, 1975; Fig. 10). It is striking that so few Ingoldian forms have been detected in waters with high osmotic values. The possible terrestrial nature of these fungi is discussed below.

4. Climate and Distribution

Nilsson's (1964) classification of "aquatic Hyphomycetes" has been corrected and updated in the light of recent research (Table I). Most of the information traditionally originated from temperate regions, the tropical mycoflora still remaining largely unexplored. With regard to distribution, too few records exist to warrant any generalizations (Table I). Furthermore, species that appear to be widespread may have been misidentified, as it has become a custom to report them on the basis of identification of conidia in foam, which can be difficult in small or sigmoid forms.

The following discussion may serve as an explanatory note for Table III, in which the most outstanding taxonomic characters are listed. Although Table III may be used as a preliminary key to the identification of Ingoldian fungi, its main purpose is to draw attention to the taxonomic features currently in use, or which may need critical evaluation.

5. Pigmentation

Most Ingoldian fungi have colorless mycelia. However, their colony pigmentation in pure culture is strikingly varied and taxonomically useful if growth

---

Fig. 9. Conidia of Ingoldian fungi. (A) Phialoconidium of *Heliscus (Nectria) lugdunensis*. Clove-shaped conidia like this are produced under water. (B) Phialoconidum of *H. lugdunensis*. Conidia of this shape are produced in air. (C) *Clavariopsis (Massarina* sp.*) aquatica*. Note the percurrent proliferation of the conidiophore lacking a conidium. (D) Phialoconidia of *Lemonniera aquatica*. (E) *Tetracladium marchalianum*. (F) *Lunulospora curvula*. Note the ring of detachment scars near the apex of the conidiophore. (G) *Anguillospora longissima* (*Massarina* sp.). A collapsed separating cell is seen to the left. The conidiophore to its right shows percurrent proliferation. (H) *Pyricularia (Massarina) aquatica*. Note the tubercles around the equator of the spore. The conidiophores show percurrent proliferation. (Photographs by P. F. Sanders.)

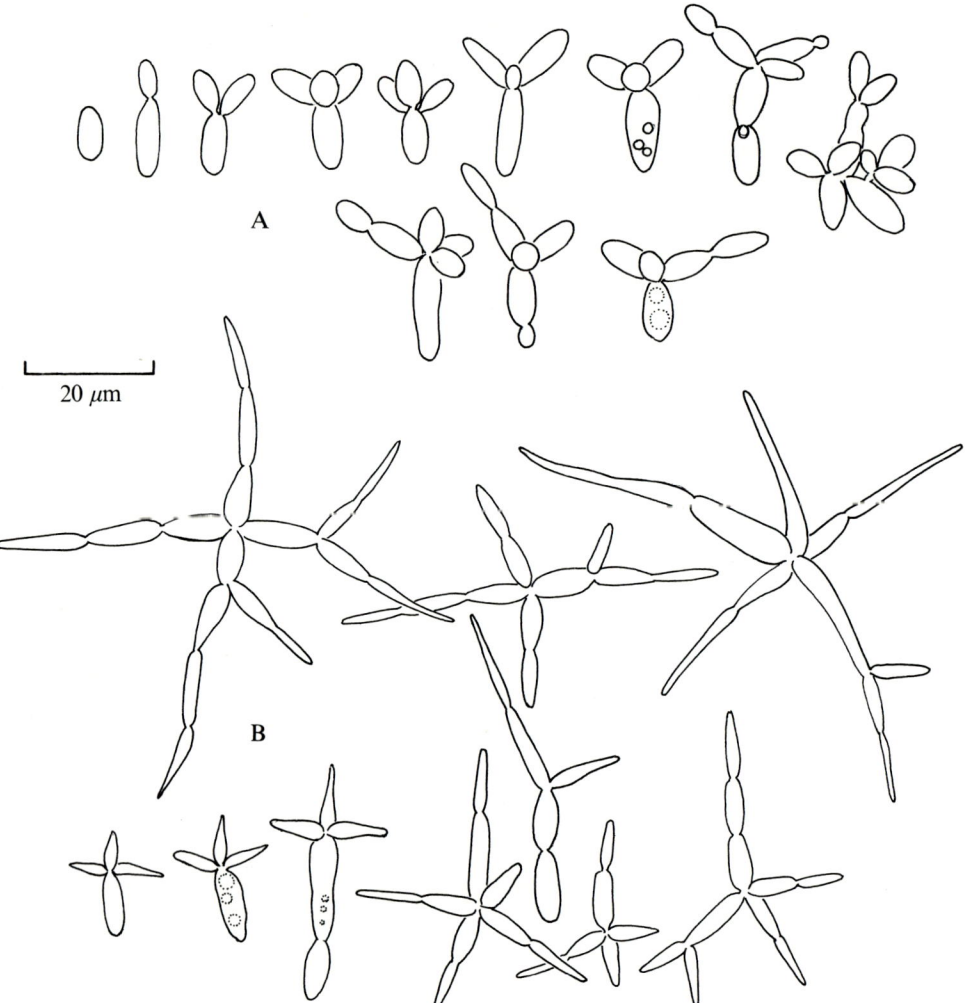

Fig. 10. *Candida aquatica.* Effect of nutrition on colony morphology. (A) Cells grown on mineral salts agar plus 0.5% pectin; (B) cells grown on mineral salts agar plus 0.5% cellulose. (After Webster and Davey, 1975.)

conditions (e.g., medium and temperature) are standardized. Most information in this respect is based on cultures grown on 2% malt agar. Species with dark mycelia (some with slightly pigmented, thin-walled conidia) are numerous and may represent a gradual adaptation to more terrestrial habitats. It is interesting that their sporulation in semisubmerged pure cultures frequently takes place at or above water level. Some typically dematiaceous Hyphomycetes are worth men-

TABLE III

Important Taxonomic Characters (Pure Culture)[a]

Sporodochial conidiomata present: 6, 10, 15, 40, 71, 74, 93
Synnematal conidiomata present: 13 (*Anguillospora pulchella*)
Anamorph subhyaline to brown: 21 (conidiophore), 30, 31, 39, 40, 93, 99, 126, 127, 129, 139, 153
Conidiophores poorly differentiated (micronematous, integrated): 75–77, 79, 124–127, (129), 136, 137, 151, 153, 154
Conidiophores dichotomously branched: 1 (*Actinospora megalospora*)
Conidiophores branched at apex (or penicillate): 19, 20, 27, 36, 46, 51–53, 59–62, 67–71, 72?, 82–87, 89, 91, 95, 101, 136, 138, 147
Conidiogenous cell
  Monoblastic (with single conidiogenous locus)
    Percurrent: 6, 8?, 11, 21, (22), (27), 28, 38, 39, (41), 46, 52, 53, 63, 65?, 66?, 89?, 92, 100?, 114?, 129, 137, 146, 155?
    Sympodial: 16–20, 22, 24, (27), (28), 39, (41), 46, (47), 54, 63, 75, 80, 93–96, 102, 105, (113), 117, 123–126, 131, 133, 138, 147, 153–156, 160
    Retrogressive: 1 (*Actinospora megalospora*)
    Phialidic: 2–5, 23, 31–36, 40, 59–62, 71, 72, 81–87, 90, 91, plus spermatial states
  Polyblastic successive: 17–20, 22, 27, 63, 66–70, 88, 89, 97, 100, 101?, 103, 109?, 118?, 134, 136, 138, 147, 148, (150), (157), 160, 165
Conidium:
  Cells distinctly swollen (moniliform to doliiform): 19, 48, 56, 65, 112, 125, 126, 134, 135, 154, 157
  With clamp connections: 73, 74 (dikaryons only), 128
  Amero- or didymosporous: 41, 51, (90), (91), (92), (113), 114, (134), (135), 158 and spermatial states.
  Phragmosporous: 62, (71), (72), (79), 96, (102), (113), 114, (138), (146), (160)
  Scolecosporous
  Falcate, sigmoid or helical: (2), 6–15, 23, 52, 53, 59–61, (89), (90), 93–95, 97, 103–105, (115), (116), 123, 124, (165)
  Geniculate: 16, 37, 98
  Staurosporous
    Main body swollen to globose: 1, 21, 22, 27, 80, 81, 83, 90–92, 99
    Main body elongated and curved: 24, 32, 48–50, 67–70, (79), (100), 150, 162, 163, 165
    Main body capitate: 54–56, 58
    Main body crozier-like: 24–26, 150, 151
    Branches curved: 4, 18, 19, (48), 67–70, 84, 122, 125, 127, 130, 134, 152, (160), 162, (164), 165
    Branches arranged asymmetrically: 46, 67–70, 80, 100, 112, 131–134, 145, 149, 151, 164, 165
    Branches on concave side of parent arm: 68–70
    Branches radiating equidistantly: 1, 21, 22, 54, 56–58, 80–87, (90), 91, 92, 112
    Branches are short protuberances: 29, 34, 35?, 40, 92, 106–113, 135
    Branches apical, dichotomous: 152 (*Triposporina ceranoica*)
    Branches, apical, coronate: (4), 17–20, 31, 33, 34, 35?, 36, 40, 75–77, 99, (131–133)
    Branches successive, apical or lateral: 17–19, 20 (underwater), 65, 66, 74 (underwater), 80, (141), 145, 153–156
    Branches apical and lateral: 4?, 24–26, 32, 38, 39, 81, 112, 134
    With caudal branch (semiaxial or percurrent): (1?), 8, 13, 15, 23, 27, 38, 39, (93), 94, (97), 139

(continued)

TABLE III (continued)

Branches opposite or more or less verticillate, arranged symmetrically: 2–5, 42–44, (46), 47, (48), 85, 130, 153–156
Secondary branches usually present: 42, 45–48, 68, 69, 100, 101, 149, 157–164
Branches budding out (base usually constricted): 17–22, 38, 39, 42–50, 54–58, 63, 64, 66, 67–70, 75–77, 80, 125, 126, 129, 134–138, 144, 146, 153–156, 160, 162–165
One or more branches at least partly seta-like: 25, 32, 36, 68–70, 131–133
Conidia polymorphic (shape, number or location of branches variable): 17, 20, 90, 91, 125, 126, 131, 133, 152
Conidium released by:
  Separating cell: 10?, 11, (12?), 16, 54?, 74, 88, 89, 106, 111?, 130, 142?, 157
  Violent discharge: 28, 92, 126
  Leaving a membranous sheath: 38 (*Culicidospora aquatica*)
  Secondary conidia present: 68–70, 138?, 160?, *Entomophthora*
Arthrospores present: 33 (*Clavatospora longibrachiata*)
Reported as plant pathogens: 93, 131?, 133, 145
Metasclerotia (= "chlamydospores") present: 3, 6, 20?, 22, 23, 28, 30, 41, 51, 60, 61, 68, 71, 72, 80, 82–87, 102, 124, 130?, 131, 142, 155?, 157, 164
Sclerotia present: 10, 28, 53, 69, 71, 72, 82–87, 90, 91, 102, 125, 157
Accessory state present (probably spermatial):
  Phialidic hyphomycete: 6, 12, 15, 18, 43, 47, 51, 61, 71, 94, 119, 141, 146, 150, 161
  Spermogonial: 11, 22, 27, 28, 113
Conidial cells multinucleate: 1, 22
With dolipore septa: 49, 50?, 142?
Isolated from brackish or sea water: 30, 35, 99, 157

a This table can be used like a synoptic key to help identify cultures of amphibious hyphomycetes. The numbers correspond to those of the species in Table I. ( ), not always; ?, doubtful.

tioning, because they are also capable of underwater sporulation, e.g., *Anavirga, Casaresia,* and *Tetraploa* (Ingold, 1975b; Descals and Sutton, 1976).

Conidiophore aggregations into sporodochia and synnemata are much more typical of terrestrial species and may also be adaptations to drier habitats. Their formation is significantly affected by environmental conditions such as temperature (Wolfe, 1976; Webster and Descals, 1979). The majority of the Ingoldian fungi sporulate effusely or in small patches on natural substrates, although extensive carpets of conidia can often be seen on the bottom of a dish directly under a submerged leaf.

6. The Conidiogenous Cell

The conidiogenous cell is discernible or discrete in phialidic species and in a few others, e.g., species with moniliform cells. It is usually apical but sometimes intercalary (*Varicosporium elodeae, Polycladium equiseti*) and may be recognized by the presence of detachment scars. The type of sporulation or proliferation seems to be somewhat dependent on environmental conditions. Sporulation is mostly monoblastic, but *Articulospora* and several other anamorph-genera have fasciculate conidia (polyblastic successive) appearing sequentially at the same level but from different conidiogenous loci (Sutton, 1975). Phialide-like conidiogenous cells occur in seven anamorph-genera, but successive sporulation has not been demonstrated in all species. Phialides may also proliferate directly into further phialides either sympodially (Marvanová, 1977b) or percurrently (Marvanová, 1972). Successive retrogressive transformation of conidiogenous cells into conidia is known only in *Actinospora megalospora*.

7. Conidium Initiation

Conidium initiation is normally defined as blastic or thallic. Derivatives of these (e.g., entero- or holo-) are terms which are avoided here because they require the detection of basal septa, conidiogenous cells, or conidial initials, and the distinction between inner and outer wall layers, which is not always possible with the light microscope, especially when observing the first-formed conidium (e.g., *Anguillospora, Dendrospora*). The process of initiation frequently takes place as a gradual differentiation of an undefined portion of the conidiogenous cell apex. The formation of a basal septum during this process is probably not of any fundamental significance, as there is evidence that the cross-wall is perforated and does not interfere with protoplasmic translocation (Descals *et al.*, 1976). The basal septum is usually recognized only at the time of conidium secession, although in a few anamorph species (*Varicosporium elodeae, Gyoerjfiella* spp.) the conidium initial is discrete from its inception.

8. Conidial Development

A few amero-, didymo-, or phragmosporous species have been traditionally included among the Ingoldian fungi, some belonging to characteristically terrestrial anamorph-genera. What are probably spermatial states appear occasionally under water, forming phialidic amerospores especially in pure culture. Scolecosporous forms, however, are much more typical of aquatic habitats, with conidia curving in varying degrees, giving rise to falcate, helical, or even geniculate shapes.

Staurosporous (tetraradiate) species are the classic representatives of Ingoldian fungi. Conidial development here has been described in many ways, sometimes in great detail. The shape, arrangement, and sequence of appearance of the conidial components have been given taxonomic status by Ingold (1942) and later workers. These criteria have sometimes been interpreted too rigidly, with the consequent unnecessary proliferation of anamorph-genera (58 anamorph-genera for 153 anamorph-species) and a confusing synonymy. The artificiality of some assemblages is now being confirmed by the discovery of teleomorphs.

9. Conidium Release

Conidium release usually occurs by median splitting of a basal septum. Irregular breakage of a basal cell is probably the case only in some dematiaceous anamorph-genera (e.g., *Casaresia*). Separating cells release conidia after their collapse (a process not yet well understood and probably affected by environmental conditions). Conidia formed by budding from a narrow point are abstricted at the site of initiation. Individual branches can also break off and behave as separate propagules (e.g., *Varicosporium*). In *Culicidospora aquatica* there is a membranous trumpetlike wall left attached to the secession scar (Fig. 4G), strongly reminiscent of that in *Endophragmia* and other anamorph-genera (Descals, 1978). They may be analogous to the collarettes of phialides after release of the first conidium.

Observation of the spent conidiophore after conidium release by natural means can be very helpful in the interpretation of processes of initiation, proliferation, and release.

10. Secondary Conidia

Formation of secondary conidia, germination by repetition, is known in a few species (characteristic of *Gyoerffiella*) and may in some cases be dependent on external conditions or aging.

11. Accessory States (Pleomorphic Anamorphs)

Accessory states are suspected to have a spermatial role and can be either phialidic Hyphomycetes or spermagonia. Phialidic states may sometimes occur directly on the other anamorph, as in *Anguillospora crassa*, *Tricladium giganteum*, and *T. splendens*.

12. Metasclerotia

Metasclerotia, swollen cells often referred to as chlamydospores but not detachable, frequently aggregate into sclerotial bodies and occur in pure culture in a number of species (e.g., *Lemonniera*). They may also be an indication of storage or drought-resistant structures occurring in nature.

C. Possible Significance of the Tetraradiate Shape

Tetraradiate propagules seem especially common in organisms growing in aquatic environments. These organisms have diverse affinities, as the list of organisms above shows (Table II). The main conclusion emerging from a study of this partial list is that tetraradiate propagules have evolved independently many times in water. These ideas have been elaborated by Ingold (1966, 1975b).

He has proposed two possible explanations for the repeated evolution of tetraradiate propagules. "The first is that a spore of this kind is slower to settle than a spherical or oval one of the same mass and is therefore given more time for effective dispersal by water movements before sinking to the bottom.... The second major possibility is that a tetraradiate spore behaves as a little anchor which readily catches on an appropriate substratum. This problem of anchorage may be a very real one in the turbulent conditions of the bed of a stream."

Both these ideas have been tested experimentally. Webster (1959a) followed the sedimentation of spores of a range of aquatic Hyphomycetes of different shapes by timing their rates of fall in a narrow tube at 15°C. There are difficulties inherent in this technique, because the illumination required may induce convectional currents within the tubes. This possible source of error was ignored. However, no correlation was found between spore shape and sedimentation rate. That is, it was *not* found that spores of tetraradiate shape were consistently slower to settle than spores of more conventional shape. The rates of sedimentation ranged from 0.187 mm/s for *Clavatospora stellata* to 0.046 mm/s for *Tricladium angulatum,* with a mean value for 16 species tested of 0.1 mm/s. These values are very low in relation to the flow rates of streams in which such fungi abound, which are commonly on the order of 1000 mm/s and may exceed this. It therefore seems unlikely that in a turbulent stream differences in sedimentation rate, even if they could be correlated with spore shape, are an adequate explanation of the tetraradiate form.

There is better experimental support for the idea that tetraradiate spores are more effectively trapped on underwater objects than spores of more conventional shape. Using a "water tunnel" in which spores at known concentration were circulated at different velocities over cylindrical glass rods coated with collodion, Webster (1959a) showed that several tetraradiate spores had a higher trapping efficiency than spores of other shapes. When a mixture of tetraradiate and sigmoid spores was circulated, the tetraradiate spores were trapped in higher numbers. The reason advanced is that, when tetraradiate spores are impacted on an underwater surface they make contact with the surface with the tips of three arms. The same effect is seen when tetraradiate spores are allowed to settle on a glass surface such as the bottom of a petri dish (Fig. 11). The three arms in contact with the surface quickly form adhesive pads or appressoria, from which germ tubes later emerge. The fourth arm projecting into the water usually fails to form a germ tube. There is experimental evidence that the presence of appressoria improves the ability of a spore to remain attached to a substrate when exposed to wash-off by water flow.

Bandoni (1974, 1975) has advanced an alternative explanation for the significance of tetraradiate propagules. He believes that these spores may be adapted to dispersal in aqueous films trapped between layers of terrestrial leaf litter and has shown (Bandoni, 1972, 1979) that spores of this shape are common in such

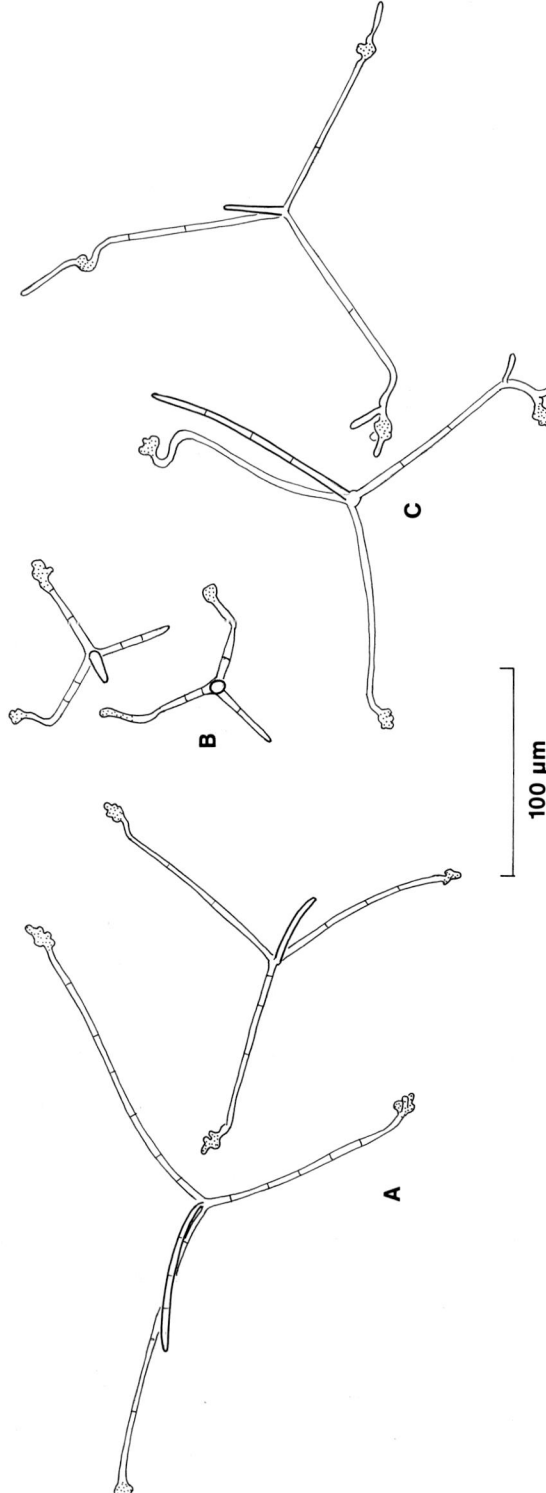

Fig. 11. Germinating conidia of Ingoldian fungi after 9 h in contact with a glass slide. Two or three of the arms in contact with the surface have formed appressoria (stippled). The fourth arm, projecting away from the surface, produces neither appressoria nor germ tubes, but germ tubes may develop from the tips of the arms in contact with the surface. (A) *Tetrachaetum elegans*; (B) *Lemonniera terrestris*; (C) *Lemonniera aquatica*.

situations in certain parts of the world. Numerous examples of tetraradiate or staurosporous conidia occur in other terrestrial habitats.

The amphibious zygomycete *Entomophthora conica* provides a beautiful example suggesting that the tetraradiate form is an adaptation to aquatic conditions. The fungus, a parasite of Diptera (flies) and Trichoptera (caddis flies), can be found on dead insects on wet wood and stones close to streams (Gustafsson, 1965). When dead infected insects are incubated in air in moist chambers, branched conidiophores emerge between the segments of the exoskeleton and develop crescent- or horn-shaped primary conidia (Fig. 12). These conidia are capable of germinating in various ways. For ease of reference the cornute primary conidia are given the symbol 1. There are two types of secondary conidia, balloon-shaped violently projected conidia (2), and tetraradiate conidia of two types (3 and 4). Type 3 is attached centrally, while type 4 has a terminal attachment point. Germination can thus be described by a series of symbols, e.g., 1-1, 1-2, 2-3. A comparison of the different modes of germination following incubation of a sample of primary conidia in air or water for 4 h (Table IV) shows that in air 57% of primary conidia formed balloon-shaped conidia (type 2), whereas in water only 10% of type-2 conidia and 58% of type-3 and -4 conidia were seen.

It is even more interesting to compare the kind of conidia which develop when infected insects are incubated in air or submerged in water. In air, all the primary conidia are of the cornute type, but in water all of them are tetraradiate and of type 4 (Descals and Webster, 1981). We speculate that the tetraradiate type of conidium is adapted to underwater infection of an aquatic larval state, but whether infection is external or internal is as yet unknown. The balloon-shaped secondary conidia which are violently projected we believe to be adapted to infecting the terrestrial adults. It is known that other amphibious *Entomophthora* species have tetraradiate propagules, e.g., *E. rhizospora* Thaxter (1888), which attacks caddis flies (Descals and Webster, 1981).

D. Possible Significance of the Sigmoid Spore Shape

Similar evidence can be presented to suggest that the sigmoid type of spore has evolved many times in aquatic organisms. Table II gives examples of sigmoid conidial forms of diverse affinity. There are also a number of submerged Ascomycetes with sigmoid ascospores. Some aquatic plants produce needle-shaped or sigmoid pollen. As yet, however, no clear ideas have emerged which explain the significance of sigmoid spores in water. A number of sigmoid spores seem to be capable of attaching themselves by their tips to solid substrates and, when such spores are allowed to settle in a petri dish, they appear to be capable of "standing on end." Germ tubes, however, are not confined to the ends of the spores.

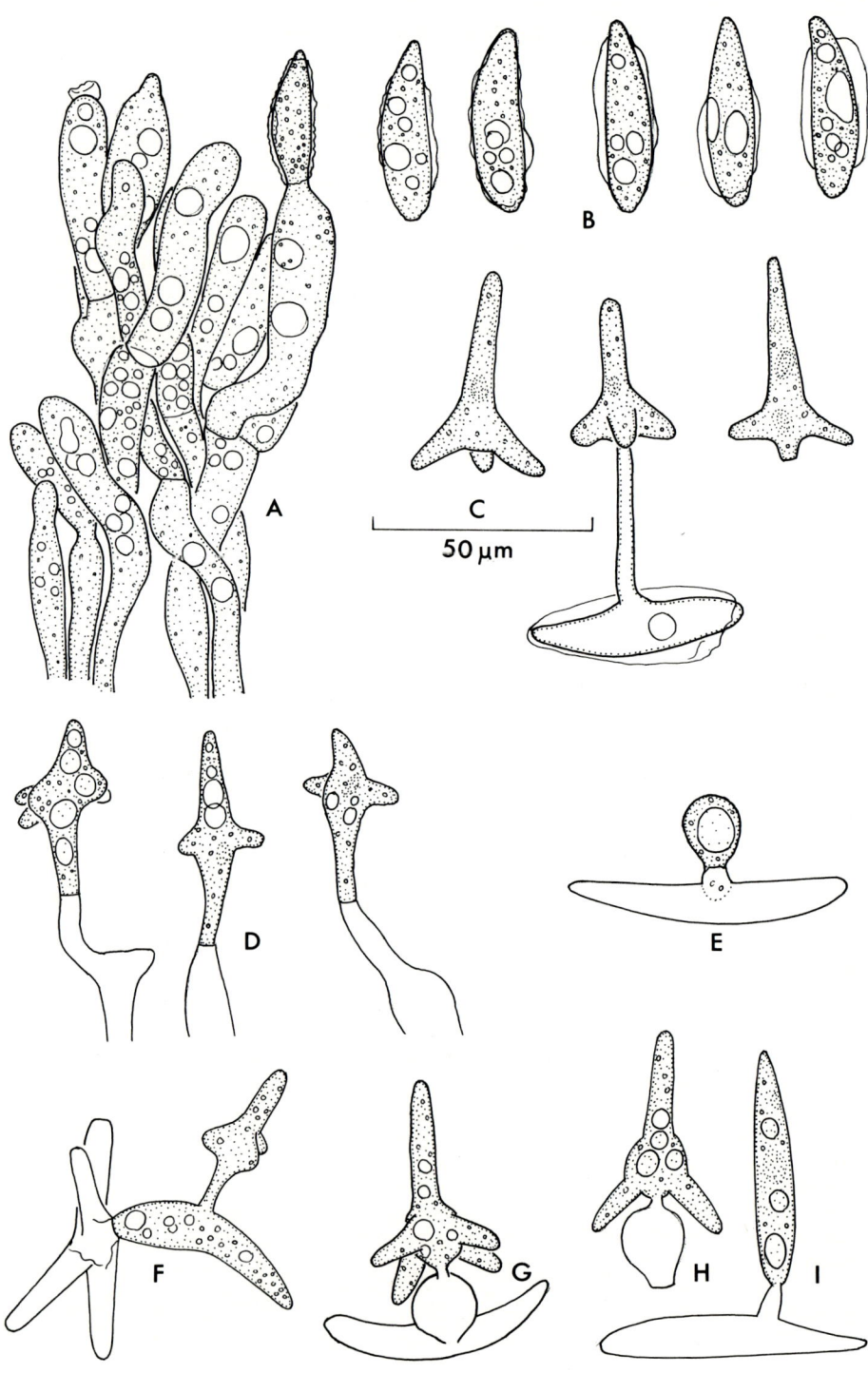

Fig. 12.

TABLE IV

Comparison of Mode of Development of *Entomophthora conica* Conidia as seen 1 h after Collection or after a Further 4 h Incubation in Air or Water[a]

Conditions	Development type[b]						Empty
	1-1	1-2	1-3	2-3	3 and 4	Ungerminated	1
1 h after collection	0	7	0	0	0	48	0
4 h later in air	0	32	2	0	0	22	1
4 h later under water	0	10	7	25	26	4	5

[a] From Webster *et al.*, 1978.
[b] For explanation of types, see text.

E. Ecology

1. Terrestrial or Aquatic?

Although Ingold's studies have concentrated attention on the freshwater habitat, other workers have shown that many of the spore forms encountered in streams can also be found in terrestrial habitats (Bandoni, 1972, 1979; Park, 1974b; Sanders and Webster, 1978). There is no evidence to suggest that such fungi are present on attached leaves, but Bandoni (1979) has shown that water collected from "stem flow" of certain trees in British Columbia may contain conidia of *Gyoerffiella* and *Volucrispora*. Gönczöl (1976b) in Hungary had earlier reported that leaves submerged in water in hollows of beech trees may bear *Articulospora tetracladia*, *Alatospora acuminata*, and *Tricladium* sp. Bandoni (1979) showed that repeated washing of "nonaquatic litter samples" in 30 changes of sterile water followed by plating on a weak medium containing yeast extract, sorbose, and the antibacterial antibiotic tetracycline permitted the isolation of 11 species of "aquatic" Hyphomycetes. It is likely that findings of this

Fig. 12. *Entomophthora conica*. (A) Conidiophores from an infected insect incubated in air. Note the cornute primary conidum to the right. (B) Detached primary conidia (type 1). The conidia are "double-walled" and the thin outer wall persists as an irregular envelope. (C) Germination of a primary conidium under water to form a tetraradiate secondary conidium (type 3). These conidia are centrally attached to their conidiophore, resembling the conidia of *Lemonniera*. Germination of this type is symbolized as 1-3. (D) Conidiophores from an infected insect incubated in flowing water. The conidiophores have formed tetraradiate conidia (type 4) which are attached terminally to their conidiophores, resembling *Heliscus*. (E) Primary conidium (type 1) germinating in air to produce a balloon-shaped, violently discharged secondary conidium (type 2). Germination of this type is symbolized as 1-2. (F) 4-1-3 germination (in flowing water). (G) 1-2-3 germination (in flowing water). (H) 2-3 germination (in flowing water). (I) 1-1 germination (in air).

kind are related to the prevailing climate. Bandoni (1979) wrote, "The climate of the area in which the isolations were made could, in part, be responsible for the presence of aquatic Hyphomycetes in upland litter samples. Maximum angiosperm leaf decay occurs during the months of November through May. During these months, rainstorms are frequent and the total average rainfall for the period is about 88 cm. The temperatures are cool, but commonly remain above freezing. The precipitation, and an abundance of overcast days, contribute to maintenance of saturation of litter for prolonged periods. However, widespread occurrences of aquatic Hyphomycetes . . . in decaying plant material on land suggest that saturation of the litter layer is not necessary for their development."

The presence of characteristic conidia in a litter sample is not in itself proof of the presence of an active mycelium. Frequent isolation of mycelia from washed litter fragments, however, provides stronger, though not conclusive, evidence of activity. It is of interest to examine whether leaves known to contain active mycelium can continue to support active growth in terrestrial habitats and whether such mycelia can extend to adjacent litter. These aspects have been studied by Sanders and Webster (1978) at sites in Devon, England, with a mean annual rainfall of 760 mm, distributed fairly evenly throughout the year. Sterilized leaf disks of *Quercus petraea* (oak) were colonized in the laboratory with a range of species of aquatic fungi, washed, and then placed in nylon mesh bags which were buried in the litter layer at two woodland sites. Some of the disks were sandwiched between two sterilized oak leaves. At intervals of a few weeks disks were removed and tested for the presence of aquatic Hyphomycetes by placing them in shallow dishes of water to induce sporulation. The sterilized leaves enclosing colonized disks were tested for the presence of aquatic Hyphomycetes. Leaves infected naturally were also collected from streams, and their species contents determined by incubation in water. Such leaves were also buried in woodland litter. The periods of survival detected using these techniques varied from 12 weeks in the case of *Pyricularia aquatica* to over 52 weeks for *Clavariopsis aquatica* and *Articulospora tetracladia*. Only 2 species of 10 tested were shown to be capable of sporulation on previously noninfected leaves: *A. tetracladia* and *Varicosporium elodeae*. Under more severe conditions of desiccation, where colonized leaves or disks were placed in litter bags 1 m above the ground, even shorter periods of survival were detected, less than 17 weeks. Since most of the aquatic Hyphomycetes examined, with the exception of *Lemonniera* spp., seem not to produce sclerotia, it is thought that survival is by means of mycelium embedded in leaf tissues. These studies have a bearing on possible dispersal methods. The ubiquitous occurrence and worldwide distribution of Ingoldian fungi in suitable habitats implies an effective method of transmission between unconnected water courses. While aerial dispersal of ascospores and basidiospores is clearly a possibility for species with teleomorphs, dispersal in wind-blown leaves is another.

We can conclude that these fungi are not exclusively aquatic. Whether they are present and active in terrestrial habitats is in part dependent on the climate. In our view the aquatic environment is the *preferred* habitat, the one in which they can be regularly and consistently found. Their occurrence in drier habitats is more sporadic and dependent on chance.

2. Seasonal Periodicity

Although Ingoldian fungi can be collected throughout the year in continuously flowing streams, their relative abundance is related to availability of litter in the water. Thus in temperate climates there is an enormous increase in spore concentration coinciding with the peak of deciduous tree litter deposition. In a study of spore concentration throughout the year in the River Creedy, a lowland river in Devon, England, Iqbal and Webster (1973b) filtered 250- to 500-ml samples through 8-μm Millipore filters. The bulk of leaf litter was deposited during the months of September and October. In October and November the aquatic hyphomycete spore concentration rose to 7000–8300/liter. In later studies (J. Webster, unpublished) values of 20,000–30,000 spores/liter were detected in the larger River Teign in November. During the following summer spore numbers in the River Creedy fell to very low levels, sometimes below the concentrations detectable by this technique. Despite these low levels there was obviously sufficient inoculum available to allow a build-up when suitable substrates were available in sufficient quantity. Despite the scarcity of leaf material in the streams in the summer, there is an addition of substrate in the spring in the form of bud scales, catkins, and so on. There are also more enduring substrates such as woody branches and roots lying in or near the river.

By the same technique it is possible to follow the concentration of spores of individual species. In some cases (e.g., *Clavariopsis aquatica* and *Flagellospora curvula*) the spore concentrations mirror the total spore concentration, suggesting a close relationship to substrate availability. In some species distinctive patterns of abundance were noted, indicating the importance of other factors controlling spore concentration. Two particularly interesting species were *Lunulospora curvula* and *Tricladium chaetocladium* (erroneously identified as *T. gracile* in the original paper; see Table I). *Lunulospora curvula* was only detected by filtration in the River Creedy from August to November. *Tricladium chaetocladium* showed an entirely different peak of abundance and was only detected from December to April. An obvious explanation might be that *L. curvula* has a higher optimum temperature for growth and sporulation than *T. chaetocladium* (Webster *et al.*, 1976). Stream temperatures during the period of study varied from 18°C in August to 3°C in January. Although some evidence has been obtained that *L. curvula* has a higher optimum temperature for growth (20°C) than *T. chaetocladium* (15°C), these values are higher than is usual for most British streams. It was found that *L. curvula* had an optimum sporulation temperature of 25°C, while that for *T. chaetocladium* was about 15°–20°C. When these fungi were grown in competition with each other in mixed culture, the temperature optima for sporulation by both fungi were lowered. Under these conditions *Tricladium chaetocladium* had an optimum temperature near 5°C, while *L. curvula* had an optimum near 10°C. These findings are in accordance with the known distribution of *L. curvula* which, according to Ingold (1975a), "has a world-wide distribution, being particularly abundant in tropical countries." There may of course be other explanations of the scarcity of this fungus after November. Iqbal (1972a) suggested that *L. curvula* was most common in the River Creedy on decaying leaves of *Alnus glutinosa* and that this relatively soft leaf was quickly decomposed or eaten, as compared with tougher leaves of some other trees such as *Quercus*.

In comparison with the tree-bordered River Creedy, streams which drain Dartmoor in Devon, England, are treeless in their upper reaches. The vegetation through which they flow is dominated by moorland plants such as *Vaccinium*, *Erica* and *Calluna* (Ericaceae), *Nardus*, *Molinia* and *Agrostis* (Gramineae), *Eriophorum* and *Carex* (Cyperaceae), and *Juncus* (Juncaceae). Despite the absence of tree litter such streams carry a flora of Ingoldian fungi, albeit a distinctive one, including fungi such as *Tricladium giganteum* (Willoughby and Minshall, 1975) and *Dendrospora juncicola*. The concentrations of spores are consistently lower than in tree-lined streams (Iqbal and Webster, 1977). Where a moorland stream flows at lower levels between wooded banks, there is an immediate increase in the concentration of these fungi.

3. Factors Influencing Sporulation

Compared with the enormous range of literature available on the physiology of nutrition and sporulation in many groups of terrestrial and aquatic fungi, the Ingoldian fungi have been less well studied. Thornton (1963, 1965) investigated eight species and showed that most were capable of utilizing a wide range of simple and polymer carbohydrates and amino acids which occur naturally in abscised leaves. Inorganic sources of nitrogen such as nitrate and ammonium were readily utilized except by *Volucrispora aurantiaca*. Three of the species tested appeared to be prototrophic for vitamins, while others had shorter lag periods for growth in the presence of pantothenic acid, biotin, or inositol.

Most common Ingoldian fungi form conidia under submerged conditions, but a few can form conidia freely when woody substrates are incubated in moist chambers. Fungi which fruit readily under such conditions include *Heliscus lugdunensis*, *Tricladium splendens*, and *Anguillospora crassa*. In agar culture most species do not form conidia on the dry agar surface, but there are exceptions. Immersion of culture strips in water will in many cases result in sporulation within 24–48 h. Because many Ingoldian fungi live in turbulent water, Webster (1959a), in an attempt to simulate the well-aerated turbulent conditions of the natural environment, subjected immersed culture pieces to forced aeration by compressed air and obtained very large numbers of spores. The effects of forced aeration on sporulation have been studied quantitatively by Webster and Towfik (1972) and Webster (1975). For a number of Ingoldian fungi there is a direct relationship between aeration rate and intensity of sporulation. Under the conditions used, in which culture disks were forcibly aerated in water in glass bottles, spores were swept out of suspension by the bubble stream and deposited as a "scum" above the line where the bubbles broke the surface. Such spores remain ungerminated so long as aeration is continued, but interruption of aeration is quickly followed by germination. It was thought unlikely that the effects of varied aeration rates on sporulation were mediated by differences in dissolved oxygen concentration, because only slight differences were detected in cultures aerated between 100 and 1000 ml air/min even though these differences in aeration rate resulted in large differences in sporulation level. Substitution of air–nitrogen mixtures for air, which resulted in lower dissolved oxygen levels, did not reduce the level of sporulation. Two lines of evidence suggested that the enhanced level of sporulation was the result of turbulence. First, a comparison of sporulation levels in cultures aerated through a single orifice (producing vigorous turbulence) with those in cultures aerated through multiple orifices (causing lower turbulence) showed significantly higher sporulation in the more turbulent cultures. Second, the numbers of spores produced in stirred aerated cultures were significantly higher than in unstirred cultures, but the differences between stirred and unstirred cultures were less when the aeration rate was increased.

If one culture disk produces more spores than another in a given time, there are two likely explanations. Either spore development is more rapid or more spores are produced per unit area of culture. There is evidence that both effects operate. In cultures aerated at 1000 ml/min, spores were detected 2–6 h earlier than in cultures aerated at 100 ml/min. Estimates of the numbers of conidiophores per square millimeter showed significantly greater density of conidiophores at the higher aeration rates. This is thought to be due to a morphogenetic effect of turbulence on the branching of conidiophores.

Sanders and Webster (1980) studied the effects of water flow rate on conidium production in a number of aquatic Hyphomycetes. Small culture strips on 1% malt extract agar 10 × 1.5 mm were placed in specially designed flow cells of the same dimensions as a microscope slide. Sterile water was allowed to flow over the culture surface at linear flow rates up to about 3 cm/s at volume flow rates varying from zero up to 200 cm^3/h. These linear flow rates were low compared with estimates of flow rates in the beds of streams. Following dye movement, stream bed flow was found to vary from 3 to 50 cm/s. Spore output was estimated by collecting the outflow from the flow cells on fiberglass filters. The 20 species studied responded in two different ways to increasing flow. A group of 8 species including *Anguillospora longissima, Heliscus lugdunensis,* and *Lemonniera ter-*

restris produced more conidia when flow was increased from 0 to 5 cm³/m, but no significant increase when flow rates were further increased. A second group of 9 species including *Anguillospora crassa, Clavariopsis aquatica,* and *Lemonniera aquatica* continued to increase spore production when the flow rate was increased from 5 to 200 cm³/h. The increased spore production was, as in the aeration experiments, found to be associated with earlier spore formation and with a greater density of conidiophores. When a small culture strip of 1% malt agar was placed "upstream" of the observation chamber, some depression in sporulation level was found. This suggests that a possible effect of water flow is to remove nutrients which might induce vegetative rather than reproductive growth.

4. Leaf Colonization and Animal Nutrition

During recent years it has become appreciated that aquatic fungi play a vital intermediary role in energy flow in stream ecosystems (Kaushik and Hynes, 1971; Bärlocher and Kendrick, 1973a,b, 1974, 1975, 1976, 1979; Willoughby, 1974; Berrie, 1976). Most streams receive the bulk of their carbon input not from attached macrophytes or algae within the streams, but from adjacent terrestrial plants which shed litter into them. It has been estimated that 50–90% of the energy available to streams may be derived from such allochthonous sources. In temperate climates the litter consists largely of autumn-shed leaves. In the period of leaf fall and in the months immediately afterward there is a build-up in the population of aquatic invertebrates such as *Gammarus* which feed on the decaying leaves (Berrie, 1976). The water-soluble materials present in the leaves are fairly quickly leached out, and the most important insoluble residues, cellulose and lignin, are largely unavailable to the animal populations because they lack the appropriate enzyme systems in their gut to digest them. When leaves are immersed in stream water, the protein content declines for several days, but after a week or two, depending on temperature, it rises steeply. In laboratory experiments in which leaf disks were immersed in stream water enriched with added nitrate and phosphate, the addition of antifungal or antibacterial antibiotics depressed the increase in protein content as compared with that in controls lacking antibiotic. It was therefore inferred that the protein increase was of microbial origin. It has been concluded that the fungi make a greater contribution to the fungal biomass than bacteria. Streams contain low concentrations of inorganic nitrogen in forms such as nitrate. The microbial population, growing at the expense of the carbon present in the leaf tissue, can absorb inorganic nitrogen from the stream water, which is continually replenished as water flows over the leaf, and transforms it into microbial protein. Although the normal terrestrial leaf-inhabiting microflora, such as species of *Alternaria, Cladosporium, Aureobasidium,* and *Epicoccum,* remain active in the leaves in water, at lower temperatures down to 0°C the Ingoldian Hyphomycetes become relatively more important as colonizers of leaf tissue. The activities of the microbial populations which decompose leaves in water are important in several ways. Not only do they bring about an increase in protein content, but they cause a softening and comminution of the leaf tissue because they possess the necessary enzymes to bring about tissue disintegration (Suberkropp and Klug, 1979). They thus render leaf tissue more palatable and nutritious to aquatic invertebrates than uncolonized leaf material. In feeding experiments in which the amphipod *Gammarus pseudolimnaeus* was fed on colonies of Ingoldian Hyphomycetes such as *Anguillospora, Clavariopsis,* or *Tricladium* (grown on malt extract), the animals put on more than twice the dry weight they did when fed on elm or sugar maple leaves. On fungal diets the daily consumption of dry matter per animal was about $1/10$ of the consumption on leaf diets (Bärlocher and Kendrick, 1973a). It is tempting to extrapolate from these findings to other animals, but Marcus and Willoughby (1978) found that the isopod *Asellus aquaticus* grew better on oak leaves collected from a lake than on a pure culture of *Lemonniera aquatica.* In studies on *Gammarus pulex* Willoughby and Sutcliffe (1976) showed that the mean interval between molts (an index of growth rate) was shorter on diets of decaying elm and oak leaves than on diets of the aquatic Hyphomycetes *Tricladium giganteum* and *Clavariopsis aquatica.*

One other aspect of leaf colonization by Ingoldian Hyphomycetes in relation to palatability should be mentioned. If aquatic invertebrates are presented with a choice of species of tree leaf material in the form of disks, certain leaves are preferred, but the development of microflora on the leaves can affect the preference. Bärlocher and Kendrick (1973b) showed that, when *Gammarus pseudolimnaeus* was presented with leaf disks bearing a negligible microflora as sole food supply, the preference was ash > maple > oak. Suitable choice of a fungal inoculum could reverse this preference. In nature it is therefore possible that kinds of fungi which colonize leaves in streams are just as important as the nature of the leaf itself in determining food preferences, thus controlling the turnover of leaf detritus.

III. AEROAQUATIC CONIDIAL FUNGI

Park (1972), in a thought-provoking discussion on the ecology of heterotrophic microorganisms in fresh water, used the term ''indwelling'' for organisms with a high degree of ecological adaptation to water. ''This relates to the extent to which a heterotrophic micro-organism is able to maintain itself in an aquatic environment. The assumption is made that indwelling micro-organisms are fully adapted, *i.e.* that they are able to maintain their biomass at their appropriate site and in their appropriate season at a more or less constant level from year to year utilizing substrata and nutrients that become available there.''

Fisher (1977a) has defined the term "aeroaquatic" in terms of Park's ideas as follows: "Aeroaquatic fungi are indwelling organisms characterized by the production of purely vegetative mycelium in substrata under water and by the formation of conidia with a special flotation device, formed only when the substrate on which the fungus is growing is exposed to a moist atmosphere." A large number of conidial fungi, some with ascomycete and some with basidiomycete teleomorphs, and others with no known teleomorphs, conform to this definition. The habitats within which they can be found with most certainty are stagnant pools, sluggish ditches, or muddy creeks on leaves and twigs lying on the mud surface. This is not, however, their exclusive habitat. By suitable techniques they can be shown to be present in soil, in peat, on rotting wood in streams, and so on (Gönczöl, 1976b; Fisher, 1978; Abdullah and Webster, 1981a). The taxonomy of the group is based mainly on the publications of Linder (1929, 1931), Glen-Bott (1951, 1955), van Beverwijk (1951a,b, 1953, 1954), Moore (1955), Hennebert (1968), and Tubaki (1975a,b).

A. Techniques for Study

The techniques for studying aeroaquatic fungi are simple (Fisher, 1977b). Leaves and twigs dredged from suitable habitats are collected in plastic bags and, after rinsing in clean water, incubated in moist chambers for several days or weeks preferably at a low temperature (10°–15°C). Within a few days the characteristic glistening buoyant conidia develop on the moist surface of the substrate, projecting into the air. Cultures can readily be prepared by picking off conidia and transferring them to dilute agar media, e.g., 0.1% malt extract agar, with added antibiotic to minimize bacterial contamination. Most aeroaquatic fungi fruit well on weak agar media. Small agar plugs can be used to inoculate autoclaved leaf disks (e.g., *Fagus*) under liquid. Daily shaking by hand aids colonization, presumably by mycelial fragments. After a few weeks' growth under these conditions, the leaf disks can be incubated in moist chambers (a petri dish lined with moist filter paper) and sporulation will take place within a few days. Fisher used small (3.6 mm diameter) *Fagus sylvatica* (beech) leaf disks in quantitative experiments. Various numbers of such small disks were submerged in the mud layer of ponds and left for several weeks. The disks were recovered, rinsed, and incubated for 12 days at $12 \pm 2°C$ on moist filter paper. Sporulation of aeroaquatic fungi was used as an index of colonization. When leaf disks collected from ponds were subjected to a 7-day period of incubation in aerated distilled water before being placed on moist filter paper, the proportion of disks bearing spores was greater than in untreated disks. Disks from one pond which were treated by incubating in aerated distilled water yielded eight species of aeroaquatic fungi, whereas only four species had been identified in the previous 12 months on whole leaves taken directly from the habitat. This technique therefore provides an improved method for assaying the species present in a given habitat.

B. Common Anamorph-Genera

Many, but not all, aeroaquatic fungi have helically coiled conidia. Flat, watch-spring-like spirals are found in *Helicosporium* and *Helicomyces* (Fig. 13).

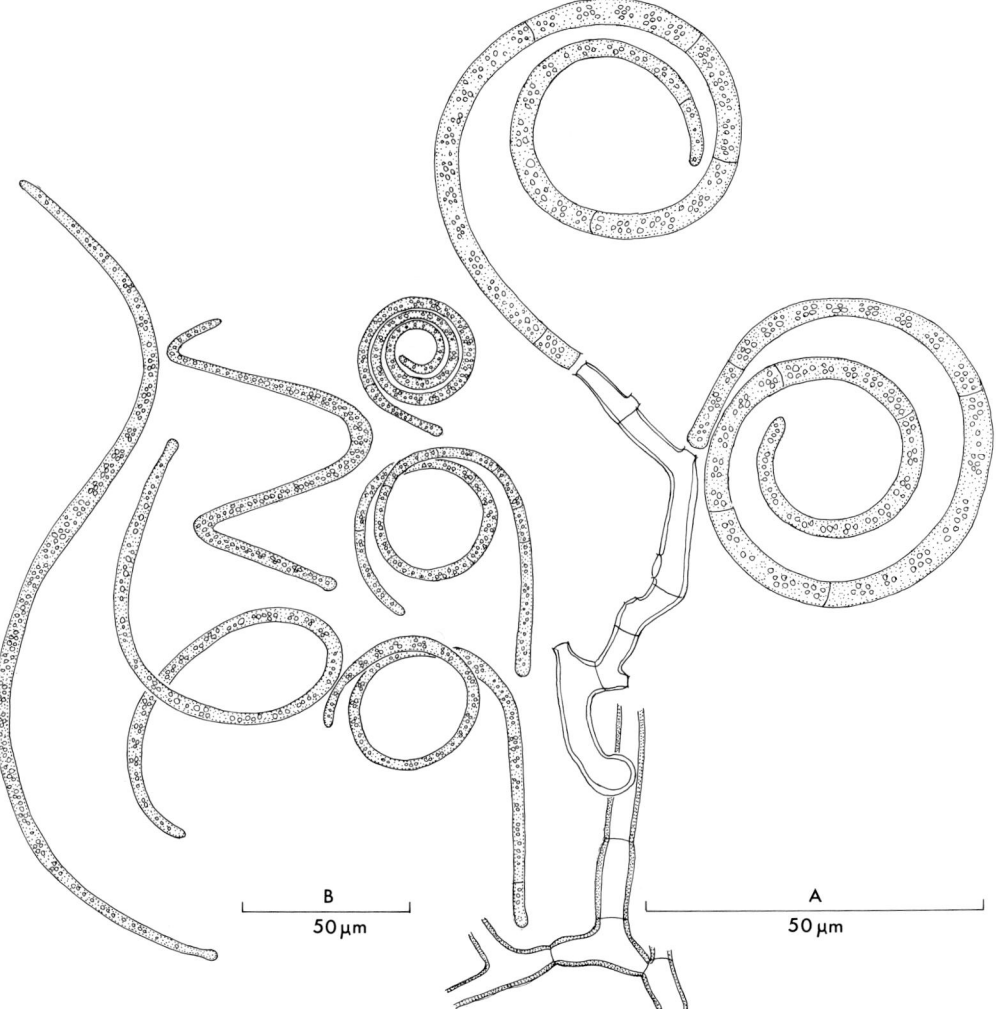

Fig. 13. *Helicomyces* sp. (A) Conidiophore and two conidia. (B) Detached conidia. Note that the conidia may unwind and become sigmoid.

Three-dimensional, barrel-shaped spores formed by the tight winding of a spiral hypha are found in *Helicoon* and *Helicodendron* (Fig. 16, B-H). Conidia of the last two enclose air as they develop at the air–water interface at the surface of the moist substrate. Their spores are virtually unsinkable in water, although by addition of detergents such as 0.1% Brij 35 (polyoxyethylene lauryl ether, BDH Chemicals, Ltd., Poole, England) the spores can be induced to sink while still

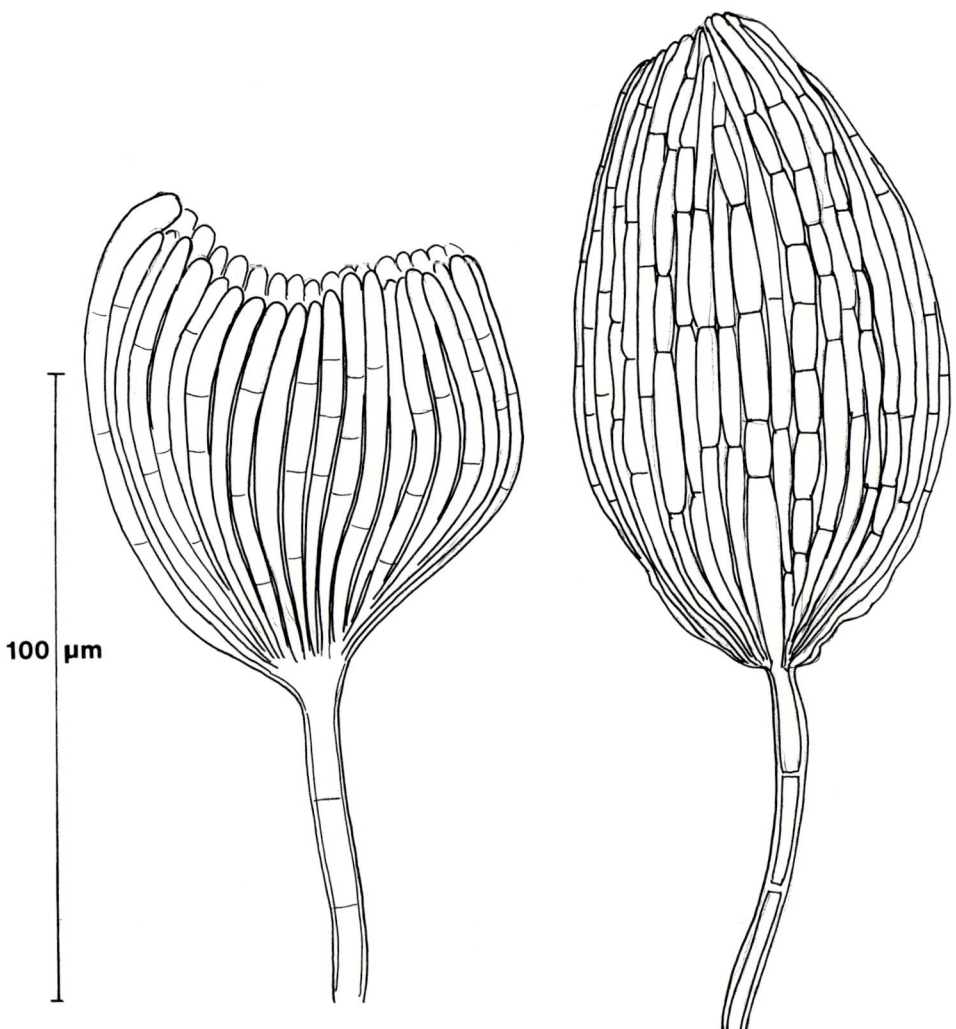

Fig. 14. Propagules of the tropical aeroaquatic fungus *Cancellidium applanatum*.

retaining viability. It is fascinating to see in other aeroaquatic fungi how the device of a nonwettable, buoyant propagule has evolved independently in a variety of different ways. In *Clathrosphaerina* (Fig. 16A) the propagule is a hollow, clathrate sphere formed by repeated dichotomous branching and incurving of the branches of the conidium initial. In *Candelabrum spinulosum* a more flattened arrangement of dichotomies ending in vertical spiny lobes which touch each other (Fig. 17E) achieves a similar result. In *Peyronelina* (Fig. 15) a morula of fertile globose cells is enclosed in a basket of incurved spiny arms. In *Aegerita candida* a cluster of globose cells with a clamp connection at the septum separating each cell from the one beneath it (Figs. 15A, 17G) encloses air. These clamped structures are the dikaryotic propagules of *Bulbillomyces farinosus* (Bres.) Jülich (1974) and are extremely common on rotting wood in wet places, sometimes accompanied by the basidial state. *Subulicystidium longisporum* (Pat.) Parm. (Jülich, 1974) is another basidiomycete with buoyant dikaryotic propagules, in this case surrounded by a fringe of white nodulose setae (Fig. 17H). Again air is trapped between the globose cells which make up the propagule. A similar arrangement is found in *Aegerita viridis,* which lacks clamps, and in the bilobed propagules of *Beverwijkella pulmonaria* (Fig. 15F). *Spirosphaera floriforme* (Fig. 15D) makes a buoyant propagule by trapping air between a large number of short, incurved branches. In *Cancellidium applanatum,* first found on balsa wood test blocks submerged in a Japanese lake (Tubaki, 1975b) but since found on leaves dragged from ponds, streams, and swamps in Malaysia, the propagule is shaped like a flattened wine glass (Webster and Davey, 1981). It develops as a series of parallel contiguous fingerlike septate hyphae (Figs. 14 and 17C, D). Air is trapped in the bowl and, when the the leaf is flooded, the propagule breaks away from its uniseriate stalk. An undescribed aeroaquatic collected from Malaysia in the same habitats as *Cancellidium* has a club-shaped transversely septate spore beset with studs derived from fragments of the outer wall (Figs. 17A, B). Air is trapped between the studs, giving the spore a greenish appearance. The spore floats away when flooded. This fungus has been described as *Fusticeps bullatus* gen. et sp. nov. (J. Webster, unpublished).

Fig. 15. Propagules of some aeroaquatic fungi. (A–C) *Peyronelina glomerulata.* (A) Propagule with incurved arms. (B) Propagule under moist conditions with extended arms. (C) High-power drawing showing the origin of the arms and two detached globose cells. (D) *Spirosphaera floriforme,* two immature propagules. (E) *Aegerita candida* (conidia of *Bulbillomyces farinosus*), developing propagule. The arrow points to a clamp connection. (F) *Beverwijkella pulmonaria.* (A, B, D, and F) to the same scale. (A–C) after Fisher *et al.,* 1977.

Fig. 16. Propagules of aeroaquatic fungi. (A) *Clathrosphaerina zalewskii.* (B) *Helicodendron giganteum.* (C) *Helicodendron giganteum.* (D) *Helicodendron giganteum* seen from above. As seen from above, the direction of coiling of the gyres of the spiral is clockwise. (E) *Helicoon pluriseptatum* seen from below. The coiling is counterclockwise. (F) *Helicoon pluriseptatum* seen from above. (G) *Helicodendron triglitziense.* Note the fibrillar coating to the wall. (H) *Helicondendron tubulosum.*

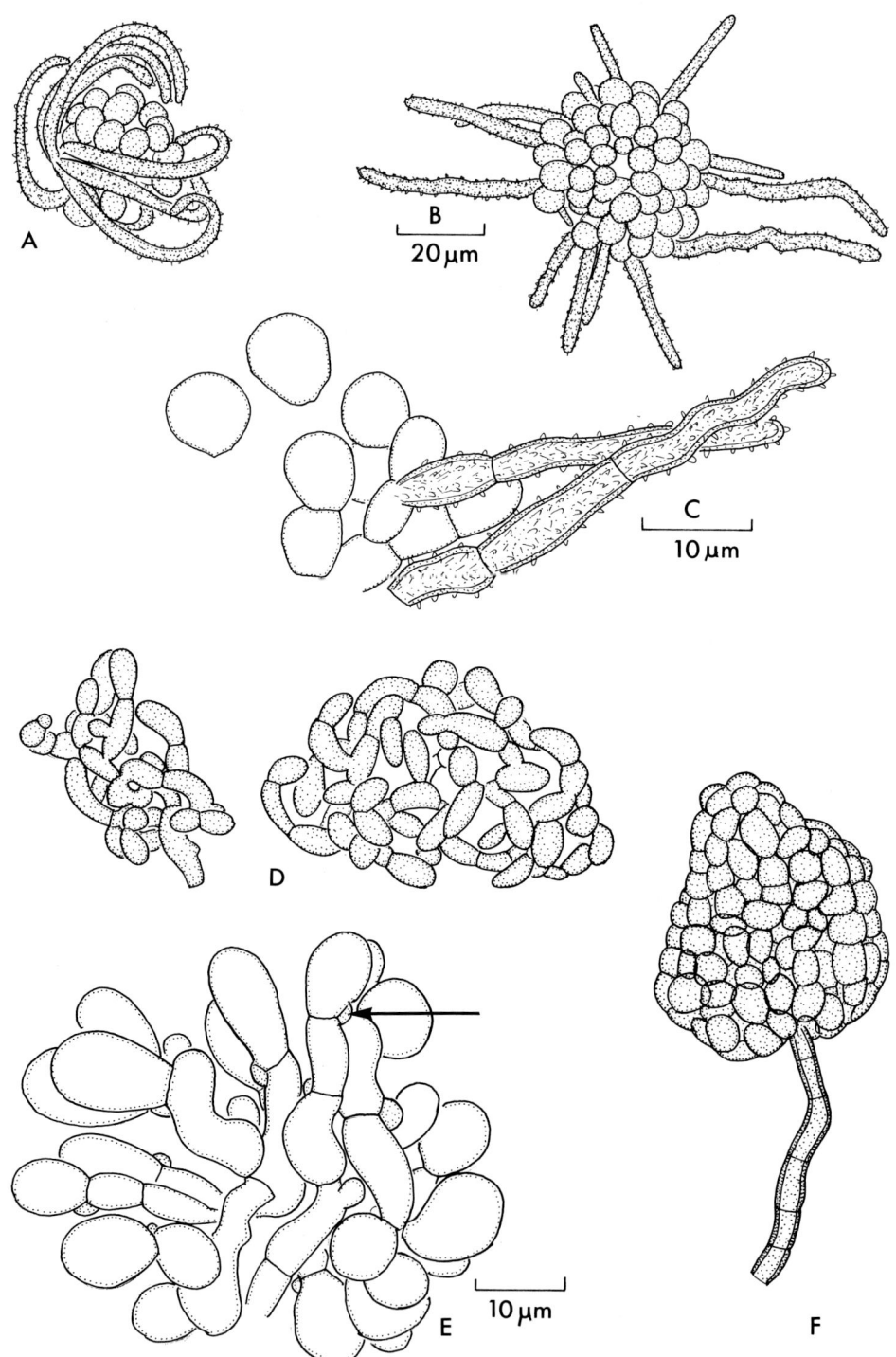

Fig. 15. Caption on page 339.

Fig. 16. Caption on page 339.

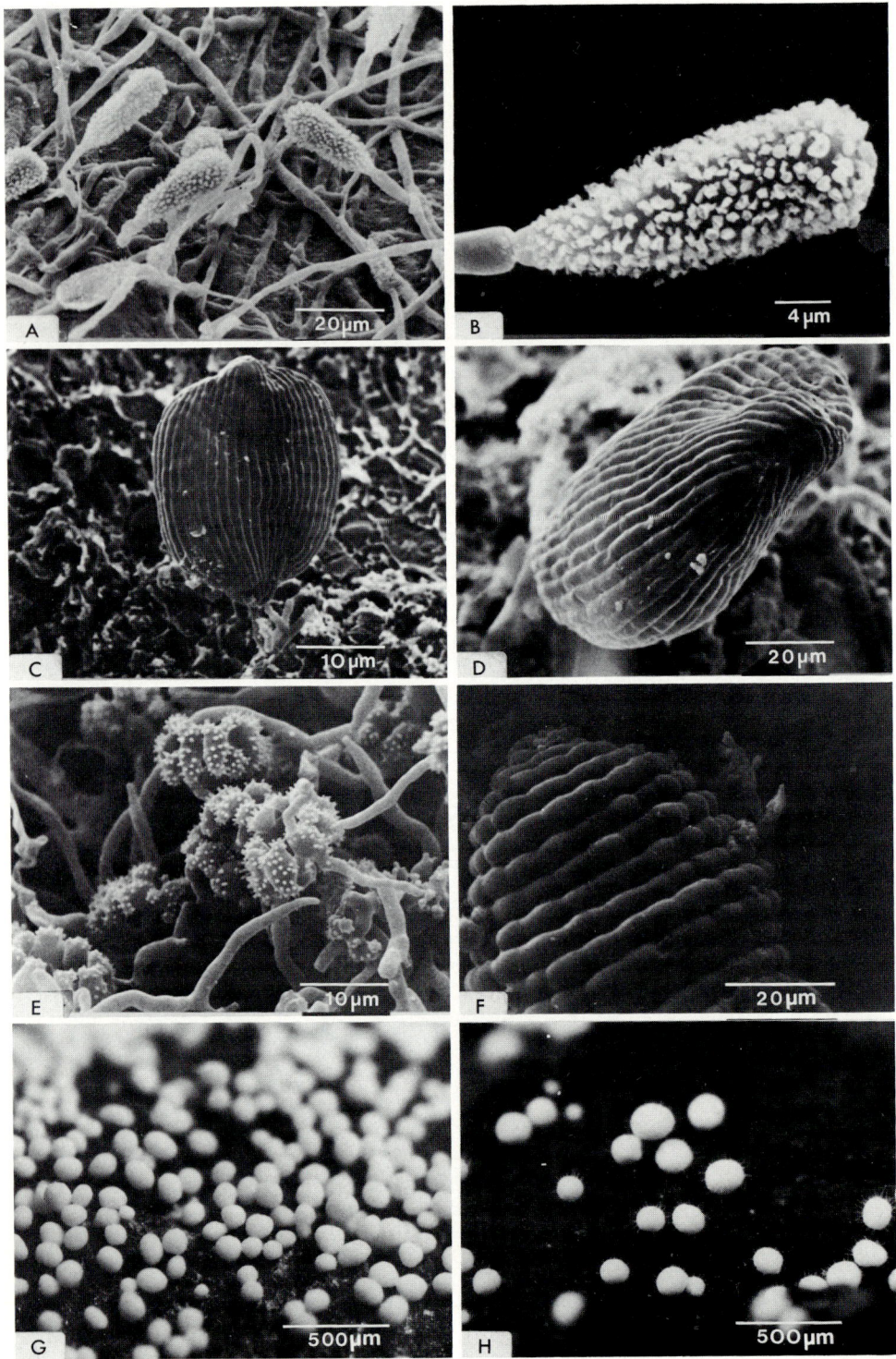

Fig. 17.

C. Affinities of Aeroaquatic Fungi

The aeroaquatic habit has evolved separately among Basidiomycotina and Ascomycotina. Mention has been made above of two basidiomycetes with buoyant propagules. The basidial states which develop on moist wood in air produce air-dispersed ballistospores. *Clathrosphaerina zalewskii* is the anamorph of an inoperculate discomycete with minute apothecia, *Hyaloscypha zalewskii* (Descals and Webster, 1976). Abdullah and Webster (1981b) have shown that *Helicodendron tubulosum* has a *Lambertella* teleomorph, and S. K. Abdullah (unpublished) has shown that an *Aegerita*-like fungus (propagules lacking clamps) has a *Hyaloscypha* teleomorph. It is therefore possible that some of the anamorph-genera of aeroaquatic fungi are more closely related than their conidium morphology leads one to suppose. It is interesting that several species, including *Helicodendron giganteum*, *H. conglomeratum*, *H. tubulosum*, *H. triglitziense*, *S. floriforme*, and *Candelabrum brocchiatum*, produce phialoconidia. These have not yet been shown to germinate. We speculate that they may be spermatia and, if suitable techniques can be developed, it may be possible to use them in the induction of teleomorphs. We have no evidence that the phialoconidia play any role in dispersal under water. The affinities of *Helicosporium* are with *Tubeufia* and *Ophionectria*. For example, *H. phragmitis*, reported from reed swamp plants such as *Glyceria* as well as from rotting wood in water, is the anamorph of *Tubeufia helicomyces* (Webster, 1951).

D. Ecology

It is clear that aeroaquatic fungi are an "ecological group" rather than a natural one. They share a series of similar habitats and have adopted common strategies for growing vegetatively beneath water and developing a dispersal phase in air. It is likely therefore that they will have a number of similar physiological adaptations.

1. Substrates

There appears to be little specificity for substrates. We have collected a wide range of aeroaquatics (over 30 species in Britain) on fruits, leaves, and wood from deciduous trees and *Pteridium* petioles. Pine needles tend to have a more

---

Fig. 17. Propagules of aeroaquatic fungi. (A and B) *Fusticeps bullatus* club-shaped conidia. Air is trapped between the roughened extensions of the spore wall. (C) *Cancellidium applanatum* propagule seen from the side. (D) *Cancellidium applanatum* propagule seen from above. The propagule is a closed, flattened sac. (E) *Candelabrum spinulosum* propagules, made up of a cylinder of dichotomously branched spiny branches. (F) *Helicodendron giganteum* propagule bearing phialophores. (G) *Aegerita candida* (=*Bulbillomyces farinosus*) propagules. (H) *Aegerita tortuosa* (=*Subulicystidium longisporum*) propagules.

specialized flora, including *Helicodendron fractum* and *H. hyalinum* (Abdullah *et al.*, 1979). Wood piles in streams and rivers are a good source of material, and wood test blocks of pine and beech submerged in rivers have yielded a number of aeroaquatic forms. These include *Helicoon sessile* and *Peyronelina glomerulata* when recovered and incubated for periods of up to 3 months (Kane, 1978; Jones, 1979). It is known that some species are cellulolytic. Fisher *et al.* (1977), who tested nine species for their ability to bring about weight loss from filter paper, showed that all had this capacity. *Helicodendron tubulosum, H. conglomeratum, Helicoon fuscosporum,* and *Aegerita* sp. were particularly active. When inoculated onto leached, sterilized *Alnus* leaves in flasks on a shaker, five species of aeroaquatic fungi (*Aegerita candida, Beverwijkella pulmonaria, Helicodendron giganteum, H. intestinale,* and *H. triglitziense*) caused significant weight loss in the absence of mineral supplements, but it was suggested that external supplies of mineral nutrients would hasten the process of leaf degradation (P. J. Fisher and J. Webster, unpublished).

2. Survival in Water

Conditions at the mud surface in ponds will vary greatly, depending on the size and depth of the pond, its catchment area, the availability of leaf litter, nutrient content, exposure to light, and so on. At sites where drainage water accumulates, outflow is limited, and leaf litter is plentiful the water layers near the mud surface, especially during the summer, may be totally depleted of oxygen for several months at a time. The black sulfide-rich silt and the smell throughout the year of hydrogen sulfide, which would readily combine with oxygen if it were available, indicate permanent oxygen depletion a few millimeters below the surface. Fisher (1977a,b) and Fisher and Webster (1979b) have shown that under these conditions a number of aeroaquatic fungi can survive for months and possibly for longer. It is likely that many survive in the mycelial state because relatively few appear to form chlamydospores or sclerotia. *Helicodendron westerdijkiae* is an exception, and forms chlamydospores.

3. Colonization and Growth under Water

The ability of aeroaquatic fungi to colonize beech leaves under semianaerobic conditions was tested and compared with a better aerated setup in the laboratory. Leaf detritus from a eutrophic pond was separated from silt and homogenized in a vegetable blender. Sixty unsterilized beech leaf disks were mixed with this field inoculum and quartz sand in 5-cm petri dishes and placed 15 cm below the water level in the pond during a period of low oxygen content. A similar batch of disks was incubated under aerated water in the laboratory. The disks were recovered 1 month later and placed on moist filter paper for 12 days at about 12°C to induce sporulation. Under these conditions none of the disks which had been incubated in the field produced spores but, after a further 7 days' incubation at 20°C in distilled water, being shaken daily to ensure aeration, spores were produced by *Helicodendron triglitziense* on a proportion of the disks. In contrast, the disks incubated under aerated water in the laboratory produced spores of 8 species of aeroaquatic fungi in proportions varying from 5 to 46 out of 60 disks. Fisher has concluded that, although aeroaquatic fungi may be able to survive periods of oxygen depletion, they are unable to colonize effectively unless adequate oxygen is available.

Placing nonsterile beech leaf disks close to homogenized infected leaf material, while obviously artificial, is probably not very different from the way in which leaf material is colonized in nature. Because the conidial propagules do not sink readily the likely sources of inoculum are leaf-to-leaf contact on the pond bottom, mycelial fragments floating in the water, or ascospores or basidiospores in species with teleomorphs. We believe that the first two sources are the most likely.

While black leaves which have been in the water for several months produce plentiful conidia of aeroaquatic fungi after suitable incubation, there is no evidence to suggest that they are preferentially colonized as compared with recently abscised leaves. Fisher (1979) compared the colonization of black beech leaves, previously submerged in a eutrophic pond for several months, with that of brown, freshly abscised leaves. He exposed batches of 60 nonsterile disks of each type of leaf to a mycelial inoculum of aeroaquatic fungi grown in beech leaf decoction at two different field sites, a eutrophic pond and a small oligotrophic lake. At the eutrophic site the brown leaf disks were more readily colonized than the black disks by all fungi, but at the oligotrophic site brown and black disks were colonized equally well by *Clathrosphaerina zalewskii* and *Helicodendron giganteum*. He used a colonization index (CI) to assess the competitive saprophytic ability of an aeroaquatic fungus in a particular situation:

$$CI = \frac{\text{mean number of black leaf disks colonized}}{\text{mean number of brown leaf disks colonized}}$$

A high value of CI, approaching 1.0, as found for *C. zalewskii* and *H. giganteum* at the oligotrophic site, shows an equally effective capacity to colonize both types of leaf disks along with the resident microflora. It is of interest that both these species were indigenous to the oligotrophic site. Lower values of CI, e.g., *H. giganteum* 0.4 and *C. zalewskii* 0.06, at the eutrophic site where neither species was indigenous, indicate a reduced capacity to colonize black leaf material. The reasons for these differences are probably complex but may reflect differing oxygen availability at the two sites.

Similar experiments were done with 6-mm lengths of freshly abscised brown needles of *Pinus sylvestris* compared with leaves which had been submerged in nylon bags at the eutrophic site for 18 months (leached needles). None of the leaf segments were colonized at the eutrophic site, and only the leached segments were colonized at the oligotrophic site by *Helicodendron fractum*. This fungus appears to be particularly common on pine needles. Whether its inability to colonize freshly abscised leaf segments is due to the presence of some inhibitory substance is not known.

The effects of varying gas mixtures on mycelial growth of four species of *Helicodendron* have been studied by Fisher and Webster (1979a). The fungi were grown in tubes of beech leaf decoction in atmospheres of air, oxygen, carbon dioxide, and mixtures of these gases. The best growth for all fungi was in liquid equilibrated with air (160 mm oxygen partial pressure). Lowering the oxygen content to 10 mm oxygen and 750 nitrogen reduced dry weights to 24–66% of the air control. In mixtures of 160 mm oxygen and 70 mm carbon dioxide dry weights of 42–92% of air control values were obtained. Whether any of these fungi are capable of growth, i.e., production of additional dry weight, under strictly anaerobic conditions is currently under investigation. Present indications are that under strictly anaerobic conditions growth is negligible (J. I. Field, unpublished). Whether aeroaquatic fungi are able to compete successfully with other microbes in their aquatic environment may possibly be related to their ability to grow at lowered oxygen tensions, but at present we have no evidence to support this. It seems more likely, however, that it is their ability to survive under these conditions which explains their common occurrence in their habitat.

4. Conditions Favoring Sporulation

The effects of different gas regimes on sporulation by 12 species of aeroaquatic fungi in light and darkness have been studied by Fisher and Webster (1978). Autoclaved beech leaf disks were inoculated with mycelial fragments and after moist incubation for 1 month at 20°C were rinsed in sterile distilled water and placed on sterile, moist filter paper at about 10°C in closed petri dishes surrounded by different gaseous atmospheres in the dark for 30 days. Dishes incubated in light in air were used as controls. Comparison of air-dark and air-light treatments showed that in five species (*Helicodendron conglomeratum, H. giganteum, H. triglitziense, H. tubulosum*, and *Spirosphaera floriforme*) light markedly stimulated sporulation. The effect was most marked in *H. tubulosum*, where no sporulation was found on dark-incubated disks. In six species, however (*Clathrosphaerina zalewskii, Helicodendron fuscum, H. fuscosporum, H. hyalinum, H. westerdijkiae*, and *Helicosporium* sp.), all disks bore spores irrespective of light or dark incubation. The responses to different gas regimes were less clear-cut, and a spectrum of response was noted. A particularly interesting difference was a comparison between incubation in 760 mm nitrogen-dark and air-dark. One group of species (*C. zalewskii, H. conglomeratum, H. fuscosporum, H. fuscum, H. giganteum*, and *H. westerdijkiae*) sporulated on fewer disks (sometimes on none) in an atmosphere of nitrogen as compared with air. When the nitrogen atmosphere was replaced by air during a further period of 30 days, all these species responded by sporulating on an increasing number of disks. It is presumed that for these species a period of aerobic growth is necessary for sporulation. In a second group of species (*Helicodendron hyalinum, Helicodendron fractum* and *Helicosporium* sp.) the numbers of disks bearing spores were the same in nitrogen or air. These results emphasize the lack of uniformity in the physiological behavior of this group.

Much remains to be learned about the physiology of sporulation in this group of fungi, and, in particular, the way in which transfer from water to air will induce the formation of conidia. Possibly, in this respect, these fungi are no different from *Penicillium* and *Aspergillus,* usually regarded as terrestrial fungi but capable of mycelial growth when submerged. Morton (1961) has written, "In the species of *Penicillium* and *Aspergillus* examined . . . the most powerful and consistently effective stimulus to sporulation was found to be exposure of mycelium to aerial conditions." *Penicillium griseofulvum* can be induced to form submerged conidiophores by manipulation of culture conditions, e.g., by the addition of Ca^{2+}. It would be interesting to see if the aerial stimulus to sporulation can be replaced in aeroaquatic fungi, but the nature of their propagules makes it doubtful if they could function properly if formed under water.

5. Survival on Land

When a leaf-filled pond or ditch dries out, the carpet of leaves is exposed. Under these conditions, leaves and twigs and other debris may bear sporulating colonies of aeroaquatic fungi. Such leaves may blow away from the original habitat. It is of interest to study the survival of aeroaquatic fungi out of water. Fisher (1978) has studied survival in the form of infected leaf material, mycelium, and spores in the field and under desiccation in the laboratory. Sterilized homogenized beech leaf mash was inoculated in the laboratory with eight species. After 7-weeks' growth the mash was air-dried for 7 days. Some of the material was placed on garden soil. At intervals, 3- to 5-g samples were recovered and tested for the presence of the inoculated fungi. Some material was also held over silica gel in a desiccator. Mycelium, grown in beech leaf decoction, was harvested, washed, placed on dry filter paper and air-dried at 20°C. To test the survival of spores beech leaf disks bearing conidia were stored over silica gel, and their percent germination tested at intervals. Survival in leaf mash on garden soil was of variable duration. Three species of *Helicodendron* (*H. tubulosum, H. conglomeratum,* and *H. giganteum*) survived for 11 months, *H. triglitziense* for 8 months, and *H. hyalinum* and *H. ellipticum* for 2 months. Survival of mycelium in the desiccator was also variable, but *H. triglitziense* was particularly impressive, being able to survive for 276 days. No resting structures were observed. Spores survived in the desiccator for periods up to 10 days, but in no case could viability be demonstrated after 20 days. These experiments show a capacity for existence out of water. It is likely that some species are better adapted to a terrestrial existence than others.

ACKNOWLEDGMENTS

We are indebted to Dr. P. J. Fisher and Dr. S. K. Abdullah for permission to use scanning electron micrographs of aero-aquatic fungi (Figs. 16 and 17) and to Dr. P. J. Sanders for scanning electron micrographs of Ingoldian fungi (Fig. 9). All three have also contributed to helpful discussions. We thank the numerous authors whose figures we have traced (Figs. 1-8). We also acknowledge gratefully technical help from Mr. R. A. Davey and Mr. M. S. Alexander, and assistance in typing from Mrs. Penny Legg, Miss Joan Vaughan, and Mrs. Barbara Vermeulen.

REFERENCES

Abdullah, S. K., and Webster, J. (1981a). Occurrence of aero-aquatic fungi in soil. *Trans. Br. Mycol. Soc.* (in press).
Abdullah, S. K., and Webster, J. (1981b). *Lambertella tubulosa* sp. nov. teleomorph of *Helicodendron tubulosum*. *Trans. Br. Mycol. Soc.* (in press).
Abdullah, S. K., Fisher, P. J., and Webster, J. (1979). Two new species of aero-aquatic Hyphomycetes. *Trans. Br. Mycol. Soc.* **72,** 324-329.
Abdullah, S. K., Descals, E., and Webster, J. (1981). Teleomorphs of three aquatic Hyphomycetes. *Trans. Br. Mycol. Soc.* (in press).
Alasoadura, S. O. (1968a). *Flabellospora crassa* n.gen., n.sp., an aquatic hyphomycete from Nigeria. *Nova Hedwigia* **15,** 415-418.
Alasoadura, S. O. (1968b). *Flabellospora verticillata,* a new species of aquatic hyphomycete from Nigeria. *Nova Hedwigia* **15,** 419-421.
Alasoadura, S. O. (1968c). Some aquatic hyphomycetes from Nigeria. *Trans. Br. Mycol. Soc.* **51,** 535-540.
Anastasiou, C. J. (1962). Fungi from salt lakes. I. A new species of *Clavariopsis. Mycologia* **53,** 11-15.
Bandoni, R. J. (1972). Terrestrial occurrence of some aquatic hyphomycetes. *Can J. Bot.* **50,** 2283-2288.
Bandoni, R. J. (1974). Mycological observations on the aqueous films covering decaying leaves and other litter. *Trans. Mycol. Soc. Jpn.* **15,** 309-315.
Bandoni, R. J. (1975). Significance of the tetraradiate form in dispersal of terrestrial fungi. *Rep. Tottori Mycol. Inst.* **12,** 105-113.
Bandoni, R. J. (1981). Aquatic hyphomycetes from terrestrial litter. *In* "Fungal Ecology" (D. T. Wicklow and G. C. Carroll, eds.). Dekker, New York (in press).
Barghoorn, E. S., and Linder, D. H. (1944). Marine fungi: Their taxonomy and biology. *Farlowia* **1,** 395-467.
Bärlocher, F., and Kendrick, B. (1973a). Fungi in the diet of *Gammarus pseudolimnaeus* (Amphipoda). *Oikos* **24,** 295-300.
Bärlocher, F., and Kendrick, B. (1973b). Fungi and food preferences of *Gammarus pseudolimnaeus. Arch. Hydrobiol.* **72,** 501-516.
Bärlocher, F., and Kendrick, B. (1974). Dynamics of the fungal population on leaves in a stream. *J. Ecol.* **62,** 761-791.
Bärlocher, F., and Kendrick, B. (1975). Assimilation efficiency of *Gammarus pseudolimnaeus* (Amphipoda) feeding on fungal mycelium or autumn-shed leaves. *Oikos* **26,** 55-59.
Bärlocher, F., and Kendrick, B. (1976). Hyphomycetes as intermediaries of energy flow in streams. *In* "Recent Advances in Aquatic Mycology" (E. B. Gareth Jones, ed.), pp. 435-446. Elek Science, London.
Bärlocher, F., and Kendrick, B. (1981). The role of aquatic hyphomycetes in the trophic structure of

streams. *In* "Fungal Ecology" (D. T. Wicklow and G. C. Carroll, eds.). Dekker, New York (in press).
Berrie, A. D. (1976). Detritus, micro-organisms and animals in fresh water. *In* "The Role of Terrestrial and Aquatic Organisms in Decomposition Processes" (J. M. Anderson and A. Macfadyen, eds.), pp. 323-338. Blackwell, Oxford.
Bonorden, H. F. (1851). "Handbuch der allgemeiner Mykologie." Schweizerbart. Stuttgart. 336 pp.
Crane, J. L. (1968). Freshwater hyphomycetes of the northern Appalachian highland including New England and three coastal plain states. *Am. J. Bot.* **55**, 996-1002.
Crane, J. L., and Shearer, C. A. (1977). *Rogersia,* a later name for *Filosporella. Mycotaxon* **6**, 27-28.
Deighton, F. C. (1972). *Mycocentrospora,* a new name for *Centrospora* Neerg. *Taxon* **21**, 716.
Dennis, R. W. G. (1964). Remarks on the genus *Hymenoscyphus* S. F. Gray, with observations on sundry species referred by Saccardo and others to the genera *Helotium, Pezizella or Phialea. Persoonia* **3**, 29-80.
Descals, E. (1978). Taxonomic studies of freshwater hyphomycetes and related fungi. Ph.D. Thesis, University of Exeter, England.
Descals, E., and Sutton, B. C. (1976). *Anavirga dendromorpha* n.sp. and its *Phialocephala* phialidic state. *Trans. Br. Mycol. Soc.* **67**, 269-274.
Descals, E., and Webster, J. (1976). The *Hyaloscypha* perfect state of *Clathrosphaerina zalewskii. Trans. Br. Mycol. Soc.* **67**, 525-528.
Descals, E., and Webster, J. (1978). *Miladina lechithina* (Pezizales), the ascigerous state of *Actinospora megalospora. Trans. Br. Mycol. Soc.* **70**, 466-472.
Descals, E., and Webster, J. (1980). Taxonomic studies on aquatic hyphomycetes. II. The *Dendrospora* aggregate. *Trans. Br. Mycol. Soc.* **74**, 135-158.
Descals, E., and Webster, J. (1981). Variations in asexual reproduction of *Entomophthora* species on aquatic insects. *Trans. Br. Mycol. Soc.* (in press).
Descals, E., Nawawi, A., and Webster, J. (1976). Developmental studies in *Actinospora* and three similar aquatic hyphomycetes. *Trans. Br. Mycol. Soc.* **67**, 207-222.
Descals, E., Webster, J., and Dyko, B. J. (1977). Taxonomic studies on aquatic hyphomycetes. I. *Lemonniera* de Wildeman. *Trans. Br. Mycol. Soc.* **69**, 89-109.
de Wildeman, E. (1893). Notes mycologiques. Fascicle II. *Ann. Soc. Belge Microsc.* **17**, 35-68.
de Wildeman, E. (1894). Notes mycologiques. Fascicle III. *Ann Soc. Belge Microsc.* **18**, 135-161.
de Wildeman, E. (1895). Notes mycologiques. Fascicle VI. *Ann Soc. Belge Microsc.* **19**, 193-206.
Dudka, I. O. (1974). (Ukrainian aquatic Hyphomycetes) (in Ukrainian). *Acad. Sci. Ukr., R.S.R.M.G. Holodny Bot. Inst.* Naukova Dumka, Kiev. 240 pp.
Dyko, B. J. (1978). New aquatic and waterborne hyphomycetes from the southern Appalachian mountains in the United States. *Trans. Br. Mycol. Soc.* **70**, 409-416.
Dyko, B. J., and Sutton, B. C. (1978). *Filosporella,* an earlier name for *Coeloanguillospora. Mycotaxon* **7**, 323-326.
Fisher, P. J. (1977a). Ecological studies on aero-aquatic hyphomycetes. Ph.D. Thesis, University of Exeter, England.
Fisher, P. J. (1977b). New methods of detecting and studying saprophytic behaviour of aero-aquatic hyphomycetes from stagnant water. *Trans. Br. Mycol. Soc.* **68**, 407-411.
Fisher, P. J. (1978). Survival of aero-aquatic hyphomycetes on land. *Trans. Br. Mycol. Soc.* **71**, 419-423.
Fisher, P. J. (1979). Colonization of freshly abscised and decaying leaves by aero-aquatic hyphomycetes. *Trans. Br. Mycol. Soc.* **73**, 99-102.
Fisher, P. J., and Webster, J. (1978). Sporulation of aero-aquatic fungi under different gas regimes in light and darkness. *Trans. Br. Mycol. Soc.* **71**, 465-468.
Fisher, P. J., and Webster, J. (1979). Effect of oxygen and carbon dioxide on growth of four aero-aquatic hyphomycetes. *Trans. Br. Mycol. Soc.* **72**, 57-61.

Fisher, P. J., and Webster, J. (1981). Ecological studies on aero-aquatic hyphomycetes. *In* "Fungal Ecology" (D. T. Wicklow and G. C. Carroll, eds.). Dekker, New York (in press).

Fisher, P. J., Sharma, P. D., and Webster, J. (1977). Cellulolytic ability of aero-aquatic hyphomycetes. *Trans. Br. Mycol. Soc.* **69**, 495-520.

Glen-Bott, J. I. (1951). *Helicodendron giganteum* n.sp. and other aerial-sporing hyphomycetes of submerged dead leaves. *Trans. Br. Mycol. Soc.* **34**, 275-279.

Glen-Bott, J. I. (1955). On *Helicodendron tubulosum* and some similar species. *Trans. Br. Mycol Soc.* **38**, 17-30.

Gönczöl, J. (1976a). *Monotosporella tuberculata*, a new species of aquatic hyphomycetes from Hungary. *Nova Hedwigia* **27**, 493-500.

Gönczöl, J. (1976b). Ecological observations on the aquatic hyphomycetes of Hungary. II. Observations on biotopes of aquatic hyphomycetes in south-west Hungary. *Acta Bot. Acad. Sci. Hung.* **22**, 51-60.

Gönczöl, J. (1976c). *Clavatospora flagellata* sp. nov., an aquatic hyphomycete from Hungary. *Acta Bot. Acad. Sci. Hung.* **22**, 355-360.

Goos, R. D. (1970). "*In vitro*" sporulation in *Actinospora megalospora*. *Trans. Br. Mycol. Soc.* **55**, 335-337.

Greathead, S. K. (1961). Some aquatic hyphomycetes in South Africa. *J. S. Afr. Bot.* **27**, 195-228.

Grove, W. B. (1884). New or noteworthy fungi. *J. Bot. Br. Foreign* **22**, 195-201.

Gupta, P. C., and Gandhi, S. K. (1979). *Tricladium* leaf spot, a new disease of sorghum. *Z. Pflanzenkr. Pflanzenschutz.* **86**, 287-289.

Gustafsson, M. (1965). On species of the genus *Entomophthora* Fres. in Sweden. I. Classification and distribution. *Lantbrukhoegsk. Ann.* **31**, 103-212.

Haskins, R. H. (1958). Hyphomycetous fungi: *Volucrispora aurantiaca*, *V. ornithomorpha*, *Tricellula curvata*, with the genus *Tricellula* emended. *Can. J. Microbiol.* **4**, 273-285.

Hennebert, G. L. (1968). New species of *Spirosphaera*. *Trans. Br. Mycol. Soc.* **51**, 13-24.

Hudson, H. J. (1961). *Heliscus submersus* sp. nov., an aquatic hyphomycete from Jamaica. *Trans. Br. Mycol. Soc.* **44**, 91-94.

Hudson, H. J., and Ingold, C. T. (1960). Aquatic hyphomycetes from Jamaica. *Trans. Br. Mycol. Soc.* **43**, 469-478.

Hudson, H. J., and Sutton, B. C. (1964). *Trisulcosporium* and *Tetranacrium*, two new genera of Fungi Imperfecti. *Trans. Br. Mycol. Soc.* **47**, 197-203.

Hughes, S. J. (1958). Revisiones hyphomycetum aliquot cum appendice de nominibus rejiciendis. *Can J. Bot.* **36**, 727-836.

Ingold, C. T. (1942). Aquatic hyphomycetes of decaying alder leaves. *Trans. Br. Mycol. Soc.* **25**, 339-417.

Ingold, C. T. (1943a). Further observations on aquatic hyphomycetes of decaying leaves. *Trans. Br. Mycol. Soc.* **26**, 104-115.

Ingold, C. T. (1943b). *Triscelophorus monosporus* n.gen., n.sp., an aquatic hyphomycete. *Trans. Br. Mycol. Soc.* **26**, 148-152.

Ingold, C. T. (1944). Some new aquatic hyphomycetes. *Trans. Br. Mycol. Soc.* **27**, 35-47.

Ingold, C. T. (1952). *Actinospora megalospora* n.sp., an aquatic hyphomycete. *Trans. Br. Mycol. Soc.* **35**, 66-70.

Ingold, C. T. (1958). New aquatic hyphomycetes. *Trans. Br. Mycol. Soc.* **41**, 365-372.

Ingold, C. T. (1959). *Polycladium equiseti* gen. nov., sp. nov.: An aquatic hyphomycete on *Equisetum fluviatile*. *Trans. Br. Mycol. Soc.* **42**, 112-114.

Ingold, C. T. (1966). The tetraradiate aquatic fungal spore. *Mycologia* **58**, 43-56.

Ingold, C. T. (1974). *Tricladium chaetocladium* sp. nov., an aquatic hyphomycete from Britain. *Trans. Br. Mycol. Soc.*, **63**, 624-626.

Ingold, C. T. (1975a). An illustrated guide to aquatic and waterborne Hyphomycetes (Fungi Imperfecti) with notes on their biology. *Freshwater Biol. Assoc., Sci. Publ.* **30**, 1-96.

Ingold, C. T. (1975b). Hooker Lecture 1974; Convergent evolution in aquatic fungi: The tetraradiate spore. *Biol. J. Linn. Soc.* **7**, 1-25.
Ingold, C. T. (1976). The morphology and biology of freshwater fungi excluding Phycomycetes. *In* "Recent Advances in Aquatic Mycology" (E. B. Gareth Jones, ed.), pp. 335-357. Elek Science, London.
Ingold, C. T., Dann, V., and McDougall, P. J. (1968a). *Tripospermum camelopardus* sp. nov. *Trans. Br. Mycol. Soc.* **51**, 51-56.
Ingold, C. T., McDougall, P. J., and Dann, V. (1968b). *Volucrispora graminea* sp. nov. *Trans. Br. Mycol. Soc.* **51**, 325-328.
Iqbal, S. H. (1971). New aquatic hyphomycetes. *Trans. Br. Mycol. Soc.* **56**, 343-352.
Iqbal, S. H. (1972a). Some observations on aquatic hyphomycetes. Ph.D. Thesis, University of Exeter, England.
Iqbal, S. H. (1972b). New aquatic hyphomycetes. *Trans. Br. Mycol. Soc.* **59**, 301-307.
Iqbal, S. H. (1974a). New aquatic hyphomycetes. *Biologia (Lahore)* **20**, 1-10.
Iqbal, S. H. (1974b). *Pseudoanguillospora,* a new genus of hyphomycetes. *Biologia (Lahore)* **20**, 11-16.
Iqbal, S. H. (1974c). *Scorpiosporium minutum* gen. nov., sp. nov., an aquatic hyphomycete. *Biologia (Lahore)* **20**, 17-21.
Iqbal, S. H., and Webster, J. (1973a). The trapping of aquatic hyphomycete spores by air bubbles. *Trans. Br. Mycol. Soc.* **60**, 37-48.
Iqbal, S. H., and Webster, J. (1973b). Aquatic hyphomycete spora of the River Exe and its tributaries. *Trans. Br. Mycol. Soc.* **61**, 331-346.
Iqbal, S. H., and Webster, J. (1977). Aquatic hyphomycete spora of some Dartmoor streams. *Trans. Br. Mycol. Soc.* **69**, 233-241.
Jones, E. B. G. (1981). Observations on the ecology of lignicolous aquatic hyphomycetes. *In* "Fungal Ecology" (D. T. Wicklow and G. C. Carroll, eds.). Dekker, New York (in press).
Jones, E. B. G., and Sloof, W. (1965). *Candida aquatica* sp. nov., isolated from water scums. *Antonie van Leeuwenhoek* **32**, 223-228.
Jones, E. B. G., and Stewart, R. J. (1972). *Tricladium varium,* an aquatic hyphomycete on wood in water-cooling towers. *Trans. Br. Mycol. Soc.* **59**, 163-167.
Jülich, W. (1972). Monographie der Athelieae (Corticiaceae, Basidiomycetes). *Wildenowia Beih.* **7**, 283 pp.
Jülich, W. (1974). The genera of the Hyphodermoideae (Corticiaceae). *Persoonia* **8**, 59-97.
Kane, D. (1978). The effect of sewage effluent on the growth of microorganisms in the marine environment. Ph.D. Thesis, C.N.A.A., Portsmouth Polytechnic, England.
Kaushik, N. K., and Hynes, H. B. N. (1971). The fate of the dead leaves that fall into streams. *Arch. Hydrobiol.* **68**, 465-515.
Kegel, W. (1906). *Varicosporium elodeae,* ein Wasserpilz mit auffallender Konidienbildung. *Ber. Dtsch. Bot. Ges.* **24**, 213-216.
Kohlmeyer, J., Schmidt, I., and Nair, N. B. (1967). Eine neue *Corollospora* (Ascomycetes) aus dem Indischen Ozean und der Ostsee. *Ber. Dtsch. Bot Ges.* **80**, 98-102.
Kol, E. (1928). Ueber die Kryovegetation der Hohen-Tátra. *Folia Cryptogam., Szeged* **1**, 614-622.
Kuzuha, S. (1973). Two new species of aquatic Hyphomycetes. *J. Jpn. Bot.* **48**, 220-224.
Linder, D. H. (1929). A monograph of the helicosporous Fungi Imperfecti. *Ann. Mo. Bot. Gard.* **16**, 227-388.
Linder, D. H. (1931). Brief notes on the Helicosporeae with descriptions of four new species. *Ann. Mo. Bot. Gard.* **18**, 9-16.
Marcus, J. H., and Willoughby, L. G. (1978). Fungi as food for the aquatic invertebrate *Asellus aquaticus. Trans. Br. Mycol. Soc.* **70**, 143-146.
Marvanová, L. (1968). *Lemonniera centrosphaera* sp. nov. *Trans. Br. Mycol. Soc.* **51**, 613-616.
Marvanová, L. (1972). Concerning *Calcarispora hiemalis. Ceska Mykol.* **26**, 230-232.

Marvanová, L. (1975). Concerning *Gyoerffyella* Kol. *Trans. Br. Mycol Soc.* **65**, 555–565.
Marvanová, L. (1977a). *Taeniospora gracilis* gen. et sp. nov. *Trans. Br. Mycol. Soc.* **69**, 146–148.
Marvanová, L. (1977b). Two new *Alatospora* species. *Arch. Protistenkd,* **119**, 68–74.
Marvanová, L. (1979). *Cylindrotrichum helisciforme* sp. nov., a water-borne hyphomycete. *Trans. Br. Mycol. Soc.* **73**, 368–369.
Marvanová, L., and Iqbal, S. H. (1973). *Pleuropedium tricladioides* gen. et sp. nov. *Antonie van Leeuwenhoek* **39**, 401–408.
Marvanová, L., and Marvan, P. (1963). Einige Hyphomyzeten aus den fliessenden Gewässern des Hrubý Jeseník. *Acta Mus. Silesiae, Ser. A* **12**, 101–118.
Marvanová, L., and Nilsson, S. (1971). Validation of aquatic hyphomycete names. *Trans. Br. Mycol. Soc.* **57**, 531–542.
Marvanová, L., Marvan, P., and Růžička, J. (1967). *Gyoerffyella* Kol. 1928, a genus of the hyphomycetes. *Persoonia* **5**, 29–44.
Matsushima, T. (1971a). Mycological reports from New Guinea and the Solomon Island. 7. Some interesting Fungi Imperfecti. *Bull Natl. Sci. Mus., Tokyo* **14**, 460–480.
Matsushima, T. (1971b). "Microfungi of the Solomon Islands and Papua, New Guinea." Shionogi Res. Lab., Shionogi & Co. Ltd., Kobe.
Matsushima, T. (1975). "Icones Microfungorum a Matsushima Lectorum." Matsushima, Kobe, Japan.
Meyers, S. P., and Kohlmeyer, J. J. (1965). *Varicosporina ramulosa* gen. et sp. nov., an aquatic hyphomycete from marine areas. *Can. J. Bot.* **43**, 915–921.
Miura, K. (1972). Notes on filamentous fungi from Japan, Nos. 7, 8. *J. Jpn. Bot.* **47**, 65–70.
Miura, K. (1974). Stream spora of Japan. *Trans. Mycol. Soc. Jpn.* **15**, 289–308.
Miura, K., and Kudo, M. Y. (1970). An agar medium for aquatic hyphomycetes. *Trans. Mycol. Soc. Jpn.* **11**, 116–118.
Miura, K., and Kudo, M. Y. (1971). Two new species of filamentous fungi *from Japan. J. Jpn. Bot.* **46**, 39–46.
Miura, K., and Okano, S. (1979). *Lateriramulosa quadriradiata,* a new aquatic hyphomycete. *J. Jpn. Bot.* **54**, 204–210.
Moore, R. T. (1955). Index to the Helicosporae. *Mycologia* **47**, 90–103.
Morton, A. G. (1961). The induction of sporulation in mould fungi. *Proc. R. Soc. London, Ser. B* **153**, 548–569.
Nawawi, A. (1973a). *Clavatospora filiformis* sp. nov., an aquatic hyphomycete from Malaysia. *Trans. Br. Mycol. Soc.* **61**, 390–392.
Nawawi, A. (1973b). Two clamp-bearing aquatic fungi from Malaysia. *Trans. Br. Mycol. Soc.* **61**, 521–528.
Nawawi, A. (1973c). A new species of *Flabellospora* from Malaysia. *Malays. J. Sci.* **2**, 55–58.
Nawawi, A. (1974a). Two new *Varicosporium* species. *Trans. Br. Mycol. Soc.* **63**, 27–31.
Nawawi, A. (1974b). Two new *Tricladium* species. *Trans. Br. Mycol. Soc.* **63**, 267–272.
Nawawi, A. (1974c). A new *Campylospora. Trans. Br. Mycol. Soc.* **63**, 603–606.
Nawawi, A. (1975a). *Triscelophorus acuminatus* sp. nov. *Trans. Br. Mycol. Soc.* **64**, 31–34.
Nawawi, A. (1975b). Another hyphomycete with branched conidia. *Trans. Br. Mycol Soc.* **64**, 243–246.
Nawawi, A. (1976a). A new genus of hyphomycetes. *Trans. Br. Mycol. Soc.* **66**, 20–23.
Nawawi, A. (1976b). *Condylospora* gen. nov., a hyphomycete from a foam sample. *Trans. Br. Mycol. Soc.* **66**, 363–365.
Nawawi, A. (1976c). Another new *Flabellospora. Trans. Br. Mycol. Soc.* **66**, 543–547.
Nawawi, A. (1976d). *Filosporella* gen. nov., an aquatic hyphomycete. *Trans. Br. Mycol. Soc.* **67**, 173–176.

Nawawi, A., Descals, E., and Webster, J. (1977a). *Leptosporomyces galzinii*, the basidial state of a clamped branched conidium from freshwater. *Trans. Br. Mycol. Soc.* **68**, 31-36.

Nawawi, A., Webster, J., and Davey, R. A. (1977b). *Dendrosporomyces prolifer* gen. et sp. nov., a basidiomycete with branched conidia. *Trans. Br. Mycol. Soc.* **68**, 59-63.

Nilsson, S. (1962a). Some aquatic hyphomycetes from South America. *Svensk. Bot. Tidskr.* **56**, 351-361.

Nilsson, S. (1962b). Second note on Swedish freshwater hyphomycetes. *Bot. Not.* **115**, 73-86.

Nilsson, S. (1964). Freshwater hyphomycetes: Taxonomy, morphology and ecology. *Symb. Bot. Ups.* **18**, 1-130.

Nimura, H., and Suzuki, S. (1962). A list of the aquatic hyphomycetes in Japan. *J. Jpn. Bot.* **37**, 30-32.

Nolan, R. A. (1972). The aquatic hyphomycete *Geniculospora inflata* from Labrador. *Mycologia* **64**, 1169-1174.

Park, D. (1972). On the ecology of heterotrophic micro-organisms in fresh water. *Trans. Br. Mycol. Soc.* **58**, 291-299.

Park, D. (1974a). *Tricladium terrestre* sp. nov. *Trans. Br. Mycol Soc.* **63**, 179-183.

Park, D. (1974b). Aquatic hyphomycetes in non-aquatic habitats. *Trans. Br. Mycol. Soc.* **63**, 183-187.

Petersen, R. H. (1960). *Culicidospora*, a new genus of aquatic, aleuriosporous hyphomycetes. *Bull. Torrey Bot. Club* **87**, 342-347.

Petersen, R. H. (1962). Aquatic hyphomycetes from North America. I. Aleuriosporae (Part 1), and key to the genera. *Mycologia* **54**, 117-151.

Petersen, R. H. (1963a). Aquatic hyphomycetes from North America. II. Aleuriosporae (Part 2) and Blastosporae. *Mycologia* **55**, 18-29.

Petersen, R. H. (1963b). Aquatic hyphomycetes from North America. III. Phialosporae and miscellaneous species. *Mycologia* **55**, 570-581.

Ranzoni, F. V. (1953). The aquatic hyphomycetes of California. *Farlowia* **4**, 353-398.

Ranzoni, F. V. (1956). The perfect stage of *Flagellospora penicillioides*. *Amr. J. Bot.* **43**, 13-17.

Saccardo, P. A. (1880). Conspectus generum fungorum Italiae inferiorum. *Michelia* **2**, 1-38.

Sanders, P. F. (1979). Experimental studies on aquatic hyphomycetes. Ph.D. Thesis, University of Exeter, England.

Sanders, P. F., and Webster, J. (1978). Survival of aquatic hyphomycetes in terrestrial situations. *Trans. Br. Mycol. Soc.* **71**, 231-237.

Sanders, P. F., and Webster, J. (1980). Sporulation responses of some aquatic hyphomycetes to flowing water. *Trans. Br. Mycol. Soc.* **74**, 601-605.

Shaw, D. (1972). *Ingoldiella hamata* gen. et sp. nov., a fungus with clamp connexions from a stream in North Queensland. *Trans. Br. Mycol. Soc.* **59**, 255-259.

Sinclair, R. C., and Morgan-Jones, G. (1979a). Notes on hyphomycetes. XXIX. *Obstipispora chewaclensis* gen. et sp. nov. *Mycotaxon* **8**, 152-155.

Sinclair, R. C., and Morgan-Jones, G. (1979b). Notes on hyphomycetes. XXXII. Five new aquatic species. *Mycotaxon* **9**, 469-481.

Singh, N. (1972). *Pyramidospora constricta* sp. nov., a new aquatic hyphomycete. *Trans. Br. Mycol. Soc.* **59**, 336-339.

Singh, N. (1976). *Pyramidospora herculiformis* sp. nov., a new aquatic hyphomyete from Sierra Leone. *Trans. Br. Mycol. Soc.* **66**, 347-350.

Spegazzini, C. (1918). Notas micológicas. *Physis (Buenos Aires)* **4**, 281-295.

Suberkropp, K., and Klug, M. J. (1981). The degradation of leaf litter by aquatic hyphomycetes. *In* "Fungal Ecology" (D. T. Wicklow and G. C. Carroll, eds.), Dekker, New York (in press).

Subramanian, C. V., and Lodha, B. C. (1964). Two interesting hyphomycetes. *Can. J. Bot.* **42**, 1057-1063.

Sutton, B. C. (1975). Hyphomycetes on cupules of *Castanea sativa*. *Trans. Br. Mycol. Soc.* **64**, 405-426.
Svrček, M. (1972). *Miladina* gen. nov., eine neue Gattung für *Peziza lechithina* Cooke. *Ceska Mykol.* **26**, 213-216.
Thaxter, R. (1888). The Entomophthoreae of the United States. *Mem. Boston Soc. Nat. Hist.* **4**, 133-201.
Thornton, D. R. (1963). The physiology and nutrition of some aquatic hyphomycetes. *J. Gen. Microbiol.* **33**, 23-31.
Thornton, D. R. (1965). Amino acid analysis of fresh leaf litter and the nitrogen nutrition of some aquatic hyphomycetes. *Can. J. Microbiol.* **11**, 657-662.
Tubaki, K. (1957). Studies on the Japanese hyphomycetes. III. Aquatic group. *Bull. Natl. Sci. Mus., Tokyo* **41**, 249-268.
Tubaki, K. (1958). Studies on Japanese hyphomycetes. V. Leaf and stem group with a discussion of the classification of hyphomycetes and their perfect stages. *J. Hattori Bot. Lab.* **20**, 142-244.
Tubaki, K. (1966). An undescribed species of *Hymenoscyphus*, a perfect stage of *Varicosporium*. *Trans. Br. Mycol. Soc.* **49**, 345-349.
Tubaki, K. (1975a). Notes on the Japanese hyphomycetes. VI. *Candelabrum* and *Beverwijkella* gen. nov. *Trans. Mycol. Soc. Jpn.* **16**, 132-140.
Tubaki, K. (1975b). Notes on the Japanese hyphomycetes. VII. *Cancellidium*, a new hyphomycete genus. *Trans. Mycol. Soc. Jpn.* **16**, 357-360.
van Beverwijk, A. L. (1951a). Zalewski's "*Clathrosphaera spirifera.*" *Trans. Br. Mycol. Soc.* **34**, 280-290.
van Beverwijk, A. L. (1951b). *Candelabrum spinulosum*, a new fungus species. *Antonie van Leeuwenhoek* **17**, 278-284.
van Beverwijk, A. L. (1953). Helicosporous hyphomycetes. I. *Trans. Br. Mycol. Soc.* **36**, 111-124.
van Beverwijk, A. L. (1954). Three new fungi: *Helicoon pluriseptatum, Papulaspora pulmonaria,* and *Tricellula inaequalis. Antonie van Leeuwenhoek* **20**, 1-16.
Vanev, S. (1976). A new species of the genus *Gyoerffyella* (in Russian). *Fitologiya* **4**, 46-50.
Vischnevskaja, Z. A. (1955). De fungus aquaticis ante in regionis leningradensis non inventis. *Bot. Mat. Otd. Spor. Rast. Bot. Inst. V. L. Komarova* **10**, 148-150.
von Höhnel, F. (1912). Fragmente zur Mykologie. XIV. Mitteilung, Nr. 719-792. *Sitzungsber. Akad. Wiss. Wien, Math.-Naturwiss. Kl., Abt.* 3 **121**, 339-424.
Webster, J. (1951). Graminicolous pyrenomycetes. I. The conidial state of *Tubeufia helicomyces*. *Trans. Br. Mycol. Soc.* **34**, 304-308.
Webster, J. (1959a). Experiments with spores of aquatic hyphomycetes. I. Sedimentation and impaction on smooth surfaces. *Ann. Bot. (London)* [N. S.] **23**, 595-611.
Webster, J. (1959b). *Nectria lugdunensis* sp. nov., the perfect state of *Heliscus lugdunensis. Trans. Br. Mycol. Soc.* **42**, 322-327.
Webster, J. (1959c). *Tricellula aquatica* sp. nov., an aquatic hyphomycete. *Trans. Br. Mycol. Soc.* **42**, 416-420.
Webster, J. (1961). The *Mollisia* perfect state of *Anguillospora crassa* Ingold. *Trans. Br. Mycol. Soc.* **44**, 559-564.
Webster, J. (1965). The perfect state of *Pyricularia aquatica. Trans. Br. Mycol. Soc.* **48**, 449-452.
Webster, J. (1975). Further studies of sporulation of aquatic hyphomycetes in relation to aeration. *Trans. Br. Mycol. Soc.* **64**, 119-127.
Webster, J., and Davey, R. A. (1975). Sedimentation rates and trapping efficiency of cells of *Candida aquatica. Trans. Br. Mycol. Soc.* **64**, 437-440.
Webster, J., and Davey, R. A. (1981). Two aero-aquatic hyphomycetes from Malaysia. *Trans. Br. Mycol. Soc.* (in press).
Webster, J., and Descals, E. (1979). The perfect states of waterborne hyphomycetes from fresh

water. *In* "The Whole Fungus" (W. B. Kendrick, ed.), pp. 419-451. National Museums of Canada, Ottawa.
Webster, J., and Towfik, F. H. (1972). Sporulation of aquatic hyphomycetes in relation to aeration. *Trans. Br. Mycol. Soc.*, **59**, 353-364.
Webster, J., Moran, S. T., and Davey, R. A. (1976). Growth and sporulation of *Tricladium chaetocladium* and *Lunulospora curvula* in relation to temperature. *Trans. Br. Mycol. Soc.* **67**, 491-495.
Webster, J., Sanders, P. F., and Descals, E. (1978). Tetraradiate aquatic propagules in two species of *Entomophthora*. *Trans. Br. Mycol. Soc.* **70**, 472-479.
Willoughby, L. G. (1974). Decomposition of litter in fresh water. *In* "Biology of Plant Litter Decomposition" (C. H. Dickinson and G. J. F. Pugh, eds.), Vol. 2, pp. 659-681. Academic Press, New York.
Willoughby, L. G., and Archer, J. F. (1973). The fungal flora of a freshwater stream and its colonization pattern on wood. *Freshwater Biol.* **3**, 219-239.
Willoughby, L. G., and Minshall, G. W. (1975). Further observations on *Tricladium giganteum* Iqbal. *Trans. Br. Mycol. Soc.* **65**, 77-82.
Willoughby, L. G., and Sutcliffe, D. W. (1976). Experiments on feeding and growth of the amphipod *Gammarus pulex* (L.) related to its distribution in the River Duddon. *Freshwater Biol.* **6**, 577-586.
Wolfe, C. C. (1976). The distributional history of the biota of the southern Appalachians. IV. (Algae and Fungi). Aquatic hyphomycetes of the southern Appalachians. *Bull., Water Resour. Res. Cent., Va. Polytech Inst. State Univ.* 243-264.

12

Distribution and Ecology of Conidial Fungi in Marine Habitats

Jan Kohlmeyer

I.	Introduction	357
II.	Morphology and Dispersal of Marine Conidial Fungi	360
III.	Biology of Marine Conidial Fungi	361
	A. Marine Fungi as Parasites	361
	B. Marine Fungi as Saprobes	362
	C. Marine Fungi as Symbionts	368
IV.	Distribution of Marine Conidial Fungi	368
	A. Geographical Distribution	368
	B. Vertical Distribution	369
	References	370

I. INTRODUCTION

Since Saccardo (1880) described the first conidial fungus (*Camarosporium roumeguerii* Sacc.), found on the salt marsh plant *Salicornia herbacea* L., more than 50 species of indigenous marine imperfect fungi have been recorded (Kohlmeyer and Kohlmeyer, 1979). These fungi occur as parasites, saprobes, or symbionts on a wide variety of substrates in oceans and estuaries. Hosts or substrates include marine algae, submerged parts of tropical mangrove trees and temperate salt marsh halophytes, dead wood and other cellulosic substances, and even animal chitin and conchyolin. Conidial fungi constitute approximately one-fourth of all described marine filamentous fungi, with Ascomycetes (149) embracing the largest and Basidiomycetes (4) the smallest number of species.

Fungi from marine habitats are separated into obligate and facultative marine species, the former being restricted to the marine environment and the latter occurring also in freshwater or terrestrial localities, or both (Kohlmeyer and Kohlmeyer, 1964). The manner of collection is of utmost importance for the decision whether a fungus found in a marine milieu is active in a particular habitat or whether it occurs there merely in a dormant, inactive state. As discussed at length elsewhere (Kohlmeyer and Kohlmeyer, 1979), pour plate techniques for marine sediments or water performed in the laboratory may result in the isolation of terrestrial fungi derived from dormant propagules. The latter are prevented from germinating by a mycostatic principle that occurs under natural conditions in seawater. This chapter includes only fungi known to be active in the marine environment, that is, those that have been observed in a growing or fruiting stage on marine substrates (Table I). Keys for identification and full descriptions of all species are provided in a recent monograph (Kohlmeyer and Kohlmeyer, 1979).

TABLE I

List of Accepted Marine Conidial Fungi and Genera with Doubtful Marine Isolates

Hyphomycetes
  Agonomycetales
    Agonomycetaceae
      Papulaspora halima Anastasiou
  Hyphomycetales
    Moniliaceae
      Blodgettia bornetii Wright
      Botryophialophora marina Linder
      Clavatospora stellatacula Kirk
      Varicosporina ramulosa Meyers & Kohlm.
    Dematiaceae
      Alternaria spp. (doubtful marine isolates)
      Asteromyces cruciatus Moreau & Moreau
      Cirrenalia fusca I. Schmidt
      Cirrenalia macrocephala (Kohlm.) Meyers & Moore
      Cirrenalia pseudomacrocephala Kohlm.
      Cirrenalia pygmea Kohlm.
      Cirrenalia tropicalis Kohlm.
      Cladosporium algarum Cooke & Massee (also doubtful marine isolates of *Cladosporium*)
      Clavariopsis bulbosa Anastasiou
      Cremasteria cymatilis Meyers & Moore
      Dendryphiella arenaria Nicot

TABLE I (*Continued*)

    Dendryphiella salina (Sutherland) Pugh & Nicot
    Dictyosporium pelagicum (Linder) G. C. Hughes
    Drechslera halodes (Drechsler) Subramanian & Jain
    Humicola alopallonella Meyers & Moore
    Monodictys pelagica (Johnson) E. B. G. Jones
    Orbimyces spectabilis Linder
    Periconia abyssa Kohlm.
    Periconia prolifica Anastasiou
    Sporidesmium salinum E. B. G. Jones
    Stemphylium triglochinicola Sutton & Pirozynski (also doubtful marine isolate of *Stemphylium*)
    Trichocladium achrasporum (Meyers & Moore) Dixon
    Zalerion maritimum (Linder) Anastasiou
    Zalerion varium Anastasiou
  Tuberculariales
    Tuberculariaceae
      Allescheriella bathygena Kohlm.
      Epicoccum spp. (doubtful marine isolates)
      Tubercularia pulverulenta Spegazzini
Coelomycetes
  Sphaeropsidales
    Sphaerioidaceae
      Ascochyta salicorniae Magnus
      Ascochytula obiones (Jaap) Diedicke
      Camarosporium metableticum Trail
      Camarosporium palliatum Kohlm. & Kohlm.
      Camarosporium roumeguerii Saccardo
      Coniothyrium obiones Jaap
      Cytospora rhizophorae Kohlm. & Kohlm.
      Diplodia oraemaris Linder
      Macrophoma spp.
      Phialophorophoma litoralis Linder
      Phoma laminariae Cooke & Massee
      Phoma marina Lind
      Phoma suaedae Jaap (probably other marine spp. of *Phoma* occurring)
      Rhabdospora avicenniae Kohlm. & Kohlm.
      Robillarda rhizophorae Kohlm.
      Septoria ascophylli Melnik & Petrov
      Septoria thalassica Spegazzini
      Stagonospora haliclysta Kohlm.
      Stagonospora sp.
    Excipulaceae
      Dinemasporium marinum Nilsson
  Melanconiales
    Melanconiaceae
      Sphaceloma cecidii Kohlm.

II. MORPHOLOGY AND DISPERSAL OF MARINE CONIDIAL FUNGI

Marine and freshwater fungi are often morphologically adapted to life in the aquatic environment. The surface of ascospores of numerous marine species is increased by appendages. Conidia of many aquatic anamorphs have a tetraradiate shape (Ingold, 1975, 1976) which decreases the settling rate, assists in keeping the spores suspended in water, and increases the efficiency of impaction on a substrate. Whereas conidial fungi with tetraradiate conidia are commonly found in fresh water, this shape is rarer among marine species. *Varicosporina ramulosa* (Fig. 1) occurs on decaying plant material on sandy beaches of the tropics and subtropics. The branched conidia of this species can be found in foam along the seashore. *Asteromyces cruciatus* is a hyphomycete from similar habitats in temperate zones. The dispersal unit consists of a central bulbous conidiogenous cell bearing several ellipsoidal cells (conidia?) on thin denticles (Cole, 1976). Conidia of two other hyphomycetes, *Orbimyces spectabilis* and *Clavariopsis bulbosa*, are composed of radiating arms connected to a globose basal cell. A helicoid

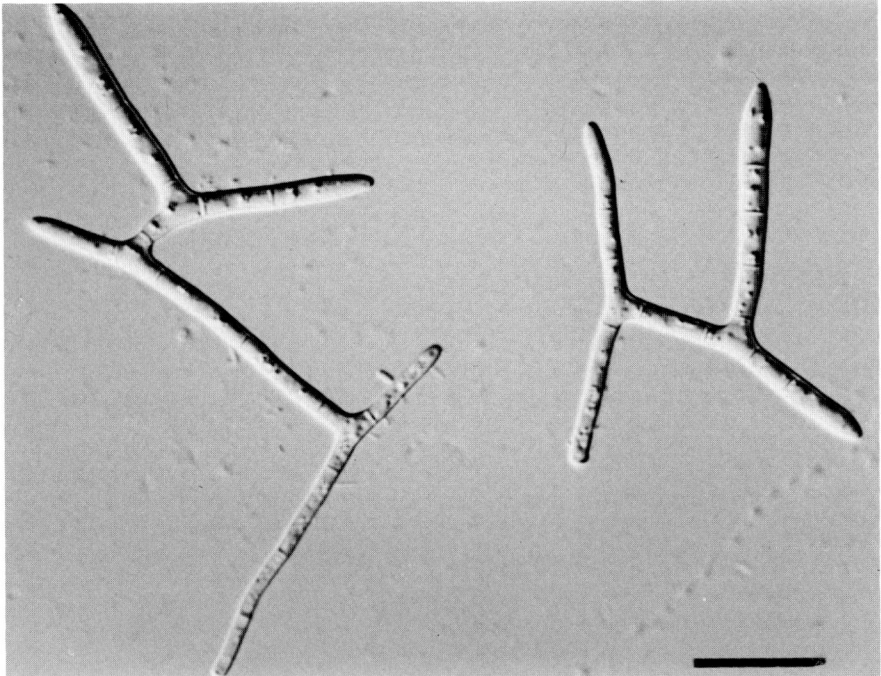

Fig. 1. Conidia of *Varicosporina ramulosa*. The left one is attached to the conidiophore. Nomarski interference contrast light microscopy; from incubated leaves of *Zostera marina*. North Carolina, Herb. J. K. No. 3907. Bar = 20 μm.

shape, as found in conidia of *Zalerion maritimum* (Cole, 1976) may also be advantageous for distribution and impaction in an aquatic environment.

Among the Coelomycetes, *Robillarda rhizophorae* has ellipsoidal conidia with radiating apical appendages. Muriform conidia of *Camarosporium metableticum* bear a caplike gelatinous appendage at each apex, which probably facilitates attachment of the spores to a substrate.

Information on conidial release and the dispersal of marine Fungi Imperfecti is limited. Conidia of most marine hyphomycetes are readily severed from the conidiogenous cell and washed away by currents. In Sphaeropsidales, conidia form beads or cirrhi on the ostioles of exposed pycnidia when the substrates, such as mangrove roots and pilings, dry out during low tide. Conidial release has not been observed under submerged conditions, but it is likely that conidia are dispersed by the water as soon as they reach the pycnidial ostiole. In some species, e.g., *Ascochyta salicorniae, Camarosporium roumeguerii,* and *Phoma laminariae,* the ostiolar canal is lined with gelatinous material or mucilage which probably prevents penetration of water into the venter.

It has long been known that conidia of freshwater Hyphomycetes are trapped in the scum and foam of streams and lakes (Ingold, 1942). Marine foam was first examined by Kohlmeyer (1966) who observed that propagules of marine fungi also accumulated between the air bubbles. Spores found in foam along the seashore indicate which species of fungi are common on decaying matter in the sand, e.g., the hyphomycetes *Asteromyces cruciatus* and *Varicosporina ramulosa.* Conidia trapped in the foam are deposited on washed-up material such as wood, algae, and leaves of sea grasses and then germinate and penetrate the substrate. Thus the entrapment of conidia in foam is an effective mechanism of bringing the propagules in contact with a substrate. Conidia in the foam have not germinated when just collected, but germination starts within a few hours after they have come to rest.

III. BIOLOGY OF MARINE CONIDIAL FUNGI

The number of parasitic and symbiotic marine Fungi Imperfecti is small, and most representatives of this group are saprobes on dead organic matter along the shores of oceans and estuaries.

A. Marine Fungi as Parasites

1. Parasites in Algae

The only described parasitic conidial fungus is *Sphaceloma cecidii,* which belongs to the Melanconiales. This fungus occurs as a hyperparasite in galls caused by members of *Haloguignardia* (Ascomycetes) in brown algae

(Kohlmeyer, 1972). Host plants are *Cystoseira osmundacea* (Menzies) C. Agardh, *Halidrys dioica* Gardner, *Sargassum fluitans* Börgesen, *S. natans* (L.) J. Meyen, and unidentified species of *Sargassum*. Galls induced by the ascomycete develop as subglobose or ellipsoidal outgrowths of stipes, vesicles, or blades of the hosts and contain immersed ascomata and spermogonia of *Haloguignardia* spp. The acervuli of *Sphaceloma cecidii* usually develop near the fruiting bodies, often closing their ostioles and rupturing the outer cell layers. Hyperparasite-free galls are brown in color, but infected galls appear black because of the dark-colored acervuli. They produce large numbers of one-celled, hyaline conidia which are ellipsoidal, truncate at the base, and provided with a gelatinous appendage which probably serves as a device for attaching the conidium to the substrate. *Sphaceloma cecidii* has not been cultured, and no information is available on its dispersal, mode of infection, or physiology. The fungus appears to have a wide distribution, as it has been found in Australia, California, and the Sargasso Sea.

2. Parasites in Higher Plants

Parasitic marine conidial fungi are rare or absent in salt marsh plants of temperate zones. Gessner and Goos (1973b) reported a *Phoma* sp. possibly causing leaf browning in *Spartina alterniflora* Loisel., although this fungus is probably not marine in the strict sense. Submerged parts of mangrove trees in the tropics and subtropics are occasionally infected by marine coelomycetes. The sole known parasitic fungus is *Cytospora rhizophorae*, which causes a dieback in young plants and seedlings of red mangroves, *Rhizophora* spp. (Kohlmeyer and Kohlmeyer, 1971). Lobed pycnidia of the parasite are embedded under the cortex of the host and release conidia from a neck that breaks through lentciles or bark. Under dry conditions the conidia are discharged as yellowish or orange cirrhi or beads. When diseased parts of red mangroves are submerged, conidia are washed away by water currents.

Rhabdospora avicenniae occurs on live, submerged pneumatophores of black mangroves, *Avicennia* spp. Since the pycnidia are embedded in dead cork cells of the roots, apparently without damaging the host, the fungus may be considered a saprobe rather than a parasite.

B. Marine Fungi as Saprobes

The majority of marine conidial fungi are saprobes in organic substrates of permanently submerged or intertidal habitats. Wood has been used most often to observe and isolate marine fungi, whereas other materials, such as abundantly available algae and sea grasses, have been more or less neglected.

1. Saprobes on Wood

Among 107 species of higher fungi found on wood in marine habitats, 29 have been identified as conidial anamorphs. The remainder belong mostly to the

Ascomycetes (Kohlmeyer and Kohlmeyer, 1979). Wood-decaying fungi are the most studied marine fungi, and much information on their morphology, physiology, ecology, and distribution has become known since Barghoorn and Linder (1944) published a pioneering paper on this subject.

Terrestrial and marine Ascomycetes and conidial fungi cause a decay termed soft rot in the outer layers of wood. Hyphae grow in the less lignified middle layer of the secondary cell wall; they follow the orientation of the cellulose microfibrils and cause spirally arranged chains of cavities which appear as holes in cross sections of the wood. Such micromorphological symptoms of wood decay can be reproduced in the laboratory by pure cultures of *Cirrenalia macrocephala, Cremasteria cymatilis, Humicola alopallonella, Monodictys pelagica,* or *Zalerion maritimum* (Jones, 1971).

Wood-inhabiting or lignicolous fungi are collected in nature on so-called intertidal wood, i.e., fixed substrates in the intertidal zone, or on loose driftwood washed up on the beach (Hughes, 1968). Marine conidial fungi do not readily sporulate under submerged conditions in nature, but conidia-bearing mycelia are often found on wood near the high-water mark. Wood samples from marine habitats incubated in damp chambers in the laboratory may yield indigenous marine fungi; however, terrestrial contaminants may develop as well, and isolations must be made with a critical mind. Most marine mycologists use wood panels, submerged for certain periods of time, to obtain lignicolous fungi (Meyers and Reynolds, 1958). This method is rather selective, and additional sampling of naturally occurring wood, such as driftwood and pilings, should be used in ecological studies to observe the majority of species present in a particular area (Höhnk, 1954, 1955; Kohlmeyer, 1959, 1960).

Among the enzymes produced by marine conidial fungi, the occurrence of cellulase has been tested most often by physiologists and biochemists. Barghoorn and Linder (1944) first grew isolates of marine fungi on wood flour and regenerated cotton cellulose and found poor growth of *Phialophorophoma litoralis,* whereas *Zalerion maritimum* (*sub Helicoma salinum*) and most of the ascomycetes showed vigorous development. However, the behavior of a fungus in pure culture does not necessarily give clues to its activity in the natural environment. Using *Humicola alopallonella,* as well as 35 terrestrial lignicolous fungi, Nilsson (1974) suggested that growth on cellulose agar was not a valid measure of cellulase activity. Other researchers have used sodium carboxymethyl cellulose to demonstrate the production of cellulase (e.g., Meyers and Reynolds, 1959). Schaumann (1974) concluded that only the activity of β-1,4-glucanases of the cellulase complex could be determined with the methods employed.

Another technique for testing the wood-degrading ability of a certain species is the determination of weight losses of the substrate. Marine conidial fungi employed by Henningsson (1976) to demonstrate loss in weight of softwoods and hardwoods were *Monodictys pelagica, Trichocladium achrasporum,* and *Zalerion maritimum.* The highest loss after 3 months was caused by *Z. maritimum* in oak (12.2%). After 6 months, *M. pelagica* caused a loss in dry weight of 25.9% in birch wood. Leightley and Eaton (1977), using *Alternaria maritima, Cirrenalia macrocephala, Culcitalna achraspora, Humicola alopallonella,* and *Z. maritimum,* found weight losses between 1.4 and 4.1% in pine (*Pinus sylvestris* L.) after 3 months. Losses in beech (*Fagus sylvatica* L.) measured between 1.7 and 11.3%, with *H. alopallonella* causing the greatest decomposition. Weight losses in nature, where several fungal species and bacteria may attack the wood at the same time, are considerably higher. Jones and Irvine (1971) and Byrne and Eaton (1972) determined weight losses of 27.6% and about 20% in beech and pine, respectively, after 40–42 weeks of exposure.

Little is known of the preference of marine conidial fungi for certain wood species. Jones (1968) has reported that *Cirrenalia macrocephala* and *Humicola alopallonella* occur preferentially on pine wood (*Pinus sylvestris*). Data on frequencies of imperfect fungi in marine habitats are also rare. Divergent results of studies made by different researchers may be explained by the different substrates used and by environmental differences in the test areas. In a worldwide test conducted by Jones *et al.* (1972), *C. macrocephala* was found on 45% of the wood panels. *Zalerion maritimum* occurred on 32%, *Dendryphiella salina* on 16%, and *H. alopallonella* on 13% of the wood blocks. In a study conducted in the North Sea, Schaumann (1975) collected *C. macrocephala* on about 16%, *Z. maritimum* on 14%, and *H. alopallonella* on 8% of 344 wood specimens.

2. Saprobes on Bark and Mangrove Seedlings

Nine conidial fungi have been reported from bark in marine habitats: *Cirrenalia macrocephala, C. pygmea, Dictyosporium pelagicum, Humicola alopallonella, Phialophorophoma litoralis, Rhabdospora avicenniae, Trichocladium achrasporum,* and *Zalerion maritimum.* Of these species, only *R. avicenniae* is restricted to bark. The others occur on alternate substrates as well (Kohlmeyer and Kohlmeyer, 1979). Bark consists mostly of dead cork cells containing suberin, which makes these tissues rather resistant to microbial decay. No information is available on the mechanism by which marine fungi break down bark.

Mangrove bark appears to be particularly well protected against microbial attack by high amounts of tannin. Wood-inhabiting fungi penetrate the wood of submerged mangrove roots or trunks only after the bark has been damaged. Newell (1976) made a thorough study of the fungal colonization of seedlings of red mangrove (*Rhizophora mangle* L.) in Florida estuaries. About 80% of the viviparous seedlings, while still attached to the tree, bear 16 common phylloplane hyphomycetes, most of which belong to the genera *Alternaria, Aureobasidium, Cladosporium,* or *Pestalotia.* After the seedlings drop into the water, the terrestrial species persist for about 2 months, and some facultative marine species appear. The subepidermis is penetrated by fungi only after the second to fifth month of submergence, and a *Pestalotia* sp. is most common at this stage. The first marine fungi appear after 5 or 6 months, among them *Cytospora rhizophorae* and *Zalerion varium*. The cortex becomes heavily invaded by hyphae and fungal pseudoparenchyma, and in a last stage of succession representatives of *Papulaspora*, *Penicillium*, and *Trichoderma* appear on the senescent or dead seedlings. Among the total of 84 species isolated by Newell (1976), there were only 15 obligate marine fungi. Obligate and facultative marine conidial fungi found in this study included *Cirrenalia pseudomacrocephala*, *Culcitalna achraspora*, *Cytospora rhizophorae*, *Dendryphiella salina*, *Papulaspora halima*, *Periconia prolifica*, *Robillarda rhizophorae*, and *Zalerion varium*. The attack by microorganisms causes a marked increase in nitrogen content in the seedlings, and an improvement in nutritional value of the substrate for detritivores can be expected.

3. Saprobes on Higher Plants in Temperate Salt Marshes

Although salt marshes are important producers of detritus (Odum and De La Cruz, 1967), which supports complex food webs in estuaries, the role of microorganisms, in particular the fungi, in the breakdown of the halophytes is not fully understood. A recent compilation of higher filamentous fungi occurring on submerged parts of salt marsh plants lists 33 ascomycetes, 23 conidial fungi, and 1 basidiomycete (Kohlmeyer and Kohlmeyer, 1979). It shall suffice to give some examples of the imperfect fungi growing on two major hosts. Genera of Fungi Imperfecti found on *Salicornia* spp. are *Ascochyta*, *Camarosporium*, *Dendryphiella*, *Phoma*, and *Tubercularia*. On *Spartina* spp. the following genera occur: *Alternaria*, *Asteromyces*, *Cirrenalia*, *Dendryphiella*, *Dictyosporium*, *Drechslera*, *Epicoccum*, *Monodictys*, *Phoma*, *Stagonospora*, and *Stemphylium*.

Few ecological studies on fungi from salt marsh halophytes have been carried out. Dickinson (1965), Dickinson and Pugh (1965a,b,c), and Dickinson and Morgan-Jones (1966) examined the mycota of root, leaf, and propagule surfaces of *Halimione portulacoides* (L.) Aellen in England. *Dendryphiella salina* and *Ascochytula obiones*, in addition to some terrestrial fungi, were the initial colonizers. Moribund stems of the same host are invaded by the coelomycetes *Camarosporium obiones* and *Coniothyrium obiones*. Saprobic fungi usually appear on halophytes after the fleshy leaves and stems dry out. For example, pycnidia of *Ascochyta salicorniae* and *Camarosporium roumeguerii* develop first under the epidermis of the lower, dry branches of *Salicornia* spp. Eventually the fruiting bodies are spread over the whole desiccated plant.

In a survey of fungi occurring on standing *Spartina alterniflora*, Gessner and Goos (1973b) found the following marine conidial fungi: *Cirrenalia macrocephala*, *Dendryphiella arenaria*, *D. salina*, *Dictyosporium pelagicum*, and *Monodictys pelagica*. Using litter bags filled with *S. alterniflora*, the same authors (Gessner and Goos, 1973a) observed *M. pelagica* and *Phoma* sp. by microscopic examination on decaying plants at the end of the experiment. In dilution plates of the same grass a large number of terrestrial conidial fungi were isolated besides the marine species *Dendryphiella arenaria, D. salina,* and *Dictyosporium pelagicum.* The small number of marine fungi, in contrast to the many terrestrial species isolated by the pour plate technique, indicates the presence of dormant terrestrial fungi on the plants' surface in nature. The pour plate technique is apparently selective for terrestrial species.

4. Saprobes on Submerged Leaves

Shed leaves of sea grasses and mangroves, which occur in great volume in estuaries, are transported by currents to the bottom of the ocean (Menzies *et al.,* 1967). Leaves in the marine environment are mainly decomposed by conidial and lower fungi. Ascomycetes are uncommon, and basidiomycetes appear to be absent from submerged leaves.

Fell and Master (1973) and Fell *et al.* (1975) studied fungal colonization on red mangrove leaves (*Rhizophora mangle*). Initially, some of the terrestrial leaf-inhabiting fungi, e.g., *Pestalotia* sp., persist on the substrate after the leaves are submerged. Lower fungi are the first aquatic invaders, accompanied by a few hyphomycetes. After one or two weeks obligate marine fungi appear, e.g., *Cirrenalia macrocephala, Dictyosporium pelagicum, Varicosporina ramulosa,* and *Zalerion varium.* Most of the lower fungi have disappeared after 3 weeks. A total of 43 conidial fungi, most of them ubiquitous saprobes, were isolated in these studies. Within 6 weeks, unless they accumulated in areas protected against wave action, the mangrove leaves were reduced to detritus.

Experiments in temperate waters of British Columbia yielded mainly conidial fungi on submerged leaves of *Alnus rubra* Bong., *Arbutus menziesii* Pursh, and *Prunus laurocerasus* L. (Anastasiou and Churchland, 1969; Churchland and McClaren, 1973). *Zalerion maritimum* occurred most commonly and was also a pioneer species.

Leaves of *Zostera marina* L. washed up on the beaches of North Carolina are invaded by higher fungi, among them the conidial forms *Alternaria* sp., *Phoma* sp., and *Varicosporina ramulosa* (Kohlmeyer, 1966). In the tropics, leaves of the turtle grass *Thalassia testudinum* Banks ex König are attacked by *Dendryphiella arenaria, Varicosporina ramulosa,* and ascomycetes (Meyers and Kohlmeyer, 1965; Meyers *et al.,* 1965). Incubated leaf segments of *T. testudinum,* yielded additional hyphomycetes: *Aspergillus, Cephalosporium (Acremonium),* and *Penicillium* spp. (Meyers *et al.,* 1965). It has not been proven, however, that these fungi actively grow on the leaves under submerged conditions.

5. Saprobes on Algae

Saprobic conidial fungi have been found on decomposing Phaeophyta and on a few Rhodophyta of drift plants along the seashore. Only 10 imperfect fungi and 8 ascomycetes have been recorded on such substrates (Kohlmeyer and Kohlmeyer, 1979). The majority of these fungi belong to Hyphomycetes (*Asteromyces, Cladosporium, Dendryphiella, Epicoccum, Stemphylium, Varicosporina* spp.) and include only three coelomycetes (*Phoma laminariae, Phoma* sp., and *Stagonospora haliclysta*). Hyphae of *Cladosporium algarum* pervade decaying thalli of *Laminaria* spp., forming a dense layer of conidiophores on the surface, while partly immersed pycnidia of *Phoma laminariae* develop nearby.

There is little information available on the breakdown of algal components by conidial fungi. The only studies in this field are by Chesters and Bull (1963), who demonstrated substantial laminarinase activity in *Dendryphiella salina,* and by Tubaki (1969), who cultured *D. arenaria* and *Varicosporina ramulosa* on laminarin.

6. Saprobes on Animal Products

Records of conidial fungi on animal substrates are rare. An anamorphic fungus, tentatively identified as *Periconia prolifica,* was found to cause black spots on the chitinous exoskeleton of *Carcinus maenas* L. in England (Alderman, 1976). Hyphae, conidia, and aleuriospores are also present in gills and other tissues of this crab.

Hyphomycetes often develop on the inner surface of empty calcareous tubes of shipworms in wood submerged in tropical waters. Conidia of *Cirrenalia pygmea* and *Humicola alopallonella* are found on the surface of these tubes. *Periconia prolifica* is found superficially and immersed in the calcareous substrate, which becomes soft and brittle (Kohlmeyer, 1969a). The organic matrix of the tubes, conchyolin, probably serves as an organic nutrient for the growth of these fungi. Hyphae and conidia of *P. prolifica* also occur in and on dead mites attached to empty shipworm tubes (Kohlmeyer and Kohlmeyer, 1977).

7. Saprobes in Marine Sediments and Soil

As discussed in Section I, isolations of fungi made with traditional plating techniques cannot give clues to the activities of the organisms in nature. Hundreds of fungi have been isolated from marine sediments and waters, but only a few of the conidial forms among them (*Asteromyces cruciatus, Dendryphiella arenaria,* and *D. salina*) have been observed growing *in situ*. It has been argued (Kohlmeyer and Kohlmeyer, 1979) that the majority of the fungi isolated from marine sediments and waters derive from dormant propagules and not from actively growing species. Culture methods may exclude most indigenous marine organisms.

I will not list the large number of conidial fungi isolated by plating techniques, because many may not be true marine species. The relevant literature has been compiled by Kohlmeyer and Kohlmeyer (1979). Investigators using standard isolation techniques in the future ought to consider the "mycostatic factor" present in seawater *in situ* (Tyndall and Kirk, 1973) and determine the ecological role of the isolated fungi in the natural environment. Without such knowledge, routine isolations of fungi with plating methods are meaningless.

C. Marine Fungi as Symbionts

Among symbiotic marine fungi there is only one conidial form, namely, *Blodgettia bornetii*. Anastomosing hyphae form a network inside the wall of *Cladophora* spp. (Feldmann, 1938) found in the tropics and subtropics. Hyaline chlamydospores are also produced between the outer and inner wall layer of the host. This marine hyphomycete does not penetrate the host cells but remains restricted to the outer walls and is absent from the cross walls and apexes of the algae. The association between *Cladophora* spp. and *B. bornetii* is considered a mycophycobiosis, that is, a symbiosis between a marine macroalga and a systemic fungus (Kohlmeyer and Kohlmeyer, 1979). Examples of these symbiotic relationships are usually found in the upper intertidal zone where they are regularly exposed to desiccation. It is probable that the alga–fungus association, like terrestrial lichens, has evolved resistance to rigorous environmental conditions.

IV. DISTRIBUTION OF MARINE CONIDIAL FUNGI

A. Geographical Distribution

The most important environmental factor controlling the geographical distribution of marine fungi appears to be temperature. Hughes (1974) established five temperature-determined biogeographical zones for the distribution of marine fungi: arctic, temperate, subtropical, tropical, and antarctic. In view of the small number of distributional data available, most marine fungi cannot be placed as yet in definite zones. Observations on *Asteromyces cruciatus* and *Varicosporina ramulosa* indicate that the former require temperate and the latter tropical-subtropical zones (Kohlmeyer, 1971). Their distributions do not overlap; *A. cruciatus* was found from New Jersey northward, and *V. ramulosa* from North Carolina southward. The dividing line is most probably at Cape Hatteras, which also represents the boundary between northern and southern species of marine algae and animals. Some cosmopolitan species, such as *Humicola alopallonella* and *Trichocladium achrasporum* (anamorph of *Halosphaeria mediosetigera* Cribb & Cribb), appear to have less distinct temperature requirements, as they occur all around the world.

Host-specific fungi are usually found throughout the range of their respective host plants. Introduction of certain plants into new areas may result also in introducing their parasites. Red mangroves, *Rhizophora mangle*, were brought to Hawaii from Florida in 1902. Apparently, *Cytospora rhizophorae* and the ascomycete *Keissleriella blepharospora* Kohlm. & Kohlm. accompanied the seedlings to their new habitat (Kohlmeyer, 1969b).

Intensive collecting of marine fungi has so far been restricted to Europe and North America. The tropics in general have been little explored, and information on the occurrence of fungi in the Indian Ocean is particularly scarce.

B. Vertical Distribution

Investigations on the vertical distribution of marine fungi have been made only by Schaumann (1968, 1969), who observed the occurrence of ascomycetes and conidial fungi on stationary wood in the Weser estuary and Helgoland. At the Helgoland site, *Cirrenalia macrocephala, Dictyosporium pelagicum,* and *Monodictys pelagica* grew near the mean high-tide line. In the Weser estuary the same fungi developed from below the mean low-tide zone to above the mean high-tide mark. Definite zonation patterns have not been demonstrated so far in temperate areas. Also, observations of mangrove roots in the tropics did not show clear vertical distributions of marine fungal species (Kohlmeyer, 1969c).

There is a distinct separation between marine and terrestrial fungi along standing roots and stems of mangroves and along salt marcs plants (e.g., *Spartina* spp.). Marine fungi develop in the lower, submerged parts of the plants up to the high-water mark. Terrestrial fungi grow in the upper, nonimmersed parts of the hosts. An overlapping between the two groups of fungi may occur at the high-water line.

Actively growing conidial fungi occur in all parts of the oceans, from the shore to depths of several thousand meters. The mycota found below about 500 m is different from that above this line (Kohlmeyer and Kohlmeyer, 1979). Only a few collections of fungi have been made in the deep sea. Among the five indigenous species described from deep-sea habitats, two were anamorph-taxa, *Allescheriella bathygena* and *Periconia abyssa* (Kohlmeyer, 1969d, 1977), and both were wood decomposers. The first species occurred at 1722 m, while the second was found at 3975 and 5315 m. Higher deep-sea fungi have not been isolated, and information on their behavior in culture is lacking. Possibly, these species require, or at least can tolerate, low temperatures and high pressures.

ACKNOWLEDGMENTS

Support by grants from the U.S. National Science Foundation (DEB74-18539) and the Brown-Hazen Fund of the Research Corporation is gratefully acknowledged. Mrs. Erika Kohlmeyer assisted in many ways in the preparation of this chapter.

REFERENCES

Alderman, D. J. (1976). Fungal diseases of marine animals. In "Recent Advances in Aquatic Mycology" (E. G. B. Jones, ed.), pp. 223-260. Wiley, New York.
Anastasiou, C. J., and Churchland, L. M. (1969). Fungi on decaying leaves in marine habitats. *Can. J. Bot.* **47,** 251-257.
Barghoorn, E. S., and Linder, D. H. (1944). Marine fungi: Their taxonomy and biology. *Farlowia* **1,** 395-467.
Byrne, P. J., and Eaton, R. A. (1972). Fungal attack on wood submerged in waters of different salinity. *Int. Biodeterior. Bull.* **8,** 127-134.
Chesters, C. G. C., and Bull, A. T. (1963). The enzymic degradation of laminarin. I. The distribution of laminarinase among microorganisms. *Biochem. J.* **86,** 28-31.
Churchland, L. M., and McClaren, M. (1973). Marine fungi isolated from a Kraft pulp mill outfall area. *Can. J. Bot.* **51,** 1703-1710.
Cole, G. T. (1976). Conidium ontogeny in marine hyphomycetous fungi: *Asteromyces cruciatus* and *Zalerion maritimum. Mar. Biol.* **38,** 147-158.
Dickinson, C. H. (1965). The mycoflora associated with *Halimione portulacoides*. III. Fungi on green and moribund leaves. *Trans. Br. Mycol. Soc.* **48,** 603-610.
Dickinson, C. H., and Morgan-Jones, G. (1966). The mycoflora associated with *Halimione portulacoides*. IV. Observations on some species of Sphaeropsidales. *Trans. Br. Mycol. Soc.* **49,** 43-55.
Dickinson, C. H., and Pugh, G. J. F. (1965a). The mycoflora associated with *Halimione portulacoides*. I. The establishment of the root surface flora of mature plants. *Trans. Br. Mycol. Soc.* **48,** 381-390.
Dickinson, C. H., and Pugh, G. J. F. (1965b). Use of a selective cellulose agar for isolation of soil fungi. *Nature (London)* **207,** 440-441.
Dickinson, C. H., and Pugh, G. J. F. (1965c). The mycoflora associated with *Halimione portulacoides*. II. Root surface fungi of mature and excised plants. *Trans. Br. Mycol. Soc.* **48,** 595-602.
Feldmann, J. (1938). Le *Blodgettia confervoides* Harv. est-il un lichen? *Rev. Bryol. Lichénol.* **11,** 155-163.
Fell, J. W., and Master, I. M. (1973). Fungi associated with the degradation of mangrove (*Rhizophora mangle* L.) leaves in south Florida. *Estuarine Microb. Ecol., Pap. Belle W. Baruch Symp. Mar, Sci., 1st, 1971* pp. 455-465.
Fell, J. W., Cefalu, R. C., Master, I. M., and Tallman, A. S. (1975). Microbial activities in the mangrove (*Rhizophora mangle*) leaf detrital system. In "Biology and Management of Mangroves" (G. Walsh, S. Snedaker, and H. Teas, eds.), Vol. 2, p. 661-679. University of Florida, Gainesville.
Gessner, R. V., and Goos, R. D. (1973a). Fungi from decomposing *Spartina alterniflora*. *Can. J. Bot.* **51,** 51-55.
Gessner, R. V., and Goos, R. D. (1973b). Fungi from *Spartina alterniflora* in Rhode Island. *Mycologia* **65,** 1296-1301.
Henningsson, M. (1976). Degradation of wood by some fungi from the Baltic and the west coast of Sweden. *Mater. Org., Beih.* **3,** 509-519.
Höhnk, W. (1954). Studien zur Brack- und Seewassermykologie. IV. Ascomyceten des Küstensandes. *Veroeff. Inst. Meeresforsch. Bremerhaven* **3,** 27-33.
Höhnk, W. (1955). Studien zur Brack- und Seewassermykologie. V. Höhere Pilze des submersen Holzes. *Veroeff. Inst. Meeresforsch. Bremerhaven* **3,** 199-227.
Hughes, G. C. (1968). Intertidal lignicolous fungi from Newfoundland. *Can. J. Bot.* **46,** 1409-1417.

Hughes, G. C. (1974). Geographical distribution of the higher marine fungi. *Veroeff. Inst. Meeresforsch. Bremerh., Suppl.* **5,** 419-441.
Ingold, C. T. (1942). Aquatic Hyphomycetes of decaying alder leaves. *Trans. Br. Mycol. Soc.* **25,** 339-417.
Ingold, C. T. (1975). Convergent evolution in aquatic fungi: The tetraradiate spore. *Biol. J. Linn. Soc.* **7,** 1-25.
Ingold, C. T. (1976). The morphology and biology of freshwater fungi excluding Phycomycetes. *In* "Recent Advances in Aquatic Mycology" (E. B. G. Jones, ed.), pp. 335-357. Wiley, New York.
Jones, E. B. G. (1968). The distribution of marine fungi on wood submerged in the sea. *In* "Biodeterioration of Materials" (A. H. Walters and J. J. Elphick, eds.), pp. 460-485. Elsevier, Amsterdam.
Jones, E. B. G. (1971). The ecology and rotting ability of marine fungi. *In* "Marine Borers, Fungi and Fouling Organisms of Wood" (E. B. G. Jones and S. K. Eltringham, eds.), pp. 237-258. Organization for Economic Cooperation and Development, Paris.
Jones, E. B. G., and Irvine, J. (1971). The role of fungi in the deterioration of wood in the sea. *J. Inst. Wood Sci.* **29,** 31-40.
Jones, E. B. G., Kühne, H., Trussell, P. C., and Turner, R. D. (1972). Results of an international cooperative research programme on the biodeterioration of timber submerged in the sea. *Mater. Org.* **7,** 93-118.
Kohlmeyer, J. (1959). Neufunde holzbesiedelnder Meerespilze. *Nova Hedwigia* **1,** 77-99.
Kohlmeyer, J. (1960). Wood-inhabiting marine fungi from the Pacific Northwest and California. *Nova Hedwigia* **2,** 293-343.
Kohlmeyer, J. (1966). Ecological observations on arenicolous marine fungi. *Z. Allg. Mikrobiol.* **6,** 94-105.
Kohlmeyer, J. (1969a). The role of marine fungi in the penetration of calcareous substances. *Am. Zool.* **9,** 741-746.
Kohlmeyer, J. (1969b). Marine fungi of Hawaii including the new genus *Helicascus*. *Can. J. Bot.* **47,** 1469-1487.
Kohlmeyer, J. (1969c). Ecological notes on fungi in mangrove forests. *Trans. Br. Mycol. Soc.* **53,** 237-250.
Kohlmeyer, J. (1969d). Deterioration of wood by marine fungi in the deep sea. *Am. Soc. Test. Mater., Spec. Tech. Publ.* **445,** 20-29.
Kohlmeyer, J. (1971). Annotated check-list of New England marine fungi. *Trans. Br. Mycol. Soc.* **57,** 473-492.
Kohlmeyer, J. (1972). Parasitic *Haloguignardia oceanica* (Ascomycetes) and hyperparasitic *Sphaceloma cecidii* sp. nov. (Deuteromycetes) in drift *Sargassum* in North Carolina. *J. Elisha Mitchell Sci. Soc.* **88,** 255-259.
Kohlmeyer, J. (1977). New genera and species of higher fungi from the deep sea (1615-5315 m). *Rev. Mycol.* **41,** 189-206.
Kohlmeyer, J., and Kohlmeyer, E. (1964). Synoptic Plates of Higher Marine Fungi, 2nd ed. Cramer, Weinheim.
Kohlmeyer, J., and Kohlmeyer, E. (1971). Marine fungi from tropical America and Africa. *Mycologia* **63,** 831-861.
Kohlmeyer, J., and Kohlmeyer, E. (1977). Bermuda marine fungi. *Trans. Br. Mycol. Soc.* **68,** 207-219.
Kohlmeyer, J., and Kohlmeyer, E. (1979). Marine Mycology: The Higher Fungi. Academic Press, New York.
Leightley, L. E., and Eaton, R. A. (1977). Mechanisms of decay of timber by aquatic microorganisms. *Br. Wood Preserv. Assoc. Annu. Conv.* pp. 1-26.

Menzies, R. J., Zaneveld, J. S., and Pratt, R. M. (1967). Transported turtle grass as a source of organic enrichment of abyssal sediments off North Carolina. *Deep-Sea Res.* **14,** 111-112.

Meyers, S. P., and Kohlmeyer, J. J. (1965). *Varicosporina ramulosa* gen. nov. sp. nov., an aquatic Hyphomycete from marine areas. *Can. J. Bot.* **43,** 915-921.

Meyers, S. P., and Reynolds, E. S. (1958). A wood incubation method for the study of lignicolous marine fungi. *Bull. Mar. Sci. Gulf Caribb.* **8,** 342-347.

Meyers, S. P., and Reynolds, E. S. (1959). Growth and cellulolytic activity of lignicolous Deuteromycetes from marine localities. *Can. J. Microbiol.* **5,** 493-503.

Meyers, S. P., Orpurt, P. A., Simms, J., and Boral, L. L. (1965). Thalassiomycetes. VII. Observations on fungal infestation of turtle grass, *Thalassia testudinum* König. *Bull, Mar. Sci.* **15,** 548-564.

Newell, S. Y. (1976). Mangrove fungi: The succession in the mycoflora of red mangrove (*Rhizophora mangle* L.) seedlings. *In* "Recent Advances in Aquatic Mycology" (E. B. G. Jones, ed.), pp. 51-91. Wiley, New York.

Nilsson, T. (1974). The degradation of cellulose and the production of cellulase, xylanase, mannanase and amylase by wood-attacking microfungi. *Stud. For. Suec.* **114,** 1-61.

Odum, E. P., and De La Cruz, A. A. (1967). Particulate organic detritus in a Georgia salt marsestuarine ecosystem. *In* "Estuaries" (G. H. Lauff, ed.), Publ. No. 83, pp. 383-388. Am. Assoc. Adv. Sci., Washington, D. C.

Saccardo, P. A. (1880). Fungi gallici. Series II. *Michelia* **2,** 39-135.

Schaumann, K. (1968). Marine höhere Pilze (Ascomycetes und Fungi imperfecti) aus dem Weser-Ästuar. *Veroeff. Inst. Meeresforsch. Bremerhaven* **11,** 93-117.

Schaumann, K. (1969). Über marine höhere Pilze von Holzsubstraten der Nordsee-Insel Helgoland. *Ber. Dtsch. Bot. Ges.* **82,** 307-327.

Schaumann, K. (1974). Experimentelle Unterschungen zur Produktion und Aktivität cellulolytischer Enzyme bei höheren Pilzen aus dem Meer- und Brackwasser. *Mar. Biol.* **28,** 221-235.

Schaumann, K. (1975). Ökologische Untersuchungen über höhere Pilze im Meer- und Brackwasser der Deutschen Bucht unter besonderer Berücksichtigung der holzbesiedelnden Arten. *Veroeff. Inst. Meeresforsch. Bremerhaven* **15,** 79-182.

Tubaki, K. (1969). Studies on the Japanese marine fungi, lignicolous group (III), algicolous group and a general consideration. *Annu. Rep. Inst. Ferment., Osaka* **4,** 12-41.

Tyndall, R. W., and Kirk, P. W., Jr. (1973). Factors in seawater affecting spore germination in marine lignicolous fungi. *Va. J. Sci.* **24,** 136.

13

The Aerobiology of Conidial Fungi

J. Lacey

I.	Introduction	373
II.	Methods of Study	374
	A. Spore-Trapping Techniques	374
	B. Isolation Media	376
	C. Choice of Spore Trap	376
III.	The Aerial Environment	378
	A. Structure of the Atmosphere	378
	B. The Indoor Aerial Environment	379
IV.	Populations of Airborne Conidia	380
	A. Outdoor Populations	380
	B. The Air Spora Indoors	393
V.	Dispersion of Airborne Conidia	396
	A. Liberation and Takeoff	396
	B. Dispersal	399
	C. Deposition	404
VI.	Implications of an Air Spora	406
	A. Human and Animal Disease	406
	B. Plant Pathology	408
	C. Biodeterioration and Biodegradation	409
VII.	Conclusion	409
	References	410

I. INTRODUCTION

The air we breathe is seldom free of fungus spores, but their number and type vary with the time of day, season, weather, geographical location, and nearness of large spore sources. Conidia make a large contribution to this air spora, mostly originating from living and dead vegetation outdoors and from stored products indoors. The air spora comprises spores in the process of dispersal following their liberation by means characteristic of each species and prior to deposition on new substrates either by physical processes or by rain washing.

The study of aerobiology is then important in understanding the distribution and ecology of fungi and also relates to their interactions with plants, with man, and with their products. Many spores will be killed by exposure to the elements. Of those that survive, some may be harmless to other forms of life, others cause diseases of plants, humans, and animals, and yet others cause deterioration of foodstuffs and manufactured products.

Much of our knowledge of the behavior of airborne spores comes from study of the epidemiology of plant disease and of infectious disease and allergy in man. It has been stimulated during the last 25 years by the development of new methods of trapping spores, by collaboration among engineers, physicists, microbiologists, hygienists, allergists, and pathologists, and by the synthesis made by P. H. Gregory in his book, *Microbiology of the Atmosphere* (1973). Here this knowledge will be related to the study of conidial fungi.

II. METHODS OF STUDY

A. Spore-Trapping Techniques

In order to study populations of airborne particles and their dispersal, they must first be trapped, classified, and counted. These three processes provide a succession of problems that have long exercised the minds of physicists, engineers, microbiologists, and allergists. The result has been a large number of devices, many developed for industrial hygiene use, each with its own characteristics but few well suited to trapping spores. The range of instruments available has been described by Wolf *et al.* (1959) and Gregory (1973), and only the most useful and widely used will be discussed here.

Different spore-trapping techniques use a variety of processes. Sedimentation methods are widely employed by allergists, but recently the advantages of impaction techniques have been recognized. Other traps may utilize impingement into liquids, filtration, cyclones, electrostatic charges, or thermal precipitation.

Sedimentation has been widely used to trap spores. *Settle plates* are inexpensive and easy to use but suffer from severe limitations. They are not volumetric and can only give qualitative results. The rate of deposition of spores varies with their terminal velocity, which is a function of the square of their radius and causes large spores to be sampled, effectively, from a greater volume of air than small ones. Trapping efficiency varies with wind speed and turbulence because of aerodynamic effects such as edge drift and shadowing. Edge drift results in increased spore deposits behind the leading edge and in front of the trailing edge of the dish at low wind speeds (1.1 and 1.7 m/s), whereas shadowing results in poor deposition above 3.2 m/s unless the dish is placed at the bottom of a tube sunk below a horizontal surface. *Gravity slides* suffer drawbacks similar to those of settle plates, with almost zero trapping efficiency at 1.7 m/s, increasing at greater wind speeds because of turbulent deposition of spores on both upper and lower surfaces.

Vertical cylinders 5 mm in diameter, covered with cellophane dipped in glycerine or sugar jelly, offer many of the advantages of sedimentation methods without some of the disadvantages. Trapping efficiency increases with increasing wind speed and particle size. The choice of diameter is a compromise between efficiency of trapping small spores and losses from large spores blowing off (Ramalingam, 1968).

The use of *impactor traps,* particularly those that operate continuously, has revolutionized our concept of the air spora. In principle, traps of this type are similar in that air is drawn through rectangular or circular jets and then deflected at right angles by the trapping surface behind. Particles with sufficient momentum are thrown out of the airstream and impact on the trapping surface. As jet velocity increases, smaller particles will be impacted. Impactor traps allowing microscopic assessment include cascade impactors and various continuously operating traps which give time discrimination (Hirst, 1952; Kramer and Pady, 1966; Brown and Jackson, 1978a; Stedman, 1979a), whereas isolation in culture is possible with a slit sampler (Bourdillon *et al.*, 1941; Brown, 1970) and an Andersen sampler (Andersen, 1958). Trapping efficiencies up to 98% may be obtained in the slit sampler with a jet 0.3×28 mm using unit density spheres 1 μm in diameter. The automatic volumetric spore traps of Hirst (1953) and others were mostly designed for large-spored plant pathogens and have jets often 2×14 mm that trap *Lycopodium* spores with an efficiency of 62–94% depending on the wind speed. Trapping is most efficient when the jet is sampling isokinetically with the wind speed or at a stagnation point in a hemispherical baffle (May, 1967). Under still conditions, the intake should be fitted with a bell-shaped mouth and directed upward. The outdoor efficiency of the Hirst and similar traps is further decreased by their slow response to changes in wind direction (May *et al.*, 1976). However, changes in the outdoor air spora are so great that this is of relatively little importance. With Rotorod and Rotoslide samplers the trapping surface is moved to the spore at high speed (Perkins, 1957; Ogden and Raynor, 1967). Effectively sampling at 120 liters/min, Rotorod traps particles efficiently down to 12 μm in diameter. To prevent overloading of the small trapping surfaces, short-period or intermittent operation is necessary, perhaps with the trapping surfaces protected by swing shields when not in use (Raynor and Ogden, 1970).

In liquid impingers (Greenburg and Smith, 1922; May and Harper, 1957) air is accelerated through a capillary tube jet, which also acts as a limiting orifice, into a liquid. Even single bacterial cells 0.5–1 µm in diameter are trapped efficiently. The liquid can be examined microscopically or diluted and plated. A preimpinger may be used to prevent deposition of large particles in the inlet tube (May and Druett, 1953). A modified version has been used to collect splash-dispersed conidia of *Leptosphaeria (Phoma) nodorum* (Faulkner and Colhoun, 1977). Particle size discrimination may be obtained with a *multistage impinger* designed to simulate deposition in the human respiratory tract giving a 50% cutoff with 6-µm diameter particles between the first two stages and with 3-µm particles between the second and third (May, 1966).

Cyclones are probably most useful when spore concentrations are small or when large quantities of spores are needed for antigenic studies. Sampling rates can be up to 1000 liters air/min. Trapping characteristics vary with design and air velocity (Errington and Powell, 1969).

Filtration may also be used in some situations, although it is difficult to operate isokinetically. Filters allow both microscopic and cultural assessment and have been used to estimate occupational exposure to fungus spores in cork factories (Lacey, 1973). However, some membrane filters may be inhibitory to some microorganisms (Al-Diwany *et al.*, 1978).

B. Isolation Media

The choice of culture media for spore trapping is wide and may depend in part on which organisms are of particular interest. The suitability of different media has been discussed by Rogerson (1958), Lacey and Dutkiewicz (1976), and Burge *et al.* (1977b). Care should be taken when using rose bengal in media, because fungitoxic substances are produced during exposure to light (Kramer and Pady, 1961).

C. Choice of Spore Trap

Different spore traps yield different types of information, and none gives a complete picture of the air spora. Microscopic assessment is the least selective, allowing all spores to be counted and classified, but identification to the species level is seldom possible and often even genera cannot be distinguished. However, the illustrations provided by Gregory (1973) supplemented by local collections of reference material of different fungi may help. Isolation in culture gives greater precision of identification, but numerical estimates seldom coincide with microscopic counts (Burge *et al.*, 1977a; Batchelder, 1977), being limited by spore viability and the ability of the fungus to grow on the chosen medium under the incubation conditions used. Also, the spore form obtained in culture may not be that which was deposited and can be counted microscopically. Colonies of *Phoma* and *Fusarium*, for instance, may have grown from ascospores of *Leptosphaeria, Nectria,* or other genera (Ganderton, 1968; Burge *et al.*, 1977a).

Some of the differences can be seen in Table I, drawn from the results of about 200 studies in different parts of the world. Microscopic studies, particularly those utilizing inertial traps, record many more ascomycetes, basidiomycetes, rusts, smuts, and *Erysiphe* spores than those utilizing culturing, partly because of the time of sampling and partly because these spores do not grow in culture or produce only sterile mycelia or alternative spore forms. In contrast, culture methods allow separation of *Aspergillus, Penicillium,* and other genera with small spherical spores, which can be identified to the species level, and yeasts, which probably account for some of the ballistospores on microscope slides.

TABLE I

Comparison of the Results of Spore Trap Studies Obtained by Different Methods

	Mean percent of total catch		
Spore type	Settle plate, 112 series	Gravity slide, 41 series	Inertial trap, 44 series
Conidial fungi			
Cladosporium	32.3	45.9	33.4
Alternaria	11.1	11.8	4.4
Curvularia	2.3	2.0	1.2
Aspergillus	6.1	2.6	3.3
Penicillium	9.9		
Acremonium	0.9		
Aureobasidium	3.4	0.6	[a]
Botrytis	0.8	0.1	0.2
Epicoccum	1.9	0.7	0.7
Erysiphe		0.1	0.2
Fusarium	2.8	3.5	0.6
Helicosporium[c]			0.7
Helminthosporium	1.6	3.1	1.0
Monilia sitophila	0.5		
Nigrospora	0.3	0.7	0.5
Phoma and conidia of other pycnidial fungi	1.6		0.3
Pithomyces	0.1	0.1	0.2
Stemphylium	1.3	2.1	0.3
Zygosporium			0.7
Other fungi			
Myxomycetes		0.2	0.5
Lower fungi	1.4	2.1	0.1
Ascomycetes	0.6	1.7	14.7
Basidiomycetes[b]		1.1	21.7
Rusts		4.3	0.8
Smuts		4.6	3.7
Yeasts	6.3		

[a] Forming <0.05% of catch.
[b] Including ballistospores of *Sporobolomyces* in microscopic assessments.
[c] Possibly coiled ascospores (Allitt, 1979).

Overall, more categories of spores have been recognized on inertial trap slides than on gravity slides.

There is much value in combining at least two trapping methods in an investigation, one allowing microscopic assessment of the total air spora and the other identification of the predominant types in culture. The trapping methods chosen also need to fit the investigation. The objectives must be defined and a number of questions answered, for instance, whether spot samples are adequate or time discrimination is required, the levels of contamination expected, whether total spore counts, particle counts, or size-graded counts are required, the distinctiveness of the spores or colonies, the resources and facilities available, the levels of skill required and available. Once these questions have been answered, the choice of possible spore traps will be much clearer. Whenever possible, impaction traps, either suction or whirling-arm, should be used because of their greater efficiency, whereas for assessing the total air spora a 24-h operation and time discrimination are essential to detect diurnal periodicity.

III. THE AERIAL ENVIRONMENT

A. Structure of the Atmosphere

The atmosphere surrounding the earth is essentially a series of concentric shells of differing physical and gaseous composition, but only the innermost of these, the troposphere, is of significance to airborne microbes. This zone comprises 80% of the mass of the atmosphere and is composed of 78% nitrogen, 21% oxygen, 0.03% carbon dioxide, and trace amounts of other inert and pollutant gases. It extends to an altitude of 17 km at the equator, 6–8 km at the poles, and in middle latitudes varies between 7 and 13 km between low- and high-pressure areas. The troposphere contains all the phenomena that make up our weather, for instance, convection, cloud formation, and precipitation. It also provides the environment for airborne spores. The gaseous components form the supporting medium, air movement assists dispersal, precipitation aids deposition, and radiation, temperature, and humidity influence survival.

Air movements vary in scale from turbulent eddies a few meters or less across to frontal systems thousands of kilometers long and hundreds wide, with jet streams in the upper troposphere and stratosphere rapidly transporting air around the world. All scales have relevance to the dispersal of airborne particles, but they are not uniformly distributed through the troposphere which can be separated into five distinct zones (Gregory, 1973):

1. *Laminar boundary layer*. A microscopically thin layer of still air surrounds the earth's surface and all objects projecting from it. Adjacent is a layer of air flowing in streamlines without turbulence and increasing in speed linearly with height. Thickness varies with wind speed and surface roughness from < 1 mm to

10 cm on a cloudy day to >10 m on a still night. Extremes of heating and cooling occur in this zone, and spores have to cross it before they can become airborne.

2. *Local eddy layer*. Surface roughness leads to stationary eddies froming behind roughness or in depressions.

3. *Turbulent boundary layer*. Objects projecting through the laminar boundary layer create eddies which break away downwind to give lateral and vertical components to the forward motion of the air. Flux of momentum decreases linearly with height and, because of the mixing caused by turbulence, temperature, and wind velocity change linearly with the logarithm of height. The likelihood of turbulence can be calculated using Reynolds' formula, $Re = dv/k$, where d is a characteristic dimension of the object, e.g., height (cm), v is wind velocity (cm/s), and k is the kinematic viscosity of air (under average conditions, 0.14 cm^2/s). Turbulence occurs when the Reynolds' number, Re, exceeds 2000. The thickness of this layer increases with increasing wind speed and is thickest (to 150 m) on hot, sunny days and least on clear, calm nights.

4. *Transitional zone*. Through this zone to 500–1000 m, turbulence decreases with altitude and diurnal changes disappear. The top of this layer is the highest to which particles may be carried by turbulence.

5. *Convective layer*. Airborne particles can only be lifted into this zone by convection which occurs when air is warmed close to the ground so that the temperature decreases more rapidly than the adiabatic lapse rate (10 cm/km for dry air but nearly halved when the air is saturated). Bubbles of warm air may rise from an area 1.25 km^2 every 6–15 min in summer, carrying many microorganisms. The bubbles expand as they rise to 3–15 km, depending on the temperature gradient and water content of the air unless a temperature inversion, with warm air above cool, is encountered.

Precipitation may both aid the liberation of spores (Section V,A) and contribute to their deposition (Section V,C). The form of precipitation and its intensity depend on many factors, but raindrops are usually 1–2 mm in diameter, although they may be only 0.2–0.5 mm diameter in drizzle and up to a maximum of 5 mm from convective clouds. However, such large raindrops may be uncommon in temperate regions. In rainfall accumulating at a rate of 5 mm/h only 1 drop in 300,000 is 4 mm in diameter (Bent, 1950). Although such rainfall occurred 100 times in a year at Rothamsted, it usually lasted less than 5 min (Stedman, 1979b).

The survival of airborne conidia may be affected by other features of the aerial environment. Ultraviolet radiation, especially at high altitudes, may damage spores, while freezing and desiccation may provide some protection against this. However, the mechanism is little understood.

B. The Indoor Aerial Environment

Although the aerial environment indoors may be considered a walled-in portion of the outdoors, it differs in patterns of air movement, humidity, temperature, buoyancy, and possibly also in gas composition. Air movement results from ventilation, convection, and buoyancy, but air does not move continually over a surface as outdoors. Instead fresh air mixes with stale, and a mixture of stale and fresh air is displaced. A spore cloud is thus diluted and not completely removed by ventilation. Outside air movement influences movement indoors by forcing air through cracks on the windward side and sucking it out on the leeward, with the direction of airflow changing as the wind direction changes. In contrast, artificial ventilation tends to stabilize flow, but whether ventilation is natural or artificial it occurs at a few points, not over wide areas, and flow is usually turbulent. Similarly heat is applied at only a few points. Occupants also contribute to turbulence both through their movement and from convection currents over the body surface drawing air up from floor level (Daws, 1967; Lewis *et al.*, 1969).

Spore concentrations in a room may decrease through exchange with outdoor air, by sedimentation under still conditions, and through death. Usually the decrease is logarithmic. If the room is likened to a large stirred sedimentation chamber, air changes decrease the load in the ratio $1/e^n$, where e is the base of Naperian logarithms. This can be expressed in terms of "equivalent ventilation turnovers" or "air changes" and partitioned to account for losses by different means (Bourdillon *et al.*, 1948). Artificial ventilation can rapidly circulate spores through a building, but convection can also be very effective, carrying spores from first- to fourth-floor halls within 5 min and into rooms within 20 min (Christensen, 1950).

IV. POPULATIONS OF AIRBORNE CONIDIA

A. Outdoor Populations

Reports on the air spora in different parts of the world are numerous, but the information that can be derived from them is restricted by the method of trapping used, the time and period of sampling each day, and the period over which the study was continued. About 200 of these reports have been used to construct Tables I and II. Table I compares results from different traps. Because the general picture is similar, only results from inertial traps (Table II) have been grouped into climatic regions derived from the map of world climate and food potential in the *The Times Atlas of the World* (Anonymous, 1975). Most culture records were obtained by exposing open petri dishes for a short period at the same time or times daily, microscopic assessments from slides exposed for 24 h in rain shelters (e.g. Durham, 1946), and inertial trap records from continuously operating suction traps together with a few from intermittent Rotorod samples.

Care is necessary in interpreting these tables. Authors vary in their skill in identifying cultures and spore types, and the different trapping methods have their limitations and advantages (Section II,A); but they do not prevent general conclusions from being drawn, which will be discussed later in this section. A major discrepancy between trapping methods is in the representation of ascospores and basidiospores. No basidiospores and few ascospores are detected by culturing (8% of colonies nonconidial); a few more are caught on gravity slides (14% nonconidial), but the relatively high trapping efficiency of inertial spore traps results in over 40% of the catch being of the nonconidial type. Conversely, the relative importance of conidial fungi is decreased by the more efficient trapping methods, although counts of individual species on culture plates may also be decreased if spores are aggregated or dead or if colonies are inhibited by other antagonistic species. Thus, allowing for nonconidial fungi, *Cladosporium* provided 57% of the conidia on inertial trap slides, 53% on gravity slides, but only 35% of the colonies on settle plates. *Aspergillus* and *Penicillium* provide a larger proportion of the catch by culturing (18% of conidial fungi) than by other methods (3.0% on gravity slides, 5.6% with inertial traps) where identification is difficult and other taxa with small spherical spores may also be included. The proportion of the catch formed by large *Alternaria* and *Helminthosporium* conidia is less in inertial traps than on gravity slides, perhaps because the slow response to wind direction changes decreases the sampling efficiency of the former (May et al., 1976).

1. Characteristics of Air Spora Outdoors and Regional Variations

The air outdoors is seldom free of fungus spores. However, their number and type vary with time of day, weather, season, geographical location, and the presence of local spore sources. Similarly, there are variations in the relative abundance of conidia, ascospores, basidiospores, yeasts, rust, and smut spores in the air.

Total numbers of spores in the air may vary from fewer than 200 to more than 2 million/m^3 air, but the average number on a daily basis is usually about 10,000–20,000/m^3, with peak concentrations only rarely exceeding 200,000/m^3 for short periods, perhaps of a few hours only. However, these peak concentrations are usually the consequence of conditions favoring the formation and liberation of numerous ascospores (e.g., *Didymella exitialis;* Frankland and Gregory, 1973), ballistospores of *Sporobolomyces* (Gregory and Sreeramulu, 1958), or basidiospores. Conidial fungi have seldom been recorded in such large numbers under normal conditions.

Numbers of spores in the air tend to be larger in temperate and tropical regions than at high latitudes and in deserts. In polar and subarctic regions the peak spore season is short, with a peak of less than 10,000/m^3 air, and ascospores and basidiospores predominate, providing nearly 70% of the catch (Rantio-Lehtimäki, 1977). Fungal colonization in deserts is limited by high temperatures

TABLE II

Relative Abundance of Different Spore Types Caught in Inertial Traps in Different Climatic Regions[a]

	Polar		Cooler humid			Warmer humid			Dry		Tropical humid	
	Tundra, 2 series	Subarctic, 1 series	Continental cool summer, 2 series	Continental warm summer, 2 series	Marine west coast, 9 series	Mediterranean, 3 series	Humid subtropical, 7 series	Steppe, 9 series	Desert, 2 series	Savanna, 6 series	Rain forest, 1 series
Spore type												
Conidial fungi												
Cladosporium	15	16	43	65	33	18	39	18	57	41	23
Alternaria	[b]		1	3	1	9	15	13	3	3	[b]
Curvularia						[b]	2	8	[b]	3	[b]
Aspergillus/ Penicillium	3	3	[b]	3	2	[b]	1	1	18	7	
Arthrinium				1		[b]			[b]		
Aureobasidium					[b]	[b]		[b]	[b]		
Botrytis			[b]	1	1	[b]	1	[b]	[b]	[b]	
Deightoniella											[b]
Epicoccum			[b]		[b]	1	4	[b]	[b]	[b]	1
Erysiphe					1	1	[b]	[b]	[b]	[b]	

Fusarium	7							
Helicosporium[a]				2		5	b	
Helminthosporium		b	b	b	1	3		1
Memnoniella							5	2
Nigrospora						1		2	b
Periconia							1	1
Pithomyces		b		1	1	b		b
Pseudotorula						1		
Conidia of pycnidial fungi						3		
Stemphylium				b	1	2		
Other fungi								
Myxomycetes	4	b			1			b
Lower fungi		b		b	b	b		2
Ascomycetes	24	45	b	14	21	6	3	2	28
Basidiomycetes[c]	34	22	16	43	16	10	15	2	17	36
Rusts	b	b	b	4	2	2	b	b
Smuts	b	1	3	2	18	4	2	4	1

[a] Classified after *The Times Atlas of the World Comprehensive Edition* (Anonymous, 1975). Most records are from automatic volumetric spore traps; the remainder are from intermittently operated Rotorods.
[b] Forming less than 0.5% of the catch.
[c] Including ballistospores of *Sporobolomyces*.
[d] Possibly coiled ascospores (Allitt, 1979).

and sparse vegetation: weekly mean spore concentrations in Kuwait seldom exceed 700/m^3 and are often less than 400/m^3 (Davies, 1969b).

A remarkable feature of the air spora is the ubiquitous occurrence of some elements which usually provide the predominant spore type in daytime, while species characteristic of particular climatic regions are often present in relatively small numbers, individual species contributing less than 1% to the total air spora. Gregory (1973) has likened the frequency of distribution of individuals of different species in an air spora to a series of the logarithmic or lognormal type.

Cladosporium is the most numerous conidial fungus in the air spora of most climatic regions of the world, often predominating to the extent that its numbers determine the magnitude of the total air spora during the day. At night, especially in cooler climatic regions, it is replaced by ascomycetes, basidiomycetes, and *Sporobolomyces*. Only in warm, dry steppe regions or in hot, dry seasons in other zones is *Cladosporium* exceeded in the daytime air spora by other fungi, usually *Alternaria* (Sandhu et al., 1964; Agarwal et al., 1974; Hariri et al., 1978), but occasionally also by *Curvularia* (Tilak and Srinivasulu, 1967) or *Helminthosporium* (Gupta et al., 1960).

Cladosporium often accounts for more than half the air spora and at peak times, when concentrations may reach 240,000 spores/m^3 air, they may form 93% of the total (Gregory and Sreeramulu, 1958; Sreeramulu and Ramalingam, 1966). However, mean daily concentrations are usually much smaller, and on the order of 5000 spores/m^3 air. Thus, in Derby, England, during the period April to November the mean daily concentrations in the years 1970-1977 varied only between 3261 and 6532 spores/m^3 (F. A. Jackson and H. M. Brown, personal communication). Many other records fall within this range (e.g., Gregory and Hirst, 1957; Hamilton, 1959; M. E. Lacey, 1962). *Cladosporium* spores are often airborne in aggregates with an average of 1.3-3.6 spores in each dispersal unit (Hyde and Williams, 1953; Davies, 1957; Harvey, 1967).

The second most abundant spore type is usually *Alternaria,* although it may give mean daily spore concentrations up to only 150/m^3 air and commonly only 50/m^3 (Gregory and Hirst, 1957). Peak concentrations of 6000 spores/m^3 have been found over short periods of sampling. In tropical areas, *Curvularia* and sometimes *Nigrospora* make large contributions to the air spora with peak concentrations of 4000-9000 spores/m^3, although mean concentrations, like those of *Alternaria,* are often 50/m^3 or less (Sreeramulu and Ramalingam, 1966).

Aspergillus and *Penicillium* spores are widespread in the air, although they can only be identified with certainty on culture plates. *Penicillium* predominates in most regions, but it is replaced by *Aspergillus* in the humid tropics. In the United Kingdom, *Penicillium* accounts for 2.5-13% of catches, and *Aspergillus* for only 0.9-3.0% (Hyde and Williams, 1949; Richards, 1954; Hudson, 1969).

Hudson identified 14 species of *Aspergillus,* with *A. amstelodami, A. fumigatus, A. repens,* and *A. versicolor* most common. Similarly in the United States (Kansas) 35 *Penicillium* species formed 6% of the catch on settle plates, and 23 *Aspergillus* species 5% (Kramer *et al.,* 1960). *Penicillium oxalicum, A. niger,* and *A. amstelodami* were predominate. In contrast, 5.6% of colonies isolated in Mysore, India, were formed by 37 *Aspergillus* species and only 1.7% by *Penicillium; A. nidulans, A. ruber, A. awamori,* and *A. flavus* accounted for 60% of the *Aspergillus* catch (Rati and Ramalingam, 1976). Large numbers may even occur in desert regions, forming 35% of the total catch at Fresno, California, and <6% in Kuwait (Anonymous, 1976; Davies, 1969b). Thermophilic and thermotolerant fungi may occur in the air, even in temperate regions in winter, with a maximum occurrence of 14 spores/m^3, 69% of which were *Aspergillus fumigatus.* Also frequently isolated were *Talaromyces thermophilus, Humicola (Thermomyces) lanuginosa,* and *Mucor pusillus* (Hudson, 1973), all species common in moldy, heated hay (Gregory *et al.,* 1963).

As previously pointed out, regional differences mostly refer to relatively minor components in the air spora. However, there is a general trend of increasing variety of fungi trapped, by whatever method, in passing from cooler to warmer climatic zones. Conidial fungi provide the largest proportion of spores in desert regions and are also the largest component of the air spora in continental warm summer humid, steppe, and savanna zones, on the basis of published inertial spore trap records (Table II). Characteristically *Alternaria, Helminthosporium, Fusarium,* and *Aspergillus* are most numerous in dry steppe and desert zones, *Nigrospora* and *Curvularia* in tropical humid savanna regions extending into humid, subtropical steppe and rain forest zones, while *Monilia, Periconia, Pithomyces, Zygosporium, Deightoniella,* and *Memnoniella* are among those that become most numerous in rain forest zones.

Enumerating spores is only one way of characterizing the air spora. An alternative is to consider the volume contribution of different taxa to the total. This can give a quite different picture of the relative abundance of different fungi in the air spora and one very relevant to the allergenic potential (Table III). The importance of large spores (e.g., *Alternaria, Curvularia, Helminthosporium, Epicoccum*) present in small numbers is enhanced compared with that of more numerous small spores (e.g., *Cladosporium, Aspergillus, Lacellinopsis*). The apparent dominance of fungal spores over pollens also disappears when volume is considered (Hyde and Adams, 1960; Reddi, 1974).

In addition to spores, hyphal fragments, mostly conidiophores, are widespread in the air. They may be 5–100 μm long, averaging 20 μm, and number up to 600–1800/m^3 air, with the largest numbers occurring in the growing season (Pady and Kramer, 1960; Pady and Gregory, 1963; Sinha and Kramer, 1972). Most are dematiaceous, belonging to *Cladosporium* and *Alternaria* species.

TABLE III

Comparison of Numerical and Volume Incidence of Different Spore Types

Spore type	Cardiff, Hirst suction trap[a]		Anakapelle, 5-mm glass rod[b]		Visakhapatnam, 5-mm glass rod[b]	
	By no.	By vol.	By no.	By vol.	By no.	By vol.
Alternaria	17.9	27.5	0.6	0.3	0.8	0.6
Aspergillus			27.9	0.5	26.0	0.8
Botrytis	1.3	3.9				
Cladosporium	21.7	21.3	63.3	2.0	47.9	2.5
Curvularia			7.7	21.4	6.5	30.5
Epicoccum	0.4	5.7				
Helminthosporium			1.8	6.2	1.6	9.2
Lacellinopsis			28.7	4.0	0.4	0.1
Nigrospora			11.1	18.3	5.5	15.4
Periconia			2.3	0.2	1.6	1.8
Trichoconis			0.7	5.3	0.7	8.5
Ascospores	17.9	27.5	0.6	0.3	0.8	0.6
Basidiospores[c]	56.3	35.6	2.8	2.1	2.6	3.4
Rust urediniospores			3.6	13.7	0.8	5.0
Ustilago	2.2	1.9				

[a] Hyde and Adams, 1960.
[b] Reddi, 1974.
[c] Including ballistospores of *Sporobolomyces* and *Tilletiopsis*.

2. Local Variation

Besides variation among climatic zones, the air spora may vary over quite small areas depending on the amount and type of vegetation, the local microenvironment, and human activity. Thus M. E. Lacey (1962) found 2.6 times more spores in a valley, close to a stream, than on a nearby exposed hill. However, although there were 5 times more ascospores and 3 times more basidiospores, there were only 1.4 times more conidia.

Grassland and agricultural crops are major sources of spores, their numbers increasing up to harvest. Despite large differences in the species composition of grass swards, the fungal flora remains remarkably constant, dominated by *Cladosporium* (Lacey, 1975a). At mowing, haymaking, and harvest, large numbers of spores became airborne, sometimes giving concentrations exceeding 10^9 spores/m^3 (Sreeramulu, 1958; Darke *et al.*, 1976).

Such spore clouds will be carried away by the wind and diluted by diffusion and deposition (Section V,B,C), so that concentrations decrease with distance from the source. This could lead to wide variations in the air spora at different locations, but the most marked variations found have been between rural and city sites and between coastal and inland sites. Trapping at eight sites within 60 km of Derby, England, Brown & Jackson (1978b) found *Cladosporium* to vary only between 83 and 125% of the Derby figure and *Alternaria* between 54 and 102%, although at some sites counts as small as 25% were obtained for some spore types. However, counts in London were less than 50% of those at Rothamsted, 40 km to the north (Hamilton, 1959), and in the center of Cardiff counts were 30–55% of those at a nearby rural site (Harvey, 1967). On-shore prevailing winds also lead to low spore counts. For instance, mean daily spore counts in Cardiff in 1958 were about 2000/m^3, while close to London they were 11,300–29,700/m^3 (Adams, 1964; M. E. Lacey, 1962). Similarly catches at Point Lynas, Anglesey, were only 26% of those at Derby (Brown and Jackson, 1978c). The difference between inland and coastal sites was smaller when the prevailing wind blew from inland.

3. Spore Viability

Airborne spores are not necessarily viable, and it has been found that the viability of some types varies through the day, perhaps depending on the intervals between spore formation, liberation, and sampling, radiation intensity, and degree of desiccation. Thus Pathak and Pady (1965) found the viability of *Alternaria* spores to average 80% through the day, while that of *Cercospora* was 70–90% and that of *Cladosporium* 20–30% when the numbers of both were greatest. However, Kramer and Pady (1968) failed to detect such periodicity, although average viabilities were similar (*Cercospora* 58%, *Cladosporium* 42%), while Davies (1957) reported that 84–96% of *Cladosporium* dispersion units (average size, two spores) germinated unless there were heavy soot deposits. Ali *et al.* (1976) suggested that the survival of airborne fungus spores may be related to their sensitivity to light. In Egypt, darkly pigmented *Alternaria, Stemphylium, Rhizopus,* and *Epicoccum* predominate when solar radiation is greatest, while *Cladosporium, Aspergillus niger,* and *Penicillium* with less pigmentation are most numerous when radiation is least. This could partly be an effect of temperature, however (Section IV,A,5).

Some airborne hyphal fragments may also retain their viability (average 16–20%; Kramer and Pady, 1968; Pady and Gregory, 1963). Viability was correlated with viability of *Cladosporium* spores, and some fragments rapidly produced secondary conidiophores with up to 20 conidia.

4. Diurnal Periodicity

Growth cycles of fungi, methods of spore liberation, and changes in meteorological conditions combine to give characteristic diurnal changes in spore concentrations. The daytime maxima exhibited by most anamorphic fungi usually relate to the conditions necessary for spore liberation (see Section V,A), but turbulence, wind speed, convection, and temperature inversions will all

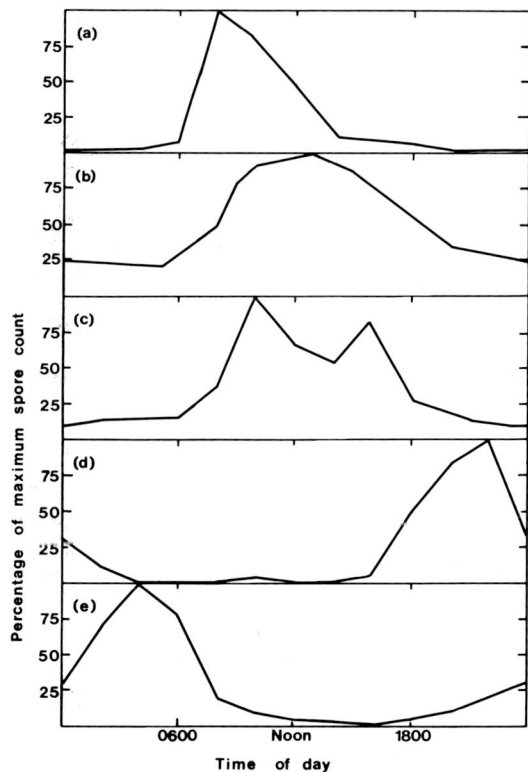

Fig. 1. Mean diurnal periodicities of airborne spores illustrating different patterns of release. (a) Early morning pattern, *Nigrospora* spp.; (b) midday pattern, *Cladosporium* spp.; (c) double-peak pattern, *Tetraploa* spp.; (d) postdusk pattern, tetraspore type; (e) night pattern, *Pyricularia grisea*. (After Hirst, 1953; Meredith, 1962a,b; Sreeramulu and Ramalingam, 1966; Shenoi and Ramalingam, 1975.)

affect the rate of dilution of spore clouds and consequently the concentrations of spores in the air. Periodicities of many spores have been determined by estimating concentrations at 2-h intervals using continuously operating traps.

Five patterns have so far been recognized (Fig. 1; Shenoi and Ramalingam, 1975). Conidial fungi demonstrating these different patterns are listed in Table IV.

a. Daytime Maxima

i. Postdawn Pattern (Fig. 1a). This is demonstrated by species with liberation mechanisms (hygroscopic movements, water rupture) depending on rapidly changing or decreasing relative humidity leading to maxima between 0700 and 1000 h.

ii. Midday Pattern (Fig. 1b). Species with this pattern may have spores

TABLE IV

Diurnal Periodicities of Airborne Conidia[a]

Postdawn pattern (maximum concentration, 0700–1000)
  Cercospora spp.   *Phaeotrichoconis* sp.
  Cordana musae   *Polythrincium trifolii*
  Corynespora cassiicola   *Trichoconis padwickii*
  Deightoniella torulosa   *Zygophiala jamaicensis*
  Epicoccum spp.   *Zygosporium oscheoides*
  Nigrospora spp.
Midday pattern (maximum concentration, 1000–1600)
  Alternaria spp.   *Penicillium*
  Aureobasidium   *Periconia*
  Beltrania   *Periconiella*
  Botrytis   *Pithomyces*
  Cladosporium   *Pseudocercospora*
  Curvularia   *Spegazzinia*
  Dendryphiella   *Stemphylium*
  Erysiphe   *Tetraploa*
  Helminthosporium   *Torula*
  Memnoniella
Double-peak pattern (maximum concentrations, 0800–1000 and 1400–1800)
  Alternaria   *Helminthosporium*
  Cladosporium   *Memnoniella*
  Curvularia   *Periconia*
  Epicoccum   *Pithomyces*
  Fusidium   *Tetraploa*
Postdusk pattern (maximum concentration, 2000–2200)
  Spegazzinia   *Ustilaginoidea virens*
  Tetraspore-type
Night pattern (maximum concentration, 0200–0400)
  Fusarium   *Pyricularia oryzae*

[a] Data from Hirst (1953), Cammack (1955), Meredith (1962a,b, 1966a,b), Pady *et al.* (1967, 1969), Sreeramulu (1959, 1962, 1970), Sreeramulu and Ramalingam (1962, 1966), Sreeramulu and Vittal (1966a, b), Shenoi and Ramalingam (1975), and Sreeramulu *et al.* (1971).

released by deflation or mechanical disturbance of the substrate resulting from the increased temperature, wind speed, and turbulence found at midday. Sometimes, spores may be released by wind following weakening of their attachment to the conidiophore by earlier hygroscopic movements (e.g., *Alternaria porri;* Meredith, 1966a).

iii. Double-Peak Pattern (Fig. 1c). The reasons for two daytime maxima remain unexplained, although it could be a variant of the midday pattern resulting from dilution of the midday spore cloud by intense convection. Species with this pattern are often classified in the midday pattern in different regions or years (Table IV).

b. Nighttime Maxima

i. Postdusk Pattern (Fig. 1d). A few species give maxima between 2000 and 2200 h. The determinants of this pattern are not known, although an active spore release mechanism could be operated by increasing humidity.

ii. Night Pattern (Fig. 1e). Nighttime maxima from 0200 to 0400 h are characteristic of ascospores and basidiospores but of few conidia. High relative humidity may perhaps be required for both spore formation and liberation. Active spore release mechanisms are usual, and in *Pyricularia oryzae* liberation follows the bursting of a cell at the base of the conidium (Ingold, 1964).

Although the general pattern of diurnal periodicity for a given spore type is followed in different regions, the timing of maximum concentrations may vary slightly with season, location, and climate. Thus peak concentrations of *Nigrospora* and *Cladosporium* in Nigeria occurred 2 h earlier in the dry season than in the wet (Cammack, 1955), while in India there was a 2–4 h difference between Visakhapatnam and Mysore attributed to the later occurrence of quick-drying conditions at Mysore (Shenoi and Ramalingam, 1975). Perhaps diurnal periodicities might be better defined in terms of spore liberation mechanisms and meteorological conditions than time of day.

5. Effects of Weather and Season

Weather conditions affect both the growth and sporulation of fungi and the liberation, dispersal, and deposition of their spores. Consequently they affect the numbers and types of spores in the air. Hamilton (1959) found effects of dew point, relative humidity, temperature, and wind on the numbers of some fungi. *Alternaria, Botrytis, Cladosporium, Helminthosporium, Polythrincium,* and *Aureobasidium* were all increased by periods with a high dew point, and *Botrytis* by high relative humidity, while the temperatures giving the largest numbers of different spore types are listed in Table V. Increasing wind speed decreased concentrations of *Alternaria* and *Cladosporium*, although the former increased with increased gustiness. However, large numbers of *Erysiphe graminis* conidia were correlated with high wind speed, dry leaves, high temperature, and low relative humidity, although the onset of wind was more important than its continuation (Hammett and Manners, 1971).

The air spora is profoundly affected by rain. At first, the tap and puff of raindrops falling on dry vegetation increases *Cladosporium, Alternaria, Erysiphe,* and other spore types (Hirst and Stedman, 1963). These are then replaced by characteristic damp-air types, predominantly ascospores (Table VI), while prolonged rain removes most spores from the air.

Weather conditions and the cycle of plant growth interact to give large sea-

TABLE V

Temperatures Favoring Highest Counts of Different Conidial Types[a]

Temperature (°C)	Spore type
10–12	*Penicillium*
16–18	*Entomophthora*
21–23	*Cladosporium, Dicoccum, Erysiphe, Helicomyces, Periconia*
24–26	*Alternaria, Polythrincium, Aureobasidium, Torula*
27–29	*Botrytis, Epicoccum, Helminthosporium, Macrosporium*

[a] After Hamilton (1959).

sonal differences in the air spora. In temperate regions, maximum spora concentrations occur during the growing season, with peaks in June–August dominated by *Cladosporium* and *Sporobolomyces* (Gregory and Hirst, 1957; M. E. Lacey, 1962). Within this overall pattern, individual spore types, especially plant pathogens, show well-defined seasons, with *Erysiphe* conidia most numerous in June–July, *Polythrincium* in August–September, and *Helminthosporium* in July–August, but with *Penicillium* in cities often most numerous in winter (Hamilton, 1959).

Seasonal trends in the tropics are less marked than in temperate regions, although spores are usually least numerous in the dry season. At Visakhapatnam, mean spore concentrations were 4650/m^3 air in the dry season (March–June),

TABLE VI

Effect of Rain on Airborne Spore Concentrations in a Jamaican Banana Plantation[a]

Spore type	Spore concentration (no./m^3)	
	Before rain	During rain
Alternaria	310	30
Cladosporium	640	110
Curvularia	240	0
Deightoniella	40	0
Nigrospora	20	10
Periconiella	50	0
Zygosporium	360	130
Ascospores	510	242,840
Basidiospores	180	70

[a] After Meredith (1962a).

$9730/m^3$ in the rainy season (July–October), and $11,500/m^3$ in the cold season (November–February) (Sreeramulu and Ramalingam, 1966). Similar trends were found at Mysore (Ramalingam, 1971; Shenoi and Ramalingam, 1976). Sreeramulu and Ramalingam (1966) recognized spore types occurring independently of crop growth and others whose occurrence was correlated with stages in the development of local vegetation and crops. In the first category, *Fusarium* was correlated with rainfall while, in the second, the incidence of *Cladosporium* and *Nigrospora* correlated with similar growth stages of each of the two rice crops, while *Deightoniella torulosa* occurred on rice only during the rainy season. *Aspergillus* species were common throughout the year, but *Alternaria* concentrations were greatest at the end of the hot season as the first rains occurred. Similarly with *Sorghum,* some pathogens were associated with the earlier crop (*Sphacelotheca sorghi*), some with the later crop (*Cercospora*), and others with both (*Helminthosporium turcicum*) (Shenoi and Ramalingam, 1976).

6. Spore Populations at High Altitudes

Upward movement of spores results from atmospheric turbulence and convection. Its extent has been studied by trapping from balloons and airplanes, and results have been summarized by Gregory (1973). Concentrations are greatest close to the ground where the air spora is continuously replenished by the liberation of new spores. Spore concentrations decrease logarithmically with height in unstable air unless temperature inversions prevent upward movement. However, if inversions are formed within a spore cloud, spore concentration increases immediately below the inversion while the original decrease with height continues above. A stable layer in contact with the ground leads to a larger concentration above because spores are deposited from within the layer. Deposition occurs continuously close to the ground and, if there is no replenishment, as over the sea, the bottom of the concentration profile may become eroded, giving maximum spore concentrations at an altitude depending on the distance from the source and the rate of deposition (Hirst and Hurst, 1967).

Upward movement is greatest when there is the most convection (Heise and Heise, 1949; Fulton, 1966a). Microbial concentrations found by Fulton (1966a) averaged $200/m^3$ (c. 50% fungi) at 690 m altitude, $60/m^3$ at 1600 m, and $30/m^3$ at 3127 m (both ca. 10% fungi) from 1200 to 1800 h when convection was greatest, but only 45, 25, and $23/m^3$, respectively, from 0600 to 1200 h. A temperature inversion at 1500 m probably contributed to the large difference between the two lower altitudes. Spore types trapped were similar to those at ground level.

Few spores have been found in the stratosphere. There is a little interchange through the tropopause, and much of the airborne dust above this is probably extraterrestrial and meteoric in origin. However, $0.03-0.14$ microorganisms/m^3 have been trapped at a 20-km altitude, about half of which were fungi (Meier, 1936; Bruck, 1967).

B. The Air Spora Indoors

The numbers and types of fungus spores in indoor air depend on air exchange with the outside and the presence of indoor spore sources. Without such sources, the air spora will consist of the same species as outdoors in the same proportions, but their numbers are usually smaller (Table VII; Richards, 1954; Ackermann *et al.*, 1969; Owen and Baker, 1970; Hirsch *et al.*, 1978; Levetin and Horowitz, 1978; Levetin and Hurewitz, 1978). The air is seldom devoid of spores, and even in "clean" rooms up to 25 spores/m^3 have been found (Favero *et al.*, 1966), while up to 2300 spores/m^3 of the potentially pathogenic *Aspergillus fumigatus* have been found in hospitals (Nobel and Clayton, 1963; Solomon *et al.*, 1978).

Spores may come from many sources within buildings. They may come from fungi growing in condensation on walls and paintwork and on food, or spores may accumulate in house dust and grow if the humidity is high enough (Davies, 1960). They may then become dispersed by human activity within the building such as sweeping, bed-making, and building repair work, giving spore concentrations, dominated by *Penicillium,* 17 times greater than before disturbance (Maunsell, 1954). Air-conditioning systems can rapidly disperse spores through a building as well as act as sources themselves, especially if they incorporate humidifiers. In one system *Aspergillus fumigatus* grew in dust accumulations in the ducts, causing aspergillosis in exposed people (Wolf, 1969). Cold mist

TABLE VII

Comparative Counts of Airborne Spores Indoors and Outdoors

	Tulsa, Oklahoma		Cardiff, Wales	
	(% of plates)		(no. colonies/626 plates)	
Colony type	Outdoors	Indoors	Outdoors	Indoors
Alternaria	75.8	43.7	44	
Arthrinium	13.8	7.2	2	1
Aspergillus	6.3	5.6	204	53
Aureobasidium	68.0	29.0	588	25
Botrytis	1.1		160	29
Chrysosporium		1.4	105	24
Cladosporium	78.3	41.9	3097	463
Epicoccum	38.1	20.5	152	9
Fusarium	3.1	1.4	9	5
Helminthosporium	21.8	11.6		
Penicillium	18.9	21.1	668	194
Phoma	17.7	3.3	199	6
Sterile mycelia	30.7	20.7	1171	349

a After Levetin and Horowitz (1978), Levetin and Hurewitz (1978), and Richards 1954.

humidifiers may become colonized by such fungi as *Penicillium, Aspergillus, Verticillium, Geotrichum,* and *Phialophora hoffmanii,* as well as yeasts, bacteria, and amebas. American homes with hydronic heating systems had more airborne *Gliocladium* than those with forced-air systems, more *A. fumigatus* where there were pets, and more *Alternaria* and *Monilia sitophila* where there were wool carpets (Hirsch and Sosman, 1976).

By far the largest numbers of airborne conidia found indoors occur when moldy organic substrates are handled. Deteriorated stored products can form vast spore sources producing concentrations up to $10^9/m^3$ air when they are disturbed. These are often associated with occupational lung disease, although actinomycetes as well as fungi may be implicated. Characteristically, the fungi involved all produce abundant conidia that are easily liberated when the substrate is disturbed, becoming airborne in large numbers. The predominant conidial fungi associated with different organic substrates are listed in Table VIII from which it can be seen that workers in many different industries can be exposed to spores. The air spora in many of these situations has been described (Barkai-Golan, 1961; Stallybrass, 1961; Lacey, 1972, 1973, 1974, 1977, 1980; Kotimaa, 1977; Pohjola *et al.,* 1977).

The numbers and types of spores in the air depend not only on the substrate involved but also on storage conditions, degree of disturbance, ventilation, and other factors. The type of molding in hay and grain depends on the water content at which it is stored, aeration, and the consequent degree of spontaneous heating.

TABLE VIII

Predominant Conidial Fungi in Organic Substrates and Associated Air Spora

Substrate	Fungi
Hay	*Aspergillus fumigatus, A. nidulans, A. versicolor, A. glaucus* group, *Humicola lanuginosa*
Cereal grains	As hay plus *Penicillium* spp., *Thermoascus (Paecilomyces) crustaceus*
Malting barley	*Aspergillus clavatus, A. fumigatus*
Straw	As hay plus *A. terreus, A. flavus, Cladosporium* spp., *Penicillium* spp.
Mushroom compost	*Scytalidium thermophilum, Humicola grisea, A. fumigatus, Penicillium* spp., *Doratomyces stemonitis*
Maple logs	*Cryptostroma corticale*
Cork	*Penicillium frequentans, P. granulatum, Aphanocladium album, Monilia sitophila*
Red wood sawdust	*Aureobasidium pullulans, Graphium* sp.
Cheese	*Penicillium casei*
Sugarcane bagasse	*A. fumigatus, A. niger, Penicillium* spp., *Chrysosporium thermophilum, Paecilomyces varioti*
Cotton	*A. niger, A. ochraceus, A. versicolor, A. nidulans, Fusarium moniliforme, Cladosporium* spp.
Citrus fruits	*Penicillium digitatum, P. italicum*

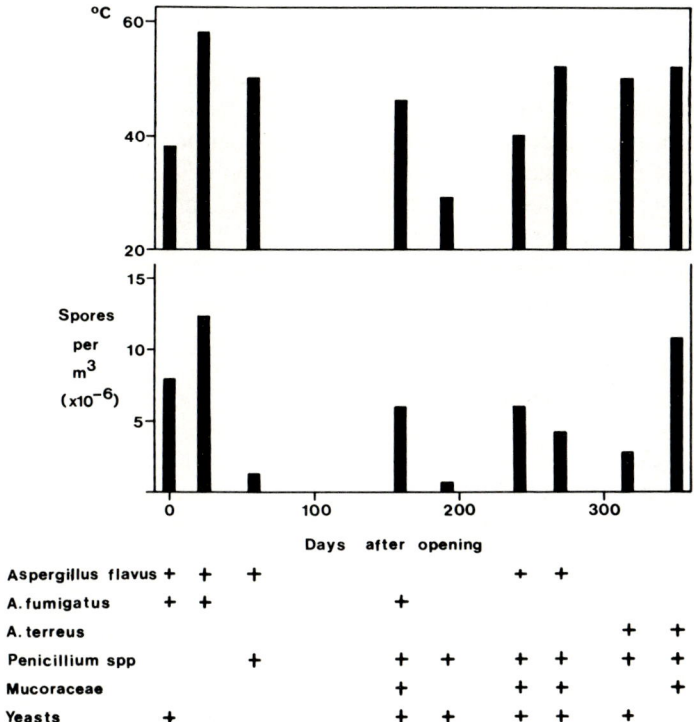

Fig. 2. Numbers and types of spores in the air of a moist barley silo at different times during unloading. °C indicates temperature of uppermost grain in silo.

For instance, hay stored with less than 25% water content molds predominantly with members of the *Aspergillus glaucus* group and *Wallemia sebi,* while at 29% water content, with heating to 40°C, these species are replaced by *A. versicolor* and *Scopulariopsis brevicaulis,* and at 32% with heating to 65°C by *A. fumigatus* and thermophilic fungi (Gregory *et al.,* 1963; Festenstein *et al.,* 1965). A similar pattern can be observed in grain, but *Penicillium* species tend to be more abundant than in hay, and with restricted aeration leading to high carbon dioxide concentrations, *P. roqueforti* may predominate. In unsealed, top-unloading, moist barley silos, after the top 30 cm which molds during the storage period has been removed, molding of the grain is related to the rate of unloading, water content, and spontaneous heating, and the changes that occur are reflected in the air spora (Fig. 2; Lacey, 1971).

Concentrations of airborne spores are usually greatest with heavy molding, vigorous disturbance, and little ventilation. For instance in cattle sheds when moldy hay is being fed (Baruah, 1961), in moist barley silos when grain is being unloaded (Lacey, 1971), in cork factories when moldy cork is being processed (Lacey,

1973), and in particle-board mills when a "mattress" is being formed with moldy bagasse (Lacey, 1974). The rate at which spore concentrations decline depends on spore size, the consequent rate of sedimentation, and the amount of air movement and ventilation. In most situations there is enough air movement and disturbance to ensure that some spores remain airborne, and in a small shed, about 9% of fungus and 12.5% of actinomycete spores were still airborne 20 min after shaking moldy hay (Lacey and Lacey, 1964). If exhaust ventilation is to remove spores effectively from an environment, it must be situated close to their source. Although this may be feasible in a factory, it is impracticable where the source is extensive, as on farms.

V. DISPERSION OF AIRBORNE CONIDIA

Dispersal of conidia through the air involves three stages: (1) *Liberation and takeoff,* enabling the conidium to overcome the adhesive forces holding it to the conidiophore, cross the laminar boundary layer, and enter the turbulent boundary layer; (2) *dispersal* in air currents from the source to other parts of the troposphere; and (3) *deposition* on surfaces prior to germination and growth. This sequence of events is the same for all fungi, although many have adaptations that assist these processes.

A. Liberation and Takeoff

Many fungi are well adapted to airborne dispersal by having tall conidiophores projecting into or through the laminar boundary layer or by particular liberation mechanisms which forcibly project the spores through this layers. The various liberation methods are listed in Table IX and have been reviewed by Ingold (1971) and Gregory (1973), so that their characteristics will only be summarized here.

1. Passive, Dry Methods

Shedding under gravity is useful only for spores elevated on vegetation or sporophores, but its importance is uncertain as it is impossible to eliminate all air currents in experiments. Convection currents caused by temperature gradients of at least 1°C/cm height of glass cylinders were sufficient to lift conidia of *Botrytis cinerea* and *Monilia sitophila,* but although deflation may be important in liberating conidia of many fungi, it is difficult to separate from mechanical disturbance. The minimum wind speed required to remove conidia varies with species between 0.4 and 2.0 m/s. Numbers released increase with increasing wind speed and turbulence and decreasing relative humidity, but wind speeds as high as 0.5 m/s are uncommon in cereal crops. Therefore most of the release of

TABLE IX

Classification of the Liberation Mechanisms of Fungus Spores

Liberation mechanism	Dry	Moisture-requiring
Passive	Shedding under gravity	Mist pick-up
	Shedding in convection currents	Bubble scavenging
	Blowing away (deflation)	Droplet launching
	Mechanical disturbance	Bellows mechanisms[a]
	Insect transmission	Rain tap and puff
		Rain splash
		Drip splash
		Splash cup[a]
Active	Hygroscopic movements	Squirt-gun mechanisms[a]
	Water rupture	Squirting mechanisms
		Rounding of turgid cells
		Ballistospore discharge[a]

[a] Not found in conidial fungi (but note that a splash-cup mechanism is employed by conidial *Phaeotrametes decipiens*).

Erysiphe graminis conidia may result from the shaking and knocking of leaves and stems (Bainbridge and Legg, 1976). Insects may carry fungus spores as casual contaminants, but in some fungi the life cycle is adapted for insect transmission. The *Pesotum* anamorph of *Ceratocystis ulmi,* the cause of Dutch elm disease, forms in the brood chambers of bark beetles that carry it to other trees, while the conidial stage of *Claviceps purpurea* is mixed with nectar attractive to insects.

2. Passive, Moisture-Requiring Mechanisms

Release of *Cladosporium* and *Verticillium albo-atrum* conidia may be aided by mist droplets carried in air currents that when dry are insufficient to release spores (Davies, 1959). Aquatic fungi may become airborne in water droplets released as waves break or as bubbles rise to the surface and burst. Organic material, including spores, may also become concentrated on the surface of the bubbles as they rise (Blanchard, 1973).

A number of processes require raindrops for spore liberation. Falling on a dry stem at terminal velocity, drops spread radially over the surface at high speed, initially about 70 m/s, covering a radius of about 2 cm within 2 ms. A puff of air preceding the water film disturbs the laminar boundary layer and, with the vibration from the drop striking the stem, projects spores into the turbulent boundary layer (Hirst and Stedman, 1963). Falling into a thin film of water, a raindrop produces 100–5000 satellite droplets 5–2400 μm in diameter. Their total volume is similar to that of the incident drop, but they are composed equally of water from the drop and from the surface film. Such splashing may be important in the liberation of many slimy-spored fungi such as *Colletotrichum lindemuthianum* and *Fusarium* species. The processes involved have been described by Gregory *et al.* (1959) and Stedman (1979b). Drip splashes from vegetation never reach terminal velocity and so may be larger than 5 mm without fragmenting, but the total volume of droplets produced on impact may be less than the volume of the incident drop, especially if the drop falls on a surface that is not rigid, is coated with wax, or is inclined from the horizontal. Droplets are deposited in a concentric pattern with numbers decreasing with distance from the source up to 2 m, unless there is a wind, when the pattern may be displaced and droplets may travel 16 m or more. Spores may occur in water films, in discrete droplets on nonwettable surfaces, as suspended drops, or concentrated at sites where runoff may occur. They may be suspended and redeposited several times in rain, as well as falling in drip splashes, while the smallest droplets may evaporate, leaving the spores to be dispersed dry.

3. Active, Dry Mechanisms

The twisting of conidiophores in response to falling or rapidly changing relative humidity, as often occurs in the early morning, may be sufficient to release conidia of *Botrytis cinerea* and *Alternaria porri* from organic connection with the conidiophore, allowing subsequent deflation (Jarvis, 1962; Meredith, 1966a). Loss of water from conidiophores may also lead eventually to the rupture of tensile water and the formation of a gas bubble accompanied by a sudden movement which jerks the spore free. Often the conidiophore wall is thickened unevenly, causing distortion with drying and emphasizing the movement as the original shape is restored (Fig. 3a). This mechanism has been found in *Deightoniella torulosa*, *Zygosporium oscheoides*, *Helminthosporium turcicum*, and other anamorphic fungi (Meredith, 1961, 1962c, 1963, 1965) and leads to a characteristic postdawn release pattern.

4. Active, Water-Requiring Mechanisms

The bursting of a turgid cell causes the discharge of many Ascomycetes and lower fungi but has been found in only two conidial anamorphs, *Nigrospora sphaerica* and *Pyricularia oryzae*. In *Nigrospora* (Fig. 3b), the contents of an ampulliform basal cell are thought to be under pressure and discharged through a small channel when the conidiophore is ripe, projecting the conidium several centimeters (Webster, 1952). The whole of the basal cell is thought to disintegrate in *Pyricularia* (Ingold, 1964). In a number of other groups the sudden rounding of turgid cells causes discharge, but this method is thought to occur in only a few conidial fungi, e.g., *Epicoccum purpurascens* (Fig. 3c) and *Arthrinium cuspidatum* (Webster, 1966).

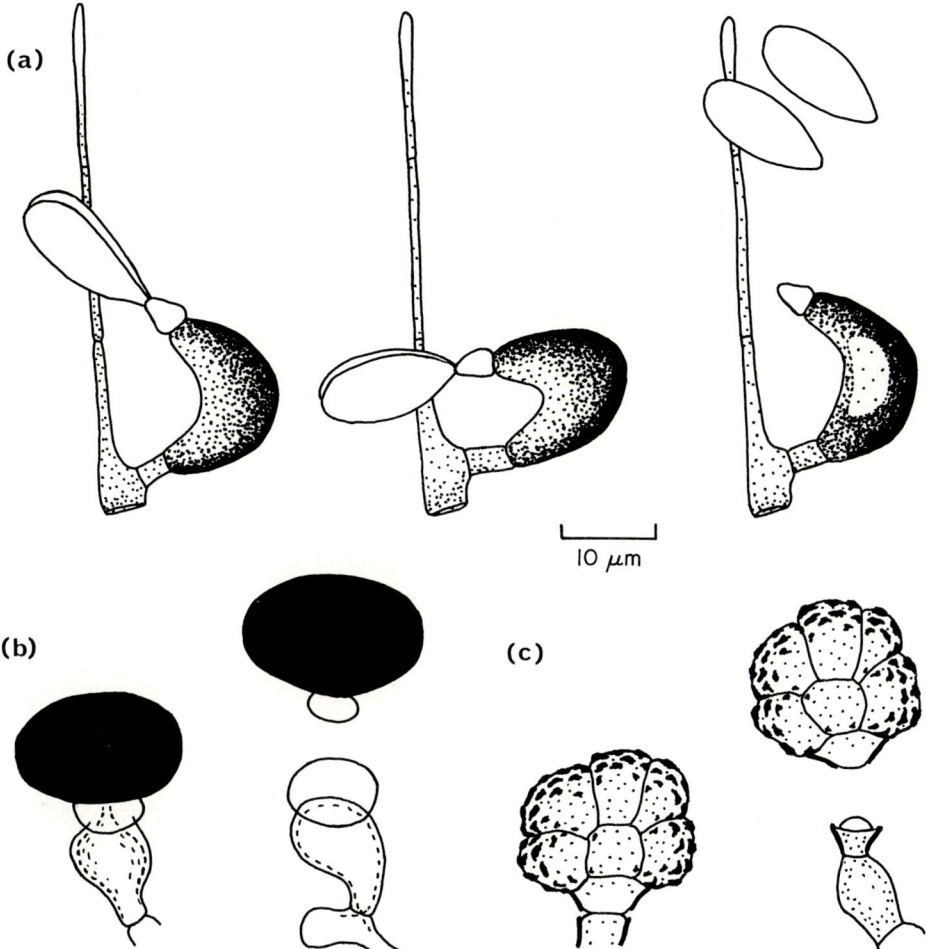

Fig. 3. Spore liberation mechanisms in conidial fungi. (a) *Zygosporium oscheoides* showing increasing curvature during drying and sudden straightening following rupture of tensile water. (After Meredith, 1962c.) (b) *Nigrospora sphaerica*. The black conidium is carried on a small supporting cell above the ampulliform cell which supplies the liquid jet for discharge. The discharged spore shows an attached drop of cytoplasm. (After Webster, 1952.) (c) *Epicoccum pupurascens* Conidium discharge by rounding of turgid cells. (After Webster, 1966.)

B. Dispersal

Two aspects of dispersal must be considered: the fate of individual spores and the behavior of groups or clouds of spores. Both aspects are closely related depending on the size, shape, roughness, density, and electrostatic charging of individual spores and air movement, turbulence, air viscosity, layering, convection, wind gradients near the ground, and patterns of atmospheric circulation in the environment. Spores are heavier than air and so tend to sediment, but this is counteracted by upward movements of the air resulting from turbulence and convection. Turbulence dilutes spore clouds according to the theories of eddy diffusion. The interplay of all these factors determines the distance a spore or spores will travel, subject to the limitations imposed by such a phenomenon as precipitation.

1. Sedimentation

The specific gravity of spores is 1.1–1.2, so that they sink in air, reaching terminal velocity within one spore diameter. The relationship between size and terminal velocity for smooth spheres 1–50 μm in diameter in viscous fluids is described by Stokes's law. This law states that

$$V_t = \frac{2}{9} r^2 g \frac{(\rho_p - \rho_a)}{\eta} \qquad (1)$$

where V_t is the terminal velocity (cm/s), r the radius of the sphere (cm), g the acceleration due to gravity (cm/s/s), ρ_p the density of the sphere (g/cm³), ρ_a the density of the medium (air, 1.27×10^{-3} g/cm³), and η the viscosity of the medium (air, 1.8×10^{-4} cm/s at 18°C). Usually ρ_a is negligible compared with ρ_p, allowing simplification to

$$V_t = 2r^2 g \rho_p / 9 = 0.0121 r^2 \qquad (2)$$

where r is the radius in micrometers.

Experimental determinations of the terminal velocities of many spores give values close to those predicted by Stokes's law (Gregory, 1973). Values range from 0.003 cm/s for *Coccidioides immitis* to 2.0–2.8 cm/s for *Helminthosporium sativum*. Estimates differ slightly, because spore dimensions may vary within a range characteristic of the species and also because the degree of hydration of constituent colloids, occurrence of intracellular gas bubbles, and size vary with relative humidity. Drag on spores is increased by surface roughening which creates eddies or a layer of stationary air around the spore and by deviation from the ideal sphere. A dynamic shape factor (α) has been proposed (Chamberlain, 1967) to allow prediction of the behavior of nonspherical spores. For cylinders with length and breadth equal, $\alpha = 1.06$, while for cylinders four times longer than broad, $\alpha = 1.32$. Gregory (1945) obtained similar results by considering cylinders as spheres with a diameter equal to the geometric mean of the long and short axes. The aggregation of spores similarly decreases terminal velocity, because not only is drag increased but also air may be trapped, decreasing the bulk density. Four spheres attached in line give $\alpha = 1.56$–1.58.

2. Wind

Wind speed usually greatly exceeds terminal velocity, at least at the standard height, 10 m above ground level, at which measurements are made. Mean wind speed in England is about 5 m/s, with 90% of hourly means exceeding 1 m/s and 50% exceeding 3 m/s. However, close to the ground wind speed is much less, decreasing by 20% between 10 m and 2 m above ground and to 8–16% of the 2-m value in the lower half of vegetation canopies (Scott, 1978) or less than 0.5 m/s in cereal crops (Bainbridge and Legg, 1976). With laminar airflow, spores would travel along a trajectory determined by height of liberation, terminal velocity, and wind speed, but this ignores turbulence and convection which add vertical components to air movement.

3. Diffusion of Spore Clouds

Eddies caused by friction at the earth's surface and convection vary in scale from 1 cm across, through those recognized as changes in wind direction, to large-scale motions in cyclones and anticyclones. Two-thirds have periods of less than 5 s and many last less than 1 s. Their activity can be observed in the smoke eddying from a chimney. Depending on their size, eddies dilute the smoke plume as it travels downwind, spreading it both vertically and horizontally, or move it bodily. Temperature gradients and atmospheric stability may lead to greater horizontal diffusion than vertical, while looping may result from parts of the plume being lifted by convection and other parts being carried down by compensatory air currents. Fanning occurs when temperature inversions prevent upward diffusion.

Eddy diffusion is complex and not fully understood, but Sutton's theory has been used to predict diffusion of spore clouds over short distances (Gregory, 1973). Sutton's formula relates the standard deviation of diffusing particles (σ), a diffusion coefficient (C), time and wind speed which together equal distance (x), and turbulence (m), so that $\sigma^2 = \frac{1}{2} C^2 x^m$.

C decreases with height as conditions for eddy formation become less favorable, while m is 1.24 in very stable non-turbulent wind to 2.00 under extremely turbulent conditions. Normal overcast conditions with a steady wind give $m = 1.75$, but the smoothing of random elements over a long period may cause m to increase to 2.00. Diffusion occurs over the whole distance of travel, but its rate is not necessarily constant. Wind speed increases and turbulence decreases with height, horizontal and vertical diffusion rates differ, wind shear may cause the top of the cloud to travel faster than the base and diffuse down in advance, concentrations may become patchy in widely dispersed clouds distant from their source because large eddies are too infrequent to smooth out variations, and particles are lost by deposition. As a consequence of these factors and because C and m refer only to neutral overcast conditions, Sutton's theory is less useful for distances greater than 1 km. Gregory

(1945) modified the theory to allow for turbulent deposition of spores, and for further discussion reference should be made to Gregory (1973), Tyldesley (1967), and Pasquill (1974).

4. Buoyancy

The upward movement of turbulence may be supplemented by convection currents which carry spores vertically from heated vegetation to cloud level.

5. Other Factors

The effect of the small electrostatic charges demonstrated on spores is probably negligible unless they are closer to another body than 1 mm. Movement of spores may also occur down a temperature gradient, away from a hot surface toward a cold one, and away from light.

6. Spread of Airborne Conidia

The spread of airborne particles has to be considered on two scales: short-distance spread relevant to the spread of fungi within a crop and the epidemiology of plant disease, and long-distance spread concerned with the spread of fungi from field to field and area to area. Because it is difficult to find and recognize small numbers of spores a long distance from their source, most studies have been on short-distance spread.

a. Short-Distance Spread. The spread of fungus spores from local sources results in characteristic dispersal gradients. Plotting spore or lesion numbers against distance on linear scales gives hollow curves (Fig. 4a), with numbers decreasing rapidly at first and then gradually flattening out and becoming almost parallel to the distance axis. The comparison of such curves is difficult, because the point of inflection varies with the scales used for the axes and the slope of the

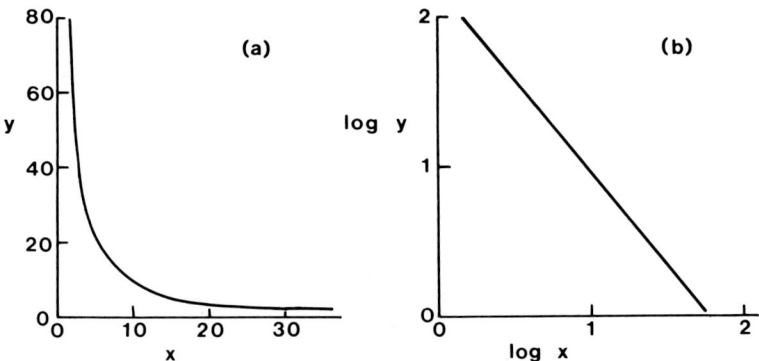

Fig. 4. Dispersal gradients. (a) Typical dispersal gradient from a point source plotted on linear scales, e.g., number of infections (y) against distance (x). (b) Dispersal gradient plotted on double logarithmic scales. (After Gregory, 1968.)

curve changes from point to point. However, plotting the same data on logarithmic scales allows a straight line to be fitted to the points (Fig. 4b), which may be characterized by the slope b, equal to the tangent of the angle the line makes with the x axis, where y is amount of disease or number or concentration of spores at distance x from their source. Values of b calculated for *Phytophthora infestans* infection gradients vary between -1.2 and -3.3, depending on the source geometry, and other dispersal gradients fall within a similar range. Line, strip, and area sources give flatter curves than point sources, since infection declines less rapidly with distance. Within a crop, impaction also helps to remove spores, causing steeper gradients, but unexpected results sometimes occur. For instance, *Erysiphe graminis* conidia are deposited more rapidly than expected, probably because spores stick together (Bainbridge and Stedman, 1979). Further discussion of dispersal gradients may be found in Gregory (1968).

b. Long-Distance Spread. A spore cloud diffusing through the atmosphere loses spores by deposition so that only 5-10% are carried away from the ground and travel long distances. Small particles travel higher and further than large ones, so that although 50% of the catch at a 700-m altitude may be fungus spores, at 1600 m the figure is only 10% (Fulton, 1966b).

As with short-distance spread, plant-pathogenic fungi have also yielded information on long-distance dispersal of fungus spores as, for instance, when spores are trapped before the disease is present in local crops. It has sometimes been possible to identify their sources by plotting wind trajectories. Thus early catches of *Puccinia graminis* urediniospores, accompanied by *Alternaria,* occurred in South Wales only when trajectories had passed over wheat-growing areas in Europe. Long travel over country producing few spores resulted in erosion of the base of the spore cloud and maximum numbers at a 600- to 900-m altitude (Hirst and Hurst, 1967).

The dispersal of spores across the North Sea has been studied by sampling at different altitudes along a transect downwind of England (Hirst and Hurst, 1967). Wind usually blows across England in ½ day, carrying spores out to sea. A series of morning samples showed three regions with large concentrations of *Cladosporium* spores close to the coast and 300 and 600 km downwind (Fig. 5). These were interpreted as having been liberated on successive days and were separated by clouds of spores typical of the night spora. The clouds became more dilute with distance from the coast, and their bases were progressively eroded.

Extension of the geographical range of fungi is often a consequence of long-distance spread. Thus *Peronospora tabacina* crossed Europe in 4 yr and *Puccinia polysora* crossed Africa in a similar period, while *Puccinia graminis* f. sp. *tritici* can travel the length of Europe or North America in a single season (Gregory, 1963; Zadoks, 1967). Modern air travel may also contribute, and spores of many species were found on the footwear and luggage of passengers arriving in Hawaii (Baker, 1966).

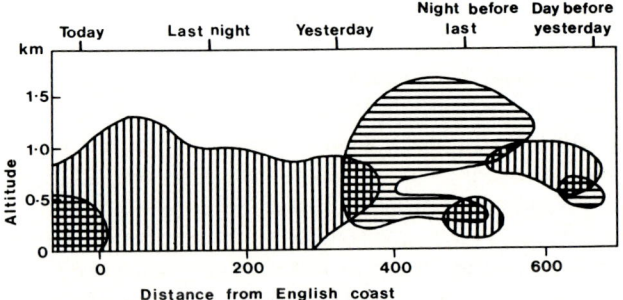

Fig. 5. Position of peak concentrations of *Cladosporium* spp. spores (>80 spores/m^3; horizontal hatching) and damp-air-spore types (>5 spores/m^3; vertical hatching) at various altitudes over the North Sea downwind of the English coast on 16 July 1964. Interpreted as windborne remnants of alternating day and night spore clouds liberated from the English land surface. (After Hirst and Hurst, 1967.)

C. Deposition

1. Deposition Outdoors

Deposition marks the end of airborne dispersal of spores, returning them to the boundary layer of plant or soil where they may colonize new substrates. Several different processes can be involved, of which sedimentation, boundary layer exchange, turbulent deposition, impaction, and rain washing are most important in the deposition of fungus spores. Most conidia are too large to be deposited by Brownian movement and electrostatic charging.

Sedimentation of spores at terminal velocity under gravity is only important in still air or at low wind speeds. Outdoors it probably only occurs within dense vegetation or under stable conditions at night. Boundary layer exchange occurs when eddies break into the laminar boundary layer, removing spore-free air and leaving spores behind to sediment under gravity. Under turbulent conditions spores are deposited at a faster rate than would be expected by sedimentation, at a rate that increases with wind speed and on both upper and lower surfaces of horizontal leaves.

One of the most important methods of deposition of dry spores is impaction. The airstream encountering a leaf or twig is deflected, but spores may be slower to respond to the change in direction because of their momentum and impact on the windward side of the obstruction. Impaction is most efficient for large spores blown fast toward small objects, and large-spored plant pathogens such as *Helminthosporium* are thus well suited for deposition in this way. Small-spored soil fungi such as *Penicillium* and *Aspergillus* must be deposited in other ways. A wind speed of 2 m/s, the usual maximum in vegetation, fails to impact spores 4–5 μm in diameter on objects only 1 mm diameter. In contrast *Erysiphe graminis* conidia are deposited on wheat leaves with 40–60% efficiency (Gregory, 1973).

Dense vegetation efficiently filters spores from the airstream by impaction and slows the airstream sufficiently to allow sedimentation. Impaction efficiency and ability to penetrate vegetation are inversely related, and many fungi compromise between dispersal and deposition in having spores about 10 μm in diameter. Spores formed within dense vegetation are unlikely to escape into the air above unless they are liberated near the upper surface. Liberated close to the soil, few spores escape (Legg and Bainbridge, 1978).

Efficient impaction is not the only essential for deposition on surfaces. Spores must be retained and not bounce off. Wet and sticky surfaces are most retentive of particles, giving results close to those predicted aerodynamically (Chamberlain and Chadwick, 1972), and wax structures on the leaf surface may also assist retention (Forster, 1977). Once deposited, larger fungus spores are more easily resuspended than small ones.

Rain washing rapidly removes spores from the air and may be the most important process in depositing small spores and the remnants of spore clouds not deposited close to their source. Spores may be impacted on raindrops, be captured by cloud droplets, or even form their nuclei. Impaction efficiency is a function of drop radius and the terminal velocity of both raindrop and spore. Terminal velocities of raindrops range from 2 to 9 m/s. *Penicillium* conidia would be collected with 15% efficiency by 2-mm diameter raindrops, but not at all by 1-mm diameter drops; while large *Erysiphe graminis* conidia would be trapped by any raindrop, reaching a maximum 80% efficiency with 2.8-mm-diameter drops.

The concentration of particles remaining in the air decreases exponentially with the length of washing, so that 60, 36, 12, and 1% of 30 μm diameter spores would remain airborne after, respectively, 15, 30, 60, and 120 min of rain falling at 2 mm/h. However, 72% of 4-μm spores would still be airborne at the end of this period (Chamberlain, 1967). The wettability of spores determines where they are carried by the raindrop and where they are deposited. Nonwettable conidia of *Aspergillus, Penicillium,* and *Cladosporium* are carried on the surface and left behind as the drop rolls across a nonwettable leaf surface, while wettable *Acremonium, Fusarium,* and *Verticillium* conidia are carried within the drop until it comes to rest.

2. Spore Deposition Indoors

The same deposition processes occur indoors as out, but less air movement allows more sedimentation, with concentrations of airborne spores decreasing in a way similar to those in a stirred settling chamber (Dimmick, 1969).

3. Spore Deposition in the Respiratory Tract

The anatomy of the respiratory tract ensures that different deposition processes operate in different parts, depositing progressively smaller particles. After the nasopharyngeal region, the lung consists of a series of bifurcating tubes decreasing in diameter but increasing in total cross-sectional area so that airflow progressively decreases from 100 cm/s in the trachea to about 1 cm/s in the bronchioles. At the end of the system are about 10^7 alveoli 0.5 μm in diameter with a total surface area of 30 m².

In nose and pharynx most deposition is by impaction, retaining spores down to 10 μm in diameter. Impaction is aided by turbulence in the trachea, bronchi, and bronchioles, the chances of impaction increasing as airway size decreases. In the deeper parts of the lung with slow air movement, sedimentation becomes important and, for particles smaller than 0.5 μm, Brownian movement. Particles 2-4 μm in diameter are deposited optimally in the alveoli. Elongated particles, such as fibers and chains of spores may be deposited by interception with the walls of the airways as they rotate in the airstream (Gregory, 1973).

VI. IMPLICATIONS OF AN AIR SPORA

Airborne conidia are thus widespread in the atmosphere and often numerous. They come mostly from living and dead vegetation and only occasionally from blown soil. They serve to distribute far and wide carbon copies of the parent organisms, contrasting with sexual spores which allow recombination and the distribution of new genotypes. The nature and properties of the air spora thus have many implications for the spread of diseases of animals and plants, for spoilage of food and other materials, and for industrial processes. The full implications are still not known, although better trapping methods and computer modeling techniques are helping to provide a better appreciation.

A. Human and Animal Disease

1. Allergy

Allergy is perhaps the most common human reaction to airborne conidia and is defined as an acquired, specific, altered capacity to react to a substance (von Pirquet, 1906). Sensitivity is acquired by exposure to an allergen, is specific to that allergen, and results in an allergic reaction not previously found in the subject. People vary in the ease with which they may be sensitized and in the type of allergic reaction produced.

About 20% of the population are atopic and easily sensitized by concentrations usual in the air spora (up to $10^6/m^3$). They react immediately on exposure in the upper airways with hay-fever-like symptoms and asthma and may become sensitive to several of the allergens to which they are exposed. The remainder of the population requires more intensive exposure (10^6–10^9 spores/m³) for sensitization, reacting only after 4-14 h in the deeper parts of the lung and producing allergic alveolitis (hypersensitivity pneumonitis) leading to breathlessness (Lacey et al., 1972; Lacey, 1975b). Sensitivity is usually only to one allergen, and the disease is characteristically occupational, often being associated with stored products (Table X).

Spores associated with immediate allergy are mostly larger than 5 μm, while those associated with delayed allergy are mostly smaller, ensuring deeper penetration into the lung (Section V,C,3). However, the cutoff between different regions of the lung is not absolute, and the volume or numbers of larger spores reaching the alveoli will depend on intensity of exposure, while aggregation of small spores, e.g., those of *Aspergillus fumigatus,* may cause deposition in upper airways.

Many common spores are allergenic (Hyde, 1972). New allergens are continually being found, so that all spores should be regarded as potentially allergenic. Most come from local sources such as agricultural crops, and numbers in the air may be boosted by human activity, especially harvesting. It is also possible that body convection currents, which form about 10% of inspired air (Lewis *et al.,* 1969), can carry large spore concentrations to the nose from ground level. However, sensitivity to conidia does not necessarily indicate that these are the usual airborne spore forms; e.g., sensitivity to *Phoma* may indicate exposure to ascospores of *Leptosphaeria* (Ganderton, 1968). Relief from airborne allergens may be found where the vegetation and climate do not favor fungal growth, e.g., in mountain valleys (Davies, 1969a).

TABLE X

Allergenic Conidial Fungi and Their Sources

Immediate allergy		Delayed allergy	
Species	Source	Species	Source
Cladosporium herbarum	Vegetation, litter, straw	*Aspergillus clavatus*	Malting barley
Alternaria alternata		*A. fumigatus*	Malting barley, hay, straw, composts
Epicoccum purpurascens			
		A. versicolor	Straw
Verticillium lecanii	Straw	*A. flavus*	Maize
Aphanocladium album		*Cryptostroma corticale*	Maple bark
Paecilomyces farinosus		*Penicillium frequentans*	Cork
Arthrinium phaeospermum	Reeds	*P. casei*	Cheese
Cladosporium fulvum	Tomato plants	*Aspergillus* spp.	Citric acid fermentation
		Penicillium spp.	
Aspergillus fumigatus	Composts, hay, stored straw	*Graphium* sp.	Redwood sawdust
		Aureobasidium pullulans	

2. Infection

Airborne transmission is important to many of the fungi causing disease in man and animals, although seldom occurring from host to host. Most pathogenic fungi are opportunistic and able to grow saprophytically in stored crops, composts, and soil, especially if these are enriched with fecal material. *Aspergillus fumigatus* from stored crops may give up to 10^8 spores/m³ and cause respiratory aspergillosis, particularly in young poultry, and mycotic abortion in cattle. It may also grow in the human lung, although the form of the disease depends on the immunological constitution of the subject (Lacey *et al.*, 1972).

Coccidioides immitis and *Histoplasma capsulatum* are far more virulent than *A. fumigatus*, and both grow in soil, *C. immitis* in desert areas of the southwestern United States and *H. capsulatum* more widely in soil enriched with poultry or bat feces. *Coccidioides immitis* becomes airborne in dust storms, during road building, and during the picking of dusty grapes and cotton, infecting a large proportion of the exposed population. In southern Arizona, 97% of Indian children and about 90% of the populations of Bakersville and Phoenix have antibodies indicating previous infection. The spores are about 2.0–2.5 μm, have a terminal velocity of only 0.003 cm/s, are difficult to wet, and can penetrate deeply into the lung where 10–100 spores are sufficient to cause infection in dogs (Converse and Reed, 1966; Levine, 1969). Spores of *C. immitis* have been detected in air only by exposing mice, but it is estimated that, for *H. capsulatum*, up to 6 spores/m³ air may be found when infected soil is disturbed (Rooks, 1954).

3. Mycotoxins

Inhalation of aflatoxins in spores of *Aspergillus flavus* from groundnuts has been suggested as the cause of lung carcinoma in exposed workers (Dvorackova, 1976). Pulmonary mycotoxicosis has also been described in agricultural workers heavily exposed to fungus spores (Emanuel *et al.*, 1975), although the symptoms could also result from activation of the alternative pathway of complement (Edwards *et al.*, 1976).

B. Plant Pathology

Much has been written on the aerobiology of plant pathogens with reference both to their local spread within crops and their long-distance spread into new crops and regions. Aerobiological studies are essential to knowledge of the epidemiology of diseases and to their forecasting. They give information on numbers of conidia of plant pathogens in the air, on the timing of their release, which can then be related to weather conditions, and on dispersal gradients. These data can then be utilized to develop mathematical models simulating disease epidemics. These aspects are further discussed in Gregory (1973), Scott and Bainbridge (1978), and van der Plank (1963).

C. Biodeterioration and Biodegradation

Airborne conidia contribute both to the spoilage of stored agricultural produce and manufactured goods and to the degradation of waste materials, composts, and manures. Spores deposited on crops before harvest may develop in storage, the predominant species depending on water content and spontaneous heating (Festenstein *et al.*, 1965). In bakeries and food shops, airborne conidia contaminating bread and other food decrease shelf life, their growth affecting appearance and nutritional value, and mycotoxins may be produced. In industry, fungal fermentations may become contaminated by undesirable fungi, and manufactured goods may deteriorate, particularly in the tropics, following the deposition of airborne conidia. Fungi, such as *Amorphotheca* (*Cladosporium*) *resinae* may grow in oil, causing blockages in pipes, corrosion, and other problems. In contrast, we rely on fungi, many deposited from the air, to degrade waste materials and pollution and maintain a clean environment.

VII. CONCLUSION

Although much is now known of the numbers and types of airborne fungi and their variations with environment, location, and substrate, there are still large areas of the world virtually unexplored aerobiologically and in others inadequate methods are still used. The characteristics of different trapping methods are now well known. Only visual methods can reveal the whole air spora and indicate categories requiring identification and further study using more specialized methods. As indicated by Gregory (1973), much could be learned from routine sampling stations distributed in different climatic and vegetation zones throughout the world, covering different topographies and altitudes, using efficient, continuously operating traps, and being aware of the different interests of cities, agriculture, and medicine. The aims would need careful definition, and visual assessments of the air spora would need to be supplemented by cultural studies and assessments of the surface deposition of spores from dry air and in precipitation. Conidial fungi form a large, often dominant part of this air spora, so that our knowledge of this group would benefit from such a study.

ACKNOWLEDGMENTS

I would like to thank Mrs. F. A. Jackson and Dr. H. M. Brown for permission to report unpublished results.

REFERENCES

Ackermann, H. W., Schmidt, B., and Lenk, V. (1969). Mykologische Untersuchungen von Aussen- und Innenluft in Berlin. *Mykosen* **12**, 309-320.

Adams, K. F. (1964). Year to year variations in the fungus spore content of the atmosphere. *Acta Allergol.* **19**, 11-50.

Agarwal, M. K., Singh, K., and Shivpuri, D. N. (1974). Studies on the atmospheric spores and pollen grains: Their role in the etiology of respiratory allergy. *Indian J. Chest Dis.* **16**, Suppl., 268-285.

Al-Diwany, L. J., Unsworth, B. A., and Cross, T. (1978). A comparison of membrane filters for counting *Thermoactinomyces* endospores in spore suspension and river water. *J. Appl. Bacteriol.* **45**, 249-258.

Ali, M. I., Salama, A. M., and Ali, J. M. (1976). Possible role of solar radiation on the viability of some air fungi in Egypt. *Zentralbl. Bakteriol., Parasitenkd., Infektionskr. Hyg., Abt. 2* **131**, 529-534.

Allitt, U. (1979). Coiled ascospores in the Hypodermataceae *Trans. Br. Mycol. Soc.* **72**, 147-151.

Andersen, A. A. (1958). New sampler for the collection, sizing and enumeration of viable airborne particles. *J. Bacteriol.* **76**, 471-484.

Anonymous (1975). "The Times Atlas of the World Comprehensive Edition," 5th ed. Times Books, London.

Anonymous (1976). "Statistical Report for 1976," Pollen and Weed Committee, American Academy of Allergy, Columbus, Ohio.

Bainbridge, A., and Legg, B. J. (1976). Release of barley-mildew conidia from shaken leaves. *Trans. Br. Mycol. Soc.* **66**, 495-498.

Bainbridge, A., and Stedman, O. J. (1979). Dispersal of *Erysiphe graminis* and *Lycopodium clavatum* spores near to the source in a barley crop. *Ann. Appl. Biol.* **91**, 187-198.

Baker, G. E. (1966). Inadvertent distribution of fungi. *Can. J. Microbiol.* **12**, 109-112.

Barkai-Golan, R. (1961). Air-borne fungi in packing houses for citrus fruits. *Bull. Res. Counc. Isr., Sect. D* **10**, 135-141.

Baruah, H. K. (1961). The air spora of a cowshed. *J. Gen. Microbiol.* **25**, 483-491.

Batchelder, G. L. (1977). Sampling characteristics of the Rotorod, Rotoslide and Andersen machines for atmospheric pollen and spores. *Ann. Allergy* **39**, 18-27.

Best, A. C. (1950). The size distribution of raindrops. *Q. J. R. Meteorol. Soc.* **78**, 200-275.

Blanchard, D. C. (1973). Bubble scavenging and the water-to-air transfer of organic material in the sea. *Adv. Chem. Serv.* **145**, 360-387.

Bourdillon, R. B., Lidwell, O. M., and Thomas, J. B. (1941). A slit sampler for collecting and counting airborne bacteria. *J. Hyg.* **41**, 197-224.

Brown, H. M. (1970). An automatic volumetric culture plate slit sampler. *In* "Airborne Transmission and Airborne Infection" (J. F. P. Hers and K. C. Winkler, eds.), pp. 57-58. Oosthoek Publ. Co., Utrecht.

Brown, H. M., and Jackson, F. A. (1978a). Aerobiological studies based in Derby. I. A simplified automatic volumetric spore trap. *Clin. Allergy* **8**, 589-597.

Brown, H. M., and Jackson, F. A. (1978b). Aerobiological studies based in Derby. II. Simultaneous pollen and spore sampling at eight sites within a 60 km radius. *Clin. Allergy* **8**, 599-609.

Brown, H. M., and Jackson, F. A. (1978c). Aerobiological studies based in Derby. III. A comparison of simultaneous pollen and spore counts from the east coast, Midlands and west coast of England and Wales. *Clin. Allergy* **8**, 611-619.

Bruck, C. W. (1967). Microbes in the upper atmosphere and beyond. *Symp. Soc. Gen. Microbiol.* **17**, 345-374.

Burge, H. P., Boise, J. R., Rutherford, J. A., and Solomon, W. R. (1977a). Comparative recoveries of airborne fungus spores by viable and non-viable modes of volumetric collection. *Mycopathologia* **61**, 27-33.

Burge, H. P., Solomon, W. R., and Boise, J. R. (1977b). Comparative merits of eight popular media in aerometric studies on fungi. *J. Allergy Clin. Immunol.* **60,** 199-203.
Cammack, R. H. (1955). Seasonal changes in three common constituents of the air spora of southern Nigeria. *Nature (London)* **176,** 1270-1272.
Chamberlain, A. C. (1967). Deposition of particles to natural surfaces. *Symp. Soc. Gen. Microbiol.* **17,** 138-164.
Chamberlain, A. C., and Chadwick, R. C. (1972). Deposition of spores and other particles on vegetation and soil. *Ann. Appl. Biol.* **71,** 141-158.
Christensen, C. M. (1950). Intramural dissemination of spores of *Hormodendrum resinae*. *J. Allergy* **21,** 409-413.
Converse, J. L., and Reed, R. E. (1966). Experimental epidemiology of coccidioidomycosis. *Bacteriol. Rev.* **30,** 678-694.
Darke, C. S., Knowelden, J., Lacey, J., and Ward, A. M. (1976). Respiratory disease of workers harvesting grain. *Thorax* **31,** 294-302.
Davies, R. R. (1957). A study of air-borne *Cladosporium*. *Trans. Br. Mycol. Soc.* **40,** 409-414.
Davies, R. R. (1959). Detachment of conidia by cloud droplets. *Nature (London)* **183,** 1695.
Davies, R. R. (1960). Viable moulds in house dust. *Trans. Br. Mycol. Soc.* **43,** 617-630.
Davies, R. R. (1969a). Climate and topography in relation to aero-allergens at Davos and London. *Acta Allergol.* **24,** 396-409.
Davies, R. R. (1969b). Spore concentrations in the atmosphere at Ahmadi, a new town in Kuwait. *J. Gen. Microbiol.* **55,** 425-432.
Daws, L. F. (1967). Movement of airstreams indoors. *Symp. Soc. Gen. Microbiol.* **17,** 31-59.
Dimmick, R. L. (1969). Stirred-settling aerosols and stirred-settling aerosol chambers. *In* "An Introduction to Experimental Aerobiology" (R. L. Dimmick and A. B. Akers, eds.), pp. 127-163. Wiley (Interscience), New York.
Durham, O. C. (1946). The volume incidence of airborne allergens. IV. A proposed standard method of gravity sampling, counting and volumetric interpolation of results. *J. Allergy* **17,** 79-86.
Dvorackova, J. (1976). Aflatoxin inhalation and alveolar cell carcinoma. *Br. Med. J.* **1,** 691.
Edwards, J. H., Wagner, J. C., and Seal, R. M. E. (1976). Pulmonary response to particulate materials capable of activating the alternative pathway of complement. *Clin. Allergy* **6,** 155-164.
Emanuel, D. A., Wenzel, F. J., and Lawton, B. R. (1975). Pulmonary mycotoxicosis. *Chest* **67,** 293-297.
Errington, F. P., and Powell, E. O. (1969). A cyclone separator for aerosol sampling in the field. *J. Hyg.* **67,** 387-399.
Faulkner, M. J., and Colhoun, J. (1977). An automatic spore trap for collecting pycnidiospores of *Leptosphaeria nodorum* and other fungi from air during rain and maintaining them in a viable condition. *Phytopathol. Z.* **89,** 50-59.
Favero, M. S., Puleo, J. R., Marshall, J. H., and Oxborrow, G. A. (1966). Comparative levels and types of microbial contamination detected in industrial clean rooms. *Appl. Microbiol.* **14,** 539-551.
Festenstein, G. N., Lacey, J., Skinner, F. A., Jenkins, P. A., and Pepys, J. (1965). Self-heating of hay and grain in Dewar flasks and the development of farmer's lung antigens. *J. Gen. Microbiol.* **41,** 389-407.
Forster, G. F. (1977). Effect of leaf-surface wax on the deposition of airborne propagules. *Trans. Br. Mycol. Soc.* **68,** 245-250.
Frankland, A. W., and Gregory, P. H. (1973). Allergenic and agricultural implications of airborne ascospore concentrations from a fungus *Didymella exitialis*. *Nature (London)* **245,** 336-337.
Fulton, J. D. (1966a). Microorganisms of the upper atmosphere. III. Relationship between altitude and micropopulation. *Appl. Microbiol.* **14,** 237-240.
Fulton, J. D. (1966b). Micro-organisms of the upper atmosphere. IV. Microorganisms of a land air mass as it traverses an ocean. *Appl. Microbiol.* **14,** 241-244.

Ganderton, M. A. (1968). *Phoma* in the treatment of seasonal allergy due to *Leptosphaeria*. *Acta Allergol.* **23**, 173.

Greenburg, L., and Smith, G. W. (1922). A new instrument for sampling aerial dust. *U.S., Bur. Mines, Rep. Invest.* **2392**.

Gregory, P. H. (1945). The dispersion of airborne spores. *Trans. Br. Mycol. Soc.* **28**, 26-72.

Gregory, P. H. (1963). The spread of plant pathogens in air currents. *Adv. Sci.* **19**, 481-488.

Gregory, P. H. (1968). Interpreting plant disease dispersal gradients. *Annu. Rev. Phytopathol.* **6**, 189-212.

Gregory, P. H. (1973). "Microbiology of the Atmosphere," 2nd ed. Leonard Hill, Aylesbury.

Gregory, P. H., and Hirst, J. M. (1957). The summer air-spora at Rothamsted in 1952. *J. Gen. Microbiol.* **17**, 135-152.

Gregory, P. H., and Sreeramulu, T. (1958). Air spora of an estuary. *Trans. Br. Mycol. Soc.* **41**, 145-156.

Gregory, P. H., Guthrie, E. J., and Bunce, M. E. (1959). Experiments on splash dispersal of fungus spores. *J. Gen. Microbiol.* **20**, 328-354.

Gregory, P. H., Lacey, M. E., Festenstein, G. N., and Skinner, F. A. (1963). Microbial and biochemical changes during moulding of hay. *J. Gen. Microbiol.* **30**, 75-88.

Gupta, K. D., Sogani, J. L., and Kasliwal, R. M. (1960). Survey of the allergenic aerial mold spores at Jaipur. *Indian J. Chest Dis.* **2**, 237-241.

Hamilton, E. D. (1959). Studies on the air spora. *Acta Allergol.* **13**, 143-175.

Hammett, K. R. W., and Manners, J. G. (1971). Conidium liberation in *Erysiphe graminis*. I. Visual and statistical analysis of spore trap records. *Trans. Br. Mycol. Soc.* **56**, 387-401.

Hariri, A. R., Ghahary, M. S., Naderinasab, M., and Kimberlin, C. (1978). Air-borne fungal spores in Ahwaz, Iran. *Ann. Allergy* **40**, 349-352.

Harvey, R. (1967). Air-spora studies at Cardiff. I. *Cladosporium. Trans. Br. Mycol. Soc.* **50**, 479-495.

Heise, H. A., and Heise, E. R. (1949). The influence of temperature variations and winds aloft on the distribution of pollens and molds in the upper atmosphere. *J. Allergy* **20**, 378-382.

Hirsch, D. J., Hirsch, S. R., and Kalbfleisch, J. H. (1978). Effect of central air conditioning and meteorological factors on indoor spore counts. *J. Allergy Clin. Immunol.* **62**, 22-26.

Hirsch, S. R., and Sosman, J. A. (1976). A one-year survey of mold growth inside twelve homes. *Ann. Allergy* **36**, 30-38.

Hirst, J. M. (1952). An automatic volumetric spore trap. *Ann. Appl. Biol.* **39**, 257-265.

Hirst, J. M. (1953). Changes in atmospheric spore content: Diurnal periodicity and the effects of weather. *Trans. Br. Mycol. Soc.* **36**, 395-393.

Hirst, J. M., and Hurst, G. W. (1967). Long-distance spore transport. *Symp. Soc. Gen. Microbiol.* **17**, 307-344.

Hirst, J. M., and Stedman, O. J. (1963). Dry liberation of fungus spores by raindrops. *J. Gen. Microbiol.* **33**, 335-344.

Hudson, H. J. (1969). Aspergilli in the air spora at Cambridge. *Trans. Br. Mycol. Soc.* **521**, 153-59.

Hudson, H. J. (1973). Thermophilous and thermotolerant fungi in the airspora at Cambridge. *Trans. Br. Mycol. Soc.* **60**, 596-598.

Hyde, H. A. (1972). Atmospheric pollen and spores in relation to allergy. I. *Clin. Allergy* **2**, 153-179.

Hyde, H. A., and Adams, K. F. (1960). Airborne allergens at Cardiff, 1942-59. *Acta Allergol.* **7**, 159-169.

Hyde, H. A., and Williams, D. A. (1949). A census of mould spores in the atmosphere. *Nature (London)* **164**, 668-669.

Hyde, H. A., and Williams, D. A. (1953). The incidence of *Cladosporium herbarum* in the outdoor air at Cardiff, 1949-50. *Trans. Br. Mycol. Soc.* **36**, 260-266.

Ingold, C. T. (1964). Possible spore discharge mechanism in *Pyricularia*. *Trans. Br. Mycol. Soc.* **47**, 573-575.
Ingold, C. T. (1971). "Fungus Spores: Their Liberation and Dispersal." Oxford Univ. Press (Clarendon), London and New York.
Jarvis, W. R. (1962). The dispersal of spores of *Botrytis cinerea* Fr. in a raspberry plantation. *Trans. Br. Mycol. Soc.* **45**, 549-559.
Kotimaa, M. (1977). Airborne spores in a mill and in a veneer factory. *Grana* **16**, 159-161.
Kramer, C. L., and Pady, S. M. (1961). Inhibition of growth of fungi on rose bengal media by light. *Trans. Kans. Acad. Sci.* **64**, 110-116.
Kramer, C. L., and Pady, S. M. (1966). A new 24 hour spore sampler. *Phytopathology* **56**, 517-520.
Kramer, C. L., and Pady, S. M. (1968). Viability of airborne spores. *Mycologia* **60**, 448-449.
Kramer, C. L., Pady, S. M., and Rogerson, C. T. (1960). Kansas aeromycology. V. *Penicillium* and *Aspergillus*. *Mycologia* **52**, 545-551.
Lacey, J. (1971). The microbiology of moist barley storage in unsealed silos. *Ann. Appl. Biol.* **69**, 187-212.
Lacey, J. (1972). Actinomycete and fungus spores in farm air. *J. Agric. Sci.* **1**(2), 61-78.
Lacey, J. (1973). Air spora of a Portuguese cork factory. *Ann. Occup. Hyg.* **16**, 223-230.
Lacey, J. (1974). Moulding of sugar-cane bagasse and its prevention. *Ann. Appl. Biol.* **76**, 63-76.
Lacey, J. (1975a). Airborne spores in pastures. *Trans. Br. Mycol. Soc.* **64**, 265-281.
Lacey, J. (1975b). Occupational and environmental factors in allergy. *In* "Allergy '74" (M. A. Ganderton and A. W. Frankland, eds.), pp. 303-319. Pitman, London.
Lacey, J. (1977). Micro-organisms in the air of cotton mills. *Lancet* **2**, 455-456.
Lacey, J. (1980). The microflora of grain dusts. *In* "Occupational pulmonary disease: Focus on Grain Dust and Health" (J. A. Dosman and D. J. Cotton, eds.), pp. 417-440. Academic Press, New York.
Lacey, J., and Dutkiewicz, J. (1976). Methods of examining the microflora of mouldy hay. *J. Appl. Bacteriol.* **41**, 13-27.
Lacey, J., and Lacey, M. E. (1964). Spore concentrations in the air of farm buildings. *Trans. Br. Mycol. Soc.* **47**, 547-552.
Lacey, J., Pepys, J., and Cross, T. (1972). Actinomycete and fungus spores in air as respiratory allergens. *Soc. Appl. Bacteriol. Tech. Ser.* **6**, 151-184.
Lacey, M. E. (1962). The summer air spora of two contrasting adjacent rural sites. *J. Gen. Microbiol.* **29**, 485-501.
Legg, B. J., and Bainbridge, A. (1978). Air movement within a crop: Spore dispersal and deposition. *In* "Plant Disease Epidemiology" (P. R. Scott and A. Bainbridge, eds.), pp. 104-110. Blackwell, Oxford.
Levetin, E., and Horowitz, L. (1978). A one-year survey of the airborne molds of Tulsa, Oklahoma. I. Outdoor survey. *Ann. Allergy* **41**, 21-24.
Levetin, E., and Hurewitz, D. (1978). A one-year survey of the airborne molds of Tulsa, Oklahoma. II. Indoor survey. *Ann. Allergy* **41**, 25-27.
Levine, H. B. (1969). Biological properties of fungal aerosols. *In* "An Introduction to Experimental Aerobiology" (R. L. Dimmick and A. B. Akers, eds.) pp. 340-346. Wiley (Interscience), New York.
Lewis, H. E., Foster, A. R., Mullan, B. J., Cox, R. N., and Clark, R. P. (1969). Aerodynamics of the human microenvironment. *Lancet* **1**, 1273-1277.
Maunsell, K. (1954). Concentration of airborne spores in dwellings under normal conditions and under repair. *Int. Arch. Allergy Appl. Immunol.* **5**, 373-376.
May, K. R. (1966). Multistage liquid impinger. *Bacteriol. Rev.* **30**, 559-570.
May, K. R. (1967). Physical aspects of sampling microbes. *Symp. Soc. Gen. Microbiol.* **17**, 60-80.

May, K. R., and Druett, H. A. (1953). The pre-impinger: A selective aerosol sampler. *Br. J. Ind. Med.* **10**, 142-151.

May, K. R., and Harper, G. J. (1957). The efficiency of various liquid impinger samplers in bacterial aerosols. *Br. J. Ind. Med.* **14**, 287-297.

May, K. R., Pomeroy, N. P., and Hibbs, S. (1976). Sampling techniques for large windborne particles. *J. Aerosol Sci.* **7**, 53-62.

Meier, F. C. (1936). Collecting micro-organisms from winds above the Caribbean Sea. *Phytopathology* **26**, 102.

Meredith, D. S. (1961). Spore discharge in *Deightoniella torulosa* (Syd.) Ellis. *Ann. Bot. (London)* [N.S.] **25**, 271-278.

Meredith, D. S. (1962a). Some components of the air spora in Jamaican banana plantations. *Ann. Appl. Biol.* **56**, 577-594.

Meredith, D. S. (1962b). Spore dispersal in *Pyricularia grisea* (Cooke) Sacc. *Nature (London)* **195**, 92-93.

Meredith, D. S. (1962c). Spore discharge in *Cordana musae* (Zimm.) Höhnel and *Zygosporium oscheoides* Mont. *Ann. Bot. (London)* [N.S.] **26**, 233-241.

Meredith, D. S. (1963). Violent spore release in some fungi imperfecti. *Ann. Bot. (London)* [N.S.] **27**, 39-47.

Meredith, D. S. (1965). Violent spore release in *Helminthosporium turcicum*. *Phytopathology* **55**, 1099-1102.

Meredith, D. S. (1966a). Spore dispersal in *Alternaria porri* (Ellis) Neerg. on onions in Nebraska. *Ann. Appl. Biol.* **57**, 67-73.

Meredith, D. S. (1966b). Diurnal periodicity and violent liberation of conidia in *Epicoccum*. *Phytopathology* **56**, 988.

Noble, W. C., and Clayton, Y. M. (1963). Fungi in the air of hospital wards. *J. Gen. Microbiol.* **32**, 397-402.

Ogden, E. C., and Raynor, G. S. (1967). A new sampler for airborne pollen: The Rotoslide. *J. Allergy* **40**, 1-11.

Oren, J., and Baker, G. E. (1970). Molds in Manoa: A study of prevalent fungi in Hawaiian homes. *Ann. Allergy* **28**, 472-481.

Pady, S. M., and Gregory, P. H. (1963). Number and viability of airborne hyphal fragments in England. *Trans. Br. Mycol. Soc.* **46**, 609-613.

Pady, S. M., and Kramer, C. L. (1960). Kansas aeromycology. VI. Hyphal fragments. *Mycologia* **52**, 681-687.

Pady, S. M., Kramer, C. L., and Clary, R. (1967). Diurnal periodicity in airborne fungi in an orchard. *J. Allergy* **39**, 302-310.

Pady, S. M., Kramer, C. L., and Clary, R. (1969). Periodicity in spore release in *Cladosporium*. *Mycologia* **61**, 87-98.

Pasquill, F. (1974). "Atmospheric Diffusion," 2nd ed. Ellis Horwood, Chichester.

Pathak, V. K., and Pady, S. M. (1965). Numbers and viability of certain airborne fungus spores. *Mycologia* **57**, 301-310.

Perkins, W. A. (1957). "The Rotorod Sampler," 2nd Semiannu. Rep., CML 186. Aerosol Lab. Dep. Chem. Chem. Eng., Stanford University, Stanford, California.

Pohjola, A., Rantio-Lehtimäki, A., and Mäkinen, Y. (1977). Spore composition in a garbage disposal plant. *Grana* **16**, 167-169.

Ramalingam, A. (1968). The construction and use of a simple air sampler for routine aerobiological surveys. *Environ. Health* **10**, 61-67.

Ramalingam, A. (1971). Air spora of Mysore. *Proc. Indian Acad. Sci., Sect B* **74**, 227-240.

Rantio-Lehtimäki, A. (1977). Research on airborne fungus spores in Finland. *Grana* **16**, 163-165.

Rati, E., and Ramalingam, A. (1965). Airborne aspergilli at Mysore. *Aspects Allergy & Appl. Immunol.* **9**, 139-149.

Raynor, G. S., and Ogden, E. C. (1970). The swing-shield: An improved shielding device for the intermittent Rotoslide sampler. *J. Allergy* **45**, 329-32.
Reddi, C. S. (1974). Volume incidence of air-borne allergens, *Indian J. Med. Res.* **62**, 1190-1194.
Richards, M. (1954). Atmospheric mold spores in and out of doors. *J. Allergy* **25**, 429-439.
Rogerson, C. T. (1958). Kansas aeromycology. I. Comparison of media. *Trans. Kans. Acad. Sci.* **61**, 155-162.
Rooks, R. (1954). Airborne *Histoplasma capsulatum* spores. *Science* **119**, 385-386.
Sandhu, D. K., Shivpuri, D. N., and Sandhu, R. S. (1964). Studies on the airborne fungal spores in Delhi—Their role in respiratory allergy. *Ann. Allergy* **22**, 374-384.
Scott, D. (1978). Air movement within six vegetations. *N.Z. J. Agric. Res.* **21**, 651-654.
Scott, P. R., and Bainbridge, A., ed. (1978). "Plant Disease Epidemiology." Blackwell, Oxford.
Shenoi, M. M., and Ramalingam, A. (1975). Circadian periodicities of some spore components of air at Mysore. *Arogya - J. Health Sci.* **1**, 154-156.
Shenoi, M. M., and Ramalingam, A. (1976). Air spora of a sorghum field at Mysore. *J. Palynol.* **12**, 54-54.
Sinha, R. J., and Kramer, C. L. (1972). Identifying hyphal fragments. *Trans. Kans. Acad. Sci.* **74**, 48-51.
Solomon, W. R., Burge, H. P., and Boise, J. R. (1978). Airborne *Aspergillus fumigatus* levels outside and within a large clinical centre. *J. Allergy Clin. Immunol.* **62**, 56-60.
Sreeramulu, T. (1958). Effect of mowing grass on the concentrations of certain constituents of the air spora. *Curr. Sci.* **27**, 61-63.
Sreeramulu, T. (1959). The diurnal and seasonal periodicity of spores of certain plant pathogens in the air. *Trans. Br. Mycol. Soc.* **42**, 177-184.
Sreeramulu, T. (1962). Some observations of the *Deightoniella* fruit and leaf-spot disease of the banana. *Curr. Sci.* **31**. 258-259.
Sreeramulu, T. (1970). Conidial dispersal in two species of *Cercospora* causing tikka leaf-spots on groundnut (*Arachis hypogaea* L.). *Indian J. Agric. Sci.* **40**, 173-178.
Sreeramulu, T., and Ramalingam, A. (1962). Notes on airborne *Tetraploa* spores. *Curr. Sci.* **31**, 121-122.
Sreeramulu, T., and Ramalingam, A. (1966). A two-year study of the air spora of a paddy field near Visakhapatnam. *Indian J. Agric. Sci.* **36**, 111-132.
Sreeramulu, T., and Vittal, B. P. R. (1966a). Periodicity in the air-borne spores of the rice false smut fungus *Ustilaginoidea virens*. *Trans. Br. Mycol. Soc.* **49**, 443-449.
Sreeramulu, T., and Vittal B. P. R. (1966b). Some aerobiological observations on the rice stackburn fungus, *Trichoconis padwickii*. *Indian Phytopathol.* **19**, 215-221.
Sreeramulu, T., Vittal, B. P. R., and Ramakrishna, V. (1971). Aerobiology of *Cercospora koepkei* Krüger causing the yellow spot disease of sugar cane. *Indian J. Agric. Sci.* **41**, 655-662.
Stallybrass, F. C. (1961). A study of *Aspergillus* spores in the atmosphere of a modern mill. *Br. J. Ind. Med.* **18**, 41
Stedman, O. J. (1979a). A seven-day volumetric spore trap for use within buildings. *Mycopathologia* **66**, 37-40.
Stedman, O. J. (1979b). Patterns of unobstructed splash dispersal. *Ann. Appl. Biol.* **91**, 271-285.
Tilak, S. T., and Srinivasulu, B. V. (1967). Air spora of Aurangabad. *Indian J. Microbiol.* **7**, 167-170.
Tyldesley, J. B. (1967). Movement of particles in the lower atmosphere. *Symp. Soc. Gen. Microbiol.* **17**, 18-30.
van der Plank, J. E. (1963). "Plant Diseases." Academic Press, New York.
von Pirquet, C. (1906). Allergie. *Muench. Med. Wochenschr.* **30**, 1457.
Webster, J. (1952). Spore projection in the hyphomycete *Nigrospora sphaerica*. *New Phytol.* **51**, 229-235.

Webster, J. (1966). Spore projection in *Epicoccum* and *Arthrinium*. *Trans. Br. Mycol. Soc.* **49**, 339–343.
Wolf, F. T. (1969). Observations on an outbreak of pulmonary aspergillosis. *Mycopath. Mycol. Appl.* **38**, 359–361.
Wolf, H. W., Skaling, P., Hall, L. B., Harris, M. M., Decker, H. M., Buchanan, L. M., and Dahlgren, G. M. (1959). Sampling Microbiological Aerosols. *U.S. Dept. Health, Educ. Welfare, Publ. Health Monogr.* No. 60.
Zadoks, J. L. (1967). International dispersal of fungi. *Neth. J. Plant Pathol.* **73**, Suppl. 1, 61–80.

14

Biogeography and Conidial Fungi

D.T. Wicklow

I.	Introduction	417
II.	The Adaptive Value of Conidia	419
	A. Vehicle of Escape for "Fugitive" Species	419
	B. Survival in an Unfavorable Environment	420
	C. Population Expansion	421
	D. Varied Dispersal Options	422
	E. Competitive Colonization	423
	F. Distribution of Genetic Variability	424
III.	Convergent Evolution	426
IV.	Controls of Species Distribution	426
	A. Barriers	427
	B. Climate	429
	C. Substrates within Ecosystems	431
	D. Competition and Predation	432
V.	Species Distribution and Environmental Scale	433
VI.	Speciation and Species Diversity	433
VII.	Abundance of Conidial Fungi in Tropical and Temperate Environments	436
VIII.	Conidial Fungi and Ecological Islands	440
	References	442

I. INTRODUCTION

Biogeography is a subject concerned with the limits and geometric structure of individual species populations and with the differences in biotas at various points on the earth's surface (MacArthur and Wilson, 1967). Traditionally, biogeographers have accumulated information about the distribution of species and higher taxa and the taxonomic composition of biotas. Because it is particularly difficult to study and interpret the fundamental processes determining species distributions (i.e., dispersal, invasion, competition, adaptation, and extinction), biogeography has, until only recently, generated little in the way of quantitative theory among students of large organisms (Cody and Diamond, 1975). Mycologists have only begun to apply biogeographical theory in explaining fungal distribution patterns (Swift, 1976).

The geographical distribution of fungi is a particularly difficult subject to approach for the following reasons (Bisby, 1933; Diehl, 1937; Hawker, 1957; Pirozynski, 1968; Baker and Meeker, 1972):

1. Large areas of the earth's surface have not been explored for fungi.
2. Where exploration for fungi has occurred, collections are often limited to taxa of specific interest to the collector.
3. The complete study of the fungus flora of a particular habitat is both difficult and laborious, and thus lists of fungi found in different habitats fail to give a complete picture of the fungal community.
4. Fungi can seldom be identified with certainty in the vegetative condition.
5. The vagaries of fruiting (sporulation) times of different fungal taxa and/or collectors' field schedules make it especially difficult to document the geographical boundaries of a fungus.

Herbarium and culture collection records, along with published lists of fungi uncovered in ecological studies or mycological forays, provide the current data base for the fungal biogeographer. One of the more ambitious projects aimed at assembling information on factors governing the distribution of microscopic soil fungi is described by Anderson *et al.* (1976). Data pertaining to the biosystematics, geographical distribution, niche specialization, dissemination, survival, reproduction, growth, physiological activities, requirements, and tolerances of 400 of the more common species of soil fungi have been entered into a computerized data bank. At present, however, the data output provides only a bibliographical list arranged in alphabetical order. Mycologists are continually finding new linkages between specific sexual and conidial morphs of Ascomycetes. According to Kendrick and Carmichael (1973), anamorph–teleomorph connections are known for about one-quarter of the 595 anamorph-genera of Hyphomycetes they recognize. Increased knowledge of the ecological status of a fungus showing either a conidial or ascosporic stage may someday provide clues as to whether another spore stage should be expected in the organism's life history and where to look for it.

The purpose of this chapter is to assess the adaptiveness of the conidial habit in determining fungal species distributions on a local and a geographical scale. A second objective is to examine different mycological studies as possible examples in which biogeographical theory might be extended to encompass the conidial fungi.

II. THE ADAPTIVE VALUE OF CONIDIA

There are numerous ways in which the conidial habit may improve an organism's fitness in a given environment. Conidia serve as vehicles through which fungi escape from or survive in unfavorable environments, offer dispersal options to fungi having more than one spore type, distribute genetic variability and/or initiate the teleomorph through the bringing together of compatible haploid mycelia, permit rapid population build-up, and are a critical factor determining an organism's competitive capacity.

A. Vehicle of Escape for "Fugitive" Species

Competitively inferior fugitive species can only survive in an ecosystem if they have a higher migration rate than their competitors and can reach patches of satisfactory habitat that are constantly being renewed (MacArthur and Wilson, 1967; Horn and MacArthur, 1972). Conidial fungi commonly adopt the life history strategy of a fugitive species as first recognized (Fig. 1) by Swift (1976). Swift has observed that initial fungal colonists of leaf surfaces rapidly become established on newly available substrate, either by persisting as inoculum on twigs or buds or by synchronizing sporulation on leaf litter to coincide with bud break. Conidia are frequently involved in this and similar colonization processes on other substrates. Brown (1958) found open sands associated with a British coastal dune succession to have constantly occurring salt-tolerant microfungal species, whereas common soil fungi in the foredunes and fixed dunes varied according to whether the particular dune system was acid or alkaline. The repeated isolation of *Pyrenochaeta* and *Tilachlidium* from the foredunes suggested to Brown that these conidial fungi were indeed "characteristic members of the microflora, tolerant of the extreme conditions and perhaps 'escapers' of intenser competition in older dune soils." Given the fact that soil microenvironments on the dune progressively change with vegetational colonization and maturation, *Pyrenochaeta* and *Tilachlidium* might well be characterized as fugitive species.

An unidentified species of *Penicillium* was found to play the role of a fugitive species in an experimental petri dish "ecosystem" consisting of itself and *Aspergillus nidulans* (Armstrong, 1976). The *Penicillium* colonies were infiltrated and eventually overrun by *A. nidulans*. However, the former also produced a greater number of daughter colonies at an earlier age than *Aspergillus*. By escaping to patches of favorable environment (agar surface), they avoided local extinction. Armstrong (1976) and Swift (1976) point out that fungi frequently engaged in colonizing episodes and having a higher population growth rate may be analogous to the so-called r-selected species characterized by MacArthur and Wilson (1967). In addition to having a high intrinsic rate of natural increase (r),

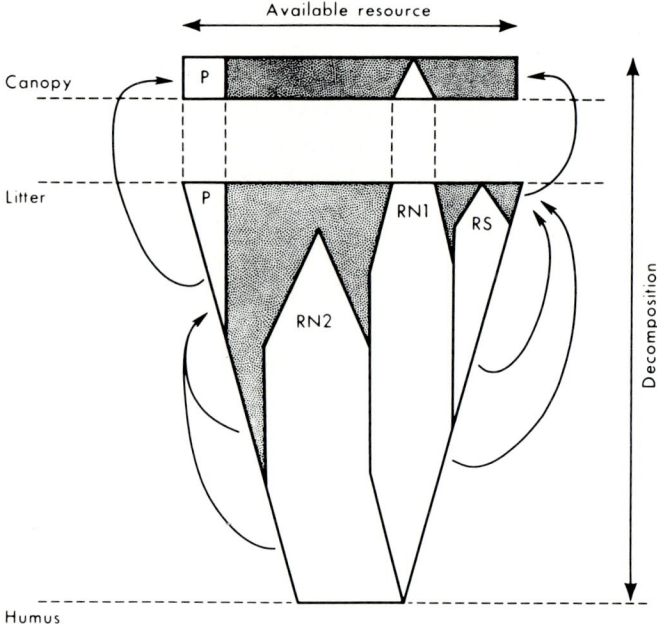

Fig. 1. Diagrammatic representation of the cyclic changes in fungal community structure on an average resource unit. Unstippled areas represent the extent of occupation of available resource by fungi. The arrows represent the dispersal of fungal propagules. P, Parasitic species; RS, resource-specific saprophytes; RN1, primary resource-nonspecific species; RN2, secondary resource-nonspecific species. (Reproduced from Swift, 1976, with permission of Blackwell Scientific Publications.)

such species are said to reach reproductive maturity earlier and to have shorter life spans and smaller body sizes than related and less prolific K-selected species; this ensures their success through effective interspecific competition (Pianka, 1970).

B. Survival in an Unfavorable Environment

The formation of resting spores or sclerotia can allow a fungus to survive without migration until favorable patches of resource become available for colonization. Conidia produced by fungi that inhabit litter or soil enable these organisms to survive in environments temporarily depleted of the substrate nutrients they require to support vegetative growth. The failure of viable conidia to germinate in soil where conditions of temperature and moisture are favorable has been attributed to the depletion of exogenous nutrients below a critical threshold concentration necessary for germination (Ko and Lockwood, 1967). Without some mechanism to restrain spores from germinating until a substrate (nutrient) becomes available, spontaneous germination would be "suicidal" (Garrett, 1970).

Spores are critical in determining the survival of fungi in desert soils. Borut (1960) observed that most of the conidial fungi isolated from samples of desert soil sporulated intensely, the optimum for sporulation being $4°-10°C$ higher than for growth. This suggests that warming trends in desert environments serve to initiate conidium formation in advance of soil drying and/or excessive sun heating. A high percentage of dark-spored conidial fungi occurs in desert soils (Nicot, 1960; Durrell and Shields, 1960; Gochenaur and Backus, 1967). Because melanized conidia appear to be more resistant to radiations than hyaline conidia, it has been suggested that the black pigments provide a selective advantage against ultraviolet light and other radiations (Durrell and Shields, 1960; Gochenaur and Woodwell, 1974). English and Gerhardt (1946) experimentally demonstrated that sensitivity of dark, olivaceous, and hyaline conidia to ultraviolet light diminished as the intensity of pigmentation increased. Significantly, the nonmelanoid dark pigment in conidia of *Aspergillus niger* does not protect against ultraviolet irradiation (Durrell and Shields, 1960). Sussman (1968) questions whether fungal melanins and other pigments provide selective advantage in protecting against ultraviolet light and other radiations, suggesting instead that melanized spores might be better able to withstand the dehydrating effects of visible light in such environments or to resist lysis by other microorganisms (Bloomfield and Alexander, 1967).

C. Population Expansion

In temperate climates there is a spring and a fall bloom of food, which should allow fungi and other organisms with large numbers of propagules and generalized dispersal mechanisms to outproduce those with smaller numbers of dispersal propagules. For example, in temperate regions where bird species frequently have to recolonize an area, the clutch size must be large enough to give the founding populations a head start (MacArthur, 1972). Similarly, through rapid production of large numbers of conidia, a fungus is able to expand its population density quickly in a given ecosystem and thereby take advantage of local or seasonal abundances of a particular resource(s). The role of conidia in expanding the realized niche of plant pathogens such as *Venturia inaequalis* or powdery mildews (Erysiphaceae) is known to all mycologists and plant pathologists (Walker, 1969; Kranz, 1974). Initial infection is commonly established by ascosporic inoculum discharged from fungal ascomata on overwintered leaves. However, rapid production of a conidial anamorph allows the pathogen to spread to other potential infection sites on the same host or to be dispersed to other host plants.

Wicklow (1973, 1975) discusses the possible role of the phialoconidial morphs of *Coniochaeta discospora* and *C. tetraspora* in accounting for the dominance of these ascomycetes in soil dilution plates following burning of a Wisconsin prairie. In this example, two competitively inferior soil ascomycetes are able to take advantage of a physical disturbance (fire) that creates temporary patches of favorable environment (Zak and Wicklow, 1978). The conidial anamorph of *Neurospora* undoubtedly performs an equivalent function following burns in tropical latitudes (Perkins *et al.*, 1976).

D. Varied Dispersal Options

Mangenot and Reisinger (1976) recognize "mechanical" and "trophical" dispersal mechanisms in conidial fungi. Mechanical dispersal is due to air or water movements that can carry liberated propagules various distances. Because of the high risk associated with dispersal by air currents or winds, selection is believed to favor forms producing large numbers of spores (Raper, 1968). Many fungi produce slimy or wet conidia that are also dispersed mechanically following contact with insects or water. The role of microarthropods in the selective dispersal of slime-spored or dry-spored conidia was examined by Brasier (1978). Mangenot and Reisinger (1976) point out that in habitats occupied by microarthropods, a sporogenous apparatus in an aerial position makes the contact easier between conidia and arthropod integuments. Moreover, small propagules or conidia with mucilaginous surfaces more easily adhere to the different parts of the animal. Trophical dispersal involves ingestion of a conidium by an arthropod, which then moves about and deposits the undigested viable conidium with its feces. Insects are frequently attracted to a spore mass by taste, as in the case of the sweet secretion accompanying the conidia of ergot (Hawker, 1957).

A fungus that is able to produce one or more conidial types in addition to the ascosporic or sexual stage increases its options for dispersal (i.e., it can use alternative vectors). Some of the more striking examples of this phenomenon are found among the sooty molds (Hughes, 1976). The sooty mold *Triposporiopsis* forcibly discharges ellipsoidal, triseptate ascospores from a perithecium bearing numerous dark, setose hyphal appendages. *Tripospermum*, the stauroconidial anamorph, is typically tetraradiate and probably serves to catch the spores on trichomes or mycelial networks covering leaf surfaces as conidia are "washed" through various strata in the canopy. Bandoni (1975) theorized that the tetraradiate spore shape may be adapted to dispersal in surface films of water on leaves. A second conidial anamorph is phialidic and produces minute, ellipsoidal conidia. This conidial type could be associated with spore dispersal via arthropod vectors. It is possible to envision a dispersal system in which airborne ascospores and insect-vectored phialoconidia distribute inoculum to upper portions of the vegetational canopy, while the tetraradiate conidium serves to disperse inoculum downward in the leaf washings. In the tropics where sooty molds are bountiful and competition for substrate is keen, such fungi may leave less to chance in determining whether their spores are deposited upon a suitable substrate. If conidia are a principal vehicle of fungal dispersal in the tropics, then we should predict that tropical fungi would have a wider range of conidial types. MacArthur (1972) recognizes a greater variety of dispersal options among other organisms in the tropics, where it is critical that an initial colonist secure the available space.

E. Competitive Colonization

Conidia may confer an advantage in competitive situations with other fungi, because they allow more efficient (rapid) colonization of substrate. The importance of inoculum abundance in the competitive colonization of the living plant or its remains is detailed by Garrett (1970). Garrett's concept of "inoculum potential" is equatable with "propagule," a term used by biogeographers and defined as "the minimum number of individuals of a species capable of successfully colonizing a habitable island" (MacArthur and Wilson, 1967). Wicklow (1973) has observed that soils from a manipulated and waterlogged plot within the Curtis prairie have an abundance of conidial fungi that produce wet or slimy spores as contrasted with nonwaterlogged plots. Competitive colonization of organic substrates incorporated into saturated surface soils may be related to the migration of conidial inoculum in soil–water films. Dry-spored conidia are not easily suspended in water films.

Most ecologists agree that one advantage of producing large seeds is that it provides a plant's progeny with an energetic head start upon germination. MacArthur (1972) observes that early successional species have turned their reproductive effort to very many light, often windblown, seeds. Trees later in the succession produce fewer, heavier seeds that contain greater reserves but are dispersed a shorter distance. No effort has been made to examine relationships between conidial size and competitive colonization for substrate. It is interesting that *Alternaria, Epicoccum, Drechslera,* and *Curvularia,* which produce some of the largest conidia known, are late successional colonists of the phylloplane. *Cladosporium* spp., which produce much smaller conidia, are among the earliest colonists of leaves (Dickinson, 1976; Flannigan and Campbell, 1977). The ability of large-spored and later invading species to colonize the phylloplane may be related to the greater endogenous resource base available to support germ tube growth and early mycelial establishment.

MacArthur and Wilson (1967) and others have suggested that temperate organisms should show a higher reproductive rate than tropical organisms, whereas tropical organisms should produce fewer progeny, survive at high densities, and be better equipped to respond to competitive pressures. For example, clutch sizes of birds are smaller in the tropics, presumably because of a shortage of resources caused in part because more parental energy must be devoted to defense against predation and interference competition (Lack, 1966; MacArthur, 1972). I am unaware of evidence that tropical fungi produce fewer propagules or vegetative fungal biomass while allocating more energy to interference competion, protection of fungi from predators, or other strategies of securing a site on a suitable substrate.

F. Distribution of Genetic Variability

Because the environment in which fungi grow is continually being altered, it is important that a fungal population (i.e., mycelium) have sufficient genetic variability to allow its continued survival in the presence of competitors. In most fungi and other organisms, the sexual process is usually responsible for reassortment of genes in different combinations and thus increased variability in succeeding generations. The conidium offers a convenient means of providing an appropriate mating-type strain necessary for inducing the sexual cycle in heterothallic fungi (Backus, 1939). Brasier (1978) presents experimental evidence that conidial transfer by mites is responsible for fertilization of *Ceratocystis ulmi* protoperithecia in nature. Moreover, when mites were the spore carriers, much larger numbers of perithecia developed in pairings of the aggressive A-type protoperithecial strain with an aggressive B-mating type strain than when the former was paired with a nonaggressive B-type strain. These results led Brasier to develop the hypothesis that mites are part of a mechanism that inhibits hybridization between the aggressive and nonaggressive strains of *C. ulmi* in nature.

Associated with the life history of a majority of fungi is a haploid vegetative phase which is generated by either sexually or asexually produced propagules. This haplophase is subjected to the immediate and rigorous effects of natural selection, except where a balanced system of genetic complementation or heterozygosity might mask genetic deficiencies (Raper, 1968). Heterokaryosis, the coexistence of genetically different nuclei in cytoplasmic continuity with one another, is an important mechanism enabling haploid fungi to survive in a changing environment. Jinks (1952) observed that the nuclear ratios of *Penicillium* heterokaryons found in nature varied in response to environmental changes. Evidence was also found that the heterokaryon could be regularly propagated and dispersed, a phenomenon of considerable importance if heterokaryosis in the absence of sexual reproduction takes on evolutionary significance (Person, 1968). Conidia provide a means of dispersing different genotypes for potential formulation of stable heterokaryons. Any conidial inoculum produced following a shift in the ratio of nuclei in a fungal mycelium would reflect these proportions of different nuclear types, and dispersal of such inoculum in advance of a gradually changing environmental front would provide greater numbers of preadapted colonists. For example, conidia from senescent leaves may be better adapted to colonize other dying or dead herbage than conidia the same organism produced on surfaces of young leaves. Such shifts in the ratios of nuclei during fungal succession on natural substrates could confer greater competitive advantage on an initial colonist, thereby expanding its realized niche.

One means of increasing the heterogeneity in a fungal population is to limit the production of asexual spores. If this is done, the major means of dispersal is through spores that result from nuclear fusion and meiosis. Conidial fungi on the other hand tend toward increasing homogeneity and the loss of sex (Fig. 2). A need for rapid population build-up and dispersibility in maintaining population size and distributing genetic heterogeneity makes a conidial stage essential to the life history of many fungi, even at the expense of a tendency toward increased homogeneity in the populations.

It has been theorized that sexual reproductive mechanisms should have an advantage in environments subject to irregular perturbations and that asexuality should be most common in stable environments, because the organism with the most fit genome should eventually replace a sexually reproducing organism with similar ecological requirements (Williams, 1975). R. P. Seifert (personal communication) predicts that one should expect to find an accumulation of species of

	Mating system	Ecological examples
Maximum genetic heterogeneity ↑	Heterothallic-ascosporic stage (outbreeder)	*Laboulbenia formicarum,* specialized parasite of ants with separate male and female thalli; ascomycetes (?) pathogenic on arctic plants
	Heterothallic-ascosporic and conidial stages (combines outbreeding and inbreeding)	*Aspergillus heterothallicus,* tropical soil; *Neurospora crassa,* burned tropical soils; *Venturia inaequalis,* pathogen of apple trees
	Homothallic-ascosporic stage (inbreeder)	Coprophilous ascomycetes with violent spore discharge (e.g., *Sordaria fimicola, Podospora anserina*)
	Homothallic-ascosporic and conidial stages (inbreeder)	Cleistothecial ascomycetes, rodent burrows, dung, birds' nests, cereal storage, arid soils (e.g., *Aspergillus nidulans, A. amstelodami*)
↓ Maximum genetic homogeneity	Asexual-conidial stage (inbreeder)	*Aspergillus niger, Penicillium chrysogenum, P. expansum,* colonists of soil, plant detritus, seeds, or fruits
	Mycelia sterilia (no dispersal propagules)	Frequently recorded from soil; prevalent in arctic tundra

Fig. 2. Genetic heterogeneity in fungal populations and the conidial habit.

conidial fungi in the tropics, where ecosystem productivity is greater and seasonality is less.

III. CONVERGENT EVOLUTION

Organisms representing differing phylogenetic ancestries but occupying similar environments may resemble each other. The closeness of the convergence might be considered the degree of selective pressure for a given morphology in a given environment. Savile (1968) has pointed out that some ascomycetous genera possess a variety of conidial anamorphs, whereas some conidial anamorph-genera have ascigerous states in different, occasionally distantly related genera. This concept is quantified by Kendrick (this volume, Chapter 2). This suggests that some conidial anamorph-genera are based on two or more convergent lineages. For example, Savile notes that the *Gloeosporium* conidial anamorph is distributed among 29 ascomycetous genera (von Arx, 1957).

The best example of convergent evolution among conidial fungi is that involving aquatic Hyphomycetes. Aquatic conidial fungi are found in abundance on decaying leaves and twigs of deciduous trees in rapidly flowing, unpolluted streams. Their spores are large and often branched or sigmoid. It has been experimentally shown that the tetraradiate type of spore is more effectively trapped on underwater surfaces than spores of more conventional shape (Webster, 1959). The tetraradiate or branched conidium has been identified with the Zygomycotina–Entomophthorales, Ascomycotina–Pezizales, Helotiales, Sphaeriales, Pseudosphaeriales, and Basidiomycotina (see Webster and Descals, this volume, Chapter 11). Because so many superficially similar spore shapes have dissimilar development or different teleomorphs, it is clear that tetraradiate or branched propagules are the result of convergent evolution (Ingold, 1975).

IV. CONTROLS OF SPECIES DISTRIBUTION

One means of obtaining insight into factors influencing patterns of species distribution involves developing explanations for the absence of a species from a given habitat sampled. Green (1971) offers three such explanations: (1) the species cannot live there because its niche does not include that point; (2) the species can live there but never has had the opportunity for zoogeographical reasons; (3) the species can and does live there, but the sample failed, by chance, to include a representative of this species. There has been very little experimental work attempted in examining the factors affecting global distributional patterns of conidial or other fungi. Before any serious effort was undertaken to quantify carefully the soil microfungal populations associated with different communities of higher plants (Warcup, 1951; Tresner *et al.*, 1954), Bisby (1943) observed that "many Fungi Imperfecti, particularly Hyphomycetes of the mould group such as those in the soil, are to be found anywhere." We now know that local microhabitat and substrate are of considerable significance in determining fungal distributions, and local habitat differences in fungal community composition are greater than continental differences (Burges, 1960; Park, 1968; Christensen, 1969; Mueller-Dombois and Perera, 1971; Wicklow, 1973; Swift, 1976).

A. Barriers

Because of advances in our knowledge of the dispersal of fungal spores in the atmosphere, it has become apparent that most fungi are not limited geographically by barriers to dispersal as is often characteristic of higher plants and animals. With the exception perhaps of conidial fungi specifically vectored by arthropods, birds, or mammals, one might anticipate that there has been a history of continuous worldwide dispersal of the detachable propagules of fungi now described and those yet to be described. However, determining the origin of geographically isolated species populations is another matter. Microfungi believed to be marine species were isolated from saline habitats in Wyoming, remote from the sea and dramatically different from marine and estuarine water with respect to salinity level or salt composition (Davidson, 1974). Whether these fungi were introduced (by migratory birds or air currents) or represent relict populations from the geological past when Wyoming was covered by ocean water is a matter for conjecture.

In an earlier review of this subject, Bisby (1943) suggested that a spore is not likely to cross an ocean or to still be alive if it does. We now have evidence showing that conidial fungi can survive long-distance dispersal across polar regions and oceans (Pady and Kapica, 1953, 1955; Pady and Kelly, 1954). In the case of insect-vectored fungal pathogens of higher plants, spore dispersal is restricted if the vector shows limited migratory ability. Historically recent introduction of the fungi causing Dutch elm disease (*Ceratocystis ulmi*) and chestnut blight (*Endothia parasitica*) to North America by humans is evidence that the Atlantic and Pacific oceans served as effective natural barriers to the spread of these insect-vectored pathogenic fungi. Strong and Levin (1975) show that the species or area regressions for fungi and insects, based on the present range of host trees in Britain, differ substantially (Fig. 3). The regular long-distance dispersal by wind transport of viable spores from fungal populations often hundreds of kilometers away and the lesser dispersibility of most insects are believed to account for this difference. Likewise, where a plant occurs sparingly, its obligate parasite may not be able to reach all of its outlying stations, especially if the fungus spores are not windborne (Diehl, 1937). Such a case was reported by Stevens (1917) for the natural distribution of *Endothia gyrosa* [=*E. parasitica*)

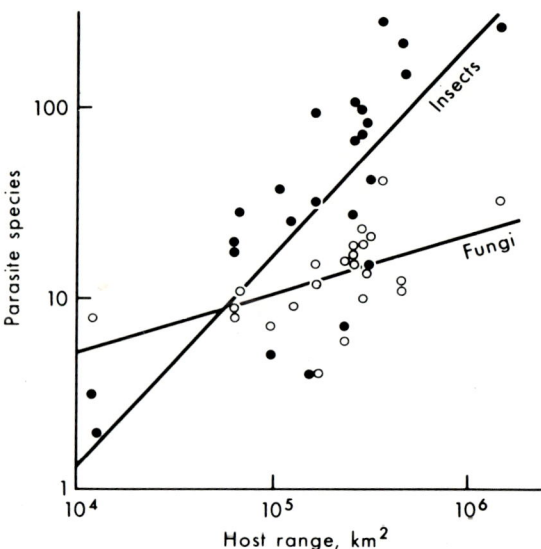

Fig. 3. A comparison of the species and area regressions for the insects and fungi of British trees. The slope of the insect curve (1.12) is significantly greater than that for the fungi (0.26) by nonparametric analysis of covariance (O.005 < P < 0.0007). (From Strong and Levin, 1975.)

on oaks. Stevens found that in the northern United States, where oaks were infrequent and thus more widely separated, the fungal pathogen was much less frequent than the host. Because the fungus was able to attack these northern hosts successfully, it is believed that there was insufficient opportunity for infection, since a forest barrier consisting of trees other than oaks prevented spread of the disease.

MacArthur (1972) notes that more species in the temperate zone seem to have ranges limited by habitat than by any other factor. At the same time, however, a given altitudinal gradient in the tropics is biologically more restrictive than in the temperate zone (Janzen, 1967). Vegetational zones near the peaks of mountains are as effectively cut off from one another as islands in the ocean. Franz (1974) isolated conidial and other microfungi from soil samples collected in different climatic zones (tropical to alpine) in central and western Nepal. No apparent trends can be recognized in the species abundance patterns of any climatic zone sampled. Although some microfungi were characteristic of only one climate zone or soil type, most of the fungi isolated were cosmopolitan in their distribution. Individual strains of a given species showed physiological adaptation (i.e., temperature tolerance) to the particular climate zone from which they were isolated.

A number of examples from the animal kingdom suggest that some taxa, including flightless birds, beetles, and grasshoppers, are able to recognize geographical barriers, since they have formed highly sedentary populations (Mayr,

1954; MacArthur and Wilson, 1967). Among the fungi, yeasts form the most sedentary populations. Since many yeasts rely on specific insect vectors, it seems reasonable to assume that fungi dispersed entirely by insects recognize the risks associated with windborne or other generalized dispersal mechanisms. It is also interesting to speculate whether the failure of numerous fungal isolates from tundra soils to produce spores of any kind (Flanagan and Scarborough, 1974) is evidence that these fungi recognize the risks of dispersal in polar regions.

B. Climate

What evidence do we have that climate influences the distribution of conidial fungi? Because climate is of primary importance in the distribution of vascular plants, it affects fungi indirectly by affecting their host and the availability of host resources (Bisby, 1943). Dearness (1923) appears to have made the initial observation that arctic regions seem to have high proportions of pyrenomycetes, frequently without conidial anamorphs and possibly living on a greater number of hosts than would be the case in temperate latitudes. Where dispersal of infective propagules is limited to the windblown ascosporic stage, it appears that fungal pathogens of arctic plants may minimize the risk associated with dispersal by being able to attack larger numbers of potential plant hosts. Savile (1968) theorizes that in arctic and alpine habitats the short summer accounts for the scarcity of pleomorphism (ascosporic and conidial morphs) in the life of Ascomycetes, noting that two spore stages are of little adaptive value if there is not enough time for sporulation, dispersal, and colonization to occur twice during the growing season. Savile (1963) has also observed that a majority of the fungal pathogens of arctic plants have a persistent perennial mycelium that remains in the crown of the host, making annual recolonization by airborne inoculum less critical. Bisby (1943) observed that, in tropical regions where a fungus does not have to survive through a cold winter, even though it generally has to be prepared for a dry season, the conidial habit may have greater importance. Morrall and Howard (1975) have shown that, although both ascospores and conidia of *Pyrenophora tritici-repentis* can initiate leaf spot disease in the Matador grassland, Saskatchewan, ascospore dispersal occurs before conidium dispersal and at the same time that the disease is increasing. This suggests that ascospores are more important as primary inoculum, with conidia being produced in abundance during the summer. Platt et al. (1977) theorized that, because *P. tritici-repentis* did not produce conidia at temperatures below 10°C in laboratory studies, this may account for low populations of airborne conidia at Matador in April, May, September, and October.

It is important to recognize that the boundaries of species distributions fluctuate greatly. The northern edge of a range is where the northernmost colony of reproducing individuals is located. The colony may become extinct randomly or with a change in climate or a shift in the balance of competition (MacArthur, 1972). *Erysiphe graminis* is able to produce only conidia and not perithecia in arctic regions. Furthermore, it does not thrive on all the species of Gramineae attacked by it further south (Lind, 1934). Jørstad (1925) and Lind (1934) theorize that *E. graminis* and *Sphaerotheca fuliginea* are divided into many host-specific *formae speciales,* each of which has its own northern limit.

Specific microclimates will often determine whether a particular conidial fungus is capable of producing maximum numbers of dispersal propagules (Ingold, 1971) and thus of influencing the outcome of competitive colonization where the energy of inoculum is critical to fungal establishment (Garrett, 1970). Lind (1934) found *Cladosporium* spp. but not *Alternaria* spp. as saprophytes on senescent or dead parts of arctic plants. This could be the result of a shortened growing season in which typically later arriving phylloplane species would be less effective colonists. Flannigan and Campbell (1977) and others have shown that *Cladosporium* spp. are among the earliest established filamentous fungi on all plant organs, whereas *Alternaria alternata* and *Epicoccum purpureum* are much later arrivals. Hill and Nelson (1976) show differences in the virulence, infection efficiency and sporulation efficiency of fungal isolates comprising a cool-environment (England, Netherlands, Scotland, and Switzerland) population and isolates comprising a warm-environment (tropical areas of Africa, Argentina, Brazil, El Salvador, Guinea, Mexico, and Spain) population of *Helminthosporium maydis* race T. Infection and sporulation efficiencies in growth chambers were greater for the cool-environment population under the cool regime (16° and 20°C) and greater for the warm-environment population under the warm regime (28° and 31°C). Their study provides some of the first experimental evidence that ecological races exist within plant-pathogenic, conidium-producing fungi.

Biotic factors associated with different climates may be important in affecting fungal distribution (Diehl, 1937). For example, *Cerebella andropogonis* Ces. is a neotropical sooty mold which, instead of existing on the honeydew of scale insects, colonizes the conidial stage (honeydew) of *Claviceps* (Ergot) and thus limits the latter's distribution in northern latitudes.

The importance of climate in the distribution of soil microfungi is also well established. Danielson and Davey (1973) surveyed *Trichoderma* populations in a variety of forest soils in the southeastern United States and Washington State and reported that *T. viride* and *T. polysporum* were largely restricted to cool temperate regions while *T. harzianum* was characteristic of warm climates. Franz (1975) found a good correlation between the temperature requirements of 154 strains of microscopic soil fungi and the climatic conditions of the site from which soil samples were collected (southwest Africa, Argentina, Canary Islands, West Germany, and Nepal). Strains of ubiquitous species were adapted to the soil temperatures characteristic of each site. Other species had a narrow temperature range and were observed only in climatic regions to which they were adapted. Gochenaur (1970) has reported that the distribution of certain fungal taxa in 29 Peruvian soil types is correlated with the pH and temperature of these soils. Conidial fungi such as *Penicillium* spp. and *Trichoderma viride* were prominent in cool, acid soils, and *Aspergillus* spp. were abundant in warm, dry, alkaline soils.

Microenvironments in nature are discontinuous in time and space because the activities of detritivore communities cause changes in the characteristics of the substrate(s) they colonize. The conidial habit is particularly suited to such fungal environments because the ability to produce and disperse large numbers of spores efficiently, coupled with the capacity for rapid growth and quick exploitation of available resources, has an understandable selective advantage under these short-term labile conditions (Stanier, 1953; Swift, 1976). For example, thermophilic conidial fungi can routinely be isolated from microhabitats in temperate latitudes and elsewhere, which develop high temperatures for only short periods (Ward and Cowley, 1972). Because they are capable of initiating the vegetative phase and producing resting spores during these transient intervals of high temperature, they are able to coexist with microorganisms that are more competitive at lower temperatures (Tansey and Brock, 1978). Some conidial fungi possess varied temperature optima for several different enzyme systems in a mycelium from a single spore. Flanagan and Scarborough (1974) showed that an isolate of *Phialophora hoffmanii* from tundra soil degraded one type of substrate only as a psychrophile and another only as a mesophile.

C. Substrates within Ecosystems

Bisby (1933) has theorized that the distribution of hosts and substrates, principally flowering plants or their remains, has more influence than climate upon distribution of fungi. He has based his theory primarily on the observation that phanerogams cultivated nearly worldwide (e.g., corn, tomatoes, potatoes) have numerous pathogens that can attack them anywhere. One can also argue that the dimensions of the abiotic environment within which one can successfully cultivate a given crop already include the fundamental niche of the pathogen and do not expand it. Even so, Bisby (1943) has also recognized that the distribution of pathogenic fungi differs from that of phanerogams in that high numbers of endemic fungi are not present where there are high percentages of endemic phanerogams. It is acknowledged that saprophytic fungi generally have a wider distribution than plant-pathogenic fungi. Whereas pathogenic fungi must adapt themselves to changes in host phenotype, saprophytic fungi are unlikely to require any adaptation to the dead remains of a new plant (Bisby, 1933). Tribe

(1957) theorized that the constant presence in soils of a diverse microfloral community indirectly supports the view that particular substrates are colonized by particular microbes. Otherwise, he argues, a few efficient microbes would have eliminated those less efficient in decomposition of organic matter. Westerdijk (1949) had earlier characterized fungal populations on different substrates as representing unique associations. It should follow then that the variety and quantity (kilograms per hectare) of individual litter types (substrates) in a given ecosystem are important in determining the dominance patterns of fungi inhabiting the organic soil horizons (Kendrick and Burges, 1962; Hering, 1965; Hudson, 1968; Wicklow and Whittingham, 1974; Swift, 1976).

The cosmopolitan distribution of most saprophytic fungi has been attributed to the fact that fungi occupy microenvironments that can occur in various ecosystems and geographical areas (Stanier, 1953; Person, 1968; Mueller-Dombois and Perera, 1971). Mueller-Dombois and Perera (1971) make the point that a worldwide system of floristic classification of higher plant communities has failed because higher plants occur in widely differing floristic provinces that vary from place to place. They argue, however, that it may be possible to characterize specific, widely separated habitats in different floristic provinces by their soil fungal communities and thus to determine biologically, in addition to physically, their similarities or differences. Applying the same data base, mycologists may some day be able to predict distributions of conidial fungi in ecosystems yet to be the subject of intensive mycological investigation.

D. Competition and Predation

The distribution and abundance of a species are ultimately determined by tolerances to extremes of physical conditions, but a species is usually limited to a smaller range of potential habitats and thus a smaller population size by interactions (e.g., competition and predation) with other organisms (Connell, 1975). There is considerable experimental evidence to show that competition for resources reduces colony or population size and the sporulating capacities of conidial fungi (Baker and Cook, 1974). However, no one has determined whether, in nature, predation suppresses individual conidial fungi from reaching a population size great enough to enable them to compete for all the resources for which they are the most efficient competitor. There is a tendency among predators to seek out the most common prey; because of this phenomenon, competition is somewhat eased and more species can coexist. Experiments should be designed to determine whether mycophagists or mycoparasites reduce the niche of certain dominant fungi, thereby enabling other less competitive species to coexist on a given substrate. Brasier (1978) theorizes that the sporulation of some conidial fungi in response to mechanical injury of the mycelium could also be interpreted as an adaptive response to predation by mites or other invertebrates.

V. SPECIES DISTRIBUTION AND ENVIRONMENTAL SCALE

The real environment in which any organism lives is a mosaic of habitats, each of which has its own structure of component subhabitats. It is critical that proper consideration be given to determination of scale in making ecological comparisons involving species numbers (Burges, 1960; Swift, 1976). Studies of fungal colonization patterns on individual litter fragments (Hudson, 1968) can be relevant to between-habitat comparisons of the fungal flora; quite often mycologists sampling the organic soil profiles in stands dominated by one or more phanerogams are likely to turn up the same prevalent fungi that occur on specific litter fragments undergoing various stages of decomposition. Whenever an effort has been made to quantify the fungal community in a given ecosystem, it has been demonstrated clearly that the number of fungal species is far greater than that of the phanerogams (Bisby, 1933, 1943; Bisby and Ainsworth, 1943; Ainsworth, 1968). Bisby (1933) has recognized that, although twice as many phanerogams as fungi have been described worldwide, in any state or country the number of species of fungi is likely to exceed the number of species of phanerogams because of the greater average range of a fungus. He also has theorized that "the smaller the area surveyed, the more that fungi outnumber phanerogams" (Bisby, 1943). Rosensweig (1973) observes that one possible reason for generally greater diversity in smaller organisms is that these organisms also have smaller resource requirements. Therefore the minimum habitat patch required to support a species is likewise small.

It is not the purpose of this chapter to detail the various surveys of conidial fungi that have been conducted thus far in different biomes. Martha Christensen is now synthesizing information pertaining to the relationships between the composition of the soil microfungal community and the composition of higher plant communities at the ecosystem level.

VI. SPECIATION AND SPECIES DIVERSITY

Species diversity, relative abundance, and population geometry vary with climate; and because such variation affects the structure, stability, and energy flow through plant–animal communities, it must also affect the rate and perhaps the mode of speciation (MacArthur and Wilson, 1967). McNaughton and Wolf (1973) produced a clear conceptual diagram of relationships between ecological and evolutionary processes in ecosystems that contribute to the evolution of community diversity (Fig. 4). A summary of possible causal factors contributing to different patterns of species diversity is provided by MacArthur (1972). Many of these concepts were advanced in an earlier paper by Hutchinson (1959).

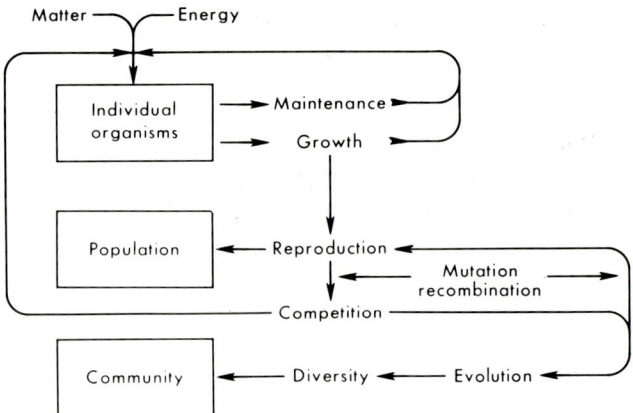

Fig. 4. A diagram of the relationships between the principal objects of study and processes of interest in ecosystems as they pertain to the evolution of community diversity.

1. "There are more species where there have been more opportunities for speciation and the presence of many species simply reflects the head start that some areas have over others." Strong and Levin (1975) assessed the influence of long geological time periods upon accumulated parasite species richness by searching for a correlation between number of associated fungal species and the relative age in Britain of host species. There was no relationship between time since the trees were introduced and number of fungal parasite species associated with these host species. Nor was there a relationship between number of fungal parasite species and the relative age in Britain of native hosts, determined as the time point where the frequency of pollen first rises to a sustained high value in the fossil record. These authors concluded that there was no justification for assuming younger host taxa to have fewer fungal pest species nor for the idea that long periods of time will add pest species to these hosts (Elton, 1966; May, 1973). Instead, evidence suggests that a great proportion of the saturation number of parasite species is accumulated by these hosts within ecological time (i.e., within several hundred years).

2. "Many species occur where fewer hazards have occurred, and . . . areas with few species have lost species through catastrophes of history." Such would more likely affect host-specific pathogenic or symbiotic fungi on locally occurring (endemic) hosts that failed to survive the catastrophe.

3. "There are more species where competitors can safely be packed closely and where the numbers of a species or its competitors will not increase with time." McNaughton and Wolf (1973) suggest that the organization of communities into guilds (Root, 1967) may be a general feature of ecosystem organization, so that diversity is increased by adding blocks of species that interact

among themselves but interact only feebly or not at all with members of other guilds. For example, entire guilds of pathogenic fungi, insect predators, or ectotrophic mycorrhizal fungi can be added to an ecosystem following initial establishment of the appropriate host tree.

4. "There are more species where the climate is benign or more stable." Environmental rigor affects productivity at the base of the food chain and therefore the number of species that can coexist (Hutchinson, 1959). The comparatively low number of microfungal species in Canadian boreal forest soils (Morrall and Vanterpool, 1968; Singh, 1976) is thought to be correlated with low numbers of boreal plant species (Lucarotti et al., 1978). Futuyma (1973) points out that observations such as the high frequency of host-specific phytophagous insects in the tropics (Janzen, 1970) contributed to the notion that many species evolve narrow niches in constant environments and that this may be one basis for high species diversity in tropical environments. Section VII will address the question of whether diversity of conidial fungi is greater in tropical environments.

5. "There are more species where the environment is complex and therefore more readily subdivided." Strong and Levin (1975) cite evidence that in Britain and the United States herbs support fewer insect species than shrubs or trees per area of distribution. They suggest that trees provide a greater diversity of niches for parasites, not because of their greater size but because of their greater morphological diversification or heterogeneity within individuals of the host plant species. Hutchinson (1959) theorized that small areas, like islands, may lack refuges for some species during unfavorable seasons or from special kinds of competition. Low-diversity areas may lack the type of environmental diversity necessary for species to cooccur in different niches at the same trophic level. An environmental mosaic may promote species diversity by allowing some organisms to occur in some areas and similar organisms of other species to occur in different areas of the same community. Hering (1965) and Swift (1976) demonstrate how different fungal guilds associated with individual litter components in a particular forest contribute to the mycological diversity associated with that ecosystem.

6. "There are more species where the environment is more productive." MacArthur (1972) suggests that, if food is in local abundance, species become more specialized in their resource requirements; but if food becomes scarce, species will have a less specialized diet. This is not necessarily the case for conidial fungi, where individuals with specific substrate requirements can await the arrival of an appropriate resource as dormant spores. In deserts where productivity is low and detritus is scattered, species numbers of conidial fungi are remarkably high, probably because of the potential for colonization of isolated organic substrates (M. Christensen, personal communication). Significantly, Gochenaur (1975) found that 75% of the soil microfungi isolated from a tropical xeric coconut grove could degrade cellulose, whereas less than one-third of the

microfungal species from a tropical mesic forest possessed this ability. She concludes that fungi in xeric habitats not only show various forms of morphological specialization protecting them from an often hostile environment but also exhibit nutritional versatility and a low level of nutrient specialization. On the other hand, she argues that the abundance and variety of organic matter in mesic soils enable microfungi with more restrictive nutritional patterns to survive.

7. "There are more species where heavy predation puts a low ceiling on the abundances of separate species, thus allowing more species to fit in." The importance of predation in providing patches for colonization by fungi has not been ascertained. Communities with more predators tend to be richer in species. We simply do not have any experimental data for conidial fungi, or any other fungal taxa for that matter, that can be applied toward answering such questions.

VII. ABUNDANCE OF CONIDIAL FUNGI IN TROPICAL AND TEMPERATE ENVIRONMENTS

The progressive increase in diversity of species from the arctic toward the equator leaves no doubt that, for most groups of organisms studied intensively, tropical environments support a greater diversity of species than temperate- or cold-zone environments (Dobzhansky, 1950). MacArthur (1972) suggests several hypotheses to explain the coexistence of increased numbers of species in a tropical community: (1) The spectrum of resources is greater; (2) the per species utilization of resources is less; (3) the niche overlap between species is greater; and (4) the dimensionality of the environment is greater (i.e., added geometric structure). Terborgh (1973) attributes the greater richness of species in tropical zones to the fact that the average area of the life zones declines as the latitude increases. Because there is more area per habitat in the tropics, species should have longer ranges and speciation should proceed more rapidly. According to MacArthur (1969), the historical accumulation of species in an area where production is greater and seasonality is less, coupled with the effects of competition and predation, should produce all the characteristics of tropical communities. Even so, Patrick (1966) found an exception to the rule of increased tropical diversity when she compared the numbers of species of major groups of organisms found in hard waters of the upper Amazon basin with those found in waters of comparable chemistry in temperate North America. The temperate waters had as many species as the tropical ones. Fungal communities may also provide exceptions. Gochenaur (1975) recorded less than one-quarter the number of fungal propagules per gram of dry soil in a mesic tropical forest than in mesic soils from several temperate forest or prairie ecosystems. She concludes that rapid cycling of the litter and the greater rate of decomposition observed in tropical plant communities probably also shortens the longevity of hyphae and dormant propagules. Gochenaur (1975) also observed that fungal species diversity and rare (low-frequency) isolates were substantially higher in a xeric coconut grove as contrasted with a mesic *Casuarina* forest. She theorizes that the relatively stable and uniform mesic habitat permits environmentally selected forms to establish themselves throughout the area; and consequently, when sampled by the dilution plate method, their high numbers mask most of the low-density taxa that may occur. The harsh and variable edaphic conditions in the xeric coconut grove are believed to keep propagule densities low and to prevent any fungal taxa from assuming dominance.

The diversity of fungi colonizing the phylloplane of living leaves of tropical forest species is apparently greater than that of temperate forest species. Runien (1961) observed that long-lived leaves in the tropics supported dense and species-rich fungal communities, the latter of which were not encountered on the shorter-lived leaves of temperate plants. Most phylloplane studies have been carried out in temperate regions and on agricultural crops (Dickinson, 1976). The few studies that have been published on tropical plants suggest that many phylloplane fungi are cosmopolitan in their distribution. Dickinson (1976) notes that, in studies of fungal colonization of wheat or barley leaves in Europe and India, the leaf floras were dominated by *Cladosporium* and *Alternaria*. *Aureobasidium* was common on leaves of either plant in Europe but was not recorded from its counterpart in India, whereas several species of *Aspergillus* and *Curvularia* that were not found at all in Europe were recorded from leaves in India. Hudson (1968) notes that, in addition to *Cladosporium herbarum* and *Alternaria tenuis,* conidial fungi initially colonizing temperate and tropical monocotyledonous leaves, the latter also support populations of *Nigrospora sphaerica* and *Curvularia lunata,* species in anamorph-genera Hudson regards as essentially tropical. Although Hudson did not attempt to contrast patterns of species abundance associated with fungal colonization, it is readily apparent from the lists of fungi assigned to successive groups of sporulating fungi that tropical monocots support more species than temperate monocots. For example, on sugarcane, group I includes 10 species, group II, 7 species, and group III, 11 species. The three groups were delimited according to the sequence in which fungi sporulated on green and senescent leaves (Hudson, 1962). The most impressive list of fungi recorded from a tropical monocot is that of Sharma (1973) who followed a succession of fungi on shoots of the grass *Setaria glauca* Beauv., beginning with their early senescence until they fell to the ground. Species were placed in one of three groups according to the timing of appearance of their sporing structures. Eight species were considered primary colonists and assigned to group I, 22 species represented the second wave of sporulating fungi, and 38 species appeared in the third group. Dematiaceous Hyphomycetes represented 73% of the conidial fungi recorded, including 69 species in 40 genera.

The sooty molds are saprophytic, dematiaceous Ascomycetes and their anamorphs that live superficially on living plants and are often associated with scale insects and other producers of honeydew (Hughes, 1976). Most sooty mold species live in mixed communities. In *Nothofagus* forests of New Zealand, black, irregular lumps of sooty molds (5 cm across) cover the trunks and branches; each lump is composed of entangled hyphae of as many as seven species. Mixed populations of sooty molds are frequently encountered growing together on leaves from the southern Pacific basin and parts of the tropics and subtropics (Hughes, 1976). To make the ecological situation even more complex, several hyphomycetes are parasitic on various sooty molds, and the phylloplane also supports bacteria, algae, lichens, mosses, and various invertebrate browsers (Hughes, 1976). There is considerable evidence that most sooty molds display no host preferences (Yamamoto, 1955). For example, Hughes reports that in New Zealand alone *Trichopeltheca asiatica* Bat., Costa & Cif. was found on over 80 different species of Filicales, Gymnospermae, Dicotyledones, and Monocotyledones; *Euantennaria mucronata* (Mout.) Hughes was recorded on over 39 different plant hosts in these same groups; and *Acrogenotheca elegans* (Fraser) Cif. & Bat., another common species, was found on 33 different hosts. Other sooty mold species are apparently restricted to one or a few related hosts (Hughes, 1976).

Although over 90% of the Hawaiian flora is endemic, only a few species comprising the community of phylloplane fungi recorded from extensive sampling of three endemic forest trees are considered potential endemics (Baker *et al.*, 1979). Most of the 150 fungal taxa isolated are widely distributed; however, the number of species isolated from individual hosts is remarkable, with the Fungi Imperfecti accounting for 90% of the species isolated from leaf surfaces. Dickinson (1976) produced a composite list totaling 76 fungal genera from the phylloplane of 35 different higher plants, and Baker *et al.* recorded 35 fungal genera from only three host species.

Data obtained in the following studies do not support the idea that diversity of conidial fungi increases in warmer latitudes. Gessner and Kohlmeyer (1976) examined and compared the filamentous fungal colonists of *Spartina alterniflora* in temperate and subtropical locations along the east coast of North America. Five of the six conidial fungi assigned to the Fungi Imperfecti were recorded from Canada to Florida. The authors concluded that temperature or salinity variations encountered over this geographical range do not appear to influence the composition of the fungal community on *Spartina* spp. Tubaki (1973) followed the microfungal colonization of two kinds of sterilized leaves (e.g., *Quercus phillyraeoides* and *Castanopsis cuspidata*) in both temperate and subtropical forest stands. The number of genera of imperfect fungi recorded from sterilized leaf packets (subtropical, 56 and 59; temperate, 60) did not differ significantly with latitude.

Data compiled by Ellis (1971) concerning the latitudinal distributions of 295 genera of dematiaceous Hyphomycetes are synthesized in Table I. At the highest latitudes (45°-60°N) 127 genera were recorded, of which only 9 have not been reported from lower latitudes. In contrast, 156 genera are listed from tropical regions (0°-15°N); 47 of these genera (30%) do not appear at higher latitudes. Lind (1934) reported that not one fungal genus among those he examined in a 10-yr study of arctic microfungi was indigenous to the northern polar regions.

Historically, there has been a greater collecting effort by European mycologists and, to a lesser extent, by North American mycologists in temperate and polar ecosystems. This is reflected in the large number of genera (211) reported between 30° and 45° N, 24% of which are exclusive to this latitudinal band (Table 1). In the temperate zone, the number of different ecosystem types and the microenvironments associated with each are more varied over a given area than in tropical latitudes (an exception being mountains in the tropics). This probably accounts to some degree for the significant (24%) number of genera found only between 30° and 45° N and not at higher latitudes or subtropical (15°-30°N) regions. At the same time, the high proportion (30%) of genera found exclusively in tropical latitudes is probably the outcome of evolution in a habitat favorable to dematiaceous fungi living on leaves, branches, and trunks of forest trees. The greater portion of the tropical genera listed by Ellis (1971) appear at distantly separated points around the globe, and there does not appear to be much evidence that conidial fungi show a wealth of endemic taxa. It will be interesting

TABLE I

Genera of Dematiaceous Hyphomycetes (295) Reported from Different Latitudes[a]

Type of conidium formation[b]	0°-15°N	15°-30°N	30°-45°N	45°-60°N
Thallic (1-12)	6	5	8	9
Blastic				
Basauxic (290-295)	6	5	4	1
Acroauxic				
Holoblastic (13-221)	111	90	148	83
Enteroblastic				
Tretic (222-247)	17	15	17	12
Phialidic (248-289)	16	16	34	22
Total[c]	156 (47)	131 (18)	211 (51)	127 (9)

[a] Data compiled from Ellis, 1971.

[b] Based on method of classification outlined by Ellis (1971, p. 11); numbers in parentheses refer to anamorph-genera cited in text.

[c] Number in parentheses refers to anamorph-genera collected from this latitude only.

to learn whether substantially larger numbers of dematiaceous genera will eventually be uncovered with increased sampling effort in tropical latitudes.

VIII. CONIDIAL FUNGI AND ECOLOGICAL ISLANDS

MacArthur and Wilson (1963, 1967) developed a theory accounting for an equilibrium in island floras and faunas, whereby the number of new immigrant species is balanced by the extinction of rarer ones. A summary of the theory (Fig. 5) shows that small and simple (habitat diversity limited) islands reach equilibrium more quickly than large or complex (habitat diversity substantial) islands. The equilibrium number is also shown to be lower for islands located far from the mainland with lower rates of immigration. A simple experimental approach in determining whether insular communities approach equilibrium consists of plotting the number of species recorded against time. Whether species numbers level off should be indicated by the resultant curve (Schoener, 1974). Equilibrium theory has applicability in interpreting differences in species abundance patterns among mainland habitats. Rosensweig (1975) observes that Wright (1941) was probably the first to put forward the hypothesis that communities have already attained their steady-state diversity. Achieving global equilibrium in species diversity also requires a balance between speciation and immigration on the one hand and extinction and emigration on the other.

No mycological studies have been specifically designed to test experimentally current biogeographical theory. Boedijn (1940) conducted a survey (1933-1934) of the fungi on the Krakatau island group located 40 km from both Java and Sumatra. During the 50 yr following a volcanic eruption in 1883, which probably left the islands of the Krakatau group sterile, 263 species of phanerogams and 61

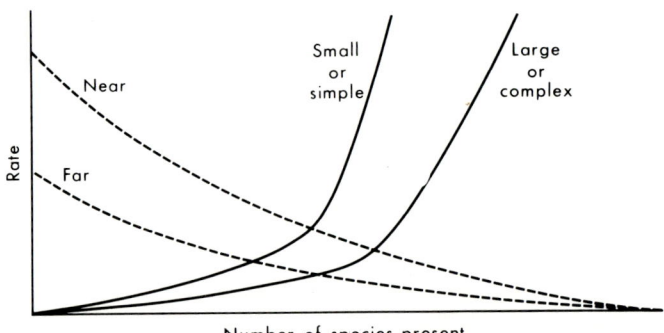

Fig. 5. The rates of immigration (falling curves) of new species not on the island and of extinction (rising curves) of species on the island are plotted for all numbers of species on an island. Where they intersect, extinction balances immigration. (From MacArthur and Wilson, 1963.)

of pteridophytes were found in gradually increasing numbers, with extinctions leaving 219 phanerogams and 52 pteridophytes in 1934 (Docters van Leeuwen, 1936). Unfortunately, immigration of fungi and lichens was not followed through the years; but Boedijn recorded 310 species of fungi, including myxomycetes, in 150 genera. Fifty-one species belonged to the Fungi Imperfecti. Bisby (1943) observed that the number of fungi would have been greater if every fungus had been found and identified. Airborne spore inoculum was considered the principal means by which fungi were dispersed to the islands (Boedijn, 1940). E. Einarsson (Museum of Natural History, Reykjavik) informs me that no studies have been attempted on the colonization by conidial fungi of the recently formed volcanic island of Surtsey.

The disinfestation of soil by steam heat or strong eradicant fungicides is used in agriculture to control soil-borne plant pathogens in greenhouses or nurseries (Baker and Cook, 1974). The treatments are often nonselective in their effects, creating islands of sterile or semisterile soil. The considerable literature on the microbiological response to such treatments should be reexamined carefully as a potential data base for experimental biogeographers. According to Kruetzer (1965), conidial fungi favored in the initial "Phycomycetes-Imperfecti" colonization stage include sugar- and amino acid-utilizing, fast-growing species of *Trichoderma, Penicillium,* and *Aspergillus*. These organisms are followed (stage 2) by degraders of complex carbohydrates including *Alternaria, Myrothecium, Aspergillus, Penicillium, Chaetomium, Sordaria,* and early basidiomycetes. Stage 3 is dominated by basidiomycetes (except for rhizosphere zones supporting simple-sugar organisms and cellulose degraders). Kreutzer (1965) considers the fourth and final stage analogous to the ecological "climax" identified with terminal successional forest stands, in that the treated soil has now become microbiologically indistinct from untreated soil.

The recolonization of four small islands of red mangrove (*Rhizophora mangle*) following extermination of existing fauna with methyl bromide was carefully monitored by Simberloff and Wilson (1969). Although the islands acquired a fauna of the original number of species within 6 months, the list of names differed from that recorded before defaunation. An equilibrium was achieved after 6-9 months; although the total number of species no longer increased, new species continued to replace earlier colonists. Immigration rates for microorganisms are exceptionally rapid. Schippers and Schermer (1966) present indirect evidence for the rapid recolonization of autoclaved soils by microbial air contaminants; more significantly, there is also evidence that with increased time for immigration the microfloral community becomes increasingly more complex, thus making it more difficult for other fungi to colonize such soils. *Verticillium albo-atrum* is a seedborne pathogen of *Senecio vulgaris* that can survive surface sterilization of the seeds it inhabits. Surface-disinfected seeds sown on autoclaved sterile soil in petri dishes produced 39.2% infected seedlings. When the same sterile soils were exposed to laboratory air for 3-10 min, 1-6 h, or 24-72 h, numbers of infected seedlings declined to 20, 14.2, and 7.5%, respectively.

Terms such as "biological balance," "pathogen-suppressive soils," and "stable microflora" are commonly used by plant pathologists when discussing cultivation practices designed to diminish the impact of a pathogen (Baker and Cook, 1974). What is clearly involved from an ecological perspective are phenomena such as competitive exclusion, niche expansion in the absence of competitors, and island equilibrium theory pertaining to microfloral species accumulation. For example, Baker (1962) and Warcup and Baker (1963) treated soil with aerated steam at temperatures (<100°C) that destroyed the pathogens and many saprophytes while leaving a microfloral community composed of *Bacillus* spp., *Streptomyces* spp., and several species in the Eurotiaceae with *Aspergillus* or *Penicillium* anamorphs. The surviving community prevented rapid and extensive recolonization by soil-borne pathogenic fungi that otherwise could spread unchecked through steam-sterilized soils and cause serious crop losses. Mycologists and plant pathologists have known for some time that successful introduction of alien fungi into samples of native or agricultural soils is particularly difficult, because in a foreign environment these fungi are less competitive than the resident fungal populations and rapidly disappear (Park, 1955).

Mycological systems may eventually prove to be ideal experimental tools for examining biogeographical theory. For example, because leaf surfaces are discontinuous habitats surrounded by a nonconducive growth medium (air), John Zak (personal communication) recommends extending biogeographical theory to explain processes of immigration and colonization of leaves by phylloplane fungi. Thus far, ecological studies on the phylloplane microflora have ignored this potentially powerful predictive theory.

REFERENCES

Ainsworth, G. C. (1968). The number of fungi. *In* "The Fungi," (G. C. Ainsworth and A. S. Sussman, eds.), Vol. 3, pp. 505-514. Academic Press, New York.

Anderson, T.-H., Bodenstein, J., and Domsch, K. H. (1976). A partially computerized documentation system for microscopic soil fungi. *Can. J. Bot.* **54**, 1709-1713.

Armstrong, R. A. (1976). Fugitive species: Experiments with fungi and some theoretical considerations. *Ecology* **57**, 953-963.

Backus, M. P. (1939). The mechanics of conidial fertilization in *Neurospora sitophila*. *Bull. Torrey Bot. Club* **66**, 63-76.

Baker, G. E., and Meeker, J. A. (1972). Ecosystems, mycologists, and the geographical distribution of fungi in the central Pacific. *Pac. Sci.* **24**, 418-432.

Baker, G. E., Dunn, P. H., and Sakai, W. S. (1979). Fungus communities associated with leaf surfaces of endemic vascular plants in Hawaii. *Mycologia* **71**, 272-292.

Baker, K. F. (1962). Principles of heat treatment of soil and planting material. *J Aust. Inst. Agric. Sci.* **28**, 118-126.

Baker, K. F., and Cook, R. J. (1974). "Biological Control of Plant Pathogens." Freeman, San Francisco, California.
Bandoni, R. J. (1975). Significance of the tetraradiate form in dispersal of terrestrial fungi. *Rep. Tottori Mycol. Inst. (Jpn.)* **12**, 105-113.
Bisby, G. R. (1933). The distribution of fungi as compared with that of phanerogams. *Am. J. Sci.* **20**, 246-254.
Bisby, G. R. (1943). Geographical distribution of fungi. *Bot. Rev.* **9**, 466-482.
Bisby, G. R., and Ainsworth, G. C. (1943). The numbers of fungi. *Trans. Br. Mycol. Soc.* **26**, 16-19.
Bloomfield, B. J., and Alexander, M. (1967). Melanins and resistance of fungi to lysis. *J. Bacteriol.* **93**, 1276-1280.
Boedijn, B. (1940). The mycetozoa, fungi, and lichens of the Krakatau group. *Bull. Jard. Bot. Buitenzorg* **16**, 358-429.
Borut, S. (1960). An ecological and physiological study on soil fungi of the northern Negev (Israel). *Bull. Res. Counc. Israel, Sect. D* **8**, 65-80.
Brasier, C. M. (1978). Mites and reproduction in *Ceratocystis ulmi* and other fungi. *Trans. Br. Mycol. Soc.* **70**, 81-89.
Brown, J. C. (1958). Soil fungi of some British sand dunes in relation to soil type and succession. *J. Ecol.* **46**, 641-664.
Burges, A. (1960). Time and size as factors in ecology. *J. Ecol.* **48**, 273-285.
Christensen, M. (1969). Soil microfungi of dry to mesic conifer-hardwood forests in northern Wisconsin. *Ecology* **50**, 9-27.
Cody, M. L., and Diamond, J. M., eds. (1975). "Ecology and Evolution of Communities." Harvard Univ. Press, Cambridge, Massachusetts.
Connell, J. H. (1975). Some mechanisms producing structure in natural communities. *In* "Ecology and Evolution of Communities" (M. L. Cody and J. M. Diamond, eds.), pp. 460-490. Harvard Univ. Press, Cambridge, Massachusetts.
Danielson, R. M., and Davey, C. B. (1973). The abundance of *Trichoderma* propagules and the distribution of species in forest soils. *Soil Biol. & Biochem.* **5**, 485-494.
Davidson, D. E. (1974). Wood-inhabiting and marine fungi from a saline lake in Wyoming. *Trans. Br. Mycol. Soc.* **63**, 143-149.
Dearness, J. (1923). Fungi. *Rep. Can. Arct. Exped.* **4C**, 1-24.
Dickinson, C. H. (1976). Fungi on the aerial surfaces of higher plants. *In* "Microbiology of Aerial Plant Surfaces" (C. H. Dickinson and T. F. Preece, eds.), pp. 293-324. Academic Press, New York.
Diehl, W. W. (1937). A basis for mycogeography. *J. Wash. Acad. Sci.* **27**, 244-254.
Dobzhansky, T. (1950). Evolution in the tropics. *Am. Sci.* **38**, 209-221.
Docters van Leeuwen, W. M. (1936). Krakatau, 1883 to 1933. *Ann. Jard. Bot. Buitenzorg* **46-47**, 1-506.
Durrell, L. W., and Shields, L. M. (1960). Fungi isolated in culture from soils of the Nevada test site. *Mycologia* **52**, 636-641.
Ellis, M. B. (1971). "Dematiaccous Hyphomycetes." Commonw. Mycol. Inst., Kew, Surrey, England.
Elton, C. S. (1966). "The Pattern of Animal Communities." Methuen, London.
English, H., and Gerhardt, F. (1946). The effect of ultraviolet radiation on the viability of fungus spores and on the development of decay in sweet cherries. *Phytopathology* **36**, 100-111.
Flanagan, P. W., and Scarborough, A. (1974). Physiological groups of decomposer fungi on tundra plant remains. *In* "Soil Organisms and Decomposition in Tundra" (A. J. Holding, O. W. Heal, S. F. MacLean, Jr., and P. W. Flanagan, eds.), pp. 159-181. University of Alaska, Fairbanks.
Flannigan, B., and Campbell, I. (1977). Pre-harvest mould and yeast floras on the flag leaf, bracts, and caryopsis of wheat. *Trans. Br. Mycol. Soc.* **67**, 485-494.

Franz, G. (1974). Mikrobiologische Untersuchungen an Böden aus Nepal. *Pedobiologia* **14**, 372–401.

Franz, G. (1975). Temperaturansprüche mikroskopischer Bodenpilze aus klimatisch und geographisch verschiedenen Standorten. *Z. Pflanzenernaehr. Bodenkd.* No. 1, pp. 73–87.

Futuyma, D. J. (1973). Community structure and stability in constant environments. *Am. Nat.* **107**, 443–445.

Garrett, S. D. (1970). "Pathogenic Root-infecting Fungi." Cambridge Univ. Press, London and New York.

Gessner, R. V., and Kohlmeyer, J. (1976). Geographical distribution and taxonomy of fungi from salt marsh *Spartina*. *Can. J. Bot.* **54**, 2023–2037.

Gochenaur, S. E. (1970). Soil mycoflora of Peru. *Mycopathol. Mycol. Appl.* **42**, 259–272.

Gochenaur, S. E. (1975). Distributional patterns of mesophilous and thermophilous microfungi in two Bahamian soils. *Mycopathologia* **57**, 155–164.

Gochenaur, S. E., and Backus, M. P. (1967). Mycoecology of willow and cottonwood lowland communities in southern Wisconsin. II. Soil microfungi in the sandbar willow stands. *Mycologia* **59**, 893–901.

Gochenaur, S. E., and Woodwell, G. M. (1974). The soil microfungi of a chronically irradiated oak-pine forest. *Ecology* **55**, 1004–1016.

Green, R. H. (1971). A multivariate statistical approach to the Hutchinsonian niche: Bivalve molluscs of central Canada. *Ecology* **52**, 543–546.

Hawker, L. E. (1957). Ecological factors and the survival of fungi. *Symp. Soc. Gen. Microbiol.* **7**, 238–258.

Hering, T. F. (1965). Succession of fungi in the litter of a lake district oakwood. *Trans. Br. Mycol. Soc.* **48**, 391–408.

Hill, J. P., and Nelson, R. R. (1976). Ecological races of *Helminthosporium maydis* race T. *Phytopathology* **66**, 873–876.

Horn, H. S., and MacArthur, R. H. (1972). Competition among fugitive species in a harlequin environment. *Ecology* **53**, 749–752.

Hudson, H. J. (1962). Succession of microfungi on aging leaves of *Saccharum officinarum*. *Trans. Br. Mycol. Soc.* **45**, 395–423.

Hudson, H. J. (1968). The ecology of fungi on plant remains above the soil. *New Phytol.* **67**, 837–874.

Hughes, S. J. (1976). Sooty moulds. *Mycologia* **68**, 693–820.

Hutchinson, G. E. (1959). Homage to Santa Rosalia or why are there so many kinds of animals? *Am. Nat.* **93**, 145–159.

Ingold, C. T. (1971). "Fungal Spores: Their Liberation and Dispersal." Oxford Univ. Press (Clarendon), London and New York.

Ingold, C. T. (1975). Convergent evolution in aquatic fungi: The tetraradiate spore. *Biol. J. Linn. Soc.* **71**, 1–25.

Janzen, D. H. (1967). Why mountain passes are higher in the tropics. *Am. Nat.* **101**, 233–249.

Janzen, D. H. (1970). Herbivores and the number of tree species in tropical forests. *Am. Nat.* **104**, 501–528.

Jinks, J. L. (1952). Heterocaryosis in wild *Penicillium*. *Heredity* **6**, 77–87.

Jørstad, I. (1925). The Erysiphaceae of Norway. *Nor. Vidensk. Akad. Skr., Math. Nat. Kl.* No. 10.

Kendrick, W. B., and Burges, A. (1962). Biological aspects of the decay of *Pinus silvestris* leaf litter. *Nova Hedwigia* **4**, 313–342.

Kendrick, W. B., and Carmichael, J. W. (1973). Hyphomycetes. *In* "The Fungi" (G. C. Ainsworth, F. K. Sparrow, and A. S. Sussman, eds.), Vol. 4, pp. 323–509. Academic Press, New York.

Ko, W.-H., and Lockwood, J. L. (1967). Soil fungistasis: Relation to fungal spore nutrition. *Phytopathology* **57**, 894-901.
Kranz, J., ed. (1974). "Epidemics of Plant Diseases: Mathematical Analyses and Modeling." Springer-Verlag, Berlin and New York.
Kreutzer, W. A. (1965). The reinfestation of treated soil. *In* "Ecology of Soil-Borne Plant Pathogens" (K. F. Baker and W. C. Synder, eds.), pp. 495-507. University of Califronia Press, Berkeley.
Lack, D. (1966). "Population Studies of Birds." Oxford Univ. Press (Clarendon), London and New York.
Lind, J. (1934). Studies on the geographical distribution of arctic circumpolar micromycetes. *K. Dan. Vidensk. Selsk., Biol. Medd.* **11**, 1-152.
Lucarotti, C. J., Kelsey, C. T., and Auclair, A. N. D. (1978). Microfungal variations relative to post-fire changes in soil environment. *Oecologia* **37**, 1-12.
MacArthur, R. A. (1969). Patterns of communities in the tropics. *Biol. J. Linn. Soc.* **1**, 19-30.
MacArthur, R. H. (1972). "Geographical Ecology: Patterns in the Distribution of Species." Harper, New York.
MacArthur, R. H., and Wilson, E. O. (1963). An equilibrium theory of insular zoogeography. *Evolution* **17**, 373-387.
MacArthur, R. H., and Wilson, E. O. (1967). "The Theory of Island Biogeography." Princeton Univ. Press, Princeton, New Jersey.
McNaughton, S. J., and Wolf, L. L. (1973). "General Ecology." Holt, New York.
Mangenot, F., and Reisinger, O. (1976). Form and function of conidia as related to their development. *In* "The Fungal Spore: Form and Function" (D. J. Weber and W. M. Hess, eds.), pp. 789-846. Wiley, New York.
May, R. M. (1973). "Stability and Complexity in Model Ecosystems." Princeton Univ. Press, Princeton, New Jersey.
Mayr, E. (1954). Change of genetic environment and evolution. *In* "Evolution as a Process" (J. Huxley, A. C. Hardy, and E. B. Ford, eds.), pp. 157-180. Allen & Unwin, London.
Morrall, R. A. A., and Howard, R. J. (1975). The epidemiology of leaf spot disease in native prairie. II. Airborne spore populations of *Pyrenophora tritici-repentis*. *Can. J. Bot.* **53**, 2345-2353.
Morrall, R. A. A., and Vanterpool, T. C. (1968). The soil microfungi of upland boreal forest at Candle Lake, Saskatchewan. *Mycologia* **60**, 642-654.
Mueller-Dombois, D., and Perera, M. (1971). Ecological differentiation and soil fungal distribution in the montane grasslands of Ceylon. *Ceylon J. Sci.* **9**, 1-41.
Nicot, J. (1960). Some characteristics of the microflora of desert sands. *In* "The Ecology of Soil Fungi" (D. Parkinson and J. S. Waid, eds.), pp. 94-97. Liverpool Univ. Press, Liverpool.
Pady, S. M., and Kapica, L. (1953). Air-borne fungi in the arctic and other parts of Canada. *Can. J. Bot.* **31**, 309-323.
Pady, S. M., and Kapica, L. (1955). Fungi over the Atlantic Ocean. *Mycologia* **47**, 34-50.
Pady, S. M., and Kelly, C. D. (1954). Aerobiological studies of fungi and bacteria over the Atlantic Ocean. *Can. J. Bot.* **32**, 202-212.
Park, D. (1955). Experimental studies on the ecology of fungi in soil. *Trans. Br. Mycol. Soc.* **38**, 130-142.
Park, D. (1968). The ecology of terrestrial fungi. *In* "The Fungi" (G. C. Ainsworth and A. S. Sussman, eds.), Vol. 3, pp. 5-39. Academic Press, New York.
Patrick, R. (1966). The Catherwood Foundation Peruvian-Amazon Expedition: Limnological and systematic studies. *Monogr. Acad. Natl. Sci. Philadelphia* **14**, 1-495.
Perkins, D. D., Turner, B. C., and Barry, E. G. (1976). Strains of *Neurospora* collected from nature. *Evolution* **30**, 281-313.
Person, C. (1968). Genetical adjustment of fungi to their environment. *In* "The Fungi" (G. C.

Ainsworth and A. S. Sussman, eds.), Vol. 3, pp. 395-415. Academic Press, New York.
Pianka, E. R. (1970). On r- and K-selection. *Am. Nat.* **104**, 592-597.
Pirozynski, K. A. (1968). Geographical distribution of fungi. *In* "The Fungi" (G. C. Ainsworth and A. S. Sussman, eds.), Vol. 3, pp. 487-504. Academic Press, New York.
Platt, H. W., Morrall, R. A. A., and Gruen, H. E. (1977). The effects of substrate, temperature, and photoperiod on conidiation of *Pyrenophora tritici-repentis*. *Can. J. Bot.* **55**, 254-259.
Raper, J. R. (1968). On the evolution of fungi. *In* "The Fungi" (G. C. Ainsworth and A. S. Sussman, eds.), Vol. 3, pp. 677-693. Academic Press, New York.
Root, R. B. (1967). The niche exploitation pattern of the blue-gray gnatcatcher. *Ecol. Monogr.* **37**, 317-350.
Rosensweig, M. L. (1973). Exploitation in three tropic levels. *Am. Nat.* **107**, 275-294.
Rosensweig, M. L. (1975). On continental steady states of species diversity. *In* "Ecology and Evolution of Communities" (M. L. Cody and J. M. Diamond, eds.), pp. 121-140. Harvard, Univ. Press, Cambridge, Massachusetts.
Runien, J. (1961). The phyllosphere. I. An ecologically neglected milieu. *Plant Soil* **15**, 81-109.
Savile, D. B. O. (1963). Mycology in the Canadian Arctic. *Arctic* **16**, 17-25.
Savile, D. B. O. (1968). Possible interrelationships between fungal groups. *In* "The Fungi" (G. C. Ainsworth and A. S. Sussman, eds.), Vol. 3, pp. 649-675. Academic Press, New York.
Schippers, B., and Schermer, A. K. F. (1966). Effect of antifungal properties of soil on dissemination of the pathogen and seedling infection originating from *Verticillium*-infected achenes of *Senecio*. *Phytopathology* **56**, 549-552.
Schoener, T. W. (1974). Experimental zoogeography: Colonization of marine mini-islands. *Am. Nat.* **108**, 715-736.
Sharma, P. D. (1973). Succession of fungi on decaying *Setaria glauca* Beauv.: A qualitative analysis of the mycoflora. *Ann. Bot. (London)* [N.S.] **37**, 203-208.
Simberloff, D. S., and Wilson, E. O. (1969). Experimental zoogeography of islands: The colonization of empty islands. *Ecology* **50**, 278-295.
Singh, P. (1976). Some fungi in the forest soils of Newfoundland. *Mycologia* **68**, 881-890.
Stanier, R. Y. (1953). Adaptation, evolutionary, and physiological: Or Darwinism among the micro-organisms. *Symp. Soc. Gen. Microbiol.* **3**, 1-14.
Stevens, N. E. (1917). Some factors influencing the prevalence of *Endothia gyrosa*. *Bull. Torrey Bot. Club* **44**, 127-144.
Strong, D. R., and Levin, D. A. (1975). Species richness of parasitic fungi of British trees. *Proc. Natl. Acad. Sci. U.S.A.* **72**, 2116-2119.
Sussman, A. S. (1968). Longevity and survivability of fungi. *In* "The Fungi" (G. C. Ainsworth and A. S. Sussman, eds.), Vol. 3, pp. 447-486. Academic Press, New York.
Swift, M. J. (1976). Species diversity and the structure of microbial communities in terrestrial habitats. *In* "The Role of Terrestrial and Aquatic Organisms in Decomposition Processes" (J. M. Anderson and A. Macfadyen, eds.), pp. 185-222. Blackwell, Oxford.
Tansey, M. R., and Brock, T. D. (1978). Microbial life at high temperatures: Ecological aspects. *In* "Microbial Life in Extreme Environments" (D. J. Kushner, ed.), pp. 159-216. Academic Press, New York.
Terborgh, J. (1973). On the notion of favorableness in plant ecology. *Am. Nat.* **102**, 283-293.
Tresner, H. D., Backus, M. P., and Curtis, J. T. (1954). Soil microfungi in relation to the hardwood forest continuum in southern Wisconsin. *Mycologia* **46**, 314-333.
Tribe, H. T. (1957). Ecology of micro-organisms in soils as observed during their development upon buried cellulose film. *Symp. Soc. Gen. Microbiol* **7**, 287-298.
Tubaki, K. (1973). Some aspects of the geographical distribution of leaf litter fungi in Japan. *Shokubutsu Byogai Kenkyu* **8**, 61-69.
von Arx, J. A. (1957). Revision der zu *Gloeosporium* gestellten Pilze. *Verh. K. Ned. Akad. Wet., Afd. Natuurkd., Reeks 2* **51**, 1-153.

Walker, J. C. (1969). "Plant Pathology," 3rd ed. McGraw-Hill, New York.
Warcup, J. H. (1951). The ecology of soil fungi. *Trans. Br. Mycol. Soc.* **34,** 376–399.
Warcup, J. H., and Baker, K. F. (1963). Occurrence of dormant ascospores in soil. *Nature (London)* **197,** 1317–1318.
Ward, J. E., and Cowley, G. T. (1972). Thermophilic fungi of some central South Carolina forest soils. *Mycologia* **64,** 200–205.
Webster, J. (1959). Experiments with spores of aquatic Hyphomycetes. I. Sedimentation and impaction on smooth surfaces. *Ann. Bot. (London)* [N.S.] **23,** 595–611.
Westerdijk, J. (1949). The concept association in mycology. *Antonie van Leeuwenhoek* **15,** 187–189.
Wicklow, D. T. (1973). Microfungal populations in surface soils of manipulated prairie stands. *Ecology* **54,** 1302–1310.
Wicklow, D. T. (1975). Fire as an environmental cue initiating ascomycete development in a tallgrass prairie. *Mycologia* **67,** 852–862.
Wicklow, D. T., and Whittingham, W. F. (1974). Soil microfungal changes among the profiles of disturbed conifer-hardwood forests. *Ecology* **55,** 3–16.
Williams, G. C. (1975). "Sex and Evolution." Princeton Univ. Press, Princeton, New Jersey.
Wright, S. (1941). The "age and area" concept extended. *Ecology* **22,** 345–347.
Yamamoto, W. (1955). On the so-called host range of sooty mould fungi. *Ann. Phytopathol. Soc. Jpn.* **19,** 97–103.
Zak, J. C., and Wicklow, D. T. (1978). Response of carbonicolous ascomycetes to aerated steam temperatures and treatment intervals. *Can. J. Bot.* **56,** 2313–2318.

Subject Index

A page number set in bold indicates a legend.

A

Acervulus, *see also* Conidioma, 49, 51, 58, 72
Adaptation, 418, 419-426
Adenylate cyclase, 126
Adhesive pad, *see* Appressorium
Aerial environment,
  indoor, 379-380
  outdoor, 378-379
    convective layer, 379
    laminar boundary layer, 378-379
    local eddy layer, *see also* Eddy diffusion, 379
    troposphere, 378
    turbulent boundary layer, 379
Aero-aquatic fungus, *see also* Propagule, aero-aquatic, 295, 296, 335-347
  study technique, 336
Aflatoxin, 408
Aggregate species, 35
Air spora
  allergy, 406-407
  and human activity, 386, 387, 395
  and plant disease, 408-409
  indoor, 393-396
  infection, 408
  in organic substrate, 394
  medium for isolating, 376
  outdoor, 380, 384-385, 393
  viability, 387
Aleuric conidium, 138
Aleuric dehiscence, 137, 138
Aleurioconidium, 12, 137, 138, 140

Aleuriospore, *see* Aleurioconidium
Ameroconidium, 44, 65, **66**, 68, 138
Amphibious fungus, *see also* Ingoldian fungus, 140
Anamorph, 22-24, 27, 29, 30, 32, 34, 145, 146, 147, 245, 258, 266, 271
  acervular, 44, 49-50, 164
  basidiomycetous, 28, 32
  coelomycetous, 43, 44, 47, 48, 49, 57, 60, 63, 78
  dimorphic genera, 39
  haploid, 90
  holomorphic, 29, 30, 86
  monophialidic, 32
  multiple, 29-30
  presence or absence, 146, 153, 156, 160
  pycnidial, 44, 47, 49, **54**, 57
  role in survival, 422
  stromatic, 47, 49, **54**
  taxa, 27, 28, 33
  teleomorph connections, 9, 16, 28, 29, 30-31, 44, 78, 140-141, 145-165, 418
Anamorph-genera, 35-37, 140-142
  double, 141
  multiple, 30
Anamorphic holomorph, 29, 30
Anastomosis, 31
Annellation, 259, **260**, 261, 262, **263**, 264
Annellide, 264
  phialide distinction, 70, 71
Antagonism, 172, 177-178, 187, 201, 204, 211, 229-230, 234-235
Anthracnose, 76

Subject Index

Appendage, extracellular, **66, 67,** 68
  mucoid, 49, **66,** 68, 69, 77
  tubular, **67,** 68
Appressorium, 173-174, 326
Aquatic fungus, 397
  Hyphomycetes, 426
Aquatic macrophyte, 334
Aquatic spore, 13
Arthroconidium, 12, 86, 91, 92, 136, 137, 140, 322
Arthrospore, see Arthroconidium
Ascal ontogeny, 109
Ascoma, 147, 267
  apothecioid, 25
  dimidiate, 26
  perithecial, 29
Ascomycetes
  bitunicate, 25-27, 32
  unitunicate, 32
Ascospore, 265
Asexuality, 27, 31, 425
Aspergillosis, 408
Assimilative stage, 135, 136

B

Ballistospore, 86, 88, 92, 95, 140
  air-dispersed, 343
Basidioma, 147
Biocontrol, 76, 234-235
Biodegradation, 409
Biodeterioration, 409
Biogeography, 417, 418, 442
Biomass
  fungal, 281-287, 289
  microbial, 286-287
Blastomycosis, 109
Blastospore, 12
Blight, 76
Budding, 147
  anamorph, 5
Bud scar, 101
Buried slide technique, 287

C

cAmp, see Cyclic adenosine monophosphate
Candidiasis, 97
Carbon compound assimilation, 86, 88

Cell growth, polarized, 101
Cellulolytic fungus, 344
Cell wall characteristics, 87
Chemical interaction, 177-178
Chestnut blight, 427
Chitin determination, 286
Chitin ring, 101
Chitin synthase, 100, 112, 123
  zymogen, 100
Chitin synthesis, 101, 123
Chlamydospore, see also Metasclerotium, 12, 138, 140, 321, 322, 324
Chromomycotic fungus, 97
Clamp connection, 91, 95, 287, 321, 339, **340**
Climate
  effect on air spora, 382, 383, 385
  effect on species, 429-431, 433, 436-440
Coenzyme Q, 87
Coiling see Hyphae
Collarette, 264
Color reaction, 87
Competition, 418, 420, 423, 432, 434
Conidial dispersal, 77, 78, 418, 420, 421-423
  barrier, 427-429
  by arthropods, 75, 422, 428
  in foam, 361
Conidial habit, 418, 419, 425, 431
Conidial melanization, 421
Conidiogenesis, see Conidium development
Conidioma, 28, 49-58, **50, 52, 53, 54, 55,** 147
  acervular, 28, 49
  apothecium-like, 57, 147
  cavity formation, 58
  cupulate, 51
  hyphomycetous, 51
  intermediate, 49
  pycnidial, 28, 49, 271
  pycnothyrial, 49, **52**
  shield-like, 158
  sporodochial, 10, 28, 49, 51, 139, 321, 322
Conidium
  blastic sympodial, 12
  amerosporous, 321
  dictyosporous
  didymosporous, 321
  enteroarthric, 107, 109
  helicoid, 360-361
  helicosporous
  holoarthric, 107
  hyaline, 421

internal, 265
phragmosporous, 321
polymorphic, 321
scolecosporous, 321
sigmoid
slimy, 422
staurosporous, 65, 67, 68, 321
tetraradiate, 323-325, 327, 360, 422
Conidium development, 12-15, 321, 322, 323-324
  acropleurogenous, 321
  annellidic, 264
  apical, 257, 258, 259
  holoblastic, 258
  lateral, 256, 257, 259, 265
  phialidic, 258, 264
Conidium initiation, 323
Conidia Vera, 12
Contact cell, see also Mycoparasite, 174
Coprophilous fungus, 178, 211, 280
Crystal violet stain, 49
Cyclic AMP, see Cyclic adenosine monophosphate
Cyclic adenosine monophosphate, 111, 126
Cysteine, 124, 125
  reductase, 125
Cystine, 125
  reductase, 125

D

dbcAMP, see Dibutyryl cyclic adenosine monophosphate
Dermatophyte, 136
Diaspore, 265
Diazonium blue B, 87
Dibutyryl cyclic adenosine monophosphate, 111
Dictyochlamydospore, 72
Dictyoconidium, 65, **67**, 68, 138
Dictyosome, 102
Didymoconidium, 65, **66,** 68, 138
Die-back, 76
Dikaryomycota, 22, 27
  an anamorph, 27
Dikaryomycotic fungus, 22
Dikaryon, 135, 146, 147
Dimorphic fungus, 135
Dimorphism, 135, 136
Direct hyphal isolation method, 284

Disjunctor cell, 137
Distoseptum, 65
Disulfide linkage, 119, 125
Disulfide reductase, 122, 124, 125, 126
Dolipore, see Septum
Dutch elm disease, 397, 427

E

Ecology,
  climax, 441
  island, 428, 435, 437-442
  of soil fungus, 278-279
  race, 430
Ecosystem, 277
  fungus role in stream energy flow, 334
Eddy diffusion, 401-402
Endemic fungus, 431, 438
Entomogenous fungus, 77
Entomopathogen, 76
Enzyme, lytic, 98
Equilibrium theory, 440, 442
Evolution, convergent, 13, 34, 426
  fungicolous fungus and host, 232-234

F

FDA, see Fluorescein diacetate
Fluorescein diacetate, 286
Fluorescein isothiocyanate, 287
Fluorescence microscopy, 101
Fluorescent-antibody method, 287
Form-genera, see Anamorph-genera
Fugitive species, 419-420
  K-selected, 420
  R-selected, 419
Fungal community, 418, 427
Fungal growth on Agaricales, 192-198, 200
  on Aphyllophorales, 198-201
  on Ascomycotina, 181, 203-224, 232
  on Basidiomycotina, 188-203, 233
  on Blastomycetes, 224
  on Chytridiales, 173
  on Chytridiomycetes, 185-186
  on Coelomycetes, 224-227
  on Coronophorales, 204
  on Cyttariales, 213
  on Deuteromycotina, 224-231, 233-234
  on Discomycetes, 213-215, 233

Fungal growth (*continued*)
 on Erysiphales, 183, 188, 204–206
 on Gasteromycetes, 202–203
 on Helotiales, 213
 on Hemiascomycetes, 203
 on Hymenomycetes, 192–198
 on Hyphomycetes, 227–231
 on Laboulbeniomycetes, 220
 on lichen, *see* Lichenicolous fungus
 on Loculoascomycetes, 215–220, 233
 on Mastigomycotina, 185–187, 232
 on Meliolales, 206–208
 on Mucorales, 173, 188, 204
 on Mycelia sterilia, 231
 on Myxomycota, 183–185
 on Oomycetes, 186, 187
 on Peronosporales, 186
 on Pezizales, 214
 on Phacidiales, 214–215
 on Plectomycetes, 203–204
 on Pyrenomycetes, 204–213, 233
 related to climate, 330
 related to water activity, 333, 334
 resource requirement, 431, 432, 433, 435
 on Sphaeriales, 208–213, 233
 on Taphrinales, 203
 on Teliomycetes, 188–192
 on Tuberales, 215
 on Uredinales, 181, 188–192
 on Ustilaginales, 192
 on Zygomycetes, 187–188, 232
Fungal guild, 435
Fungal standing crop, 286, 287
Fungal succession, 279, 281, 423, 425, 437
 on mangrove, 364
 seasonal, 151, 152
Fungicide, 441
Fungicolous fungus
 facultative, 173
 obligate, 173
 ubiquitous, 180, 183
Fungicolous habit, 173–179

G

G + C% of DNA, *see* Guanidine-cytosine content
Generic concept, 32, 33, 35–39
Genetic heterogeneity, 425
Genetic interaction, 178–179

Genetic variability, 424–426
Genome, 425
Genotype, 424
Germination, 420, 421
Germ pore, 148, 150
Germ slit, 148, 150
Germ tube, 326, 327
Glucose-stimulated soil respiration, 286
Glume blotch, 76
Glyceel, 48
Griseofulvin, 177
Guanidine-cytosine content, 87
Gummosis, 44

H

Habitat, 319
 mesic, 436–437
 moorland, 332
 salt marsh, 365, 369
 tropical forest, 436–440
 xeric, 437
Haploid mycelium, 146, 147
 compatible, 419
Haplophase, 424
Haustorium, 174–175
Helicospore, 65
Heterokaryosis, 31, 424
Heterothallic fungus, 424
Histoplasmosis, 109
Holoblastic cell, 102
Holoblastic development, 115
Holomorph, 29, 30–31, 43, 141, 159, 160
 anamorphic, 29, 30
 teleomorphic, 29, 30
Host alternation, 153
Host–parasite interrelationship, 45, 75
Hyperparasite, 361
Hyphae coiling, 174
 intrahyphal, 175–177
 moniliform, 107
 penetrating, 175
Hyphal degradation, 287
Hyphal wall, dematiaceous, 135
 moniliaceous, 135
Hyphopodia, 35
Hyphophore, 265

I

Immersion tube method, 283

Subject Index

Imperfect state, *see* Anamorph
Infection papilla, 176
Ingoldian fungus, 295, 296–308, 319–323, **326**, 331, 332–333, 360
    study technique, 296–297
Inhibition, *see* Antagonism
Inoculum
    potential, 423
    primary, 429
Inostiolate, 51
Intergeneric hiatus, 37–38
Involucre, 51

K

Karyogamy, 43

L

Leaf spot disease, 429
Lichen pycnidium, 246–255
    Lecanactis-type, 247, **248, 250,** 271
    Lobaria-type, **249,** 252, **253,** 270
    Roccella-type, 247–**249,** 250
    Umbilicaria-type, **249,** 251–252, 270
    Xanthoria-type, **249,** 252–255, **254,** 270
Lichenicolous Coelomycete
    conidiogenesis, 246, 258–264, **259, 261, 262,** 271
    conidiophore, **255**–258
Lichenicolous fungus, 172, 173, 186, 220–224, 233–234
Lignin decomposing fungus, 280, 281
Lignicolous fungus, 363
Lupinosis, 76
Lysigenous cavity, 28

M

MTOC, *see* Microtubule, organizing center
Macroconidial state, 153
Macroconidium, 136, 138, 264
Marine fungus, 358–359
    as parasite, 361–362
    as saprobe, 357, 362–368
    deep sea, 369
    dispersal, 360–361
    facultative, 358, 364
    geographical distribution, 368–369
    obligate, 358, 364
    symbiont, 368

    effect of climate, 360, 362, 367, 368, 369
    vertical distribution, 369
Meiosis, 425
Meiospore, 43
Meristem phialospore, 12
Metamorphosis, 22
Metasclerotium, *see also* Chlamydospore, 322, 324
Methionine, 126
Methylene blue stain, 49
Microarthropod, grazing on soil fungus, 289, 290
Microconidial state, 153
Microconidium, 109, 136, 137, 138, 264
    phialidic, 138
Microfilament, 101
Microfilamentous belt, 101
Microflora, leaf-inhabiting, 334–335
Micropore, 91
Microsclerotium, 32
Microtubule
    extranuclear, 99
    organizing center, 99
Microvesicle, 102
Mitoconidium, 43
Modified impression slide, 285
Monomorphic fungus, 136
Mucilage, pycnidial, 247, 252
Mushroom disease, 193, 195, 197, 198
Mycelium, 424, 429, 431
    dikaryotic, 146, 147
    perennial, 429
    vegetative, 336
Mycobiont, 172, 179
Mycoparasite, 173–174, 177, 234–235, 432
    contact, *see also* Mycotrophein, 174, 179
Mycophycobiosis, 368
Mycorrhizal fungus, 277, 279, 280
    ectotrophic, 435
Mycoses
Mycostatic factor, 358, 367
Mycotoxicosis, 76
Mycotoxin, 408
    nonvolatile, 177
    volatile, 177
Mycotrophein, 179

N

Nectrioid fungus, 39

Nematode-trapping fungus, 11
Nematode-trapping Hyphomycetes, 39
Niche, 426
  ecological, 29, 75
  expansion, 442
  fundamental, 431
  narrow, 435
  overlap, 436
  realized, 421
  specialization, 418
  theory, 282
Nutritional interaction, 179

O

Oidium, see Arthroconidium
Ontogeny, see Conidium development
Operculum, 60
Orifice cell, 249
Ostiole, 51, 60
Overgrowth, see Smothering

P

Paraphysis, 63, 147
Parasexuality, 31
Parasexual life cycle, 136
Parasite, 434
  facultative, 278, 279
  obligate, 146, 428
Parasymbiont, 172, 245, 265–266
Pathogen suppressive soil, 442
Pathogenic fungus, 433, 434–435, 442
Penetration peg, 176
Percurrent proliferation, 87, 264
Perfect state, see Teleomorph
Peridium, 47, 58, 60
Phase change, 22
Phenetic isolation, 232
Phialide, 12, 13, 15, 32, 147, **149**, 264, 321, 322, 323
Phialoconidium, 137, 138, **319**, 343
Phosphodiesterase, 126
Phragmoconidium, 65, **66, 67,** 68, 136, 137, 138
Phragmoscolecoconidium, **67,** 68
Phycobiont, 179
Phylloplane, 281, 423, 430, 438, 442
  Hyphomycetes, 364
Pigment
  carotenoid, 88, 90, 92, 94, 95

formation, 86, 87, 88, 90, 95
Pigmentation, 65, 247, 252
  ascomycetous, 150
  in taxonomic identification, 319, 320, 322
Plant pathogen, 421
  control, 441
Plasmogamy, 43
Pleoanamorph, 141, 324
Pleomorphic fungus, 43, 136
Pleomorphism, 8, 13
  anamorphic, 136, 137, 138
  scarcity, 429
  types, 137–139
Pluriseptate fungus, 148
Polyblastic ontogeny, 71
Polyphialide, 138
Polyphyletism, 25, 32
Polysaccharide synthase, 98, 99
Powdery mildew, 421
Predaceous fungus, 39, 280, 290
Predation, 424, 432
Primordium
  pseudoparenchymatous, 150
  pycnidial, 247, 252, 270
Primulin stain, 101
Prokaryotic parallel, 27
Proline, 111, 119
Propagule, 270, 271, 421–423, 430
  aero-aquatic, **340, 341, 342,** 343
  chlamydosporic, 32
  dikaryotic, 339, **342**
  dormant, 358, 367
  infective dispersal, see also Inoculum potential, 429
  sigmoid, 327, 337
  tetraradiate, 323, 324–325, 327
Pseudohyphae, 97, 103, 104, 105, 106
Pseudomycelium, 111
Pseudopycnidium, 49
Pseudotaxa, 27
Pycnidial fungus, 139
Pycnidium, see also Lichen, pycnidium cavity; Conidioma, 28, 246–255, 271
  cupulate pycnothyrioid, 49
  discoid, 164

R

Radulaspore, 12
Red rot, 76

Subject Index

455

Resting spore, 420, 431
Rhizosphere, 282
Root-inhabiting fungus, 280
Rubratoxin B, 177
Rust fungi, 147

S

S phase, 99, 100
Salt-tolerant microfungus, 419, 427
Saprophyte, 430, 432, 438
    obligate, 278, 280
    primary, 279, 281
    secondary, 279, 281
Schizogenous cavity, 28
Sclerotium, 150, 161, 322, 344, 420
Scolecoconidium, 137
Scolecospore, 65
Selective inhibition method, 286
Septum
    dolipore, 92, 322
    primary, 100, 118
    secondary, 118
Sequential morph, 150
Setae
    conidiomatal, 49, 51, 58, 68, 77
    hydrophobic, 77
    incipient, 58
Slime molds, 183
Smothering, 177
Soft rot, 363
Soil-agar film, 285, 288
Soil dilution plate method, 283, 284, 285
Soil fungus, 418, 419, 430-431, 435, 436
    in acid soil, 419, 431
    in alkaline soil, 419, 430
    saprophytic, 277, 280, 281, 288
    in tundra, 225, 429
Soil plate technique, 283
Soil section, 285, 288
Sooty mold, 422, 423, 430, 438
Speciation, 433-436
Species concept, 33-35
Spermatium, 69, 140, 146, 156, 265, 271
Spermogonium, *see* Pycnidium
Spherule, 109
    endospore development, 109
    endosporulating, 98
Spindle plaque, 99, 100
Spitzenkörper, 102
Sporangial formation, 109

Spore chimeric, 147
Spore concentration, in river foam, 297
Spore deposition, 404-406
Spore dispersal, 396-403
    diurnal periodicity, 387-390
    effect of precipitation, 390-391
    effect of wind, 401
    geographical distribution, 381-385
    gradient, 402-403
    liberation and take-off, 396-**399**
    long-distance spread, 403-404
Spore, secondary, 147
Spore trap, 376-378, 386
    Andersen sampler, 375
    automatic volumetric trap, 375
    cascade impactor, 375
    cyclone, 376
    filter, 376
    gravity slide, 375, 377
    liquid impinger, 375, 376
    rotorod and rotoslide sampler, 375, 377
    sedimentation, 374
    settle plate, 374, 377
    sticky cylinder, 375
Sporodochium, 10, 49, 51, 321, 322
Sporulation
    efficiency, 430
    role in survival, 421
    time, 418
    synchronization, 419
Stalk rot, 76
Stauroconidium, **67**, 68
Staurospore, 65
Sterigmatum, 85, 92, 95
Stokes Law, *see also* Terminal velocity, 400
Substrate
    alga, 366-367
    and fungal growth, 419, 422, 423, 431-432, 435
    bark, 364-365
    conchyolin, 357, 367
    halophyte, 357, 365-366
    leaf, 366
    mangrove, 362, 364-365, 369
    marine sediment, 367-368
    shipworm, 367
    relationship of soil fungus, 279, 283
    wood, 362-364
Sulfhydryl group, 119, 124
Sutton formula, see also Eddy diffusion, 401
Synnematous fungus, 139

T

Taxa
  heterogeneity, 24-25, 34
Teleomorph, *see also* Anamorph, teleomorph
      connections, 22, 23, 27-29, 30, 32, 43,
      145, 153, 245, 266, 270, 419
  anamorph connection, **146,** 147, 157, 161,
      309
  induction, 343
  multiple, 34
  overwintering, 140
Teleomorphic holomorph, 29, 30
Teliospore, 147, 150, 153
Terminal velocity, 400
Textura, 58, **59**
  angularis, 57, 58
  globulosa, 58
  intricata, 58
  porrecta, 58, 60
  prismatica, 58
Thallospore, 12
Thermophilic fungus, 385
Trichodermin, 177
Trichogyne, 265, 271
Trinomial nomenclature, 35

U

Urease, 87
Urediniospore, 147, 150

W

Wall biosynthesis, 100, 102, 103, 121, 126
Wall intussusception, 102
Wall lysis, 102
Wall synthesis, 102, 115, 119
Washing method, 284
White rot, 76
White ear rot, 76

XY

Xylose, 87
Yeast
  anamorphic, 85, 86, 88
  apiculate, 91
  black, 102, 107
  conidiation
    acropetal, 87, 90
    basipetal, 92, 94, 95
    enteroblastic, 87, 88, 92
    holoblastic, 87, 90
    phialidic, 88
    sympodial, 88, 92, 95
  fermentation, 86, 87, 92
    heterothallic, 88, 92
  hyphae, 86, 88, 90, 92
  mating experiment, 88, 90, 92
  pseudohyphae, 86, 88, 90
  red, 88
  teleomorph, 88, 90, 91, 95

Index to Taxa

A page number set in bold indicates a legend.

A

Abrothallus suecicus, 224
Acacia, 34
Acanthotheciella barbata, 58, **59**
Acarellina, 51
Acarocybella jasminicola, 227
Acarosporium, 58, 60, **66,** 68, 70
    sympodialis, 58, **59,** 60, **61**
Acer, 151, 296
Acetabula calyx, 214
Achyla conspicua, 186
Aciesia, 267
Acladium ellipticum, 225
Acleistomyces, 267
    rionegrensis, 269
Acontium ustilaginicola, 192
Acremoniella atra, 201, 213, 231
    melioliphila, 207
Acremoniula deightonii, 217
    sarcinellae, 217
    suprameliola, 207
Acremonium, 32, 138, 139, 140, 156, 174, 176, 182, 185, 204, 207, 222, 229, 234, 366, 377, 405
    acutatum, 230
    alternatum, 137
    arxii, 210
    bactrocephalum, 192
    butyri, 201
    byssoides, 204
    cerealis, 231
    colletotrichum-dematii, 227
    colletotrichum-truncatum, 227
    crotocinigenum, 197, 200
    domschii, 200
    exiguum, 200
    fungicola, 185
    furcatum, 197
    hyalinulum, 200, 227
    hypholomatis, 196
    incrustatum, 196
    kiliense, 231
    lichenicola, 214
    murorum, 211
    rutilum, 211
    sordidulum, 227
    strictum, 183, 197, 206, 213, 230
    tulasnei, 197
Acrodictys balladynae, 219, 229
    obliqua, 204
Acrodontium hydnicola, 200
Acrogenospora setiformis, 231
Acrogenotheca elegans, 438
Acroscyphus sphaerophoroides, 251, 257
Acrostalagmus, 278
    charceus, 231
Acrostaphylus hypoxylii, 210
Acrostaurus turneri, 215, **218**
Actinomucor elegans, 187
Actinomycetes, 27
Actinospora, 298
    megalospora, 298, 308, **309,** 321, 323
Actinoteichus, 267
Aderkomyces, 267
Aecidium clematidis, 191

Aegerita, 343, 344
  candida, **339, 343,** 344
  tortuosa, **343**
  viridis, 339
  weberi, 229
Aessosporon, 86, 88, 94, 95
Agaricus, 197, 198
  arvensis, 197
  brunnescens, 136
  campestris, 197
Agropyron, 281
Agrostis, 332
Ahmadinula, 163
Ajellomyces, 29
  capsulatus, 137, 140
Alatospora acuminata, 298, **309**
  constricta, 298, 329
  crassipes, 298
  pulchella, 298
Albizzia lebbek, 227
Alchemilla, 31
Alectoria ochroleuca, 257
Allescheriella bathygena, 359, 369
  crocea, 214
Allothyriella, 51
Alnus, 344
  glutinosa, 296, 332
  rubra, 366
Alternaria, 14, 162, 201, 230, 278, 334, 358,
    364, 365, 366, 377, 381, 382, 384, 386,
    387, 389, 390, 391, 392, 393, 394, 403,
    405, 423, 437
  alternata, 183, 187, 220, 230, 407, 430
  dendritica, 228
  longissima, 206
  maritima, 364
  olivacea, 225
  porri, 234, 389, 398
  tenuis, 230, 281, 437
  tenuissima, 192, 220
  zinniae, 234
Alysia, 267
Alysidium fuscum, 214
  resinae, 230
Amanita caesarea, 6
Amazonomyces, 267
Amblyosporium, 195
  botrytis, 195, 200, 203, 231
  spongiosum, **194,** 195
Amerodiscosiella renispora, 73, 75

Ameropeltomyces, 267
Amerosporium, **57,** 58, 60
Amoebomyces, 267
Amorphotheca resinae, 409
Ampelomyces, 204
  abramovii, 188
  quisqualis, 176, 188, 204, **205**
Ampullifera, 220
  foliicola, **221**
Anaptychia ciliaris, 257, 258
Anavirga, 322
Anconomyces, 267
Anemone, 155
Amerosporae, 10
Anguillospora, 298, 323, 335
  crassa, 227, 298, 308, 324, 332, 334
  curvula, 298
  filiformis, 298
  furtiva, 298, 308
  gigantea, 298
  longissima, 227, 298, 308, **309, 319,** 333
  pseudolongissima, 298
  pulchella, 298, 321
  rosea, 298, 308
  virginiana, 298
Angulospora aquatica, 298, **309,** 321
Anixiopsis, 29
Annellodochium ramulisporum, **209,** 210
Annellolacinia, 70
  dinemasporioides, **50,** 51, **62,** 63
Annellophora, 35
  dendrographii, **226,** 228
  sydowii, 228
Anthostoma turgidum, 210
Anthracoderma hookeri, 213
  selenospermum, 213
Aphanoascus, 29
Aphanocladium album, **184,** 185, 192, 204,
    394, 407
  meliolae, 207
Aphanomyces euteiches, 187
Aphrophora parallela, 77
Aphyllophorales, 88
Apinisia, 18
Apiocrea chrysosperma, 193
Apiosporina collinsii, 219
  morbosa, 153
Apiotrichum, 92, 94
  humicola, 94
  porosum, 94

Aposphaeria, 162
  parasitica, 203
Appendiculella calostroma, 207
Aquilegia, 27
Arachniotus, 29
Arbutus menziesii, 366
Aristastoma, 57
Armillaria mellea, 197, 198, 235
Arthonia galactites, 256
  impolita, 222
  medusula, 247
Arthoniales, 247, **259**
Arthrinium, 14, 35, 69, 70, 382, 393
  cuspidatum, 398
  phaeospermum, 407
Arthrobotryomyces, 267
Arthrobotrys, 11, 39
Arthrobotryum, 163
Arthroderma, 29
  benhamiae, 141
  curreyi, 141
  tuberculatum, 141
Articulospora, 298, 323
  angulata, 298
  atra, 298
  foliicola, 306
  moniliforma, 298
  ozeensis, 306
  tetracladia, 298, **309**, 329, 330
Asbolisia, 207
  indica, 225
Asbolisiomyces, 267
Aschersonia, 57, 76, 77, 225
  turbinata, 225
Ascobolus crenulatus, 178, 214
Ascochyta, 47, 71, 150, 161, 365
  gossypii, 60
  graminicola, 227
  lichenoides, 224
  phaseolorum, 76
  salicorniae, 359, 361, 365
Ascochytella, 208
  stegasphaeriae, 219
Ascochytopsis, 217
Ascochytula obiones, 359, 365
Ascoideaceae, 88, 90
Ascomycetes, 10
Ascomycota, 135, 140
Ascomycotina, 426
Ascophanus, 32, 214

Asellus aquaticus, 335
Aspergillus, 5, 35, 37, 160, 176, 184, 187, 201, 203, 204, 215, 230, 278, 349, 366, 377, 381, 382, 384, 385, 386, 392, 393, 404, 407, 419, 431, 437, 441, 442
  aculeatus, 227
  amstelodami, 385, 425
  awamori, 385
  candidus, 197
  cervinus, 215
  clavatus, 394, 407
  flavus, 176, 287, 385, 394, 407
  fumigatus, 136, 220, 385, 393, 394, 395, 407, 408
  glaucus, 8, 197, 394, 395
  heterothallicus, 425
  nidulans, 385, 394, 419, 425
  niger, 197, 230, 385, 387, 421, 425
  ochraceus, 394
  repens, 228, 385
  ruber, 385
  terreus, 394
  ustus, 177
  versicolor, 385
Asteridiella coffeae, 208
  fraseriana, 206
  glabra, 208
  tetracerae, 208
Asterina, 215, 217, 219
  colliculosa, 217
  contigua, 219
  dallasica, 217
  linderae, 215
  veronicae, 217
Asteroma fugax, 202
Asteromella, 69, 151, 159
  veronicae, 217, 225
Asteromyces, 365, 367
  cruciatus, 358, 360, 361, 367, 368
Asterophora lycoperdoides, 195
Asterostomella parasitica, 200
Astrabomyces, 267
Astraeus hygrometricus, 203
Athelia rolfsii, 201, 235
Atractilina asterinae, 217
  parasitica, 206, **212**, 217
Aureobasidium, 139, 334, 364, 377, 382, 389, 390, 391, 393, 394, 395, 407, 437
  pullulans, 107, 183, 227, 235, 281, 394, 407

Auricularia auricula-judae, 202
  mesenterica, 202
Avicennia, 362

B

Bacidia chlorococca, 265
  vezdae, 258
Bacillus, 442
Bactridium gymnosporangii, 190
  parasiticum, 228
Balladyna, 219
  magnifica, 219
Balladynella amazonica, 217
Balladynopsis entebbeensis, 206
  ledermannii, 219
Bartalinia, 68
Basididyma perexigua, 227
Basidiomycetes, 10, 86, 87
Basidiomycota, 135, 140
Basidiomycotina, 426
Basipetospora, 15
Beauveria densa, 139
Beltrania, 389
Berberis, 151, 152, 153
Betula, 296
Beverwijkella pulmonaria, **339**, 344
Bipolaris, 160
Bispora antennata, 228
  betulina, 228
  christiansenii, **221**, 222
Blarneya, 267
Blastomyces, **110**, 111
  dermatitidis, **108**, 109, 112, 113, 114, 115, 118, 119, 120, 123, 124
Blastotrichum puccinioides, 193
Bleptosporium pleurochaetum, **55**, **62**, 63
Blistum, 185
  ovalisporum, **184**, 185
  tomentosum, **184**, 185
Blodgettia, 267
  bornetii, 358, 368
Boletinus, 193
Boletus, 6, 198
  edulis, 198
  purpureus, 198
Bothrodiscus, 57, 77
  berenice, **56**, **64**, 65

Botryodiplodia, 225
  lecanidion, 225
  theobromae, 65, 76, 225, 227
Botryophialophora marina, 358
Botryosphaeria, 30, 161, 220
Botryosphaerostroma quercina, 227
Botryostroma eupatorii, 219
Botryotinia, 140
  fuckeliana 136, 213
Botryotrichum, 138, 140
Botrytis, 5, 14, 138, 140, 278, 377, 382, 386, 389, 390, 391, 393
  allii, 229
  cercosporicola, 228
  cinerea, 183, 281, 396, 398
  yuae, 228
Brachiosphaera jamaicensis, 298
  tropicalis, 299, **309**
Brachysporium minutum, 207
Breteldiellaceae, 31
Bremia graminicola, 186
Brettanomyces, 85, 86, 91
  bruxellensis, **89**, 91
Broomella, 16, 148, 162, 163
  acuta, 163, 187
  excelsa, 163
  montaniensis, **149**, 163
  vitalbae, 148, **149**, 163
Brunchorstia pinea, 76, 77
Bryoria capillaris, 267
Buellia canescens, 224
  punctata, 264
  punctiformis, 265
  stillingiana, 265
Bulbillomyces farinosus, **339, 343**
Bulgaria, 202, 214
Bullera, 86, 92
  alba, 92, **93**
Byssoloma subdiscordans, 247, 256
Byssophytum, 270
Byssostilbe stilbigera, 185

C

Calcarispora hiemalis, 299, **309**
Calcarisporium arbuscula, **194**, 195, 198, 200, 211, 213, 214
  pallidum, 185
  parasiticum, 174

Caliciaceae, 157
Calloriopsis gelatinosa, 206
Calluna, 332
Calonectria cephalosporii, 207
   gymnosporangii, 190
   ukolayi, 208
Caloplaca ferruginea, 252
Camaropycnis, 57
Camarosporium, 153, 155, 162, 365
   metableticum, 359, 361
   obiones, 365
   palliatum, 359
   roumeguerii, 357, 361, 365
Campylospora chaetocladia, 299, **309**
   filicladia, 299
   parvula, 299
Cancellidium applanatum, **338**, 339, **343**
Candelabrella, 39
Candelabrum brocchiatum, 343
   spinulosum, 339, **343**
Candelariella vitellina, 222
Candida, 85, 86, 88, 90, 91, 92, 95, **110**
   albicans, **89**, 90, 97, 102, 103, **104, 105, 106,** 111, 112, 113, 121, 122, 123, 124, 125
   anomala, 198
   aquatica, 319, **320**
   arabinosa, 198
   boleticola, 203
   buffonii, 198
   chodatii, 90
   ciferrii, 90
   curvata, 94
   fermenticarens, 224
   guilliermondii, 90
   humicola, 94
   krusei, 90
   lipolytica, 90
   macedoniensis, 90
   membranaefaciens, 201
   mesenterica, 90
   mogii, 90
   muscorum, 95
   obtusa, 198
   pseudotropicalis, 90
   pulcherrima, 90
   sake, 90
   santamariae, 201
   tropicalis, 88, **89**

   utilis, 90
   valida, 90
   vulgaris, 88
Candidaceae, 88
Cantharellus odoratus, 200
Capitorostrum asteridiellae, 206
Capnodium, 220
   meridionale, 217
Caprettia, 267
Carcinus maenas, 367
Carex, 154, 296
Casaresia, 322, 324, 332
   sphagnorum, **309**
Cashiella fuscidula, 147, **149**
Cassia, 31
Castanopsis cuspidata, 438
Casuarina, 437
Catenophora, 63, 70, 77
   yuccae, **55**
Cenangium ferruginosum, 213
Cephalosporiopsis, 39
Cephalosporium, 39, 278, 366
   curtipes, 192
   macrocarpus, 188
   saprolegniae, 186
Cephalotrichum, 142
   stemonitis, 136, 138, 142
Ceratobasidium, 32
Ceratocystis, 30, 112, 141, 161, 162, 211
   fagacearum, 211
   fimbriata, 180, 211
   ips, 211
   minor, 112, 211
   paradoxa, 138
   stenoceras, 185
   ulmi, 397, 424, 427
Ceratophorum, 35
Ceratopycnis, 57
Cercospora, 208, 230, 235, 287, 289, 392
   atromarginalis, 230
   carioae, 228
   chandleri, 219
   kakivora, 228
   koepkei, 228
   unamunoi, 228
   uromycestri, 190
   vestita, 229
Cercosporiella, 228
   cercosporicola, 228

Cerebella andropogonis, 430
Cetraria islandica, 251, 257
Ceuthospora, **54,** 57, **62,** 63, **66,** 68, 70
Chaetomella, 57, 58, 60
Chaetomium, 140, 156, 441
   cochliodes, 178
   elatum, 156
Chaetomonodorus, 267
Chaetophoma stromaticola, 225
Chaetophomella asterinarum, 217
   parasitica, 217
Chaetoscutula juniperi, 147
Chaetoseptoria, 57
Chaetospermum, 48, 58, 60, **61**
Chaetosphaerella, 148
   phaeostroma, 148
Chaetosphaeria, 30
Chaetospora quezeli, 76
Chaetothyriolum, 266
Chalara, 32, 39, 138
   cyttariae, 213
   fungorum, 200
   fusidioides, 220
   microspora, 200
   minima, 213
Chalaropsis, 39, 138, 141
   punctulata, 138
Chaos, 6
Cheiromycella microscopica, 230
Chiastospora parasitica, 217
Chionomyces chorleyi, 207
   meliolicola, 206, **212**
   sclerochitonis, 207
Chlamydomyces, 138, 140
Chloridium botryoideum, 230
   meliolae, 208
Chondroplea populea, **55,** 58, **59,** 76, 77
Chondropodiola falcispora, 217
Chrondrostroma, 57
Chrysalidiopsis, 68
Chrysanthemum, 154
Chrysomyxa, 153
   rhododendri, 152
Chrysosperma, 225
Chrysosporium, 29, 30, 32, 141, 393
   thermophilum, 394
Cicerbita, 154
Cicinnobella, 207, 219
   domingensis, 219
   heterothea, 229

   megastoma, 219
   sydowii, 219
   tetracericola, 219
Cicinnobolus, 204
   bremiphagus, 186
   cesatii, 204
   heraclei, 186
   sporophagus, 229
Ciliochora, 57, 58, 63
Ciliochorella, 68
Ciliophora cryptica, 208
Circosia manaosensis, 215
Cirrenalia, 365
   fusca, 358
   macrocephala, 358, 363, 364, 365, 366, 369
   pseudomacrocephala, 358, 365
   pygmea, 358, 364, 367
   tropicalis, 358
Cladobotryum, 193, 200
   apiculatum, 193
   australe, 228
   dendroides, 193, 200
   leptosporum, 200
   mycophilum, 193
   varium, 193, **194,** 200
   verticillatum, 193, 202, 213
Cladonia, 224
   chlorophaea, 257
   furcata, 224
   rangiformis, 222
   turgida, 257
Cladoniaceae, 257
Cladophora, 368
Cladosporiella uredinicola, 190
   uredinis, 190, **191**
Cladosporium, 14, 150, 156, 182, 183, 188, 229, 230, 278, 284, 290, 334, 358, 364, 367, 377, 381, 382, 384, 385, 386, 387, **388,** 389, 390, 391, 392, 393, 394, 397, 403, 404, 405, 423, 430, 437
   aecidiicola, 190
   algarum, 358, 367
   argillaceum, 185
   arthoniae, 222
   asterinae, 219
   balladynae, 219
   cladosporioides, 187, 197
   colocasiae, 186
   cyttariicola, 213
   elsinoes, 219

exoasci, 203
fuligineum, 197, 202, 211, 214
fulvum, 407
gallicola, 190, **191**
hemileiae, 190
herbarum, 183, 197, 201, 214, 215, 229, 230, 407, 437
lophodermii, 214
macrocarpum, 197, 201
oxysporum, 206
penicillioides, 228
phyllachorae, 208, **209**
resinae, 409
stromatum, 210
tuberculatum, 225
uredinicola, 187, 192
werneckii, 107
Cladotrichum opacum, 231
Clasterosporium, 35
parasiticum, 225
Clathrosphaerina zalewskii, **339,** 343, 345, 346
Clavariana aquatica, 299, **309**
Clavariopsis, 299
aquatica, 299, 308, **309,** 319, 330, 331, 334, 335
brachycladia, 299
bulbosa, 299, 308, 358, 360
Clavatospora filiformis, 299
flagellata, 299
longibrachiata, 299, **309,** 322
stellata, 299, 325
stellatacula, 299, 358
tentacula, 299
Claviceps, 430
purpurea, 151, 397
Cleistosphaeria macrostegia, 219
Clematis, 148, 155
Clitopilus prunulus, 198
Cnazonaria, 161
Coccidioides immitis, 98, 107, 109, 400, 408
Coccodiella, 210
Coccospora parasitica, 225
Cochliobolus, 153, 160, 163
lunatus, 141
miyabeanus, 176
sativus, 180, 220, 230, 235
Cochlonema megalosomum, 188
Coccostroma puttemansii, 215
Coeloanguillospora, 306
appalachiensis, 306

Coleomyces rufus, 214
Coleophoma, 63, 77
taxi, 60, **61, 66,** 68
Collema, 247, 265
bachmanianum, 265
multipunctatum, 265
tenax, 224
Colletotrichella xylostei, 157
Colletotrichum, 72, 160, 225, 227
aeciicola, 191
capsici, 227
caudatum, 68
dematium, 192, 200, 201, 227
falcatum, 76
gloeosporioides, 77, 192
lindemuthianum, 398
lini, 227
musae, 76, 77
palinhae, 185
pucciniophilum, 192
umemurai, 215
uredinophilum, 192
Colpoma quercinum, 215
Coltricia perennis, 201
Columnophora rhytismatis, 214
Coma, 69
circularis, **55**
Comatospora, 69
Combea mollusca, 247, 256, 258
Condylospora spumigena, 299, 302, **309**
Conicosolen, 266
Coniella, 49, 63, 65
diplodiella, 75, 76
eucalypticola, **53, 66,** 68
Coniochaeta discospora, 422
tetraspora, 422
velutina, 201
Coniomycetes, 7
Coniosporium, 266
aeroalgicola, 266
helminthosporii, 228
Coniothecium toruloides, 265
Coniothyrium, 162, 208
crepiniarum, 225
cyttariae, 213
dolium, 219
epimyces, 200
glabroides, 208
hookeri, 213
insuetum, 210

Coniothyrium (continued)
  massariae, 219
  minitans, 213, 231
  obiones, 359, 365
  occultum, 208
  tuberculariae, 229
Conotrema urceolatum, 258
Coprinus heptemerus, 178
Cordana musae, 389
Cordyceps, 30
  dipterigena, 210
Coremiella cuboidea, 139
Corethropsis epimyces, 195
Coriolus, 200
  versicolor, 198
Corniculariella, 58
Cornutispora, 224
  lichenicola, **223**, 224
  limaciformis, 76, 214
Corollospora pulchella, 308
Corticium, 198, 200
Corynascus fergusii, 195
  sepedonium, 141
Corynespora cassiicola, 389
Coryneum, 65
  brachyurum, 225
Crataegus, 158
Cremasteria cymatilis, 358, 363
Creonecte biparasitica, 189
Cristella confinis, 202
Crocicreomyces, 267
Cronartium, 189, 192
  comandrae, 189, 191
  fusiforme, 189, 191
  hemileiae, 191
  quercuum, 189
  ribicola, 189, 192
  strobilinum, 189
Cruciferae, 27
Cryptococcaceae, 85, 86, 88, 91, 93
Cryptococcus, 86, 92, 95
  hungaricus, 92
  neoformans, 92, **93,** 107
Cryptogene parodiellae, 217
Cryptogenella parodiella, 217
Cryptostroma corticale, 394, 407
Ctenomyces, 29
  serratus, 141
Cucurbitaria, 30, 161, 220
Cucurbitula berberidis, 217
  elongata, 220

Culcitalna achraspora, 364, 365
Culicidospora aquatica, 299, **309,** 322, 324
  gravida, 299
Cunninghamella africana, 188
  echinulata, 188
  elegans, 188
Curvularia, 141, 160, 377, 382, 384, 385, 386, 389, 391, 423, 437
  lunata, 141, 437
  ramosa, 288
Cycloschizon macarange, 220
Cylindrocarpon, 29, 138, 183, 185, 214, 278
  destructans, 229
  luteoviride, 219
  macrosporum, 206, **212,** 217
  ukolayi, 208
Cylindrocephalum stellatum, 228
Cylindrocolla fugax, 225
Cylindrotrichum, 299
  helisciforme, 299
  oligospermum, 299
Cyrta, 267
  licaniae, 269
Cystocoleus, 270
Cystoseira osmundacea, 362
Cytoplea parasitica, 210
Cytospora, 160, 225, 227
  abietis, 224
  leucosperma, 225
  prunorum, 225
  rhizophorae, 359, 362, 365, 369
Cytosporina, 44, 75
Cyttaria darwinii, 213

D

Dactylaria, 39
  dimorpha, 208
  domina-gregum, 206
  mycophila, 214
Dactylella, 186, 299, 306
  anisomeres, 186
  helminthodes, 186, 188
  microaquatica, 300, **309**
  minuta, 300
  spermatophaga, 187
  stenocrepis, 186
  stenomeres, 186
  tenuis, 186
Dactylium dendroides, 193
Darluca filum, 189
Dasyscyphus, 213

Davisiella, 208
   botryodiplodia, 225
Debaryozyma castellii, 90
   hansenii, 90
Deightoniella, 382, 385, 391
   leonensis, 208
   torulosa, 389, 392, 398
Dekkera, 91
Dematieae, 10
Dematophora necatrix, 228
Dendrodochium parasiticum, 225
   subeffusum, 222
   tenue, 188
Dendrographium atrum, 228
Dendrospora, 300, 323
   erecta, 300, **309**
   fastuosa, 300
   fusca, 300
   juncicola, 300, 332
   nana, 300
   tenella, 300
   torulosa, 300
Dendrosporomyces prolifer, 300, **309**
   splendens, 300
Dendryphiella, 185, 365, 367, 389
   arenaria, 358, 365, 366, 367
   salina, 359, 365, 366, 367
Dendryphion, 162
Dennisographium episphaeria, 210
Dermatiscum thunbergii, 258
Dermatocarpon, 253
   weberi, 258
Dermatodothis, 217
   zeylanica, 217
Desmidiospora, 138
Deuteromycota, 140
Deuteromycotina, 44
Dexteria pulchella, 204
Diachorella, 68
Dialaceniopsis landolphiae, 217
Diaporthales, 24
Diaporthe, 210
   woodii, 76
Diatrype, 210
   stigma, 210
Diatrypella favacea, 210
   quercina, 210
Dicoccum, 391
Dicotyledones, 438
Dicranidion inaequalis, 210
Dictyonema, 176

Dictyophrynella, 220
   bignoniacearum, **221**
Dictyosporium, 365
   pelagicum, 359, 364, 365, 366, 369
Didymaria acervulicola, 225
Didymaster, 267
Didymella, 30, 150, 161
   exitialis, 381
   lycopersici, 161
Didymium difforme, 185
Didymobotryum hymenaearum, 208
Didymopsis helvellae, 214
   phyllachorae, 210
   spicata, 214
Didymosphaeria winteri, 208
Didymosporae, 10
Didymosporium conglutinatum, 229
Didymostilbe obovoidea, 200
Diedickea piptadeniae, 227
   singularis, **52**
Diheterospora, 39, 138
Dikaryomycota, 22
Dilophospora, 57, 68
Dimerosporium macrocarpum, 217
Dimorphospora, 138, 141
   foliicola, 300, **309**
Dinemasporium, 51, 58, 68, 77
   aberrans, **62,** 63
   fimeti, 76
   marinum, 76, 359
   meliolicola, 208
Diozegia, 92
Diplocarpon maculatum 151
Diplococcium clarkii, 202
Diplodia, 71, 160
   castaneae, 225
   celtidigena, 227
   cydoniae, 225
   maydis, 76
   oraemaris, 359
   pinea, 75, 76, 77
   tumefaciens, 65
Diplodina geasterina, 202
   tylostomatis, 202
Diploicia canescens, 247, 258
Diplosporium morchellae, 214
Diplozythia, 57
Dipodascaceae, 88
Dipodascus, 91
Dirina ceratoniae, 224, 247, **249,** 256, 258
   repanda, 256

466 Index to Taxa

Discomycetes, 10, 25, 27
Discosia, 51, 72
Discosiella, 51, 69
Discosiopsis, 75
Discostroma, 148
   massarina, 148
   saccardiana, 148
Divinia diatricha, 206
Dolichocarpus chilensis, 247, 256
Doliomyces, 60, 68
   senegalensis, **53**
Domingoella asterinarum, 206, **212,** 217
   pycnopeltarum, 217
Doratomyces putredinis, 201
   stemonitis, 394
Dothichiza, 68, 76, 159, 160
   sorbi, **54,** 58, **59**
Dothidea, 140, 141
   ribesia, 219
Dothideales, 25, 28
Dothidella australis, 217
   derridis, 220
   symplocii, 217
Dothiomyces, 267
Dothiora, 159, 160
Dothioraceae, 26
Dothiorella parasitica, 225
   stratosa, 225
Dothiorellina quickii, 213
Dothiorina, 57
   tulasnei, **56, 64,** 65
Dothistroma pini, 76, 77
Drechslera, 150, 160, 365, 423
   halodes, 359
   poae, 230
   ravenelii, 228
   teres, 230
Dryas, 281
Durandiella, 148
Dwayabeeja, 139
   sundara, 137
Dwayalomella, 72

E

Echidnodes, 219
Echinobotryum, 142
   atrum, 138, 142
Echinoplaca, 265

Echinopodospora, 29
Ectosticta, 207
   insignis, 225
   popowiae, 219
Elaphomyces, 211
   japonicus, 215
Eleutheromycella mycophila, 76, 198
Eleutheromyces, 63, 70
   subulatus, 76, **196,** 198, 227
Elsinoe wisconsinensis, 219
Emmonsiella, 29
Endocarpon, 253, 257
Endocena, 270
Endocronartium harknessii, 190, 192
Endomyces decipiens, 198
Endomycetaceae, 85
Endomycetes, 26, 29
Endophragmia, 138, 324
   dennisii, 200
Endophragmiella, 138, 182
   canadensis, 210
   eboracensis, 210
   hughesii, 222
   pallescens, 225, **226**
Endothia gyrosa, 427
   parasitica, 427
Endozythia moravica, 217
Engelhardtiella alba, 224
Entodesmium, 160
Entomophthora, 9, 188, 322, 327, 391
   conica, 327, **329**
   rhizospora, 327
Entomophthorales, 426
Entomosporium, 151
Entyloma calendulae, 192
Ephelidium, 267
Epichloe typhina, 210
Epicoccum, 139, 334, 359, 365, 367, 377,
   382, 385, 386, 387, 389, 391, 393,
   423
   nigrum, 281
   purpurascens, 183, 197, 201, 220, 230, 398,
   399, 407, 430
Epiphytae, 7
Epistigme erodens, 208
   parmulariicola, 219
   teucrii, 208
Erica, 332
Eriocercospora balladynae, 206, **212,** 217
   olivacea, 207

Eriomycopsis, 204, 217
   biseptata, 176, 206
   bonplandii, 206
   bosquieae, 207
   englerulae, 217
   flagellata, 176, 206, **216,** 217
   meliolinae, 207
   minima, 206, 217
   minuta, **226,** 228
   paraensis, 207
   schiffnerulae, 217
Eriophorum, 332
Erysiphaceae, 421
Erysiphe, 204, 377, 382, 389, 390, 391
   cichoracearum, 176, 206
   graminis, 390, 404, 405, 430
   trifolii, 204
Euantennaria mucronata, 438
Eucalyptus, 34
Eudarluca caricis, 189, 217
Euphorbia, 152
Eurasina bondarzewiae, 198, **199**
Eurotium, 8
   herbariorum, 8
Eutypa armeniacae, 211
   flavovirens, 210
   leioplaca, 210
   velutina, 210
Eutypella prunastri, 44, 75
   stellulata, 225
Excipulaceae, 45
Exidia, 202
Exobasidium, 202
Exophiala dermatitidis, 107, 113
   jeanselmei, 107
   pedrosoi, 107
   werneckii, 107
Exosporiella fungorum, 198, **199**
Exosporium stilbaceum, 208
Exoserohilum, 160

F

Fagus, 336
   sylvatica, 336, 364
Farlowiella carmichaeliana, 230
Fibulocoela, 70
Filicales, 438
Filobasidiaceae, 86, 88, 91

Filobasidiella, 86, 92
   neoformans, 92
Filobasidium, 86, 92
   capsuligenum, 92
Filosporella annelidica, 300, 308
   aquatica, 300, **309**
Flabellospora acuminata, 300
   crassa, 300, **309**
   multiradiata, 306
   tetracladia, 300
   verticillata, 300
Flagellospora curvula, 301, **309,** 331
   fusarioides, 301
   penicillioides, 301, 308
   stricta, 301
Flahaultia hyalina, 202
Fomes, 201
Fomitopsis, 200
   pinicola, 201
Fonsecaea, 141
Fontanospora alternibrachiata, 301
   eccentrica, 301, **309**
Fuckelia, 57
Fuligo septica, 185
Fulvia fulva, 228
Furcaspora, 69, 70
   pini, **67,** 68
Fusarium, 37, 137, 139, 182, 202, 229, 278,
   279, 284, 376, 377, 383, 385, 389, 392,
   393, 397, 398, 403
   agaricorum, 197
   aquaeductuum, 227
   avenaceum, 192, 202, 213, 220, 234
   bactridioides, 192
   coeruleum, 230
   culmorum, 230, 288
   detonianum, 202
   dominicanum, 208
   epistromum, 210
   equiseti, 202
   heterosporioides, 231
   heterosporum, 231
   laboulbeniae, 220
   larvarum, 202, 220
   lateritium, 211
   lycoperdonis, 202
   moniliforme, 230, 394, 405
   mycophilum, 197
   oxysporum, 186, 188, 197, 211, 231
   peltigerae, 222

Fusarium (*continued*)
  sambucinum, 202, 211
  sclerodermatis, 202
  semitectum, 187, 230
  solani, 156, 187, 188, 230, 231
  sphaeriae, 219, 228
  sporotrichioides, 198
  trichothecioides, 192
  ustilaginis, 192
  volutella, 227
Fusicladium, 150
  crataegi, 159
  poriicola, 200
Fusidium, 228, 389
  hormiscii, 228
  hypophleoides, 201
  parasiticum, 211
Fusoma telimenellae, 210
Fusticeps bullatus, 339, **343**

G

Gabarnaudia fimicola, 214
  tholispora, 195, **209,** 211
Gaeumannomyces graminis, 211
Gammarus, 334
  pseudolimnaeus, 335
  pulex, 335
Gampsonema, 68, 70
  exile, 60, **61, 67,** 68
Ganoderma, 200, 201
  lucidum, 198, 201
Gasteromycetes, 10, 24, 28
Gaumannomyces graminis, 284
Geasteropsis conrathii, 202
Geastrum ambiguum, 202
  fornicatum, 202
  hungaricum, 202
  minimum, 202
Geastrumia, 65
  polystigmatis, **56, 64,** 65, 208
Gelasinospora tetrasperma, 136
Gelatinosporium, 58
Genicularia, 39
Geniculospora grandis, 301, **309**
  inflata, 301
Geoglossum, 213
  glabrum, 213
Geopyxis ciborium, 214

Geotrichum, 14, 86, 91, 201, 394
  armillariae, 198
  bipunctatum, 231
  candidum, 91, 107, 109, 203, 231
  cyphellae, 197
  fermentans, 91
Gibberella baccata, 211
  pulicaris, 211
Gibellula pulchra, 227
Giulia, 69
Gliocephalotrichum bulbilium, 138
Gliocladium, 182, 202, 394
  agaricinum, 197
  album, **184,** 185
  caespitosum, 210
  penicilloides, 200
  roseum, 137, 176, 177, 180, 187, 201, 211, 213, 229, 231
  virens, 177
  viride, 198
Gliocoryne, 161
Gloeosporium, 426
  roesteliaecola, 192
Gloeotrochila anthuriicola, 225
Gloiosporae, 13
Glomerella, 160, 210
  cingulata, 192
Glyceria, 343
Godronia, 30
  abieticola, 213
Gonatobotrys flava, 188
  simplex, 174, 177, 230
Gonatobotryum fuscum, 174, 211, 231
Gonatorrhodiella highlei, 174, 211
Graphium, 139, 394, 407
  flexuosum, 230
  hendersonulae, 225
  irradians, 210
Grifola frondosa, 198, 201
Groveolopsis, 65
Guignardia, 30, 75, 139, 157, **158**
  himalayensis, 157, **158**
  latemarensis, 157, **158**
  lonicerae, 157, 159
  mirabilis, 157, **158**
  xylostei, 157, **158**
Gyalectina carneolutea, 247
Gymnoascus, 29
Gymnospermae, 438

Gymnosporangium, 147, 153
  confusum, 190
  fuscum, 189
  gaeumanni, 147
  sabinae, 152
Gyoerffiella, 323, 324
  biappendiculata, 301
  entomobryoides, 306
  gemellipara, 306
  oxalidis, 306
  rotula, 301, **309**
  speciosa, 301
  tricapillata, 301

H

Haematomma cismonicum, 222
Hainesia, 51
Halidrys dioica, 362
Halimione portulacoides, 365
Haloguignardia, 361, 362
Halosphaeria mediosetigera, 368
Hanseniaspora, 91
Hansenula capsulata, 90
  jadinii, 90
Hansfordia, 160, 220, 229
  alba, 227
  parasitica, 174, 177, 220, 227
  pulvinata, 229
  triumfettae, 229
  ugandense, 220
Hansfordiella asterinarum, 217
  cupulifera, 207, 217, **218**
  diedickeae, 227
  meliolae, 207
Hansfordiellopsis, 220
  lichenicola, **221**
Haplaria melioliphila, 208
Haplographium delicatum, 228
Haplosporella, 57
Harknessia, 65, 68, 69, 70, 71
  renispora, **62**, 63
Harpographium corynelioides, 202
  rhizomorphum, 231
Harposporium, 39
Harzia, 138, 140
  acremonioides, 213
Harziella, 138, 140
  capitata, 195

Helicodendron, 338, 346, 347
  conglomeratum, 343, 344, 346, 347
  ellipticum, 347
  fractum, 344, 346, 347
  fuscosporum, 346
  fuscum, 346
  giganteum, **339, 343,** 344, 345, 346, 347
  hyalinum, 344, 346, 347
  intestinale, 344
  triglitziense, 339, 343, 344, 345, 346, 347
  tubulosum, **339,** 343, 344, 346, 347
  westerdijkiae, 344
Helicoma salinum, 363
  stigmateum, 227
Helicomyces, **337,** 391
  niveus, 227
Helicoon, 338
  elegans, 228
  fuscosporum, 344
  pluriseptatum, **339**
  sessile, 344
Helicosporium, 337, 343, 346, 347, 377, 383
  binale, 185
  brunneolum, 228
  phragmitis, 343
  vegetum, 227
Helicothyrium riyukyuense, 65
Heliscus, **329**
  lugdunensis, 297, 301, 308, **309, 319,** 332, 333
  submersus, 301
Helminthosporium, 228, 229, 230, 377, 381, 383, 384, 385, 386, 389, 390, 391, 393, 404
  conviva, 200
  delphinii, 229
  maydis, 430
  sativum, 180, 288, 400
  turcicum, 229, 392, 398
  velutinum, 231
Helotiales, 308, 426
Helvella, 214
Hemileia vastatrix, 176, 191, 192, 207
Hemisphaeriales, 26
Hemispora stellata, 228
Hendersonia, 46, 71, 155
  leptostromatis, 225
  roblediae, 219
  taphrinicola, 203

Hendersonia (continued)
    uredinophila, 191
Hendersonula, 217
    australis, 217
    monochaetiella, 219, 225
    symplocii, 217
    toruloidea, 76, 225
Heterobasidion annosum, 201
Heteroceras, 68
Heteroconium solanium, 207, **216,** 217
    tetracoilum, **209,** 210
Heterodea, 271
    muelleri, 222
Heterodermia, 258, 263
Heteropatella, 60, 70, 147, 164
    lacera, 60, **61,** 225
Heterosphaeria, 147, 164
    veratri, 164
Heterosporium tupae, 228
Hevea, 210
Hirneola, 202
Hirschioporus abietinus, 198
Hirsutella entomophila, 229
    versicolor, 229
Histoplasma, **110,** 111
    capsulatum, 107, 109, 111, 113, 114, 118, 119, 120, 122, 123, 124, 125, 137, 408
Hoehneliella, 58
    perplexa, **50,** 51, 58, **59**
Hormiactis nectriae, 210
Hormonema, 159, 160
Humicola, 39, 138, 140
    alopallonella, 359, 363, 364, 367, 368
    asteroidea, 200
    fuscoatra, 187
    grisea, 231, 394
    lanuginosa, 394
Hyalopycnis, 57
    blepharistoma, 198
Hyaloscypha dematiicola, 228
    zalewskii, 343
Hyalotiella, 63, 70, 72
    transvalensis, 60, **61, 67,** 68, 72
Hydnotrya tulasnei, 215
Hydnum, 200
    compactum, 200
Hygrophorus, 198

Hymenochaete rubiginosa, 201
Hymenomycetes, 10, 24, 39, 193
Hymenoscyphus fagineus, 308
    varicosporoides, 306
Hymenula sclerotii, 231
    socia, 225
Hyphoderma calyciferum, 200
Hyphodontia sambuci, 200
Hypholoma fasciculare, 196
Hyphopichia burtonii, 90
Hypnotheca graminis, 219
Hypocrea, 211
    austrograndis, 176
Hypocreales, 25
Hypocrella aleyrodis, 210
    turbinata, 227
Hypodermii, 10
Hypogymnia physodes, 251, 257, 258
Hypomyces, 30, 193, 211, 214
    aurantius, 193
    chrysospermus, 193, 200, 203
    ochraceus, 193
    odoratus, 193
    rosellus, 193
Hypoxylon, 210
    rubiginosum, 210
Hysteriales, 25, 26

I

Idiocercus macarangae, **62,** 63
    pirozynskii, **62,** 63
Idriella fertilis, 228
Illosporium, 220
Ingoldiella fibulata, 301
    hamata, 301, 308, **309**
Inonotus dryadeus, 198
    obliquus, 200
    radiatus, 201
Irene nuxiae, 207
Ireniopsis aciculosa, 206
    cryptocarpa, 206
Irpicomyces schiffnerulae, 215
Isaria, 9, 227
    fruticosa, 197
    meliolae, 208
Isariella auerswaldiae, 215
Issatchenkia orientalis, 90

Isthmotricladia britannica, 301
  gombakiensis, 301
  laeensis, 301, **309**
Itersonilia perplexans, 192

J

Jacobia conspicua, 210
Jaculispora submersa, 301, **309**
Japonia, 60, 68
Juncus, 332
Juniperis, 152

K

Kabatia, 157, 158
  americana, 157
  latemarensis, 157
  lonicerae, 157
  mirabilis, 157
Kabatiella, 139
Karakulinia, 150
Keissleriella blepharospora, 369
Kellermania, 58, 65, 70
  uniseptata, **53**
Kellermaniopsis, 75
Kilikiostroma, 267
Kloeckera, 86, 91
  apiculata, 86, **89**, 91
Kluyveromyces, 86
  fragilis, 90
  marxianus, 90
Kmetia exigua, 227

L

Laboulbenia blanchardi, 220
  formicarum, 425
Labridella cornu cervae, **60**
Labyrinthulomycetes, 27
Lacellinopsis, 385, 386
Lachnella alboviolascens, 197
Lactarius, 193, 195
  deliciosus, 197
Laeviomeliola cassiae, 208
Lagenidium hyphinicola, 186
Lagenomyces, 267
Lambertella, 343

Laminaria, 367
Lamproderma echinulatum, 185
Langermannia gigantea, 202
Laridospora appendiculata, 301, **309**
Lasiosphaeria, 210
  ovina, 210
  spermoides, 210
Lateriramulosa quadriradiata, 302
  uniinflata, 302
Lecanactis, **248**, **260,** 270
  abietina, 247, 258, **260,** 264, 271
  subabietina, 247, 256, 258, 271
Lecanora quercicola, 258
Lecanorales, 251, **261**
Lecanosticta, 63
  acicola, **55**
Leccinum aurantiacum, 198
Lecidea erratica, 265
Leightoniomyces, 220
Lembosia byrsonimae, 217
Lembosiodothis parmularioides, 219
Lemna minor, 31
Lemonniera, 302, 324, **329,** 330
  alabamensis, 302
  aquatica, 302, **309, 319, 326,** 334, 335
  centrosphaera, 302
  cornuta, 302
  filiformis, 302
  pseudofloscula, 302
  terrestris, 302, **326,** 333
Lentinus, 195
  edodes, 197
Lenzites, 198
Leotia lubrica, 198
Lepista nuda, 195
Lepraria, 270
Leprocaulon, 270
Leproplaca, 270
Leptosphaeria, 30, 147, 150, 153, 155, 160, 161, 162, 407
  acuta, 155, 219
  agnita, 155
  anemones, 155
  bellynkii, 155
  derasa, 217
  doliolum, 147, 150, 155
  dolium, 219
  dumentorum, 155

Leptosphaeria (*continued*)
   fallaciosa, 155
   haematites, 155
   macrospora, 147, 150, 155, 219
   maculans, 155, 227
   millefolium, 155
   nodorum, 376
   ogilviensis, 155
   polygonati, 155
   pratensis, 155
   weigeliniana, 219
Leptosporomyces galzinii, 308, **309**
Leptostroma, 51
   ahmadii, 225
Leptostromataceae, 51
Leucopenicillifera gracilis, 185
Leucostoma, 160
Libartania, 63
   laserpiti, 60, **61,** 70
   themedae, **66,** 68
Libertella peltigerae, 224
Lichenoconium, 222
   echinosporum, 222
   erodens, 222
   lecanorae, 222
   lichenicola, **223**
   pyxidatae, 222
   usneae, 222
Lichenodiplis, 224
   lecanorae, **223**
Lichenophoma, 224
Lichenosticta alcicoruaria, 224
Lichenothrix, 270
Lindquistia indica, 210
Linotexis deightonii, 206
Lobaria, **249,** 252, **253,** 270
   amplissima, **249,** 252, 257, 258, **262**
   laetevirens, 5, 9
   linita, 252
   pulmonaria, 222
Lobodirina cerebriformis, 247, 256
Loculoascomycetes, 308
Lonicera, 157, **158**
   alpigena, 157
   canadensis, 157, **158**
   coerulea, 157
   hispidula, 157
   nigra, 157

   qinquelocularia, 157
   xylosteum, 157
Lophodermium pinastri, 214
Lunulospora curvula, 302, **309, 319,** 331, 332
   cymbiformis, 302
Lycoperdon, 203
Lycopodium, 375
Lyromma, 267
   nectandrae, 269
Lysotheca suprastromatica, 217

M

Macronemeae, 10
Macrophoma, 359
   pinea, 65
   sapinea, 192
Macrophomina, 72
   phaseolina, 76, 227
Macrosporium, 391
   gemmivorum, 186
   parasiticum, 186
   ustilaginis, 192
Macrotyphula, 161
Magnoliaceae, 26
Malassezia furfur, 94
Manaustrum, 267
Marasmius, 202
Margaritispora aquatica, 302, **309**
   monticola, 302
Mariannaea elegans, 198
Marssonina, 71
   rosae, 75
Massaria, 30
   conspurcata, 219
   pupula, 217
   pyxidata, 217
Massarina, 308, 319
   aquatica, 308, 319
Massariothea, 63, 65
Mazosia phyllosema, 224
Melanamphora spinifera, **149**
Melanconiaceae, 8, 10, 45
Melanconiales, 10, 28
Melanconieae, 10
Melanconiopsis, 63
Melanconium, 65
   parasiticum, 200

Melanochaeta, 148, **149**
   hemipsila, 148, **149**
Melanogaster variegatus, 203
Melanographium citri, 228
Melanomma, 162
Melanospora, 140
Melasmia, 151
Meliola, 207, 208
   ambigua, 208
   arundinis, 207
   bidentata, 206
   byrsonimae, 208
   capensis, 176
   carissae, 208
   chlorophorae, 176
   clerodendri, 208
   clerodendricola, 206
   deinbolliae, 207
   kaduae, 208
   melanochylae, 208
   myriapoda, 208
   paulliniae, 207
   sapindacearum, 208
   soroceae, 207
   swieteniae, 207
   tabernaemontanicola, 208
   ugandensis, 208
Meliolales, 164
Meliolina cladotricha, 207
   kawandensis, 207
   mollis, 207
Melioliphila melioloides, 206
Melogramma, 162
   vagans, 219
Memnoniella, 13, 383, 385, 389
   echinata, 211, 230
Meria, 39
Meripileus giganteus, 201
Merulius, 200
   lacrymans, 200
Mesnieraceae, 31
Metabotryon connatum, 217
Metschnikowia lunata, 91
   pulcherrima, 90
Metschnikowiaceae, 88, 90
Micarea pycnidiophora, 258
   stipidata, 258
Microcalicium, 224
   subpedicellatum, 256, 258

Microdiplodia, 208
   arthurii, 225
   cenangicola, 213
   mycophaga, 210
Microlychnus, 267
Micronemeae, 10
Micropeltidaceae, 31
Micropera, 225
Microsphaera, 204
Microsphaeropsis centaureae, 213
   sarcinellae, 229
Microsporum, 137
   audouinii, 137
Microthecium, 215
Microthelia, 219
Microthyriaceae, 31
Microtyphula, 161
Microxyphiomyces, 267
   astrocaryifolii, **269**
Microxyphium alangii, 225
Miladina lechithina, 308
Milospium, 220
   graphidiorum, **223**
Mitteriella zizyphina, 229
Mnium cuspidatum, 200
Molinia, 332
Mollisia, 308
   crassa, 308
Monacrosporium melioliphilum, 208
Monilia, 138, 140, 142, 151, 385
   fungicola, 202
   sitophila, 13, 377, 394, 396
Moniliales, 28
Moniliella, 86, 92
Monilinia, 140
   fructigena, 151
Monochaetia, 77, 140
Monochaetiella, 63
   themeda, 225
Monocillium, 160, 222
Monocotyledones, 438
Monodia, 68
   elegans, 76
Monodictys, 222, 365
   pelagica, 359, 363, 364, 365, 369
Monosporiella meliolicola, 206
Monotospora, 278
   megalospora, 230
   priceana, 231

Monotosporella, 302, 306
　tuberculata, 302, **309**
Morchella, 214
　esculenta, 214
Mortierella, 278, 284
　gracilis, 280
　humilis, 280
　polycephala, 188
　ramanniana, 179
Mucedineae, 10
Mucor, 5, 9, 109, **110,** 111, 126, 179, 278, 284
　hiemalis, 188
　pusillus, 385
　racemosus, 111, 122, 126
　ramannianus, 280, 288
　rouxii, 109, 122
Mucoraceae, 395
Mucorales, 281
Mucorini, 10
Mycelia sterilia, 32
Myceliophthora fergusii, 195
　fusca, 200
　lutea, 195
Mycelium radicis-atrovirens, 287
Mycena, 195
Mycocentrospora, 302, 306
　acerina, 302, **309**
　angulata, 302
　aquatica, 302
　clavata, 302
　varians, 302
Mycogone, 138, 182
　calospora, 200
　cervina, 214
　meliolarum, 207
　perniciosa, 187, 193, **194,** 195
　peziza, 214
　rosea, 195, 214
　sporotrichi, 229
Mycohypallage, 57, 65, 68
Mycopara shawii, 217
Mycosphaerella, 30, 69, 141, 220
　berberidis, 151
　tassiana, 156
Mycosticta cytosporicola, 225
Mycotoruloideae, 85, 86
Mycotribulus, 63
Mycovellosiella, 227
Myriangiales, 25, 26
Myrioconium, 138, 140

Myrothecium, 29, 51, 441
　cinctum, 230
　gramineum, 51
　inundatum, 197, 200
　leucotricha, 51
　verrucaria, 187, 220
Myxomycetes, 10, 27, 183
Myxormia, **50,** 51
　atroviride, 51
Myxotrichum, 140

N

Nadsonia, 91
Naemosphaerella chalaroides, 229
　epimyces, 229
Nardus, 332
Nascimentoa pseudoendogena, 206
Nectria, 29, 30, 141, 147, 156, 161, 210, 211, 225, 376
　aurantiicola, 202, 220
　candicans, **184,** 185
　cinnabarina, 211, 228
　coccinea, 29, 210, 211
　haematococca, 156
　inventa, 201
　leptosphaeriae, 219
　lugdunensis, 308, 319
　magnusiana, 210
　mammoidea, 210
　myxomyceticola, 185
　penicillioides, 308
　prodigiosa, 210
　radicicola, 229
　ralfsii, 29
　violacea, 185
Nectriaceae, 29
Nectrioidaceae, 45
Nectriopsis, 185
　aureonitens, 200
　berkeleyanum, 201
Nematogonium aurantiacum, 174, **209,** 211
　niveum, 227
　parasiticum, 211, 230
Neobarclaya primaria, **62,** 63
Neocosmospora vasinfecta, 211
Neottiospora, 45, 58, 60, 69
　caricina, **53, 66,** 68
Neottiosporina, 58

Neoxenophila, 29
Nephroma laevigatum, 257
Neurospora, 156, 422
    crassa, 156, 425
    sitophila, 142, 156
Niesslia, 160
    cladoniicola, 222
Nigrospora, 377, 383, 384, 385, 386, **388**, 389, 390, 391, 392, 398
    sphaerica, 382, 398, **399,** 437
Nitschkia, 204
Nodulisporium, 14, 164, 210
    tuberum, 215
Nodulosphaeria, 160
Nothofagus, 438
Nyctalis asterophora, 195
    parasitica, 195, 197

O

Obstipipilus, 68
Obstipispora chewaclensis, 302
Octaviania asterosperma, 203
Oedium, 148
Oedocephalum, 200
    crystallinum, 228
Oidiodendron, 14
Oidium, 14
    fungicola, 225
    heveae, 204
    tingtanium, 206
Omphalina, 198
Omphalodium pisacommense, 257
Oncopodiella hyperparasitica, 210
Oncosporomyces, 267
Onygena equina, 204
Onygenei, 10
Oospora candicula, 228
    dothideae, 219
    placentiformis, 215
    tuberum, 215
Oosporidium, 86
Opegrapha, 271
Ophiobolus, 30, 160
Ophionectria, 343
Ophiostoma, 141, 156, 161, 162
    stenoceras, 156, 159
Orbilia, 308
Orbimyces spectabilis, 303, **309,** 359, 360

Ostropales, 258
Otthia, 161
Ovulariopsis, 229
Oxalis pes-caprae, 31

P

Paecilomyces, 39, 138
    viridis, 112
    carneus, 230
    crustaceus, 394
    farinosus, 228, 407
    marquandii, 198, 231
    varioti, 394
Pannariaceae, 257
Papaveraceae, 26
Papulaspora, 138, 365
    candida, 213
    dodgei, 231
    halima, 358, 365
    stoveri, 231
Parabotryon comatum, 217
Paracoccidioides, **110,** 112
    brasiliensis, 107, 112, 113, 114, 115, **116,** **117, 118,** 119, 120, 122, 123, 124, 125
Parahyalotiopsis, 58
Paraphaeoisaria alabamensis, 189
Paraphaeosphaeria, 162
Parasterina veronicae, 225
Paratrichoconis chinensis, 215, **219**
Parmelia, 224, 257
    acetabulum, 251, 257, 266
    caperata, 224
    saxatilis, 222
Parmeliaceae, 257
Passalora dendritica, 228
Patellariaceae, 25, 26
Patellina epimyces, 229
Peckiella completa, 193
Pediliospora ramularioides, 228
Pellicularia sasakii, 174, 176, 201
Peltaster, 227
Peltasterales, 51
Peltigera canina, 224
    polydactyla, 224
    rufescens, 222
    tomentosa, 256
Peltigeraceae, 257
Penicillium, 8, 14, 32, 35, 37, 139, 176, 183, 185, 187, 188, 198, 203, 211, 214, 215,

Penicillium (continued)
    278, 284, 288, 365, 366, 377, 381, 382, 384, 385, 387, 389, 391, 393, 419, 424, 431, 441, 442
    brevicompactum, 183, 198, 201, 206, 231
    casei, 394, 395, 407
    chrysogenum, 425
    citrinum, 211
    digitatum, 394
    expansum, 230, 425
    frequentans, 187, 280, 394, 407
    granulatum, 230, 394
    griseofulvum, 177, 347
    implicatum, 211
    italicum, 394
    janthinellum, 211, 231
    nigricans, 188
    oxalicum, 385
    roquefortii, 395
    rubrum, 177
    spinulosum, 213, 230
    stoloniferum, 213
    thomii, 201
    vermiculatum, 231
Peniophora, 200
    longispora, 200
Periconia, 383, 385, 386, 389, 391
    abyssa, 359, 369
    byssoides, 201, 229
    doidgeae, 207, 217
    prolifica, 359, 365, 369
Periconiella, 389, 391
    ellisii, 208
Periola hirsuta, 231
Perizomella inquinans, 208
Peroneutypa heteracantha, 210
Peronospora, 187, 192
    schleideni, 186
    tabacina, 403
Peronosporeae, 10
Pertusaria corallina, 222
Pesotum, 397
Pestalotia, 58, 65, 139, 148, 163, 364, 366
    duportii, 198
    funera, 224
    taphrinicola, 203
Pestalotiopsis, 65, 68, 72, 77
    guepini, 77
Pestalozziella, 68
Pestalozzina, 68

Petrakomyces, 60, 68, 69
Petriellidium boydii, 139
Peyronelia, 138
Peyronelina glomerata, 210
    glomerulata, **339**, 344
Peyronellaea, 72
Pezicula, 30
Pezizales, 25, 308, 426
Pezoloma, 308
Phacidiopycnis piri, **54**
Phaeoantenariella lichenicola, 224
Phaeobotryosphaeria plicatula, 225
Phaeobulgaria inquinans, 214
Phaeocytostroma, 57
    sacchari, 75
Phaeographina fulgurata, 265
Phaeolus schweinitzii, 201
Phaeophyscia orbicularis, **249**, 257, 258, 263
Phaeophyta, 366
Phaeopolynema, 63
Phaeosphaeria, 148, 162
    herpotricha, 148
Phaeotrametes decipiens, 397
Phaeotrichoconis, 389
Phaffia, 86, 92
    rhodozyma, 92
Phallomyces, 267
    palmae, **268**
Phallus impudicus, 202
Phanerochaete, 29
Phellinus, 200
Phialea albida, 213
    sordida, 213
Phialophora, 14, 138
    brevicollaris, 200
    dermatitidis, 107
    hoffmanii, 394, 431
    luteo-viridis, 201
    radicicola, 211
Phialophorophoma litoralis, 359, 363, 364
Phleospora mori, 225
Phloeosporella maculans, 225
Pholiota, 160
Phoma, 45, 46, 70, 71, 72, 76, 138, 139, 147, 153, 155, 161, 162, 362, 365, 366, 367, 376, 377, 393, 407
    agaricola, 198
    betae, 60
    colpomatis, 215
    consocians, 225

cytospora, 224
dothideicola, 219
exigua, 72, 230
gasteropsidis, 202
herbarum, 155, 202
humicola, 220
hypocrellae, 210
laminariae, 359, 361, 367
lingam, 227
marina, 359
medicaginis, 227
nodorum, 376
parasitica, 203
physciicola, 224
pinodella, 227
portentosa, 200
pusilla, 225
pyrenophoricola, 219
ramalinae, 224
scleroticola, 231
sclerotivora, 213
stemphylii, 229
suaedae, 359
taphrinae-pruni, 203
tremellae, 202
Phomopsis, 57, 208
  juniperivora, 76, 77
  leptostromiformis, 76
Phragmidium holoserica, 189
  kraussiana, 189
  periodica, 189
  rubi-idaei, 189
Phragmites, 296
  communis, 76
Phragmotrichum, 70
Phycomyces blakesleeanus, 188
Phycomycetes, 10
Phyllachora, 208, 210
  brenesii, 208
  paspalicola, 208
  pseudis, 208
  whetzelii, 210
Phyllactinia, 204
  corylea, 206
Phyllophiale, 270
Phyllosticta, 44, 45, 46, **53**, 57, 58, 74
  aecidiarum, 191
  bauhinicola, 229
  cavarae, 225
  consors, 225

  convallariae, 44
  gallicola, 215
  pivensis, 229
  sclerotialis, 231
Phyllostictina murrayae, 44, 45
Physalospora, 30, 220, 227
  montana, 210
Physcia aipolia, 224
  millegrana, 222
  semipinnata, 258, **263**
Physciaceae, 257, 271
Physconia pulverulacea, 257
Phytophthora, 187
  cactorum, 187
  cinnamomi, 187
  colocasiae, 186
  infestans, 403
  megasperma, 187
  sojae, 187
Picea, 152
Pichia, 86, 88
  chambardii, 90
  guilliermondii, 90
  membranaefaciens, 90
Piedraceae, 31
Pilaira anomala, **178**
Pilidiella, 65
Pilobolus, 188
  crystallinus, 178
Pinus sylvestris, 280, 345, 364
Pirus, 152, 158
Pistillaria, 161
Pisum, 152
Pithomyces, 377, 383, 385, 389
Pityrosporum, 86, 94
  ovale, **93,** 94
Placella, 51
Placodium, 256
Placonema, 57, 58, 65, 68, 69, 70
Plasmodiophoromycetes, 27
Plasmopara, 204
  nivea, 186
  viticola, 186
Plectomycetes, 25, 28
Plectronidiopsis, 49, 68, 70
  chilensis, **50,** 51
Plectronidium, 68, 70
  minor, **62,** 63
Plenodomus lingam, 60
Pleomassaria siparia, 225

Pleospora, 30, 39, 148, 161
   herbarum, 148
   vagans, 162
Pleosporales, 25
Plesiospora, 26
Pleurocatena acicularis, 213
Pleuropedium tricladioides, 303, **309**
Pleurophoma, 208
Pleurophomella, 57
Pleurosticta, 266
   lichenicola, 224
Poa, 31
Podosordaria leporina, 210
Podospora anserina, 425
Podosporium ugandense, 208
Podoxyphiomyces, 267
Pollaccia, 150
Polycladium equiseti, 303, **309**, 323
Polygonatum, 150
Polynema, 51, 58, 68
   ornatum, 68
Polyporus, 200
   brumalis, 198
   portentosa, 200, 201
Polyscytalum fungorum, 197
Polystigma pusillum, 208
   rubra, 151
Polystigmina, 151
Polythrincium, 390, 391
   trifolii, 389
Populus tremuloides, 281
Poria, 201
   ferrea, 200
Porina viridiseda, 264
Porocladium aquaticum, 303, **309**
Poropeltis, 51, 65
   davilliae, **52,** 68
Porosphaeria sporoschismoides, 148, **149**
Pragmopycnis pithya, **64,** 65
Prosthemium, 65
Protomycetes, 10
Protomyci, 7
Protostegiomyces, 217
   asterinarum, 217
   lembosiae, 217
Prunus, 151, 153
   laurocerasus, 366
Psammina stipitata, 222
Psathyromyces, 267
Pseudeurotium zonatum, 204

Pseudoanguillospora gracilis, 303
   prolifera, 303
   stricta, 303, **309**
Pseudocercospora, 389
   heveae, 210
   lichenum, 222
   mori, 230
   triumfettigena, 228, 229
Pseudocercosporella herpotrichoides, 204, 230
Pseudofusidium hansfordii, 227
Pseudographiella variiseptata, 213
Pseudohansfordia irregularis, 198
Pseudolachnea, 51
   bubaki, **50,** 51
Pseudomonilia, 90
Pseudonectria pipericola, 206
   tilachlidii, 195
Pseudoneottiospora, 58
   coprophila, 76
   cunicularia, 76
Pseudorobillarda, 45, 63, 69
   phragmitis, 76
Pseudoscypha, 25
Pseudosphaeriales, 26, 426
Pseudospiropes simplex, 228
Pseudothis coccodes, 219
Pseudotorula, 383
Pseudovalsa laviciformis, 225
Psoroma hypnorum, 257
Psorotheciopsis, 266
Pteridium, 343
Ptychogaster rubescens, 200
Puccinia, 190, 192
   aecidii-leucanthemi, 154
   cnici-oleracei, 191
   gaeumanni, 154
   graminis, 150, 152, 153, 403
   kuehniae, 191
   leucanthemi, 154
   mulgedii, 154
   pelargonii, 192
   polyspora, 403
   recondita, 189
   saccardoi, 191
   tatarica, 154
Pullospora, 68
   tetrachaeta, 76
Pycnidiopeltis, 51
Pycnociliospora, 267
Pycnofusarium, 73

Index to Taxa

Pycnopeltis conspicua, 219
Pycnothyriaceae, 51
Pyramidospora casuarinae, 303
   constricta, 303, **309**
   densa, 303
   fluminea, 303
   herculiformis, 303
   ramificata, 303
   stellata, 303
Pyrenochaeta, 57, 419
   collematis, 224
   geasteris, 202
   mitteriellae, 229
   unguishominis, 76
Pyrenomycetes, 10, 308
Pyrenophora, 150, 160, 219
   tritici-repentis, 429
Pyrenotrichum, 224, 234
Pyricularia, 306
   aquatica, 303, 306, 308, **309, 319,** 330
   grisea, 303, **388,** 398
   oryzae, 389, 390, 398
   submersa, 303
Pyriomyces, 267
Pyripnomyces, 267
Pythium, 174, 186, 187, 284
   ultimum, **175,** 176, 187

Q

Quercus, 296, 332
   petraea, 330
   phillyraeoides, 438

R

Racodiopsis, 270
Racodium, 270
Ramalia veronicae, 215, 225
Ramalina, 224
   siliquosa, 257
Ramalinaceae, 257
Ramaria aurea, 200
   decolorans, 200
Ramocercospora flagelliformis, 303
   foliosa, 303, **309**
Ramularia coleospori, 191
   episphaeriae, 219
   gerani-phaei, 229
Ranunculaceae, 26, 27

Ranunculus, 27
Ravenelia, 190
Redbia, 189
   elegans, 189
   pucciniicola, 189, **191**
Refractohilum, 222
Renispora, 29
   flavissima, 141
Reticularia fuliginosa, 185
Rhabdogloeopsis, 70
Rhabdospora, 147, 148, 150, 155, 162, 224
   avicenniae, 359, 362, 364
   elettariae, 229
   lecanorae, 224
   mycophaga, 219
   thallicola, 224
Rhagadolobium bakerianum, 219
Rhinocladiella anceps, 210
   epichloes, 210
Rhinotrichiella globulifera, 198
Rhinotrichum alterosum, 208
   gossypinum, 228
   macrosporum, 183
Rhizocarpon, 247
Rhizoctonia, 32
   solani, 231, 284
Rhizomorpha subcorticalis, 231
   subterranea, 231
Rhizophora, 362
   mangle, 76, 364, 366, 368, 441
Rhizopogon, 203
Rhizopus, 5, 278, 284, 387
   oryzae, 187
   stolonifera, 188
Rhododendron, 152
Rhodophyta, 366
Rhodosporidium, 86, 88, 94, 95
Rhodotorula, 86, 90, 95
   fujisanensis, 95
   glutinis, 95
   muscorum, 95
   rubra, **93,** 179
Rhopalomyces elegans, 188, 195
Rhytidhysteron, 161
Rhytisma, 215
   acerina, 151
   solidaginis, 215
Riessia minima, 228
Rileya, 57, 69
   piceae, **54,** 58, **59, 67,** 68

Robillarda, 45, 77
   geasteris, 202
   rhizophorae, 76, 359, 361, 365
Roccella, **249, 250,** 270
   hypomecha, 247
   portentosa, 247, 256
Roccellaceae, 247, 256
Rollandina, 29
Rosellinia necatrix, 228
Rubus, 31
Russula, 193
   adusta, 197
   nigricans, 195, 197, 211

S

Saccharomyces, 86, 88
   cerevisiae, 5, 98, **99,** 100, 101, 102, 115, 118, 119, 124
   exiguus, 90
Saccharomycetaceae, 86, 88, 90
Saccharomycetales, 88
Saccharomycetes, 87
Saccharomycodaceae, 88
Saccharomycodes, 91
Saccharomycopsis, 91
   lipolytica, 90
Salicornia, 365
   herbacea, 357
Salix, 296
Samukuta, 45
Santessonia namibensis, 258
Saprolegnia delica, 186
Saprolegnieae, 10
Sarcinella palawanensis, 229
Sarcinosporon, 94
   inkin, 94
Sargassum, 362
   fluitans, 362
   natans, 76, 362
Satchmopsis brasiliensis, **50,** 51
Satureia, 155
Schiffnerula, 219
   solani, 215
Schismatomma decolorans, 222
Schizoblastosporon, 86
   starkeyi-henricii, 94
Schizophyllum commune, 125
Schizothyra, 51
Schizotrichella lunata, 200
Schoenoplectus, 296

Sclerococcum, 222
   sphaerale, 222, **223**
Scleroderma citrinum, 203, 211
   verrucosum, 203
Sclerophoma, 63
Sclerospora graminicola, 187
Sclerotinia, 161
   cepivorum, 213
   libertiana, 213
   sclerotiorum, 213
   trifoliarum, 213
Sclerotium, 231
   clavus, 231
   patouillardii, 231
Scolecobasidium acanthacearum, 192
   dendroides, **226,** 228
   pusillum, 228
Scolecosporiella, 57
   sisyrinchii, **53**
Scolecotrichum clavariarum, 201
Scolecozythia valsivora, 210
Scopaphoma corioli, 198
Scopulariopsis, 14, 138, 185, 278
   brevicaulis, 137, 395
   putredinis, 139
Scorpiosporium angulatum, 303
   anomalum, 303
   gracile, 303
   gracile var. *oxyphilum*, 303
   minutum, 304, **309**
   rangiferinum, 304
Scytalidium, 140
   aurantiacum, 197
   thermophilum, 394
   uredinicola, 190
Sebacina, 32

Secotiaceae, 24
Seimatosporium, 63, 70, 148
   dilophosporum, **62,** 63
   falcatum, **50,** 51
Selenophoma donacis, 75
Selenosporella, 138
Selenozyma, 86, 91
   intestinalis, 91
   peltata, 91
Senecio vulgaris, 441
Sepedonium, 138, 200, 203
   ampullosporum, 138
   brunneum, 193

Index to Taxa

chrysospermum, 193, **196**
  curviseum, 188
  epimeliola, 208
  monosporum, 230
  mucorinum, 188
Septobasidium clelandii, 202
Septocylindrium lindtneri, 201
  morchellae, 214
  myxophagum, 185
Septogloeum, 71
  robustum, 219
Septoidium didymopanacis, 228
  lateritium, 228
Septonema trichomeriicola, 219
Septoria, 46, 47, 60, 71, 151, 192
  antirrhinorum, 225
  ascophylli, 359
  didyma, 225
  forskahleana, 227
  graminis, 227
  leptosphaericola, 219
  lycopersici, 60
  nodorum, 76
  passerini, 77
  thalassica, 359
Septoriomyces, 267
Septosporium bulbotrichum, 138
Sesquicillium microsporum, 185
Sessiliospora, 222
Setaria glauca, 437
Setomyces, 267, 270
Setosphaeria, 160
Sibirina fungicola, 195, **196**
Sigmoidea aurantia, 304
  prolifera, 304, **309**
Siphula, 270
Sirobasidium, 202
Sirodesmium rosae, 225
Sirodothis, 57
  populnea, **56, 64,** 65
Sirosperma florididana, 225
  sparsum, 229
Sirosporium gliricidiae, 227
Sistotrema, 308
Sitochora ellipsospora, 208
Sorbus, 158
Sordaria, 441
  fimicola, 211, 425
Spartina, 365, 369, 438
  alterniflora, 362, 365, 438

Spegazzinia, 389
  chandleri, 207
  meliolae, 207
Speira densa, 228
Speiropsis hyalospora, 306
  irregularis, 304
  pedatospora, 304, **309**
  simplex, 306
Spermatoloncha maticola, 206
Spermosporella, 206
  elegans, 228
  pulvinata, **205**
Sphacelia, 151
Sphaceloma ampelinum, 27
  cecidii, 76, 359, 361, 362
  fawcetti, 77
Sphacelotheca sorghi, 292
Sphaerella isariphora, 219
Sphaerellopsis, 76
  filum, 189, **190,** 192
Sphaeriacei, 8
Sphaeriales, 426
Sphaerographium microperae, 225
Sphaeromma mazosiae, 224
Sphaeronaemella fimicola, 214
Sphaeronema lichenophilum, 224
Sphaeronemeae, 8
Sphaeronemei, 8
Sphaeropsidaceae, 10, 45
Sphaeropsidales, 10, 28, 359, 361
Sphaeropsidei, 8
Sphaeropsis asiminae, 225
  sapinea, 192
Sphaerotheca, 204
  fuliginea, 430
Sphaerulomyces coralloides, 227
Spicaria valdiviensis, 228
Spilocaea, 151
  pomi, 159
Spinomyces, 267, 270
Spiropes, 206, 207, 219
  asterinae, **218,** 219
  capensis, 207
  dorycarpus, 207, 217, 228
  guareicola, 207
  melanoplaca, 207
  scopiformis, 219
Spirosphaera floriforme, **339,** 343, 346
Spondylocladiella botrytioides, 198
Sporendonema epizoum, 228

Index to Taxa

Sporhaplus rondiensis, 224
Sporidesmium, 35, 37, 228
  adscendens, **36**
  altum, **36**
  anglicum, **36**
  aturbinatum, **36**
  baccharidis, 228
  bicolor, **36**
  biseptatum, 219
  cajani, **36**
  cambrense, **36**
  eucalypti, **36**
  folliculatum, 265
  inflatum, **36**, 213
  longirostratum, **36**
  raphiae, **36**
  salinum, 359
  sclerotivorum, 213
  socium, 230
  vagum, **36**
Sporidiobolus, 86, 88, 94, 95
  johnsonii, 189
Sporobolomyces, 86, 94, 140, 377, 381, 384, 386, 391
  boleticola, 198
  coralliformis, 202
  holseticus, 202
  roseus, **93**, 94, 198
  salmonicolor, 95
Sporobolomycetaceae, 85, 86, 88, 93, 94
Sporocadus, 148
Sporocephalum peniophorae, 200
Sporocybomyces, 267
  pulcher, **269**
Sporonema nigropunctata, 74
Sporoschisma, 148
  saccardoi, 148
Sporothrix, **110**, 111, 185
  fungorum, 201
  schenckii, 107, 113, 114, 115, **117**, 119, 124, 156, 159, 185
  setiphila, 208
Sporotrichum, 197, 229
  biparasiticum, 228
  fallax, 225
  fungicola, 214
  hospicida, 219
  isarioides, 210
  meliola, 208

  phalloidearum, 202
Stachybotrys, 13
  atra, 177, 213, 230
  echinata, 137
  klebahnii, 197
Stagonopsis peltigerae, 224
Stagonospora, 71, 153, 155, 162, 208, 359, 365
  geastericola, 202
  haliclysta, 359, 367
  sandsteadeana, 224
Stauronema, 51, 68
Stegasphaeria pavonia, 219
Stegonosporium, 65
Stemonitis fusca, 185
Stemphyliopsis, 39, 138
Stemphylium, 148, 162, 186, 359, 365, 367, 377, 383, 387, 389
  anomalum, 229
  triglochinicola, 359
Stenospora uredinicola, 189
Stephanoascus ciferrii, 90
Stephanoma phaeospora, 174, 192, 229
  strigosum, 214
  tetracoccum, 213
Stephanosporium cerealis, 204
Stephosia, 267
  protii, **269**
Stereum, 200
  hirsutum, 200, 201
Sterigmatomyces, 86, 92, **93**
  haplophilus, 92
Sticta, 257
Stictaceae, 252, 257
Stigmina, 214
  septoidii, 228
Stilbeae, 10
Stilbella, 178
  erythrocephala, **178**, 183, 211, 214, 227
  tomentosa, 185
Stilbospora, 225
Stilbum, 197, 228
  mycetophilum, 201
Strasseria, 68, 69, 70
  carpophila, **66**, 68
Strasseriopsis, 57, 68
  tsugae, 60, **61**, **66**, 68
Streptomycetes, 442
Strigula elegans, 222, 264
Stromatopycnis rosetum, 225

Index to Taxa

Subulicystidium longisporum, 339, **343**
Suillus, 193
Suttoniella, 51
Sydowia, 160
Sympodiocladium frondosum, 304, **309**
Sympodiophora, 182, 200, 220, 234
   didyma, **218**
   meliolae, 207
   mycophila, 202
   pulchella, 220
   stereicola, 138, 200
   varanasiensis, 229
   venezuelensis, 227
Syncephalastrum racemosum, 187
Syringospora, 90
   albicans, 90

T

Taeniolella, 222
   delicata, **221,** 222
   hormiscii, 228
   scripta, 201
Taeniospora, **309**
   gracilis, 304, 308
   thermophilus, 385
   wortmanii, 188
Tapesia, 147
Taphrina, 203
   betulae, 203
   caerulescens, 203
   pruni, 203
Taraxacum, 31
Tauromyces, 267
Tegoa, 267
Teichospora obducens, 219
Telimenella gangraena, 210
Teloschistaceae, 252, 258
Teloschistes capensis, **249,** 252, 258, 261, 263
Teratosperma, 222
   anacardii, **221**
Tetrabrunneospora ellisii, 304
Tetrachaetum elegans, 304, **309, 326**
Tetracladium marchalianum, 304, **309, 319**
   maxilliforme, 304
   setigerum, 304
Tetranacrium, 65
Tetraploa, 322, **388,** 389

Tetrasporium asterinearum, 219
Thalassia testudinum, 366
Thamnidium, 188
   elegans, 183, 187
Thamnolia, 270
Thanatephorus, 32
   cucumeris, 231
Thelephora, 198
Thermoascus crustaceus, 394
Thermomyces lanuginosus, 385
Thermutis velutina, 256, 258
Therrya fuckelii, 214
Thielavia, 29
Thielaviopsis, 138, 141
   basicola, 138
Tholurna dissimilis, 256
Thyrsidina, 73
Tiarosporella, 69, 70
Tilachlidium, 419
   brachiatum, 195, 214
Tilletiopsis, 386
   pallescens, 202
Titaea, 74, 192
   callispora, 219, 227
   doidgeae, 207
   hemileiae, **190,** 192
   submutica, 227
   triradiata, 207
   ugandae, 217
Titaeella capnophila, 217
Tolypomyria fungicola, 201
Torula, 137, 389, 391
   darwinii, 213
   herbarum, 230
Torulaspora, 88
Torulopsis, 86, 88, 90, 95
   auriculariae, 202
   candida, 90
   domercqii, 90
   holmii, 90
   kruisii, 198
   molischiana, 90
   schatavii, 201
Toxosporiopsis, 63, 65, 69
   capitata, **66,** 68
Tracyella, 51
   aristata, **52**
Trametes versicolor, 101
Tremella, 202

Tremellini, 10
Triadelphia, 139, 141
   heterospora, 137
Tricellula aquatica, 304, **309**
   botryosa, 304
   curvatis, 306
   inaequalis, 304, 306
Tricharia, 265
Trichocladium, 138
   achrasporum, 359, 363, 364, 368
Trichoconis, 219, 386
   africana, 207, **216**
   angustispora, 207, 217
   appendiculata, 219
   capitata, 207
   caudata, 229
   englerulae, 219
   hamata, 207
   hibernica, **205**, 207
   padwickii, 389
   schiffnerulae, 217
   sigmoidea, 207
   trichiliae, 219
   viridula, 217
Trichoderma, 174, 183, 179, 187, 201, 211, 213, 230, 231, 235, 278, 365, 430, 441
   aureoviride, 230
   hamatum, 197
   harzianum, 187, 198, 201, 229, 230, 430
   longibrachiatum, 176
   polysporum, 201, 430
   pseudokoningii, 230
   viride, **175**, 176, 177, 187, 197, 201, 220, 227, 230, 231, 280, 430, 431
Tricholomopsis rutilans, 197
Trichomerium jambosae, 219
Trichopeltheca asiatica, 438
Trichophaea, 32
Trichophyton, 137
   ajelloi, 137
   mentagrophytes, 137
   rubrum, 137
   schoenleinii, 137
   terrestre, 137
Trichosporium saccardoi, 210
Trichosporon, 85, 86, 91, 92, 94
   beigelii, 92
   cutaneum, 92
   inkin, 94

Trichosporonoides, 86
Trichothecium, 15
   arrhenopum, 186
   helminthosporii, 228
   plasmoparae, 186
   polyctonum, 186
   roseum, 177, 183, 229, 230
   sublutescens, 219
Trichothyrium, 207
   asterophorum, 207
   hansfordii, 207
   reptans, 207
Tricladium, 305, 329, 335
   angulatum, 325
   attenuatum, 304
   brunneum, 305
   castaneicola, 305
   caudatum, 305
   chaetocladium, 305, 331, 332
   giganteum, 305, 324, 332, 335
   gracile, 331
   malaysianum, 305
   marylandicum, 305
   minimum, 306
   patulum, 305
   sorghicolum, 305
   splendens, 227, 305, 308, **309**, 324, 332
   terrestre, 305
   varium, 305
Trigonopsis, 86
Trimmatostroma lichenicola, 222
Trinacrium mycogonis, 210
   subtile, 225
Tripospermum, 305, 306, 422
   acerinum, 304
   camelopardus, 305, **309**
   prolongatum, 305
Triposporina, 305, 306
   ceranoica, 305, **309**, 321
   uredinicola, 189, 305
Triposporiopsis, 422
Triposporium ledermannii, 219
Triscelophorus acuminatus, 305
   magnificus, 305
   monosporus, 305, **309**
   septatus, 305
Trisulcosporium acerinum, 306
Tritirachium fungicola, 201
   isariae, 228

Index to Taxa

Trullula olivascens, 70
Truncatella truncata, 187
Tuber, 215
  gulonum, 215
Tuberacei, 10
Tubercularia, 228, 229, 365
  pulverulenta, 359
  vulgaris, 228
Tuberculariaeae, 10
Tuberculina, 189
  davisiana, 215
  maxima, 189
  persicina, 189, **190**
  phyllachoricola, 210
Tuberculispora, 206
  jamaicensis, **205,** 206
Tubeufia, 159, 162
  amazonica, 159
  cerea, 159, 227
  helicoma, 159
  helicomyces, 343
  palmarum, 159
  paludosa, 159
Tubulicinum dussii, 200
Tulasnella, 32
Tulostoma brumale, 202
  volvulatum, 202
Tympanis, 147
Tympanosporium parasiticum, 228
Typhula, 161

U

Umbilicaria, **249, 251,** 253, 254, 265, 270
  cinereorufescens, 247, 252
  crustulosa, 252, 257
Umbilicariaceae, 257
Uncinula, 204
Uniseta flagellifera, **55**
Uredinales, 136, 140
Uredinei, 10
Uredo, 190
  neocomensis, 154
Uromyces cestri, 190
  costaricensis, 189
  pisi, 152
Urtica, 155
Ustilaginales, 94
Ustilaginei, 10

Ustilaginoidea virens, 389
Ustilago, 192, 386
  avenae, 192
  hordei, 192
Ustomycetes, 26
Ustulina deusta, 200, 210

V

Vaccinium, 332
Valsa, 160
  ambiens, 210
  ceratophora, 210
Varicosporina, 367
  ramulosa, 305, **309,** 358, **360,** 361, 366, 367, 368
Varicosporium, 306, 308, 324
  aquaticum, 305
  delicatum, 305
  elodeae, 227, 297, 306, **309,** 323, 330
  giganteum, 306, 323
  helicosporum, 306
  macrosporum, 306
  trimosum, 306
Vasudevella, 63
Venturia, 16
  crataegi, 158, 159
  inaequalis, 151, 158, 159, 220, 421, 425
Vermes, 6
Vermicularia oligochaeta, 225
Vermispora grandispora, 206
Veronaea coprophila, 201
  filicina, 219
Verrucariaceae, 253, 258
Verrucariales, 253
Verrucaster lichenicola, 224
Verticilliopsis infestans, 195
Verticillium, 39, 138, 139, 182, 183, 186, 195, 278, 394, 405
  albo-atrum, 187, 397, 441
  berkeleyanum, 201
  catenulatum, 185
  chlamydosporium, 187
  dahliae, 188, 229
  epimyces, 215
  epiphytum, 228
  fungicola, 197
  fusisporum, 201
  hemileiae, 192

Verticillium (continued)
  lamellicola, 192, 197
  lecanii, 176, 192, 206, 407
  olivaceum, 201
  psalliotae, 188, 192, 195, **196,** 197, 215
  rexianum, 185
  tenerum, 201
Vezdaea aestivalis, 265
Virgaria nigra, 201
Vitis vinifera, 76
Volucrispora, 329
  aurantiaca, 306, 332
  graminea, 306
  ornithomorpha, 306
Volutella, 139
  ciliata, 178
Vouauxiella, 224
  lichenicola, 70
Vouauxiomyces truncatus, **223,** 224

W

Wallemia sebi, 395
Wardomyces, 139
  dimerus, 139
Wickerhamiella domercqii, 90

X

Xanthoria, **249,** 252, **254,** 257, 261, 270
  parietina, 222, 258
Xanthoriicola, 222
  physciae, 222

Xanthothecium, 29, 39
Xenosporium berkeleyi, 185
Xeromphalina, 195
Xerosporae, 13
Xylaria hypoxylon, 6
  oxyacanthae, 211

Y

Ypsilonia, 58, 65, 76, 77
  mirabilis, **67,** 68
Yucca, 75

Z

Zakatoshia hirschiopori, 198, **199**
Zalerion maritimum, 359, 361, 363, 364, 366
  varium, 359, 365
Zendera, 29
Zopfiaceae, 31
Zopfiella, 29
Zostera marina, **360,** 366
Zygomycotina, 426
Zygophiala jamaicensis, 389
Zygorhynchus, 278, 288
  moelleri, 187
Zygosporium, 377, 385, 391
  mycophilum, 220, 231
  oscheoides, 389, 398, **399**
Zythia, 208
  compressa, 201
  stromaticola, 219